W0227365

NUCLEAR SCATTERING

BY

K. B. MATHER, B.Sc. (Eng.), M.Sc.
Senior Research Officer
Australian Atomic Energy Commission

AND

P. SWAN, M.Sc., Ph.D.
Lecturer in Physics
University of Melbourne

CAMBRIDGE
AT THE UNIVERSITY PRESS
1958

CAMBRIDGE UNIVERSITY PRESS
Cambridge, New York, Melbourne, Madrid, Cape Town,
Singapore, São Paulo, Delhi, Tokyo, Mexico City

Cambridge University Press
The Edinburgh Building, Cambridge CB2 8RU, UK

Published in the United States of America by Cambridge University Press, New York

www.cambridge.org
Information on this title: www.cambridge.org/9780521279444

First published 1958
First paperback edition 2011

A catalogue record for this publication is available from the British Library

ISBN 978-0-521-05690-8 Hardback
ISBN 978-0-521-27944-4 Paperback

CONTENTS

CHAPTER 4

Neutron Scattering Techniques

CHAPTER 5

Determination of Beam Energy

CHAPTER 6

Low Energy Neutron-proton Scattering

CHAPTER 7

Low Energy Proton-proton Scattering

CHAPTER 8

Scattering of Neutrons and Protons by very Light Nuclei

CHAPTER 9

Non-central Forces and Spin-orbit Coupling

CHAPTER 10

High Energy Neutron-proton and Proton-proton Scattering

AUTHORS' PREFACE

This volume describes the physics of nuclear scattering, including experimental techniques, significant experimental results and interpretation in terms of nuclear forces and nuclear structure. The treatment is not strictly introductory. On the experimental side, standard equipment such as particle accelerators and detectors has been discussed only in the special context of scattering technique. The theoretical discussion assumes a general acquaintance with quantum mechanics, and with the elementary scattering formulae encountered elsewhere in relation to the scattering of electrons by atoms. Of the chapters concerned specifically with scattering, 2–4 treat general experimental method and 6–12 the theory of nuclear scattering and comparison with experimental data.

The nature of the nuclear force has been the underlying theme of many sections of the book. However, a treatment of nuclear scattering cannot be confined exclusively to this aspect, any more than a discussion of nuclear forces can be given using only the results of scattering experiments. Chapter 1 attempts to place scattering in appropriate perspective, by indicating other approaches which have proved fruitful and by summarizing the principal characteristics of nuclear forces. On the other hand, chapters 11 and 12 lay emphasis on nuclear structure, collision effects being described in terms of largely phenomenological models.

Our thanks are due to Dr C. A. Hurst, Dr E. G. Muirhead and Miss A. Werner for reading and commenting on certain sections of the manuscript, and to Dr D. N. F. Dunbar for assisting with the proof-reading. We are also indebted to the editor, Professor N. Feather, for his helpful comments on parts of the book.

K.B.M.

P.S.

MELBOURNE
July 1957

CHAPTER I

NUCLEAR SCATTERING IN RELATION TO NUCLEAR FORCES

1.1. Introduction

Analysis of the scattering which occurs when atomic nuclei collide forms one of the main approaches to the problem of nuclear forces. By choosing the least complex particles and varying the collision energy, experiments may be designed which lend themselves comparatively well to detailed analysis, with the result that this approach has provided a most profitable source of quantitative information. Nevertheless, as there are within the scope of nuclear physics other studies which bear significantly on the same problem, it seems desirable to place the work on scattering in correct perspective by reviewing the various lines of thought which have contributed to the current concepts of the nuclear force. We have attempted to do this at the outset by abstracting the major conclusions of the following chapters and integrating them into a general survey.

Present information falls, broadly speaking, into four chief categories:

(a) General observations on the systematics of nuclei, such as size, mass, neutron-proton ratio and stability. These reveal the important saturation property of nuclear matter, lead to rough estimates of the strength and range of action of the force between nucleons and suggest spin dependence and exchange properties for the force. In the same category is work on shell structure which throws light on the coupling between nucleons inside nuclear matter.

(b) The two-nucleon system has received the most complete treatment, leading to a satisfactory representation of the force between a pair of nucleons with low relative energies. Experimental material has been provided mainly by the properties of the deuteron and n-p and p-p scattering.

(c) Some success has attended attempts to extend a similar detailed analysis to rather more complex systems. This includes

I

work on the three-body problem, in particular the binding energies of ^{3}H and ^{3}He and n-d and p-d scattering. However, mathematical difficulties grow rapidly as the number of particles increases. Four-body and higher multi-body systems supplement only in a minor way our quantitative knowledge of nuclear forces.

(d) Finally, a sketch of topics in the meson theory of nuclear forces has been included.

We have adopted this grouping of material in the following review. For brevity, treatment of each subject is confined to an outline, indicating the basic ideas and listing the chief conclusions. Detailed accounts are available in several standard texts (Rosenfeld, 1948; Gamow and Critchfield, 1949; Blatt and Weisskopf, 1952; Evans, 1955; Bethe and Morrison, 1956).

1.2. Qualitative and semi-quantitative generalizations

The existence of stable nuclei proves that neutrons and protons exert attractive forces on each other. Since neutrons are uncharged, the force cannot be electrostatic, although protons within the nucleus must certainly repel one another, opposing the attraction and tending to decrease stability.

Evidence on the actual range and strength of the specifically nuclear force is forthcoming in the first place by considering the density of nuclear matter and the binding energies of nuclei. To calculate the former, information is required on the size of nuclei.

1.2.1. Nuclear radii. The chief methods of obtaining estimates of radii are the following:

(a) Total cross-sections σ for high energy neutrons are expected to approach $2\pi R^2$, made up of approximately equal contributions due to absorption and diffraction. The condition is that the de Broglie wavelength λ $(= \hbar/mv) \ll R$, where $\hbar$ is (Planck's constant)/2π. Direct determination of σ by transmission experiments (§§ 2.4, 4.3.5) leads to a value for R.

(b) Mean lives of α-active nuclei are sensitive to the kinetic energy of the α-particle and to the height of the Coulomb barrier, which in turn depends on the radius R. From measured half-lives and kinetic energies, radii of some of the heavy nuclei ($A > 208$) have been determined.

(c) The cross-section for a charged particle reaction may be regarded as a fast neutron cross-section multiplied by a penetrability factor (the transmission coefficient of potential barriers) which, as for α-decay, depends on the radius.

(d) On the assumption of equality of forces between n-n and p-p pairs (cf. §§ 1.2.4, 1.3.3), the binding energies of 'mirror nuclei' such as ^{11}C and ^{11}B should differ only in the amount of Coulomb energy required to introduce one more proton over the potential barrier. This is calculable in terms of R. For a uniform distribution of protons it is $6Ze^2/5R$, which may be equated to the maximum energy for β-decay between mirror nuclei, thus giving a figure for R.

(e) Electrons have negligible interaction with neutrons and protons except, in the latter case, through the long range Coulomb force. The scattering of electrons can therefore be used to study the charge (proton) distribution, if the energy is sufficiently high to allow penetration into the nucleus.

(f) Slow negative μ-mesons may be captured into a Bohr energy state of an atom, analogous to an electron state but associated with an orbital radius of ~ 0·005 that of the electron orbit because of the much larger meson mass. For the μ-meson the K shell lies partly within the nucleus itself. When it makes transitions to lower states after capture, characteristic electromagnetic radiation is emitted, the energy of which depends on the nuclear radius.

Measurements of R by the above or closely allied methods have been accumulated for many nuclei. It has long been recognized that the measures are not strictly equivalent and cannot be expected to yield identical values. Whereas (a) involves only nuclear forces, the remainder involve Coulomb effects in different degrees and reflect more particularly the radius of the proton distribution. The terms *nuclear force radius* and *electromagnetic radius*, respectively, have been suggested to designate the two quantities. Experiment shows the latter to be 10 or 20 % smaller than the former (for detailed discussion see Evans, 1955).

The important result for the present argument is that the measurements lead to an approximate relation $R = r_0 A^{\frac{1}{3}}$, where r_0 (interpreted as the radius of a nucleon) is 1·4–1·5 × 10^{-13} cm. for the nuclear force radius† and A is the mass number. Hence nuclear

† $r_0 \approx 1·2 \times 10^{-13}$ cm. for the e.m. radius.

matter possesses an essentially constant density. The simplest explanation is that the range of the inter-nucleon force is so short that any one nucleon interacts only with its immediate neighbours, suggesting a range $\sim 2r_0$.

1.2.2. *Binding energies.* Data on the binding energies of nuclei support the above view. If long range attractive forces between all pairs of nucleons existed, the total binding energy B (the energy required for complete separation of all nucleons) would be proportional to the number of pairs $\frac{1}{2}A(A-1)$. Actually B/A is approximately constant ($7\cdot5-8\cdot5$ MeV.) except for minor deviations among very light and very heavy nuclei—a property known as *saturation*. Each nucleon is about equally strongly bound in nuclear matter, implying interaction with a strictly limited number of other nucleons, i.e. short range forces. However, although a necessary condition, short range forces are not alone sufficient to give saturation. Increased packing of nucleons could lead to B increasing faster than A for large A.

1.2.3. *Nuclear saturation.* Various hypotheses have been put forward to explain saturation. They reduce mainly to two groups:

(i) *Exchange forces.* In addition to the ordinary radial dependence of interaction $V(r)$ (cf. § 1.3), any two nucleons may interchange space and/or spin co-ordinates. Some of the possible interactions lead to non-saturation but others have the required properties. The *Majorana* force, involving exchange of spatial co-ordinates of the two particles, is often employed in this connexion. Here, the sign of the potential associated with a given state of orbital angular momentum l is determined by l. The potential is attractive for states of even l (and hence does not affect ground state calculations which always involve $l=0$), but is repulsive for states of odd l. That this can lead to saturation is shown by considering the lightest nuclei, e.g. ^{5}He and ^{5}Li (cf. § 10.3). In fact the requirements of saturation can be met for nuclei in general by spatial exchange forces, total binding energies being greatly reduced by the repulsion between pairs of nucleons in even and odd l states, respectively.

(ii) *A repulsive core.* An alternative to exchange forces is a potential which becomes very strongly repulsive at small distances.

At present there are quantitative difficulties in the way of this idea, especially in reconciling the potential with requirements of low energy scattering.

1.2.4. *Stability of nuclei.* A plot of the neutron number N vs. proton number Z for stable nuclei exhibits several conspicuous features. There is a preference for $N \approx Z$ up to $^{40}_{20}$Ca, beyond which the average N/Z ratio exceeds 1 and rises to 1·6 among the heaviest nuclei. There is also a marked difference between nuclear abundances, according as N and Z are even or odd. (Even N, odd Z) and (odd N, even Z) nuclei are moderately and comparably abundant. (Even N, even Z) nuclei are three times as common, but (odd N, odd Z) are very rare.

The implication is that stable nuclei are built by completion of nucleon pairs of the same kind, this arrangement resulting in lower nuclear masses and greater stability. Moreover, even nuclei have zero angular momentum, in line with the requirement of the exclusion principle that nucleon pairs occupying the same energy state must have oppositely directed spins, as in the case of the α-particle.

It can also be inferred that the force between nucleons is approximately independent of their charge—known as the *charge independence hypothesis*. On quantum-mechanical grounds this would make the density of states available to neutrons the same as for protons. If low-lying states are filled first to give maximum stability, an approximately equal number of neutrons and protons is favoured, which is in agreement with observation. Among heavier nuclei, the long range Coulomb repulsion becomes relatively important and the stability condition is biased toward a higher N/Z ratio.

Evidence on the relative strengths of the bonds between nucleons comes from consideration of the stabilities of isobaric sets of nuclei, which show different characteristics according to whether A is even or odd (cf. the A vs. Z diagram (Rosenfeld, 1948)). When A is odd there is a single stable nucleus among the set, but when A is even there are frequently two and sometimes three stable nuclei, all of even Z. Each group consists of two nuclear types. Odd A can arise from (even N, odd Z) or (odd N, even Z), i.e. one unpaired nucleon remains in each case. No appreciable difference exists between the nuclear masses of these two types. On the other hand, even A can result from either (even N, even Z) or (odd N, odd Z). The nuclear masses of the former are systematically the smaller.

Presumably there are strong attractive forces between the paired like particles which depress the masses of (even, even) nuclei. In general the force between an odd proton and odd neutron must be smaller, making the occurrence of stable (odd, odd) nuclei very improbable. The four which do occur, ^{2}H, ^{6}Li, ^{10}B, and ^{14}N, are all of the type $A = 4n + 2$ (n integral), which may be thought of as

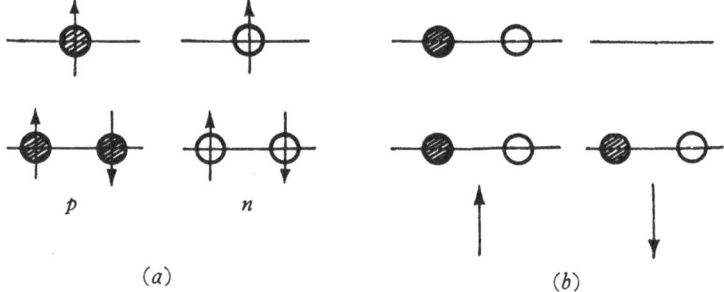

(a)   (b)

Fig. 1. Interchange of roles of spin and charge in ^{6}Li nuclei.

an α-particle core with an odd proton and neutron whose spins are aligned making the nuclear spin 1. In these cases the force between the odd nucleons must be strong, contrary to the general rule.

The fact that these (odd, odd) nuclei have $N = Z$ has suggested that roles of spin and charge may here be interchanged. Treating charge as a co-ordinate, protons and neutrons become alternative states of the nucleon, analogous to different spin states. In ^{6}Li, instead of pairing like particles with opposite spins, leaving two odd particles (fig. 1a), one can adopt a pairing of particles of like spin and unlike charge, (fig. 1b). The latter results in all particles being paired and the same can be done for the other three (odd, odd) nuclei. However, this equivalence of the roles of spin and charge only holds if the energies of states available to protons and neutrons (fig. 1a) are themselves equivalent. It breaks down because of Coulomb repulsion among heavier nuclei.

From the behaviour of certain unstable nuclei, support can be drawn for the charge-independence hypothesis referred to above.

Mirror nuclei like $^{11}C \xrightarrow{\beta+} {}^{11}B$ differ only in the interchange of a proton and a neutron. On the 'paired nucleons' concept, a pairing of two protons in ^{11}C becomes a pairing of two neutrons in ^{11}B. By measuring the maximum energy of β-decay of mirror pairs and allowing for their slightly different Coulomb energies, the strengths of the binding between nucleon pairs may be compared. The conclusion is that the n-n interaction is essentially the same as the p-p, an important result in view of the difficulty (cf. § 1.4) of obtaining direct evidence on the magnitude of the n-n force.

1.2.5. Shell structure. Various attempts have been made (Mayer, 1948; Jensen, Haxell and Suess, 1948) to explain the occurrence of 'magic numbers' in relation to nuclear structure, i.e. certain N or Z values associated with abnormal stability. A single nucleon is assumed to be acted upon by a central potential $V(r)$ and the interaction between nucleons is treated as a perturbation. Although this approach is valid in the atomic case of electrons moving in the Coulomb field of the nucleus, it seems, a priori, a poor approximation to the nuclear case and difficult to reconcile with the Bohr compound nucleus theory. However, it has been carried through with surprising success and the results have a significant bearing on the spin-dependence of the inter-nucleon force.

Assuming a particular form for $V(r)$, e.g. the oscillator potential $-V_0 + kr^2$, one can derive a set of energy levels distinguished by a principal quantum number n and an orbital angular momentum quantum number l, as in spectroscopic notation. Some of these levels have the same energy and each such degenerate group is regarded as a 'shell' which can hold a definite number of nucleons of one kind. When a shell is filled the resulting configuration possesses more than average stability. For the oscillator force the shells are: $(1s)$, $(2p)$, $(3d, 2s)$, $(4f, 3p)$, $(5g, 4d, 3s)$, ...

Since the spin of a nucleon is $\frac{1}{2}$, each l value (> 0) can give rise to two states, $j = (l \pm \frac{1}{2})$, whereupon the shells become (writing the j values as subscripts and omitting the principal quantum numbers): $(s_{\frac{1}{2}})$, $(p_{\frac{3}{2}}, p_{\frac{1}{2}})$, $(d_{\frac{5}{2}}, d_{\frac{3}{2}}, s_{\frac{1}{2}})$, ... Each state can contain $(2j + 1)$ identical nucleons, hence successive shells should close at totals of 2, 8, 20, ... particles. These numbers agree with some of the lower magic numbers but for the higher shells the theory breaks down.

Completion of the first shell in both neutrons and protons results in the particularly stable α-particle.

It is possible to account for the higher magic numbers by introducing a fresh assumption concerning the coupling between nucleons. One can distinguish two limiting coupling rules:

(i) j-j coupling. The spin s of a nucleon is oriented relative to its orbital angular momentum l to give j, the total angular momentum of the nucleon. The total angular momentum of the nucleus, J, is obtained by summing over all j but L and S are not required to be good quantum numbers.

(ii) L-S or Russell-Saunders coupling. In this case $l+s$ coupling is negligible. Individual l values couple together and likewise s values. Both the total spin S ($=\Sigma_i s_i$) and the total orbital momentum L summed over all the nucleons are constants of motion and the total angular momentum of a state is $J=L+S$. Different S values coupled to any one L value give the so-called 'spin multiplets'.

The type of coupling assumed directly affects the energy separation of the levels predicted. To arrive at the correct magic numbers some of the degeneracy in the shells must be removed. By assuming a sufficiently strong spin-orbit effect (case (i)) the $(l \pm \frac{1}{2})$ terms are split. The $(l + \frac{1}{2})$ state is depressed in energy by an amount which is small among the light nuclei but increases with increasing l and exceeds the difference of energy between shells for $l > 2$. This results in a regrouping into shells which differ from those of the former list beyond the second shell: $(s_{\frac{1}{2}})$, $(p_{\frac{3}{2}}, p_{\frac{1}{2}})$, $(d_{\frac{5}{2}})$, $(s_{\frac{1}{2}}, d_{\frac{3}{2}})$, ... Closure of shells occurs at 2, 8, 14, 20, 28, 50, 82, and 126, which are the observed magic numbers.

With certain refinements, it is also possible to account for the principal descriptive features of nuclei on the shell theory, e.g. stability, ground state spins, magnetic moments and the general pattern of the energy level schemes. The important conclusion is that j-j coupling is 'stiffer' than L-S coupling, and predominates among heavy nuclei, although there is no simple explanation of this fact. Light nuclei are probably best represented by some intermediate form of coupling, similar to that postulated in explanation of many complex atomic spectra.

1.3. The two-nucleon system

Multi-body problems cannot be solved exactly. The two-nucleon system is expected to be the most amenable to study, and also the most rewarding, subject only to the reservation that the interaction between an isolated pair of nucleons is not necessarily typical of interactions within nuclear matter. It is usual to treat the system in terms of conservative forces, described by a potential $V(r)$ which, following §1.2, must be given a form which develops rapidly below a range of $\sim 10^{-13}$ cm.

Actually, results are insensitive to the detailed shape of the potential field. Functions commonly employed are:

Rectangular (or spherical) $V(r) = -V_0, \quad r < b,$
$$= \quad 0, \quad r > b. \tag{1.1}$$

Yukawa (or meson) $\qquad V(r) = -V_0 \dfrac{\beta}{r} e^{-r/\beta}. \tag{1.2}$

Exponential $\qquad V(r) = -V_0 e^{-2r/\beta}. \tag{1.3}$

Gaussian $\qquad V(r) = -V_0 e^{-r^2/\beta^2}. \tag{1.4}$

Morse $\qquad V(r) = -V_0 [2e^{2(r_1-r)/r_0} - e^{4(r_1-r)/r_0}]. \tag{1.5}$

Equation (1.1) assumes a constant potential over a definite range b. Equations (1.2)–(1.4) are characterized by a range parameter β beyond which the potential is negligible, while equation (1.5) introduces a repulsive core into the potential at very small r. (By convention, at complete separation the nucleons are assumed to have zero energy.)

1.3.1. *The deuteron.* The ground state of the deuteron and also the process of photodisintegration and its inverse, radiative capture of neutrons in hydrogen, provided early evidence of the spin dependence of the n-p force (§§6.13, 9.7). The chief features are described approximately by treating the deuteron as a neutron and proton held together in an S state by a central potential such as one of those above.

If a rectangular well is used, a relationship can be derived involving the binding energy W of the deuteron and the well parameters:

$$\cot \kappa b = -\left[\frac{W}{V_0 - W}\right]^{\frac{1}{2}}, \tag{1.6}$$

where $\kappa = \{m(V_0 - W)\}^{\frac{1}{2}} \hbar$ and m is the mass of a nucleon. The experimental value of W is $2 \cdot 226 \pm 0 \cdot 003$ MeV. A useful approximate expression which follows for $W \ll V_0$ is

$$V_0 b^2 \approx \frac{\pi^2 \hbar^2}{4m}. \tag{1.7}$$

Other short-range potentials give similar results. It can be shown further that the n-p system has no stable excited states, either for $l = 0$ or for higher values of orbital angular momentum.

Since the ground state of the deuteron has $l = 0$, the total angular momentum J is due to spin. Nucleon spins are $\frac{1}{2}$ and that of the deuteron 1, hence the nucleon spins must be aligned and the ground state is the triplet ^3S. This suggests that the magnetic moments of neutron and proton should add to give that of the deuteron. As $\mu_p = 2 \cdot 79275$ and $\mu_n = -1 \cdot 9131$ nuclear magnetons ($e\hbar/2mc$), their sum is $0 \cdot 87965$. The measured moment is $0 \cdot 8795$, in fairly close agreement although the difference $0 \cdot 00015$ falls slightly outside experimental error ($\sim 0 \cdot 00004$). Moreover, no combination of neutron spin S_n, deuteron spin S, orbital angular momentum L and sign of the neutron moment, leads to approximately the correct deuteron moment except $L = 0$, $S_n = \frac{1}{2}$, $S = 1$ and μ_n negative.

As the deuteron exhibits a small electric quadrupole moment, its charge distribution is ellipsoidal. This could be due to the presence of states of higher l, but since it has no electric dipole moment, an $l = 1$ state cannot occur. It is estimated (§9.7) that the deuteron spends about 4% of its time in the D-state ($l = 2$). Thus, the assumed central force which gives no mixing of states of different l, should be replaced by a non-central force (chapter 9). It is then possible to explain also the discrepancy between measured and calculated magnetic moments.

1.3.2. *Low energy neutron-proton scattering.* This forms the subject matter of chapter 6. The principal results obtained in the various energy ranges for which experimental data are available are summarized below:

Low energies ($E \gtrsim 10$ MeV.). To within experimental error, scattering in the CM system† is isotropic, as expected from the short

† Frames of references are discussed in §2.5.

range character of the force. Total cross-sections are in reasonable agreement with the theoretical equation (6.19) in their dependence on the collision energy and the binding energy of the deuteron. Theoretical values are somewhat low, the discrepancy being $\sim 20 \%$ at several MeV. but much worse as the energy decreases. The triplet interaction dominates scattering except at very low energy.

Slow neutron and thermal energies. As the neutron energy decreases, the *n-p* cross-section increases much more rapidly than the theoretical, and reaches a steady value of ~ 20 b. at ~ 1 keV. This is attributed to a singlet state of the deuteron with a binding energy $|W_s| \approx 60$ keV., (§ 6.5), but of undetermined sign. This constitutes definite evidence for spin dependence of the force.

Scattering by molecular hydrogen has proved especially useful because the molecules exist in both *para*- and *ortho*-configurations. When sub-thermal neutrons (§ 4.4.2) of wavelength greater than the distance apart of the two protons interact with a molecule, coherent scattering occurs at the two centres, giving rise to interference effects. Interference is constructive and destructive for ortho- and para-hydrogen respectively. A comparison establishes that the *n-p* force is spin-dependent, the ^1S state of the deuteron is virtual (unbound) and also that the neutron spin is $\frac{1}{2}$ (§ 6.10). More accurate information has been obtained from neutron diffraction by crystals and from total reflexion by liquid mirrors (§§ 6.11, 6.12).

1.3.3. *Low energy proton-proton scattering.* This is dealt with in chapter 7. Experimentally it is much easier to investigate than *n-p* scattering, leading to better accuracy (cf. § 4.1). However, the Coulomb repulsion must be overcome before protons come within range of nuclear forces, which means that bombarding energies must be at least 100 keV. to involve detectable nuclear scattering. Differential cross-sections are now known to about 1 % from ~ 0.2 to 15 MeV. but the accuracy declines at higher energies.

To within experimental errors, low energy scattering is isotropic (*s*-wave only). The observed angular distribution can be accounted for by any short range potential, with a suitable choice of depth and range parameters (equations (1.1)–(1.5)). The Pauli principle excludes the ^3S case, so that only ^1S scattering occurs. Singlet pp and np forces are found to be closely the same, and the concept of charge independence (§ 1.2.4) draws its strongest support from this fact. *The force between any two nucleons in the same spin and orbital angular momentum state is independent of the charge on the nucleons.* Evidence that the nn force equals the pp force is the weakest link in this generalization. n-n scattering is impossible to observe. However, the non-existence of the di-neutron accords with the conclusion from analysis of p-p scattering that the di-proton should be unstable.

1.3.4. *High energy neutron-proton and proton-proton scattering.* Above about 10 MeV. states of non-zero l can be expected, about which the angular distribution should provide information. Bombarding energies at which $l=1$, $l=2$, ..., effects set in should indicate the strengths of the interactions in these states. The angular distribution of scattering should reveal the special characteristics of the potential itself.

The observed n-p distribution becomes detectably anisotropic somewhat above 10 MeV. and at 30 MeV. is markedly so. However, p-p scattering remains isotropic to ~ 350 MeV., its absolute magnitude being approximately independent of energy above ~ 100 MeV. At 1 BeV., $\sigma(30°) \approx 12\sigma(90°)$.

The analysis of high energy scattering is dealt with in chapter 10. It is found that n-p scattering cannot be explained by an ordinary force, which predicts a strong forward peak. A pure Majorana force predicts too much backward scattering and can likewise be ruled out. The most commonly used potential is of the Serber type in which ordinary and exchange forces enter with equal weight (§ 10.5). Odd orbital angular momenta do not contribute, hence the cross-section should be symmetrical about 90°.

Observed n-p cross-sections are in general agreement with this prediction up to several hundred MeV., which provides strong evidence for exchange forces in two-body interactions. Above 300 MeV. the ratio of backward to forward scattering is ~ 3 and the Serber model breaks down. Also, if theory is adjusted to fit experiment at high energy, the predicted cross-sections at ~ 40 MeV are too large. Moreover, the terms in odd l which the Serber potential excludes are specifically those required to provide the repulsive force which accounts for saturation (§ 1.2.3). In the case of p-p scattering the position is even less satisfactory. The isotropic distribution is inconsistent with a Serber potential unless further assumptions are introduced. At 300 MeV., the absolute cross-section is about twice the maximum allowed s-wave contribution. Several special models have been proposed but most of them have a somewhat arbitrary character (see § 10.1.1). None yields agreement from low to high energies and remains consistent with n-p models. Whether the charge independence hypothesis is valid at high energies is undecided (§ 10.1.2). The position may be clarified by more accurate high energy experiments, especially those involving polarized protons (§ 10.1.4 et seq.).

1.4. The three-nucleon system

Some useful information is forthcoming from a study of more complex systems; in particular investigations of systems of three nucleons provide evidence for the equivalence of nn and pp forces. ^{3}H and ^{3}He are the lightest mirror nuclei and p-d and n-d scattering are the simplest mirror reactions (see chapter 8).

1.4.1. *Binding energies of* ^{3}H *and* ^{3}He. These nuclei have binding energies of 8·49 and 7·73 MeV. respectively, the difference being consistent with Coulomb repulsion between the pp pair in the latter nucleus. Neither nucleus has any orbital angular momentum, hence the evidence is that the nn and pp forces are about equal in the ^1S state.

1.4.2. *Neutron-deuteron and proton-deuteron scattering*. Because of the large size of the deuteron, in n-d and p-d scattering waves of higher orbital angular momenta participate than is the case in n-p or p-p scattering at the same energy. This makes possible a comparison of nn and pp forces in other than s-states, which is important because the s-state does not exhibit exchange effects. The angular distributions of scattering are not isotropic but peak strongly in the forward and backward directions. There is a close resemblance between the two distributions except at small angles where Coulomb effects are present. This is consistent with equality of nn and pp forces in higher l states. Moreover, theoretical calculations are in agreement with experiment when exchange-type forces are used, but in definite disagreement with ordinary forces (cf. §8.2.3). At high energies the cross-section for n-n scattering can be derived from $\sigma_{nd} - \sigma_{np}$ (§11.1) and agrees with σ_{pp}.

Similar conclusions have been drawn from inelastic scattering and also from the similar asymmetries which have been observed at low energy in the angular distribution of particles from ^{3}H(d, n) ^{4}He and ^{3}He(d, p) ^{4}He.

1.5. The meson description of nuclear forces

The meson theory of nuclear forces attempts to explain nuclear interactions by field theoretic considerations. Thus two nucleons exert a force on each other by exchanging a virtual meson, the meson transferring the momentum exchanged. Of the many theories which have been proposed, few lead to useful results and even these are beset by mathematical and conceptual difficulties.

Yukawa's original idea was formed by analogy with electromagnetic theory, the force between nucleons being described in terms of a meson field. For the ^1S potential, meson theories give a potential energy of the form

$$V(r) = C e^{-r/\beta} / (r/\beta), \quad \beta = \hbar/mc, \quad (1.8)$$

where the range β of the Yukawa potential is inversely proportional to the meson mass m. Triplet states (^3S, ^3P, ...) have a more complicated radial structure and in general involve the non-central force component S_{12}:

$$S_{12} = 3(\boldsymbol{\sigma}_1 \cdot \mathbf{r})(\boldsymbol{\sigma}_2 \cdot \mathbf{r})/r^2 - \boldsymbol{\sigma}_1 \cdot \boldsymbol{\sigma}_2, \quad (1.9)$$

where the $\boldsymbol{\sigma}$'s are Pauli spin operators related to spin $\mathbf{S}$ by $\mathbf{S} = \frac{1}{2}\hbar\boldsymbol{\sigma}$.

1.5.1. *The π-meson spin and parity.* The π-meson has a strong interaction with nuclear matter (cross-section $\sigma \sim 1b$ as compared with $\sim 10^{-6}b$ for μ-mesons), so it is natural to identify it as the Yukawa particle. Its properties are listed below:

(1) There are three types of π-meson, π^+, π^- and π^0 carrying unit positive, unit negative and zero electronic charge respectively.

(2) By conservation of spin, the reaction $p + p \rightleftharpoons \pi^+ + {}^2$H fixes the spin of π^+ as 0. Similarly the reactions $\pi^- + p \rightarrow \gamma + n$,

$\pi^- + {}^2H \rightarrow n + n$ give the spin of π^- as 0 and its mass as $273 \cdot 3 m_e$. That π^0 has zero spin follows from the reaction $\pi^- + p \rightarrow n + \pi^0$, $\pi^0 \rightarrow 2\gamma$.

(3) The capture of π^- in deuterium $\pi^- + {}^2H \rightarrow n + n$, fixes the parity of π^- as negative, owing to the exclusion principle and the requirements of spin and parity conservation.

It is therefore believed that all three π-mesons have zero spin and odd parity. Particles with these properties are called pseudoscalar. This conclusion automatically rejects certain other possibilities—scalar (spin 0, even parity), vector (spin 1, odd parity) and pseudovector (spin 1, even parity) mesons and their associated fields.

1.5.2. *Exchange theories.* There are three types of exchange force important in the meson theory of nuclear interactions: those postulated in the '*neutral*', '*charged*' and *symmetrical* theories, respectively. The neutral theory involves exchange of π^0's only, resulting in an ordinary-type force attractive in all angular momentum states which does not lead to saturation. The charged theory uses π^+ and π^- mesons only, so that there is no force between two protons or two neutrons to second order, although there might be such a force in higher order approximations.

The most important type of exchange force is that of the symmetrical theory, which involves equal numbers of π^0, π^+ and π^- mesons. On this theory n-n, p-p and n-p forces occur and are repulsive in odd states of angular momentum, so leading to saturation. The charge-symmetry property and approximate equality of singlet n-p and p-p interactions is also obtained on this theory.

1.5.3. *Weak coupling theory.* This assumes weak coupling of a nucleon to its associated meson field. The nucleon-nucleon interaction is therefore not considered as strong. Thus the nuclear potential should be expressible as a perturbation series in powers of the interaction constant $g^2/4\pi$. However, the best form of interaction energy cannot be deduced from the experimental data owing to divergence difficulties in the theory. For example, the second order perturbation expression for the n-p potential contains a $1/r^3$ term leading to an infinite deuteron binding energy, and this necessitates arbitrary remedies such as a short range cut-off.

The procedures for renormalization of mass and charge, although successful in quantum electrodynamics, have proved inapplicable to many varieties of meson theory. However, the pseudoscalar theory with pseudoscalar coupling, hereafter denoted by $PS(ps)$, is renormalizable and fits the experimentally established pseudoscalar character of the π-meson. Nevertheless, the whole renormalization prescription still depends on expansions in powers of the coupling constant, and these show no signs of convergence after renormalization ($g^2/4\pi \sim 10$).

1.5.4. *The Feynman diagrams.* The interaction of two nucleons (including scattering) may be pictured by means of Feynman diagrams. Corresponding to the emission of a meson by one nucleon

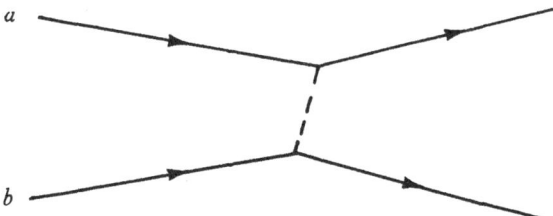

Fig. 2. Feynman diagram for interaction of two nucleons (solid lines) by exchange of one meson (dotted line). This represents the second order interaction energy.

and its absorption by another is the second order interaction energy. This is obtained by working out the change in the wave function for the whole system, to the first order in the coupling constant g, and then finding the expectation value of the total energy. Each nucleon changes the wave function by terms of order g, producing the 'self-field' of the nucleon. This gives rise to terms of order g^2 in the total energy obtained by applying the interaction energy of a second nucleon. The second order process is pictured in fig. 2, nucleon and meson lines being represented by full and dotted lines respectively.

The fourth order interaction, resulting in a contribution of order g^4 to the interaction energy, involves exchange of two mesons, so that there are two diagrams, one corresponding to each of the two possible exchange sequences (fig. 3). Higher order interactions can be pictured similarly.

However, difficulties arise, among them the divergence trouble. Fig. 4(a) illustrates two situations in which divergences occur; the first showing a meson being emitted and reabsorbed by one of the nucleons before a meson exchange occurs with the other nucleon. As the meson field round a point nucleon varies as $1/r$ for small r, the resulting contribution to the field energy of this meson cloud is infinite. Nevertheless the requirements of Lorentz and gauge invariance make it possible to deal consistently with these infinities.

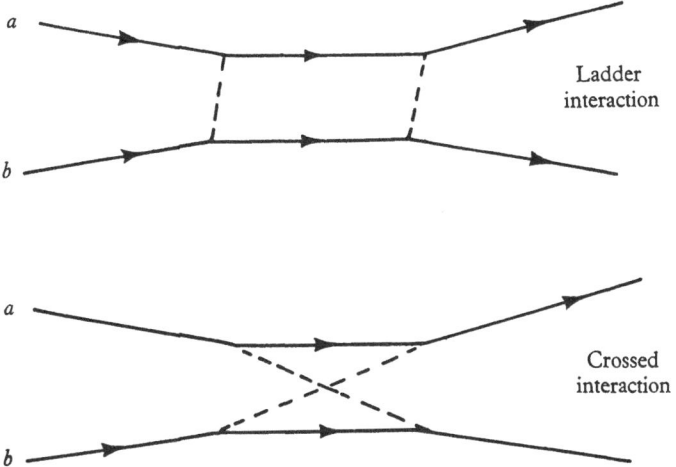

Fig. 3. Feynman diagrams for two possible fourth-order nucleon-nucleon interactions in which mesons are exchanged. They represent the fourth-order interaction energy.

They are cancelled by making infinite changes in the mass and charge of the nucleon, leaving all physically observable quantities constant.

Fig. 4(b) illustrates vacuum polarization in which a meson being exchanged ejects a proton from a negative energy state, producing a proton-antiproton pair. Although reabsorbed, the latter's interaction with the protons filling the negative energy states in a vacuum modifies electromagnetic interaction. Infinite effects result, both mass and charge having to be changed in the renormalization process, during which infinite integrals are encountered.

However, although the $2n$th order contribution to the interaction energy can be evaluated and renormalized in the $PS(ps)$ theory, the series is very probably divergent. Thus there is really no logically consistent weak coupling theory of nuclear forces.

1.5.5. *The hard core potential.* A possible solution has been suggested (Nishijima, 1951; Jastrow, 1950, 1951). Virtual states involving several mesons lead to correspondingly large momentum changes, so that two mesons simultaneously emitted give a range half that for one meson ($\Delta x \sim \hbar/\Delta p$). Similarly the $2n$th order perturbation corresponding to n mesons simultaneously emitted leads to a range roughly $1/n$ times that for one meson. Consideration of these effects and nucleon recoil suggests that second order

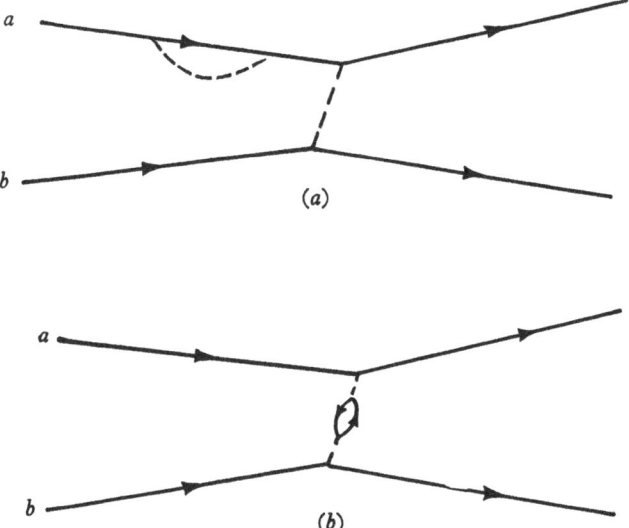

Fig. 4. (*a*) Illustration of the divergence difficulty due to mesic self energy of the nucleon. (*b*) Illustration of the divergence due to polarization of the vacuum in which a virtual meson being exchanged between two nucleons emits and reabsorbs a nucleon and an anti-nucleon.

perturbation theory should be taken seriously only for $r > r_0$, as the higher order terms predominate increasingly for smaller r.

A number of authors have worked out the second and fourth order potentials $V^{(2)}$ and $V^{(4)}$, amongst these Levy (1952 *b*), who assumed a hard core for $r < r_0$ due to heavy mesons and possible isobaric states of nucleons. This leads to approximate agreement with the deuteron binding energy and the low energy n-p scattering parameters used in the shape-independent approximation (cf. §6.7). A rough value of the deuteron quadrupole moment is also obtained.

1.5.6. *Strong coupling theory and nucleon isobars.* A nucleon is inseparably bound to its proper meson field, this having associated with it a quantized amount of angular momentum or charge and a large amount of energy. Thus excited states of nucleons should exist exhibiting charge and spin values different from those of the ground state, always subject in number to an upper limit imposed by the possible spontaneous emission of mesons. On the strong coupling theory (if $g^2/4\pi \sim 10$ is considered as corresponding to very strong coupling) isobars of the nucleon ground state should have excitation energies of the order of a few hundred MeV. or less. However, the strong coupling theory of Wentzel (1940, 1942) predicts a coupling between spin and charge, resulting in equal and opposite magnetic moments for the neutron and proton, in contradiction to experiment. The properties of the deuteron are also in conflict with a large spin inertia for the nucleon. Hence strong coupling theory is usually not taken very seriously, although nucleon isobars have been suggested as providing an explanation of possible resonances in the scattering of mesons of several hundred MeV. energy by hydrogen.

1.5.7. *Intermediate coupling theory.* There is considerable evidence that the interaction constant $g^2/4\pi \sim 10$ actually corresponds to intermediate coupling. This would still allow semi-stable nucleon isobars to exist with energies as low as 400 MeV. for the case of a very large source, and higher energies for smaller sources (Harlow and Jacobson, 1953).

Tomonaga (1947) has dealt with the stationary state of a nucleon coupled to a scalar meson field, by a method applicable to intermediate coupling, when neither the perturbation theory (weak coupling) nor the method of Wentzel (strong coupling) converges. The meson field is treated from the particle aspect, writing the Schrödinger equation for the eigenvalue problem as a number of simultaneous integral equations operating on functions ϕ_0, $\phi_1(k_1^+)$, $\phi_2(k_1^+, k_1^-)$, $\phi_3(k_1^+, k_2^+, k_1^-)$, ..., according to the number of positive and negative mesons involved. The equations are solved approximately by the Hartree method, namely by assuming the ϕ's can be written as the product of one-meson wave functions. The relative proportions of the various possible meson states as represented by the ϕ's can be found by using the Ritz variational method to minimize the expectation value of the energy of the system. The nucleon is treated as infinitely heavy so that its recoil and nucleon pair production are neglected. High momentum components are also cut off arbitrarily to secure convergence. The method agrees well with the perturbation method in the limit of weak coupling and also with Wentzel's result for strong coupling. Iteration of Tomonago's solution has further shown its accuracy to be better than 3 % under conditions of intermediate coupling. The most probable number of virtual mesons around a nucleon turns out to be about two, the number distribution roughly obeying a Poisson formula, so that processes involving more than about seven or eight mesons are not very important.

Intermediate coupling theory has been extended to the $PS(ps)$ field, using approximate methods (Mathews and Salam, 1952). For a coupling constant $g^2/4\pi \sim 10$, preliminary estimates indicate that a renormalized weak coupling expansion gives a very slowly converging picture of the meson cloud surrounding a nucleon.

CHAPTER 2

THEORETICAL CONCEPTS AND EXPERIMENTAL METHODS

2.1. Nuclear collisions

When two small bodies are subject to a mutual force they are capable of colliding with each other. A collision in this sense implies an interaction which detectably influences their velocities, their structure, or both. Fundamentally an interaction amounts to a redistribution of total energy and momentum.

The study of nuclear scattering is concerned with one type of event which takes place when atomic nuclei collide. In all practical cases in the laboratory we are concerned with the impact of a fast-moving nuclear particle a on a nucleus A which is at rest. These are customarily referred to as the *incident* (or *projectile*) particle and the *target* particle, respectively. The types of interaction which occur may be classified in various ways, the most obvious being based on considerations of conservation of kinetic energy E and conservation of particles:

(a) $\qquad a_{E_0} + A \rightarrow a_{E_1} + A_{E_2}$, where $E_0 = E_1 + E_2$.

Both particles are preserved unaltered by the event but a transfer of kinetic energy occurs, the target particle receiving kinetic energy equal in amount to that lost by the incident particle. By analogy with the collision of two macroscopic elastic bodies, this is called *elastic scattering*.

(b) $\qquad a_{E_0} + A \rightarrow a_{E_1} + A^*_{E_2}$, where $E_0 = E_1 + E_2 + W$.

Part of the kinetic energy of the projectile is used to increase the internal energy of the target particle, which is left in an excited state W and subsequently reverts to its ground state by γ-emission. No change occurs in the (N, Z) composition of either particle and the process is referred to as *inelastic scattering*.

(c) $\qquad a_{E_0} + A \rightarrow b_{E_1} + B_{E_2}$, where $E_0 + Q = E_1 + E_2$.

In the most general case, a rearrangement of the protons and neutrons which constitute the particles occurs, and the final products are different in (N, Z) composition from the reactants. However, N and Z are separately conserved in the reaction, hence the nucleons may be considered as redistributed among different nuclei whilst preserving a distinct existence. Q, which may be positive or negative, is the nuclear disintegration energy $[(m_1 + m_2) - (m_3 + m_4)]c^2$, where m_1, m_2, m_3 and m_4 are the masses of particles a, A, b and B respectively.

(d) At high energies, nucleons themselves become subject to inelastic processes and generate π-mesons. The energetics of the reaction are the same as in (c) where the particle b is now a meson, $m_3 = 274m_e$. An absolute threshold, based on the assumption that a interacts with the whole target nucleus A, is given by

$$[(m_3 + m_4)^2 - (m_1 + m_2)^2] c^2/2m_2.$$

For incident protons this threshold energy is about 290 MeV. when the target is hydrogen, 209 MeV. when it is deuterium, 181 MeV. when it is helium and 150 MeV. when it is carbon. The figures are not very different for incident neutrons. However, experimentally determined thresholds are rather higher than those given by the above formula because meson production is strictly a nucleon-nucleon process.

It is the first of these types of interaction, elastic scattering by nuclei, which forms the main subject of this monograph. The others will be introduced either in the course of the general theory which describes the whole collision process or, on the experimental side, if they compete effectively with elastic scattering so as to interfere with the measurements. However, a detailed theory of nuclear collisions reveals that elastic scattering must be regarded as a composite of two parts, *potential* and *resonance* scattering, the latter of which involves the same basic mechanism as (b) and (c), viz. the formation and subsequent disintegration of a 'compound nucleus'.

Potential scattering takes places only in the external field (Coulomb + nuclear) of the target particle. The incident particle penetrates to the surface of the nucleus but does not enter it. Resonance scattering is a phenomenon associated with penetration of the nuclear surface. To understand this, and its relationship to the other processes, calls for a brief digression on the compound nucleus model of nuclear reactions.

Bohr proposed in 1936 that a reaction of type (c) proceeds in two stages, represented by

$$a + A \rightarrow C^* \rightarrow b + B.$$

Particle a coalesces with A to form an excited compound nucleus C in which the original particles lose their identity, while the incident energy becomes shared among all the nucleons. C exists in a quasi-stationary (or 'virtual') quantum level whose excitation energy is the binding of a to A, plus the relative collision energy (equation (2.19)). The mean life of C^* is $10^{-16 \pm 3}$ sec., which is long compared with the collision time ($\sim 10^{-21}$–10^{-22} sec.). After a time of the order of its mean life, C^* disintegrates into the observed particles b and B.

In general, disintegration can occur in a variety of competing modes, depending on the excitation energy of the level and on the particular nucleus C involved. One possibility is obviously the emission of a particle of type a. If this happens without loss of kinetic energy, we have a case of elastic scattering, indistinguishable on grounds of energetics from potential scattering. Such a mechanism becomes much more probable if E_0 corresponds to one of the quantized excited levels of C, in which case a can be absorbed into C and re-emitted before any other competing process can occur. 'Resonance scattering' is the term used to describe this.

Scattering by the lightest nuclei, to which many of the following chapters are devoted, involves only the potential part, because no excited states of the compound nuclei exist. However, in the general case, both phenomena contribute simultaneously and coherently, and interference between Coulomb, nuclear and resonance scattering leads to complex effects. These will be discussed in detail in chapter 12.

It is evident from such considerations that a treatment of elastic scattering in isolation would be somewhat artificial if not impossible. Another instance of the inter-relation of nuclear processes is found in deuteron stripping (chapter 11). Though classified by energetics according to (c), potential scattering is also involved. The deuteron dissociates in the potential field of the nucleus, only one of its nucleons being captured to form a compound nucleus.

2.2. Kinematics of elastic collisions

Both the planning of experiment and its theoretical interpretation depend on a knowledge of the basic kinematics of scattering. Consider the collision illustrated in fig. 5 between an incident particle of mass m_1 and a target particle m_2 which is at rest in the laboratory. The collision scatters m_1 through an angle Θ_1 and causes m_2 to recoil at Θ_2. E_0 is the incident energy and E_1 and E_2 the energies of m_1 and m_2 respectively after collision. By applying the laws of conservation of momentum and kinetic energy, relationships can be established between the masses, angles and energies of the two particles.

The expressions are simple for low energy collisions where particle velocities are negligible compared with the velocity of light, but for high energies the relativistic treatment must be employed and results tend to become formidable. Derivations have been carried out in Appendix A. The principal results are quoted below for the non-relativistic case which applies to all low energy experiments and even at 100 MeV. is not seriously in error, for nucleon-nucleon collisions.

$$\frac{m_1}{m_2} = \frac{\sin(2\Theta_2 + \Theta_1)}{\sin \Theta_1},$$ (2.1)

$$\frac{E_1}{E_0} = \left\{ \frac{\sin \Theta_2}{\sin(\Theta_1 + \Theta_2)} \right\}^2,$$ (2.2)

$$\frac{E_2}{E_0} = \frac{m_1}{m_2} \left\{ \frac{\sin \Theta_1}{\sin(\Theta_1 + \Theta_2)} \right\}^2 \quad \text{or} \quad \frac{4 m_1 m_2 \cos^2 \Theta_2}{(m_1 + m_2)^2}.$$ (2.3)

2.3. Definition of cross-sections

The kinematics of collisions presupposes nothing about the force between the bodies, except its existence. The above expressions contain no information about the probability of scattering occurring or the angular distribution of the scattered particles. These are the features which depend upon the nature of the force and which experiments are generally designed to determine.

As is the case with other nuclear processes, the probability of scattering is specified by the cross-section. Suppose I_0 is the intensity in particles/cm.²/sec. incident in a collimated beam on a target containing N_v nuclei/cm.[3]. Attentuated by various nuclear processes during its passage through the target, the undeflected beam is reduced in intensity to I_x, which may be related to I_0 as follows. If the intensity is reduced from I to $I - dI$ by an element dx at a distance x through the target,

$$\frac{dI}{I} = -\sigma N_v dx.$$

On integration, $\qquad T = \dfrac{I_x}{I_0} = e^{-\sigma N_v x} \quad \text{or} \quad e^{-\sigma N_a}, \qquad (2.4)$

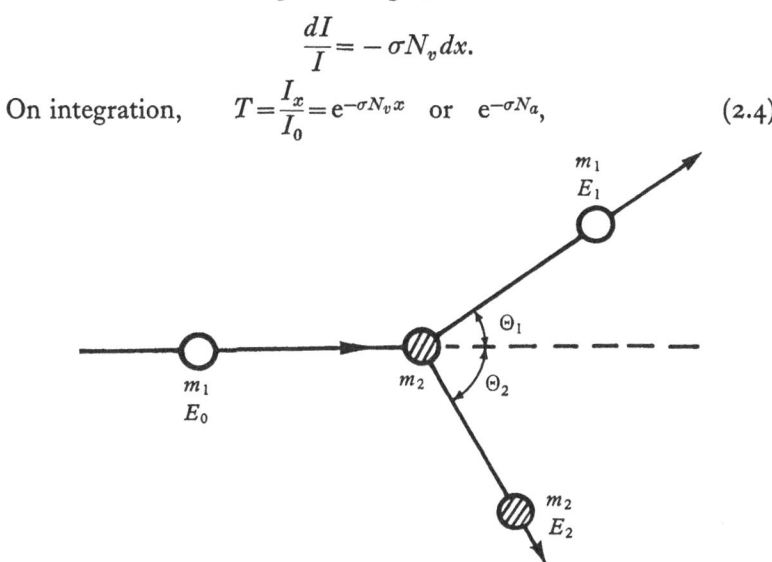

Fig. 5. Elastic collision between mass m_1 moving with energy E_0 and mass m_2 at rest in the laboratory system.

where N_a is the number of nuclei/cm.[2] or 'surface density' of the target. T is the *transmission* and the constant of proportionality σ, which has dimensions of area, is termed the nuclear *cross-section* for removal of particles from the beam. Its units are cm.[2] or *barns* ($1 \text{ b.} = 10^{-24} \text{ cm.}^2$). Classically, it is the effective area which the target nucleus must present to the bombarding particles in order to account for the diminished intensity.

Defined thus, σ covers all processes which remove particles from the parallel beam, viz. elastic and inelastic scattering and nuclear reactions. Partial cross-sections may be written for each type of event

$$\sigma = \sigma_{el} + \sigma_{in} + \sigma_r. \qquad (2.5)$$

We are chiefly concerned here with the elastic process and will use σ to designate the *total elastic cross-section*, unless otherwise stated.

The probability that a particle will be scattered out of a parallel beam is, by equation (2.4),

$$P = (1 - e^{-\sigma N_a}),\qquad (2.6)$$

or, expressed in series form,

$$P = \sigma N_a - \frac{(\sigma N_a)^2}{2!} + \dots \qquad (2.7)$$

For the thin targets generally employed in scattering, $\sigma N_a \ll 1$, and therefore

$$P = \sigma N_a,\qquad (2.8)$$

which defines σ as the probability of scattering per target nucleus/cm.^{2}.

Other common measures of probability can be related simply to σ. σN_v, which occurs in equation (2.4), is the familiar 'coefficient' of exponential attenuation laws in general. In neutron physics it is written as Σ, the *macroscopic cross-section*, and represents the total effective area of all the nuclei in 1 cm.^{3}. The reciprocal of Σ can be shown to be the *mean free path* for scattering λ. From equation (2.6), the probability of scattering per unit target thickness is

$$\frac{d}{dx}(1 - e^{-\Sigma x}) = \Sigma e^{-\Sigma x},$$

whence the mean free path is

$$\lambda = \frac{\displaystyle\int_0^\infty x\Sigma e^{-\Sigma x}.dx}{\displaystyle\int_0^\infty \Sigma e^{-\Sigma x}.dx} = \Sigma^{-1}.\qquad (2.9)$$

Numerator and denominator are standard forms, the latter being equal to unity. λ is obviously the target thickness which would reduce the incident intensity to $1/e$ by scattering.

Summarizing the above relationships, if N particles are scattered from an incident beam of N_i particles by a 'thin' scatterer,

$$P = \frac{N}{N_i} = \sigma N_v x = \Sigma x = \frac{x}{\lambda}.\qquad (2.10)$$

The definition of cross-section may be further restricted in order to include a specification of the way in which scattered particles are distributed in space. Assuming that scattering is symmetrical in azimuthal angle about the beam axis (a valid assumption in the absence of polarization effects, § 2.4) a *differential scattering cross-section* $d\sigma(\Theta)/d\Omega$ may be defined,†

$$\frac{d\sigma(\Theta)}{d\Omega} \approx \frac{N(\Theta)}{N_i N_v x \Delta\Omega}, \qquad (2.11)$$

where $N(\Theta)$ is the number of particles scattered through an angle Θ into solid angle $\Delta\Omega$. The differential cross-section (written hereafter as $\sigma(\Theta)$) is therefore the probability of scattering through angle Θ per unit solid angle per target nucleus/cm.². It is related to σ by

$$\sigma = \int_{\Omega} \sigma(\Theta)\, d\Omega = 2\pi \int_{\Theta} \sigma(\Theta) \sin \Theta \,.\, d\Theta. \qquad (2.12)$$

The principal purpose of scattering experiments is to measure the total cross-section σ and the angular distribution $\sigma(\Theta)$ vs. Θ.

2.4. Scattering experiments

The overwhelming majority of experiments have been concerned with *single scattering*, i.e. the angular distribution of particles scattered once from a single target. Although much diversity can be found in the technical details of equipment, the general arrangement has tended to standardize on several distinct forms which have proved eminently satisfactory. These are introduced here as a preliminary to more comprehensive discussion in the following chapters.

A familiar form for low energy scattering is shown in fig. 6. A beam of charged particles from an accelerator is collimated by two small slits, passed through the target gas contained in a chamber and stopped by a current collector at the far side. A detector is mounted behind a slit unit which can be rotated about the centre of the chamber in order to measure the intensity of scattering at various angles Θ. The slit unit serves three purposes. It defines the effective target thickness seen by the detector, the solid angle of acceptance of the detector and determines the angular resolution.

Adopting the standard notation for this geometry, we denote the distance from the second slit to the beam by R_0 and that between slits by h. $2b$ is the width of the first slit and the area a of the second

† In the limit as $\Delta\Omega \to 0$, this approximation becomes an identity and will be treated as such henceforth.

is the effective detector aperture. When the detector is located at angle Θ an infinitesimal area of counter aperture sees a length $2bR_0/h \sin\Theta$, which is therefore the effective target thickness x of equation (2.11). The solid angle defined by the aperture is $\Delta\Omega = a/R_0^2$. Then the differential cross-section is related to $N(\Theta)$, the number of scattered particles recorded by the detector, by

$$\sigma(\Theta) = \frac{N(\Theta)\sin\Theta}{N_i N_v G},\qquad (2.13)$$

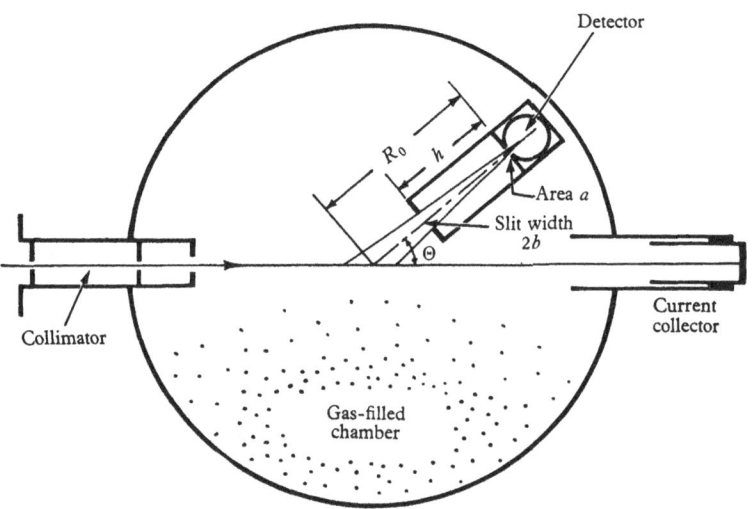

Fig. 6. Essentials of scattering equipment, showing basic geometry of detector system.

where the geometry factor $G = 2ba/R_0 h$ and N_i and N_v are as defined earlier.

This is, of course, an idealized two-dimensional version of the geometry. It neglects the finite width of the main beam and also the distortion introduced by the angular spread of the scattered particles. The latter arises because each element of the area a receives particles from an angle $\Delta\Theta = 2b/h$ (typically 1–3°). If N varies linearly with Θ, the yield averaged over an interval $\Delta\Theta$ about the central axis of the detector Θ_0 correctly measures the value of $N(\Theta_0)$, at least in the two-dimensional approximation. However, such is not the case if the dependence is non-linear, i.e. if $N''(\Theta_0)$

is finite, where the primes denote double differentiation with respect to Θ. The number of counts actually observed at Θ_0 is

$$\int_{\Theta_0-\Delta\Theta/2}^{\Theta_0+\Delta\Theta/2} N(\Theta)\,d\Theta,$$

which is not the same as $N(\Theta_0)\,\Delta\Theta$. The ratio is approximately

$$1 + \frac{N''(\Theta_0)}{N(\Theta_0)}\frac{(\Delta\Theta)^2}{24},$$

which is the factor by which observed values of the yield ($N(\Theta)$) in equation (2.13)) are too large. At small angles, where the Rutherford yield $N(\Theta)$ varies approximately as Θ^{-5}, geometrical corrections due to finite aperture are commonly several per cent, but they are usually negligible when nuclear scattering predominates.

The above is still only a rough approximation to the magnitude of expected errors. A precision experiment warrants careful attention to 'second order geometry', which calls for a detailed analysis in three dimensions. Though not unduly difficult, the treatment is too laborious to embark upon here (see Breit, Thaxton and Eisenbud, 1939; Critchfield and Dodder, 1949a).

It is important to notice that the experimental arrangement shown in fig. 6 is not limited to counting scattered particles. In general, recoil particles will be recorded also, and methods must be found for distinguishing them (cf. chapter 3). When the incident particles are neutrons, counting the recoil particles provides the only reliable means of obtaining the angular distribution. The physical information conveyed is just as complete; in fact the distribution is the same as that for scattering when referred to the centre of mass of the colliding particles (§ 2.5).

An arrangement for charged-particle scattering which is extensively employed, especially at high energies, involves the *coincidence technique* as shown in fig. 7. Two counters located at angles Θ_1 and Θ_2 form a *conjugate* pair. If a scattered particle from the target enters A, the recoil particle from the same collision will enter B. When electrical signals from the two counters are fed to a coincidence circuit, each coincidence records the scattering of a particle from a target nucleus which has the correct mass to satisfy equation (2.1). The coincidence rate is therefore independent (neglecting accidental coincidences) of scattering by other nuclei in the target material. One of the counters has an aperture of known area a located at distance R_0 from the target. This counter defines the solid angle and is referred to subsequently as the *defining* counter; the other is called the *conjugate* counter. In practice, either

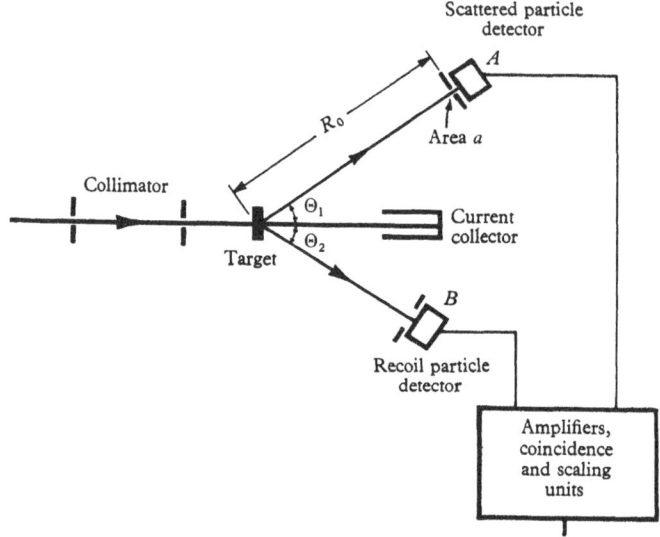

Fig. 7. Schematic diagram of arrangement of apparatus for two-particle coincidences.

scattered or recoil particles may be defined. The cross-section is evaluated from

$$\sigma(\Theta) = \frac{N(\Theta)\, R_0^2}{N_i N_a a} \qquad (2.14)$$

to the same degree of approximation as equation (2.13). N_a is the number of target nuclei/cm.^{2} which have the correct mass to allow coincidences. In principle, the coincidence technique is an elegant one, although it falls prey to a number of practical difficulties (§§ 3.5.2, 3.5.4).

A determination of the total cross-section for neutron scattering does not follow readily from any of the above procedures. The coincidence technique is inapplicable to neutron studies, while the single-counter method yields only the shape of the angular distribution unless the flux N_i is known. In some instances the latter can be found (§4.3 *et seq.*), but the most straightforward procedure is to measure the attenuation of the neutron beam in passing through a target of known composition and thickness. Such an experiment is variously described as a *transmission* or *attenuation* experiment.† The geometrical arrangement is shown in fig. 45. By equation (2.4), σ can be found if $N_v x$ and the ratio I_x/I_0 are known. Provided the incident neutron flux remains constant throughout the experiment, neither it nor the efficiency of the neutron detector enter into the determination. However, the theory holds only in the case of 'good geometry', i.e. when the source is essentially a point and distances between the three components are large compared to the sizes of scatterer and detector (§4.3.5). If σ is obtained by transmission, an angular distribution can be normalized by applying equation (2.12).

Unfortunately, the transmission cross-section is the *total* for all processes which remove neutrons from the beam (equation (2.5)). While it can be obtained with considerable accuracy, isolating the contribution due to scattering may be much more difficult, depending on the relative intensities of the partial processes. The problem is considered in §§4.3.3 and 8.3 in connexion with the break-up of the deuteron in n-d scattering, and similar difficulties arise in n-^{3}H, n-^{3}He and n-^{4}He scattering if the bombarding energy exceeds the threshold for disintegration.

Transmission studies owe their feasibility in general to two circumstances: (1) at MeV. energies elastic scattering is always the major process and may completely dominate all others, e.g. in the important n-p case competition is negligible until well beyond the π-meson production threshold at 290 MeV. (ii) Even when inelastic events are appreciable in proportion, the problem is not insoluble providing the relative intensity σ_{el}/σ_{in} can be measured. Suppose this is done in an angular distribution experiment. Then the absolute scale of σ_{el} can still be fixed by equating

$$\int_\Omega [\sigma(\Theta)_{el} + \sigma(\Theta)_{in}] \, d\Omega = \sigma.$$

† Transmission experiments are not applicable to charged particle scattering, because at energies below several hundred MeV. the range in matter is less than the mean free path for nuclear interactions. They have been used at BeV energies (e.g. Shapiro, Leavitt and Chen, 1954).

Another aspect of scattering—perhaps the most promising of recent developments—involves the use of polarized beams of protons and neutrons in order to investigate the properties of spin-dependent interactions. If polarized targets were available also, interactions could be studied as a function of the relative spin orientations of incident and target particles. Technical difficulties are formidable and very little which concerns scattering has been attempted with polarized targets. However, resonance absorption of thermal neutrons by partly polarized nuclei has been used to determine the angular momenta of levels of the compound nuclei concerned (see below). Experiments have also been carried out on aligned radioactive nuclei with the object of measuring the departure of their emitted radiation from spherical symmetry, which can lead to values for the magnetic moments of the nuclei.

Consider first the polarization of beams of particles. 'Polarization' here refers to the alignment of particle spins along a definite direction. Then

$$P = \frac{N^+ - N^-}{N^+ + N^-}, \qquad (2.15)$$

where N^+ and N^- are the numbers of particles with spins in the + and − directions respectively. In general, protons and neutrons have randomly oriented spins, but if they can be subjected to a process which favours a particular spin direction, the possibility of obtaining polarization presents itself.

The earliest and most familar case is polarization by magnetic scattering of neutrons, suggested by Bloch (1936). The intrinsic magnetic moment of the neutron interacts with a paramagnetic atom or ion in two distinct ways corresponding to parallel and anti-parallel orientations of its spin relative to the atomic spin. If a ferromagnetic substance is magnetized to saturation, the atomic spins are all aligned along the direction of the applied field. Slow neutrons traversing iron exhibit different scattering cross-sections for the two orientations. That the transmitted beam is partly polarized may be verified by sending it through a second iron slab (equivalent to the 'analyzer' of an optical polariscope) which can be magnetized in the same direction as the first, or opposite to it.

In practice, polarization by transmission suffers from a number of drawbacks† and more efficient methods are used: (a) The reflectivities of the two spin states at the surface of a magnetized mirror differ. Completely polarized beams of $\sim 10^4$ neutrons/sec. have been produced by reflexion from cobalt mirrors (Hughes and Burgy, 1949). The method is applicable at very low energies (corresponding wavelengths beyond the Bragg limit). (b) Another method which has yielded monoenergetic (0·08 eV.) polarized neutrons is based on diffraction by ferromagnetic crystals, especially of magnetite (Fe_3O_4). With an incident pile flux (§ 4.2.3) of 10^{12} neutrons/cm.^{2}/sec., the polarized intensity is $\sim 10^5$ neutrons/sec. Both methods have been described in detail elsewhere (Hughes, 1953; Feld, 1953).

All such techniques of magnetic scattering depend on the neutron energy being too small to cause spin reversal of the atomic electrons responsible for ferromagnetism, thus they are inapplicable at greater than thermal energies. In principle, any other type of spin-dependent scattering might be used in the same way if the scattering agents can be aligned in the targets (see below). However, another and more feasible approach is possible. Instead of attempting to polarize beams with respect to a preferred direction in space, it is relatively simple to obtain a high degree of polarization with respect to the plane of scattering. A scattering process or nuclear reaction in which a strong coupling exists between spin and orbital angular momentum (cf. § 1.2.5 and chapter 9) can be expected to polarize the outgoing particles (Wolfenstein, 1949). The practical condition is that $l > 0$, i.e. p or higher waves must be involved. Hence polarization effects do not occur in low energy p-p or n-p scattering, but substantial effects are observed in the scattering of nucleons by deuterons and more complex nuclei.

Single scattering from an unpolarized target does not introduce any asymmetry in the scattered intensity as a function of azimuthal angle ϕ. However, at a particular ϕ value there is a preferred spin direction if the scattering force is spin-dependent. Nucleons

† (1) Depolarization occurs when neutrons encounter imperfectly aligned domains in unsaturated iron (Halpern and Holstein, 1941). (2) The thickness of iron required to yield useful polarizations ($\sim$ several cm.) causes serious loss of intensity and also increases the average energy (beam 'hardening'). (3) Iron is transparent to neutrons of energy less than 0·005 eV., corresponding to the Bragg limit, 4·04Å. (§4.4).

34   NUCLEAR SCATTERING

scattered in direction (Θ, ϕ) have a net spin polarization perpendicular to the plane of scattering in which Θ is measured. If a
singly-scattered beam is re-scattered by a process which depends
on spin, the cross-section then varies azimuthally. This leads to
the idea of *double scattering* experiments, the first scattering causing
polarization and the second serving to analyse the partially polarized beam. The geometry is shown in fig. 8. In general, unequal

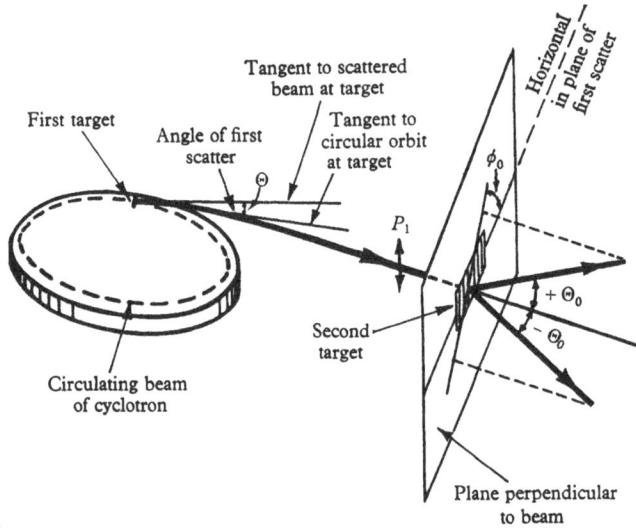

Fig. 8. Basic geometry of double scattering experiment.

numbers of nucleons will be scattered on opposite sides of the beam,
$+\Theta_0$ and $-\Theta_0$ corresponding to azimuthal angles ϕ_0 and $\phi_0 + \pi$
respectively. For a particular value of Θ_0 the asymmetry is
defined as

$$\epsilon = \frac{N(\phi_0) - N(\phi_0 + \pi)}{N(\phi_0) + N(\phi_0 + \pi)} = P_1 P_2 \cos \phi_0, \qquad (2.16)$$

where Θ_0 and ϕ_0 now refer to the co-ordinates of second scattering.
Θ_0 is measured from the incident direction on the second target
and ϕ_0 from the plane of first scattering. ϕ_0 is then simply the angle
between the two planes of scattering and hence the angle between
the two directions of polarization. P_1 is the polarization produced
by the first scattering and P_2 that which would result from the second

scattering if the incident particles were unpolarized. The asymmetry is a maximum when both scattering events are coplanar ($\phi_0 = 0$) and zero when they are perpendicular ($\phi_0 = 90°$). When first and second scatterings are equivalent, the polarization due to the first scattering can be evaluated from $P = \epsilon^{\frac{1}{2}}$.

Interpretation of double scattering experiments is discussed in §8.7 *et seq.* In principle, the azimuthally-dependent cross-section leads to the cross-section for a pure spin state. Additional information can be obtained from a *triple scattering* experiment, in which the first target is a polarizer and the third an analyser which measures the 'depolarization' which occurs at the second target.

Single scattering has been the basis of most methods of polarizing beams of fast nucleons. Strong polarizations have been produced from light nuclei of spin 0 (especially ^{4}He) following a suggestion by Schwinger for scattering from a resonance level which is split by spin-orbit coupling (Schwinger, 1946; Heusinkveld and Freier, 1952; Juveland and Jentschke, 1956a). The same principle is involved in the use of ^{4}He and ^{16}O as analysers (Adair, Darden and Fields, 1954; Levintov, Miller, Tarumov and Shamshev, 1957). Schwinger has also suggested a spin-orbit coupling mechanism for polarizing neutrons which depends on the motion of a magnetic dipole through the Coulomb field of a heavy nucleus (Schwinger, 1948). Almost complete polarization is predicted at very small angles of scattering ($\sim 1°$). Experimentally, the technique is not easy, but the effect has been detected (e.g. Longley, Little and Slye, 1952; Wilson and Voss, 1956).

Spin polarization has also been observed in the scattering of high energy protons by complex nuclei (Oxley, Cartwright and Rouvina 1954). Here, the first scattering is conveniently made to take place inside a cyclotron (fig. 8), resulting in a beam of intensity 10^3–10^5 protons/cm.^{2}/sec. about 50 ft. from the target. Polarization can be measured either by re-scattering the beam through the same angle and finding ϵ, or by degrading it to low energy and scattering from ^{4}He, or other suitable target.† With any given target, P depends on incident energy E and scattering angle Θ; in fact there appears to

† An important experimental point is that a polarized beam may be degraded by ionization loss to lower energies without significant loss of polarization. The cross-section for depolarizing collisions has been calculated as $6 \times 10^{-6}(Z^2 + 2Z)$ b. (Wolfenstein, 1949a).

be an optimum value of each. For carbon, maximum polarization occurs at about $12°$ at 290 MeV. and $25°$ at 130 MeV. Similarly, $P(\Theta)_{max}$ probably rises to a fairly flat maximum around 200 MeV., then declines again, which will limit the usefulness of polarization by scattering at very high energies. An alternative line of approach may be to effect a separation of spins first (e.g. in a large-scale atomic beam, magnetic resonance apparatus), then to accelerate the protons.

At low energies, polarized particles have been obtained from a variety of nuclear reactions although the yield is smaller than that from scattering. The direction of polarization is perpendicular to the plane of emission and its magnitude varies with bombarding energy and angle of emission of the particle. In the case of the useful d-d reaction (§4.2.4), theoretical analysis of the angular distribution indicates the presence of strong spin-orbit coupling and a p-wave contribution, hence substantial polarization of the outgoing particles is expected (Blin-Stoyle, 1951). For the protons from this reaction, P has been measured as about 30% at $E_d = 300$ keV. (Bishop, Westhead, Preston and Halban, 1952) and similar figures have been obtained for the neutrons (Longley et al. 1952; Meier, Scherer and Trumpy, 1954; Levintov, Miller and Shamshev, 1957). At $E_d = 500$ keV. P probably rises to ~50%. Neutrons from $^7Li(p, n)$ have also been employed extensively, showing polarizations of 50–60% (Adair et al. 1954; Willard, Bair, Cohn and Kington, 1956), and preliminary studies have been made of various $\left(d, \dfrac{n}{p}\right)$ stripping reactions in this connexion (e.g. Juveland and Jentschke, 1956a).

At high energies, polarized neutrons can be generated by (p, n) charge exchange reactions (§4.2.6). Intensities are comparable with those of external polarized proton beams. However, P is generally much lower, ~20% or less (Wouters, 1951). It could probably be trebled by rescattering from a second target, but this would entail altogether a triple scattering experiment, and serious trouble with intensity. Neutron double scattering experiments are consequently more difficult. ϵ is commonly only a few per cent at several hundred MeV., while the accuracy of measurement is 0·5–1%.

Finally, the various methods of polarizing or aligning target nuclei which have been proposed will be referred to briefly below:

(a) The 'brute force' method, or direct application of an intense magnetic field at low temperature (Kurti and Simon, 1935; Rose, 1949). The important parameter is $\mu B/kT$, where B is the applied field, μ the magnetic moment, k Boltzmann's constant and T the absolute temperature. In all practical cases $\mu B \ll kT$ and the fraction of nuclei polarized is

$$P = \frac{1}{3} \frac{(J+1)}{J} \frac{\mu B}{kT}, \qquad (2.17)$$

where J is the nuclear spin. Magnetic moments of odd-mass nuclei are of the order of the nuclear magneton

$$(e\hbar/2mc) = 5 \cdot 05 \times 10^{-24}\, \text{erg}\, G^{-1}.$$

To obtain about $\sim 20\%$ polarization when $J = \frac{1}{2}$, B/T needs to be $2-3 \times 10^6\, G/^\circ\text{K}$., which is not impossible if temperatures of $0\cdot1-0\cdot01^\circ\text{K}$. are used. ^{115}In, which has a large magnetic moment, was polarized by this method and the $1\cdot46$ eV. neutron capture cross-section measured as a function of the relative spin orientations (neutron polarization 80%, nuclear polarization $\sim 2\%$ (Dabbs, Roberts and Bernstein, 1955)).

(b) The field at the nucleus of certain paramagnetic ions is $10^5-10^6\, G$ due to unpaired electrons in unfilled shells. This couples electron and nuclear moments and the former may be polarized by moderate magnetic fields at low temperatures such that kT is of the order of the energy of hyperfine splitting (Gorter, 1948; Rose, 1949). A useful degree of polarization has been obtained in a number of cases, e.g. ^{55}Mn, $B = 2350\, G$, $T = 0\cdot20^\circ$ K., electron polarization 85%, nuclear polarization $\sim 16\%$ (Bernstein, Roberts, Stanford, Dabbs and Stephenson, 1954).

(c) Alignment due to interaction between the nuclear electric quadrupole moment and the field of a single crystal (Pound, 1949) may sometimes be used. No external field is required. A significant degree of alignment of ^{233}U has been achieved at 1° K. (Dabbs, Roberts and Parker, 1956).

(d) When there is strong electron-nuclear coupling, as in (b), the electrons may be aligned by the single-crystal field (Bleaney, 1951). In certain paramagnetic salts, due to the crystalline electric field producing anisotropic hyperfine splitting, the interaction between the magnetic moments of electrons and nucleus depends on their orientation with respect to a particular crystal axis as well as on their orientation relative to each other. No external magnetic field is necessary. Many nuclei have been aligned in this way. With ^{60}Co at $0.02°$ K., γ-emission following β-decay departs from spherical symmetry by more than 40% (Daniels, Grace and Robinson, 1951).

(e) Nuclear polarizations of nearly 100% are obtainable at moderate temperatures and field strengths by the 'electron spin resonance' method (Honig, 1954). This depends on the application of r.f. power to a sample when the hyperfine resonances are resolved. At a resonance corresponding to a state of nuclear magnetic quantum number m_J, electron spin relaxation produces a nuclear spin reorientation via the hyperfine coupling, thus depopulating the state. Nuclei may be polarized in a particular m_J state by sweeping through the resonances corresponding to all other m_J values in a time short compared to the nuclear relaxation time τ in the absence of r.f. For example, silicon crystals containing arsenic impurity show four resonances (^{75}As spin $= 3/2$). With an r.f. amplitude of $10^{-2}\,G$, the spin flip time under r.f. was $\sim 10^{-2}$ sec. compared with $\tau \sim 16$ min. at $4°$ K.

It will be noticed that some of the above methods produce *polarization* with spins all pointing in the same direction (when the sample has a net magnetic moment), the others merely produce *alignment* in space, parallel and anti-parallel spins being equally favoured. The latter arrangement is obviously less useful than the former for scattering experiments.

2.5. Co-ordinate systems

Scattering experiments are invariably conducted with the target particle at rest with respect to the laboratory system.† Differential cross-sections and other physical quantities introduced earlier were defined with respect to this system. However, the theoretical treatment of collisions is simplified by referring all particle motions to the system in which the centre of mass is at rest.‡ The net momentum $\Sigma p = 0$ in this frame of reference.

Throughout this text we shall use θ for the CM angle of scatter corresponding to Θ in the Lab. system; likewise ω for the solid

† Denoted hereafter by Lab.   ‡ Denoted hereafter by CM.

angle corresponding to Ω. Transformations can be effected readily from one system to the other, and the appropriate formulae have been derived in Appendix A. The principal results for non-relativistic collisions are listed below:

The kinetic energy associated with the motion of the CM of the system is

$$E_{CM} = E_0 \left(\frac{m_1}{m_1 + m_2} \right). \tag{2.18}$$

The energy associated with the relative motion of two particles is

$$E_R = E_0 \left(\frac{m_2}{m_1 + m_2} \right) \quad \text{or} \quad \tfrac{1}{2}\mu v_0^2, \tag{2.19}$$

which is the energy available in the CM system. E_0 is the incident energy and v_0 the incident velocity in the Lab., and μ is the reduced mass $m_1 m_2 / (m_1 + m_2)$. Other notation follows that of §§ 2.2 and 2.3.

Conversion of cross-sections

$$\sigma(\theta)\, d\omega = \sigma(\Theta)\, d\Omega,$$

whence
$$\sigma(\theta) = \sigma(\Theta) \left| \frac{d \cos \Theta}{d \cos \theta} \right|. \tag{2.20}$$

Two cases must be distinguished:

(i) When the cross-section for *scattering* $\sigma(\Theta_1)$ has been determined,

$$\sigma(\theta) = \sigma(\Theta_1) \left(\frac{\sin \Theta_1}{\sin \theta} \right)^2 \cos (\theta - \Theta_1). \tag{2.21}$$

(ii) When the cross-section for *recoil* $\sigma(\Theta_2)$ has been determined,

$$\sigma(\theta) = \frac{\sigma(\Theta_2)}{4 \cos \Theta_2}, \tag{2.22}$$

where $\sigma(\theta)$ is the differential cross-section in the CM system.

Conversion of angles

$$\theta = (\pi - 2\Theta_2). \tag{2.23}$$

The above formulae, or their relativistic equivalents, provide the connexion between experimentally-determined angular distributions, with which the following chapters are concerned, and the theoretical distributions of the later chapters which deal with the analysis of collision processes.

CHAPTER 3

CHARGED-PARTICLE SCATTERING TECHNIQUES

3.1. General requirements of equipment

Experiments on scattering call for four essential components: (a) an accelerator which produces a beam of ions of the required kind, energy and intensity; (b) a target composed of, or at least incorporating atoms of, the required element or isotope; (c) a beam detector to measure or monitor the ion current through the target; (d) one or more particle detectors, the purpose of which is to measure the intensity of scattering as a function of angle.

In the following sections we shall first consider various pieces of equipment designed to fill these needs. While no attempt will be made to describe their theory or detailed construction, except where these have significant bearing on the planning of the whole experiment, references are included in the text to representative accounts available elsewhere. Emphasis in all sections has been placed on the suitability of each device for precision work.

Later sections illustrate a number of complete experimental arrangements which have been employed successfully in recent years. There is, of course, a wide variety of equipment in use. Scattering has continued to receive attention at energies below 1 MeV., while, at the other end of the energy scale, experiments have been pushed well beyond 1 BeV. Inevitably, the associated techniques must be diverse and we can only hope to provide a well-balanced account of typical procedures.

3.2. Ion accelerators

The following discussion is confined to types of accelerators which have contributed substantially to the study of nuclear scattering during the last 25 years. Emphasis rests on such features as the energy range reliably covered, whether the energy is fixed or conveniently variable, the homogeneity of the beam, the available current after analysis and whether the beam is continuous or pulsed. Several reviews and bibliographies (Livingston, 1952;

Slater, 1952; Chu and Schiff, 1953; Livingston, 1954; Cushman, 1951) have been published which include further information of a general kind.

3.2.1. *High voltage sets.* These have not figured prominently in the literature of nuclear scattering. Although bombarding energies of chief interest are beyond their range they are, nevertheless, well suited to exploring the region from about 1 MeV. to a lower limit imposed by various experimental difficulties at $\sim$ 100 keV. It is rather surprising, for instance, that in spite of the many H.V. sets in existence, no systematic attempt was made until recently to locate the important minimum which occurs in p-p scattering at about 380 keV., due to interference between nuclear and Coulomb forces.†

High voltage sets make use of a variety of electrical circuits designed to produce a constant high potential from a low voltage alternating current supply (Craggs and Meek, 1954). They are generally operated at air pressure although the entire circuit can be pressurized in a steel tank for compactness (Lorrain, Béique, Gilmore, Girard, Breton and Piché, 1955). The high potential is applied directly across a long evacuated tube in order to accelerate positive ions of hydrogen, deuterium, tritium or helium.

Total ion currents are commonly of the order of 100 μA. and may be as high as several mA. in specially designed sets operating at 100–200 keV. (Peck and Eubank, 1955; cf. also § 4.2.4). The energy spread, or homogeneity, of the beam depends chiefly on the residual ripple in the high voltage. When the machine is used as a neutron generator, total current rather than homogeneity may be the prime requirement, but in charged-particle scattering it is often desirable to restrict the energy to within 0·1 %. Both the energy spread and the mean energy of the beam must then be kept within this percentage throughout a run.

It is possible to suppress the voltage ripple of a 1 MeV. set to some 50 V., subject to a long period drift of several times this figure. Alternatively, the ion source may be pulsed in phase with the voltage ripple, thereby utilizing only a narrow potential range for

† Ultimately carried out at M.I.T. with a Van de Graaff generator (Cooper, Frisch and Zimmerman, 1954).

acceleration. The latter arrangement, of course, results in a pulsed beam. Whatever steps are taken to generate monoenergetic particles, final analysis is almost always achieved by deflecting the beam with a constant magnetic field—a procedure which is, in any case, necessary in order to separate monatomic and diatomic ions. This feature, which H.V. sets have in common with other direct-type accelerators, will be discussed below.

3.2.2. Van de Graaff generators. The moving belt type of Van de Graaff accelerator has contributed virtually all the scattering data below 3 MeV., including much of the most accurate data now available. The p-p experiment of Herb and co-workers in 1939 using the Wisconsin machine has served as a model of precision for many years (Herb, Kerst, Parkinson and Plain 1939).

Electrostatic generators possess a number of conspicuous advantages to which the success of experiments conducted with them may reasonably be attributed. An external continuous beam is produced, the energy is smoothly variable up to the maximum, and stability is very good. Spread in beam energy can be confined to 0·1 % while retaining a beam current of 1–10 μA.

The most serious limitation is one of energy. When the generator is operated in dry air at atmospheric pressure the upper limit is about 1 MeV., though with a pressure tank it is considerably higher. Commercially available machines are in satisfactory operation over the range 0·5–5·5 MeV. (Foster, Dewey and Gale, 1953). The M.I.T. Van de Graaff, designed for 12 MeV., has been employed so far to 8 MeV., which appears to be somewhere near the practical limit.

Total ion currents are, typically, ∼ 100 μA. but may reach 1 mA. However, preliminary analysis is required before even approximately monoenergetic particles are available. When used to produce protons or deuterons for acceleration, the ion source forms atomic and also molecular ions, e.g. H^+ and HH^+, the relative proportions of each depending on the design of source.† Since both are accelerated through the same potential they attain equal kinetic energies, hence the energy per atom of the molecular ions is only half that of the atomic ions. This results in a 'double-

† The HHH^+ molecule, though unstable under ordinary pressure conditions, is also formed in arc sources.

energy beam'. A 1 MV. machine yields both 1 MeV. and 0·5 MeV. protons, or deuterons. Likewise, He^+ and He^{++} ions differ in energy by a factor of 2 although the latter are not very abundant. In some arrangements one beam is used for bombardment, the other to regulate the energy.

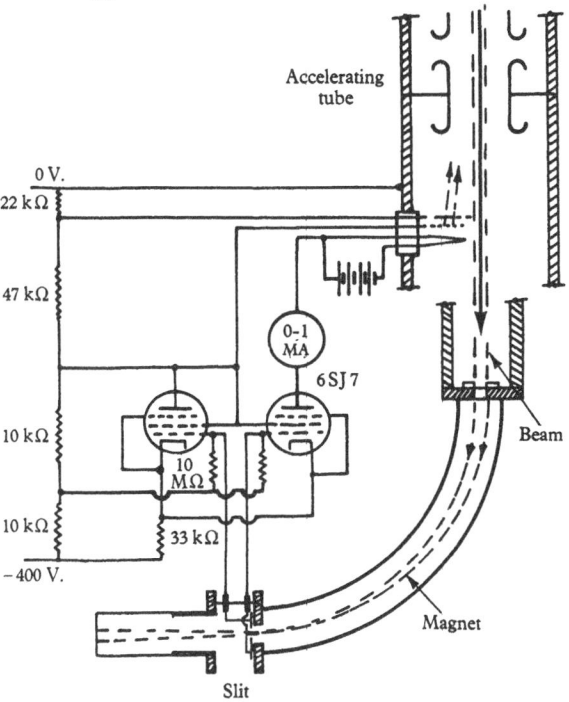

Fig. 9. Magnetic analyser and automatic potential stabilizer for Van de Graaff generator (Bennett *et al.* 1946).

The general practice is to pass the emergent beam through a magnetic field which separates the various ions according to their radii of curvature, $r = (2mE)^{\frac{1}{2}}/Be$. B is the flux density, m the mass and e the charge of the ion and E its kinetic energy, where $E = Ve$ for a machine developing a potential V. By holding the magnetic field constant and introducing a narrow defining slit at the end of the field where the beam focuses, the particle energy can be regulated to $\sim 0·1 \%$ at the same time.

The arrangement of Bennett, Bonner, Mandeville and Watt (1946) is shown in fig. 9. Here the analyser is combined with a potential stabilizer which automatically corrects within a few μsec. any tendency for the beam energy to vary. The jaws of the slit are insulated and the function of the electronic circuit is to maintain equal currents on the two jaws. An electron gun located at the lower end of the accelerating tube allows an electron beam to be accelerated up the tube to the ion source so that the potential of the

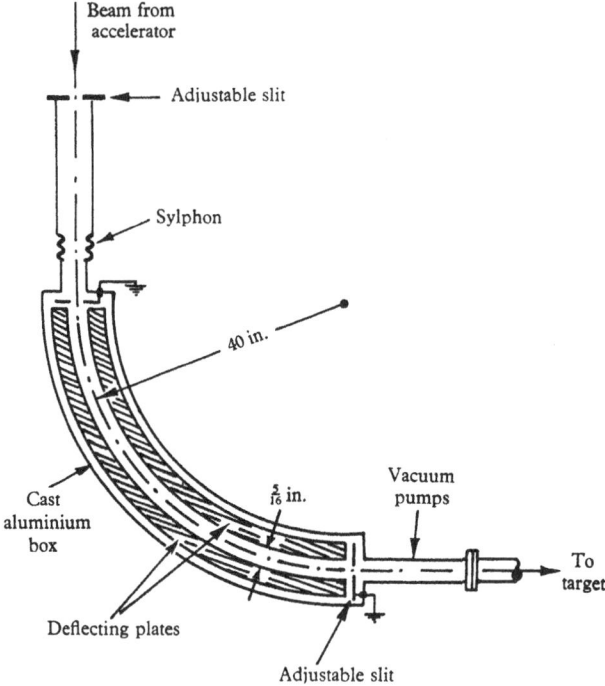

Fig. 10. Low energy electrostatic analyser (Warren *et al.* 1947).
(Not to scale.)

accelerator may be controlled by means of the electron loading, i.e. by modulating the gun. If the ion current to the lower jaw increases, the electron current up the tube increases, the potential decreases and the ion beam through the magnet returns to the centre of the slit.

A quite different system is the electrostatic analyser, developed at Wisconsin (Warren, Powell and Herb, 1947). Here the proton beam is deflected through a 90° arc by a radial electric field between curved metal plates (fig. 10). About 16 kV. is required for 1 MeV.

protons. With some modification of dimensions it has been used to about 4 MeV., but at this and higher energies magnetic analysis is more convenient. Both types of analyser are capable of a resolution of 1 in 1000. With extremely well regulated potentials, 1 in 5000 has been obtained with the electrostatic analyser although the sacrifice of current is then considerable.

Electrostatic generators give a range of energy coverage of the order of thousands of times the effective energy resolution width encountered in measuring a highly structured function, e.g. a nuclear excitation function. Also at high resolution the counting rate is low and the amount of time required per energy point is large. These two factors result in an enormous waste of time in taking data at a large number of energies. However if the accelerator target is insulated from ground potential and a modulating voltage is applied to it, it is possible to correspondingly modulate the output energy of the accelerator.

Using an a.c. potential of 50 kV. amplitude at a frequency of 10 c./sec., an electrostatic generator has covered 100 energy intervals in a single run, corresponding to an energy range of 50 keV. (Cranberg, Aiello, Beauchamp, Lang and Levin, 1957). The second factor of low counting rate has been solved for fast neutron total cross-sections by using a thick proton recoil detector as neutron detector and the pulsed-beam time-of-flight technique (§ 4.3.4) to identify the neutrons of interest (Cranberg, Beauchamp and Levin, 1957). Together with energy modulation of the beam, this enables observations at 100 energy points simultaneously.

3.2.3. *Fixed-frequency cyclotrons.* These have been the most widely used accelerators for elastic scattering between 4 and 15 MeV. Particles spiral outwards from the centre of a constant magnetic field, acquiring energy from successive impulses delivered at a frequency of the order of 10 Mc./sec. The extracted beam is therefore pulsed, having the frequency of the oscillator. The acceleration principle applies only to essentially non-relativistic particles, which imposes an upper limit on the energy that can be achieved in a fixed-frequency cyclotron.

The Oak Ridge 86 in. machine produces an internal beam of 1·5 mA. of 23 MeV. protons. Most cyclotrons are of diameter between 40 and 60 in. and accelerate particles to 5–10 MeV. If designed for an e/m ratio of 0·5 as is usual, they will accelerate HH^+, D^+ or He^{++} ions in the energy ratios 1:2:4, thereby providing a useful repertoire of scattering studies in spite of the limitation of a single operating frequency. The latter is a shortcoming which several laboratories have sought to overcome by constructing cyclotrons of variable field strength and frequency (Caro, Martin and Rouse, 1955; Fulbright, Bromley and Bruner, 1955; Taylor *et al.* 1956). The Melbourne variable-energy machine is designed to accelerate protons in the range 2–12·5 MeV. Another trend in a rather different direction has been the building of low-energy, high-current cyclotrons such as those at Brookhaven and Cornell. Here, the aim is to generate a circulating current of many mA. of protons or deuterons at 1 or 2 MeV. for neutron production (§ 4.2.4).

There is no serious difficulty in extracting a large proportion of the internal beam of a cyclotron, and external beams are frequently of several hundred μA. However, the energy spread on a target just outside the vacuum box generally exceeds 5 %, making beam analysis essential for most experiments, and the loss of much of the beam inevitable.

An attractively simple system, which, however, is unnecessarily wasteful of current, utilizes the fringing field of the cyclotron magnet to spread the beam into a spectrum of energies from which a narrow slit selects a range conforming to the requirements of the experiment. One arrangement appears in the general assembly shown in fig. 22. An external beam of 200 μA. of 10 MeV. deuterons at the target chamber resulted in a current density of 0·3 μA./cm.^{2} at the scattering chamber. With slits of 0·02 cm.^{2} area, a beam current of $\sim 6 \times 10^{-9}$ A. was obtained with an energy spread of less than 0·5 %. This is ample for single scattering work. In fact it is usual cyclotron experience to have to cut back the main beam to enable the electronic system to cope with the counting rate.†

More general practice is now to analyse the external beam with an auxiliary magnet whose focusing properties at the same time compensate in part for the angular spread of the emergent particles.

† The situation is entirely different for inelastic scattering or reaction studies where cross-sections are commonly an order of magnitude lower, and better energy and angular resolution is required. Nor does this apply to double or triple scattering.

In the Los Alamos arrangement (Curtis, Fowler and Rosen, 1949) the diverging beam was passed through a trapezoidal magnetic field which analysed and refocused it 3 m. from the magnet. The beam current obtained was $\sim 1 \cdot 3 \mu\text{A./cm.}^2$ collimated to within $\pm 0 \cdot 6°$.

Magnetic focusing has two other merits. In the first place it is sometimes convenient to be able to deflect the beam away from its initial direction and, secondly, it makes the use of a cyclotron at intermediate energies more practicable. The latter result is achieved by degrading the beam with absorbers. However, in the case of a system such as that shown in fig. 22, when foils are inserted, multiple scattering (§ 3.5.2) spreads the beam to such an extent that a large fraction is lost. For 82 mg./cm.^{2} of Al—sufficient to lower the energy of protons from 10 to 7 MeV.—the r.m.s. angle of scattering is $3 \cdot 5°$. On the other hand, the principle of alternating gradient focusing has been used successfully in similar situations (Cork and Zajec, 1953; Shull, MacFarland and Bretscher, 1954), and it could probably be applied to refocus the slowed beam in this case, especially if the foil were placed at the image of one lens pair so that the particles appeared to diverge from a point source before entering a second lens pair.

At energies of the order of 10 MeV. the shielding of scattering equipment becomes of major importance. Cyclotrons are copious neutron and γ-ray sources, especially when accelerating deuterons, and some effort should be made to reduce background counting rates. Shielding commonly takes the form of a loaded concrete wall, some 5 ft. thick for a 20 MeV. deuteron machine (Livingston, 1952), through which the primary beam must be piped. The reduction factor for fast neutron flux is then $\sim 10^{-5}$ and for slow neutrons or γ-rays $\sim 10^{-6}$.

3.2.4. *Synchrocyclotrons.* In extending the energy at which scattering cross-sections have been measured by more than an order of magnitude, these accelerators have profoundly influenced the study of nuclear forces. With them, experiments have been carried out in America, Britain and Russia between 100 and 660 MeV., while similar machines of more modest proportions have bridged the gap down to the energy of the ordinary cyclotron.

In the energy range of the synchrocyclotron, the kinetic energy of the ions becomes appreciable in comparison with their rest energy, requiring that the frequency of the accelerating electric field be progressively decreased as the ions are accelerated. To accomplish this, the frequency is modulated cyclically and the high energy beam therefore consists of ion pulses having the frequency of modulation. It follows that the duty cycle, and consequently the average beam current, are both much less than that of the fixed-frequency machine. The modulation frequency is 60–300 c./sec. for large machines and the average current on an internal probe target is $\sim 1 \mu$A. Smaller machines may operate at higher frequencies with correspondingly enhanced currents (e.g. the Amsterdam 28 MeV. deuteron cyclotron, 2000 c./sec., 25 μA.).

It is difficult to extract the beam from a synchrocyclotron efficiently and several laboratories have developed techniques for working with internal targets (§ 3.3.2). Nevertheless, methods of obtaining some external beam have been widely used (Livingston, 1952, p. 163) and in spite of the small currents available most scattering experiments have been conducted in this way.

Early Berkeley experiments made use of a pulsed electric deflector giving a proton pulse of 0·1 μsec. duration. This made the use of coincidence equipment virtually impossible since resolving time should be short compared with pulse lengths. Subsequently, in an innovation due to Leith (1950) a thin target was inserted which caused multiple scattering with a r.m.s. angle of 1 or 2°. Some of the particles developed free oscillations which enabled them to enter a magnetic channel shielded from the main field and escape from the machine (fig. 11). With this arrangement the intensity is only $\sim 1 \%$ of that given by the electric deflector (at Berkeley it yielded 10^{-12} A. in a 2·5 cm. diameter beam) but the pulses are much longer, 25–50 μsec. Most scattering work at high energy has been carried out with beam currents of 10^{-11} to 10^{-13} A.

A major improvement in beam extraction technique was introduced on the Liverpool 380 MeV. cyclotron (Moore, 1955). When the circulating beam is 1 μA. the available external beam is 3×10^{-8} A.

3.2.5. *Linear accelerators.* As their name implies, these machines accelerate ions along straight paths, a concomitant virtue of which is that the beam emerges without complication from the end of the accelerating tube. They are ideal machines for precise experiments in the energy range inaccessible to Van de Graaff generators. The first important ion accelerator in this class, operating in its own right, was the Berkeley 32 MeV. 'Linac' of the drift-tube, cavity-controlled type (Alvarez *et al.* 1955). A 10 MeV. machine

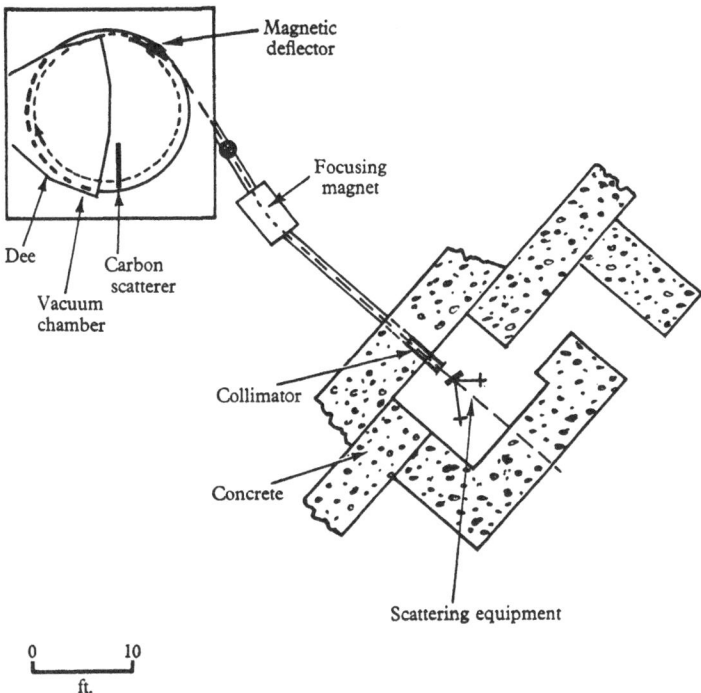

0 ____ 10
└─────┘
ft.

Fig. 11. Arrangement for beam extraction, focusing and shielding in relation to Berkeley 184-in. synchrocyclotron (Chamberlain, Sègre and Wiegand, 1951).

constructed as injector for the Bevatron (§ 3.2.6) was used in the interim for *p-p* scattering (Cork, 1955). A 70 MeV. machine, similar in design to the Linac, has been built at the University of Minnesota. In principle, the linear accelerator can achieve any energy, but in practice circular-type machines are more economical at higher energies. The final energy is, of course, fixed by the design and is not variable, although the Minnesota machine has been built in three sections, making available two intermediate energies by operating it with only one or two of the sections.

4

The beam current is pulsed but pulses are long ($600\,\mu$sec. in the case of the Linac), a circumstance which is highly important in experiments using counters in coincidence because the accidental rate is inversely proportional to the pulse length. With a repetition frequency of 15 c./sec., the duty cycle approaches 1 % and the average beam current is about $0\cdot3\,\mu$A. delivered in a parallel beam of $\frac{1}{2}$-in. diameter. The energy is constant and the homogeneity known to be better than $0\cdot3$ %. On its exit from the tube, four magnetic quadrupoles (Cork and Zajec, 1953) concentrate the beam to a circle of $\frac{3}{16}$-in. diameter about 20 ft. from the machine. 16 MeV. HH^+ molecules,† which are accelerated as well as protons, are broken by a thin 'stripper' foil and separated by a magnetic deflector (Alvarez *et al.* 1955).

3.2.6. *Proton synchrotrons.* Elastic scattering has been measured using several accelerators operating in the BeV. range. These are proton synchrotrons, essentially the same in principle as electron synchrotrons (Livingston, 1954). A constant-radius orbit is located in a magnetic field which is made to increase as the protons gain energy. Obviously, the frequency of the accelerating electric field must also increase in synchronism with the rotation frequency of the particles.

Because of the large size of the magnet and the amount of energy involved, the repetition cycle is long, 5 sec., for the Brookhaven 'Cosmotron' and Berkeley 'Bevatron' and 10 sec. for the Birmingham synchrotron. Each of these machines accelerates a pulse of $\sim 10^{10}$ protons every cycle, corresponding to a time-average current of about 3×10^{-10} A. Homogeneity in energy is better than $0\cdot1$ %.

The circulating beam of a proton synchrotron can be made to spiral inwards or outwards to strike an internal target. In the Cosmotron, the beam can be moved radially at rates between 20 and $0\cdot05$ cm./msec. Since the beam diameter at full energy is about $2\cdot5$ cm., the corresponding bursts of radiation may be as short as $0\cdot1$ msec. or as long as 50 msec. A short pulse is necessary for time-of-flight studies, a long one for counting experiments. If the pulse is 1 msec., ~ 100 counts can be recorded per pulse.

† 8 and 4 MeV. components due to particles which spend 2 r.f. cycles in each drift tube also occur.

Small external fluxes have been generated for experimental use by scattering the beam from an internal target. More recently, a method of extracting a considerable fraction of the circulating beam has been devised (Piccioni, Clark, Cook, Friedlander and Kassner, 1955). The beam is made to spiral inwards and traverse a target which reduces its energy slightly, so that after another revolution it becomes displaced several inches inwards. An electromagnet then takes charge of the beam and ejects it.

3.3. Targets

In any experiment involving dissimilar particles it is necessary to decide which will be accelerated and which shall form the target. A number of factors, such as the following, should be taken into account.

As a general rule it is better to use the lighter particle as projectile, for reasons contained in the basic mechanics of collision (fig. A. 2). The lighter particle can be scattered over the full angular range 0–180° in the Lab system while the target particle can recoil from 0–90°. By contrast, if the incident particle is heavier, its angle of scatter is confined within a cone of angle arcsin (m_2/m_1) which is 30° for d-p scattering, 19·5° for t-p and only 14·5° for α-p. Recoil particles cover the range 0–90° as before. This restriction of all scattered particles to a forward cone requires better angular resolution, more accurate setting of detector positions, congests detector and current collector and situates the detector very unfavourably in relation to slit scattering, which is unavoidably present in the forward direction.

Moreover, accelerator considerations sometimes favour using a lighter projectile. High voltage and Van de Graaff accelerators confer equal amounts of kinetic energy on all single charged particles which they accelerate. To obtain the highest relative velocity possible with a given machine the lighter particle must become the projectile. For example, in the p-d case, a 3 MV. machine produces 3 MeV. protons whose relative energy (equation (2.19)) when bombarding deuterons is 2 MeV. The same machine, in accelerating deuterons to 3 MeV. gives a relative energy of only 1 MeV. when bombarding protons.

Other factors may, however, decide in favour of the latter

arrangement in spite of its disadvantages. For instance, the coincidence technique (fig. 7) can only be applied conveniently with solid targets. While hydrogenous materials are easily found, targets containing deuterium do not come to hand so readily. Hence d-p experiments have, in fact, often been preferred to p-d experiments.

For experiments involving ^{3}H and ^{3}He, recovery is more straight-forward if these rare products are used as targets rather than accelerated through the machine, although published work by no means all conforms to this procedure.

In the following sections targets are discussed according to their most obvious classification, whether gas or solid. Liquid targets and certain special arrangements not fitting properly into either of these categories are treated separately.

3.3.1. Gas targets. In the case of expendable gases the entire chamber is filled with the gas. Scattering occurs all along the path of the beam through the chamber but a slit system placed in front of the detector defines the length of path from which scattered particles may enter the detector, i.e. the effective target thickness (fig. 6).

The constant of a target which enters into the expression for differential cross-section is the number of atoms per unit area of effective target, $N_a = N_v.x$, where N_v is the number of atoms per unit volume of target and x is the target thickness (§ 2.3). In the case of a gas target, N_v is determined for a pure gas by knowing the Loschmidt number N_L, the gas pressure p (cm. of Hg) and the absolute temperature T (° K.) pertaining at the target volume. Then

$$N_v = N_L.\nu.\frac{p}{76\cdot00}.\frac{273\cdot16}{T}, \qquad (3.1)$$

where ν is the number of atoms per molecule and $N_L = 2\cdot6870 \times 10^{19}$.

The permissible pressure is limited by two factors, (a) loss of particle energy due to ionization in the gas, (b) multiple scattering of the beam through the chamber and also of the scattered particles between the target and detector. In the latter case particles have reduced energy and are more susceptible to the effects of multiple scattering (§ 3.5.2).

In experiments below 10 MeV., the acceptable pressure is usually found to be in the region of several cm. Hg, which is rather low for precise measurement with a Hg manometer. An oil-filled U-tube manometer, which increases the head by a factor of ~ 15, is commonly substituted. The tube should be at least 0·5 in. diameter to avoid capillary effects and the low pressure arm should be continuously pumped. Suitable oils are 'Octoil S' or 'Apiezon B', which have very low vapour pressures. Their densities ρ_t are known accurately as a function of temperature (Rodgers, Leiter and Kruger, 1950; Worthington, McGruer and Findley, 1953; Beeck, 1935) and may be expressed as

$$Octoil\ S \quad \rho_t = 0\cdot 9104 + 0\cdot 00072\,(25\cdot 0 - t),$$

$$Apiezon\ B \quad \rho_t = 0\cdot 8645 + 0\cdot 00070\,(25\cdot 0 - t),$$

both correct to about $\pm 0\cdot 05 \%$. Temperature t is in ° C. and ρ_t in g./cm.³. Equation (3.1) may be re-written as

$$N_v = N_L . \nu . \frac{h\rho_t g}{1\cdot 01325 \times 10^6} . \frac{273\cdot 16}{T}, \tag{3.2}$$

where the constant in the denominator is the pressure of a standard atmosphere, g is the local value of the acceleration due to gravity, h is the head of oil of density ρ_t and T is the temperature of the gas in ° K., as before.

The head h is best measured with a cathetometer reading to $\sim 0\cdot 1$ mm. A 2 cm. Hg pressure difference (≈ 30 cm. oil) can then be read to $\sim 0\cdot 03 \%$. Gas temperature should be read to $\sim 0\cdot 1°$ C. at regular intervals throughout the run, from a thermometer sealed in through the chamber so that its bulb rests close to the target volume. A point to notice here, and one which has received insufficient attention, is that the beam produces local heating along its path through the target gas. The N_v value involved is strictly that pertaining to this heated region so that the question of gas circulation and mixing in the chamber becomes of practical importance. An error of 0·3° C. in temperature measurement corresponds to a discrepancy of 0·1 % in N_v.

At higher energies gas pressures can be increased, and different systems of handling and measurement have been evolved. A schematic diagram of the apparatus of Panofsky and Fillmore (1950) for 30 MeV. scattering at a hydrogen pressure of 1 atmosphere appears in fig. 12 (a). The gas pressure was balanced against

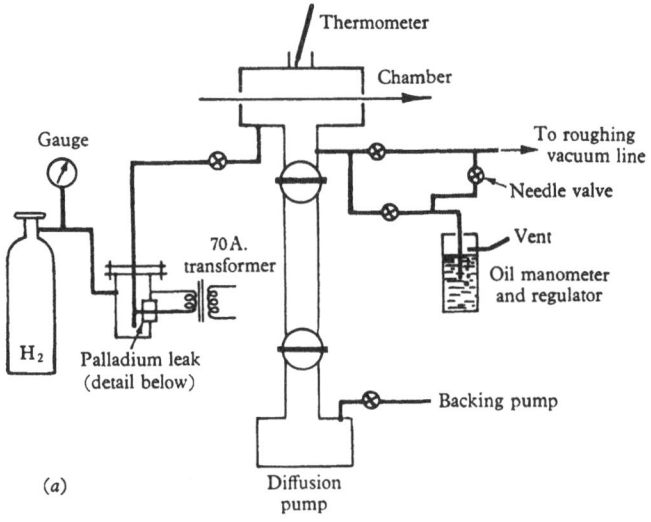

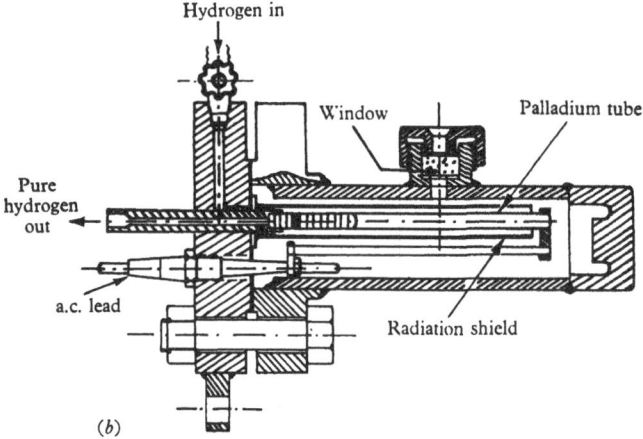

Fig. 12. (a) Gas handling system used in p-p scattering at 30 MeV. (b) Detail of palladium leak for supplying pure hydrogen (Panofsky and Fillmore, 1950).

atmospheric with a small pressure differential (~ 25 cm. oil) maintained by an oil manometer. The head of oil was measured together with atmospheric pressure read from a precision Hg barometer, the total error in pressure measurement being estimated at 0·04 %.

Above this energy, solid targets (§ 3.3.2) have been employed almost without exception. Multiple scattering becomes relatively unimportant since the r.m.s. angle is proportional to E^{-2}, and the much smaller N_v values of the gaseous state ($\sim 10^{-3}$ to 10^{-4} that of solids) would lead to targets of unwieldy size.

A special problem has been presented by recent attempts to use ^{3}H and ^{3}He targets. Because of current prices of these gases they must be used sparingly and reclaimed after completion of the experiment. Moreover, ^{3}H is radioactive and constitutes a health

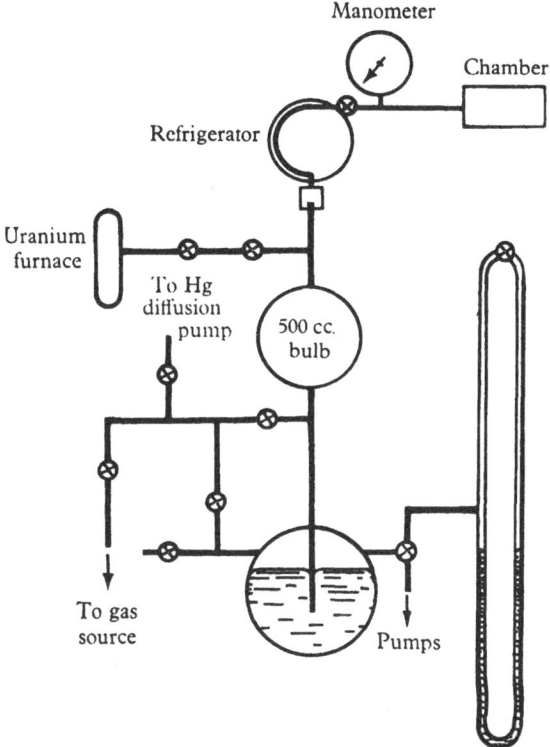

Fig. 13. Handling and recovery system for tritium gas targets
(Claassen *et al.* 1951).

hazard if released into the laboratory. A small-volume (40 cm.^{3}) scattering chamber, otherwise fairly conventional in design, was employed by Claassen, Brown, Freier and Stratton (1951), the entire chamber being filled with gas at 1·8 cm. Hg pressure. A gas-handling and recovery system such as that shown in fig. 13 will transfer efficiently the small volumes of only a few standard cm.^{3} usually available.

Tritium gas was stored as hydride in 5 g. of uranium shavings in the furnace and released to the scattering chamber by heating. A Toepler mercury pump was used to transfer it to the chamber and a refrigerator operating at about $-35°$ C. prevented mercury vapour from reaching the chamber.

^{3}H loses its purity by exchange with ^{1}H in greases, gaskets, etc., during the course of a run† and the target gas becomes a mixture of the two, with a steadily rising proportion of ^{1}H. The purity can be checked at intervals with a mass spectrometer and when con-tamination becomes too severe for use (as much as 65 % impurity in Claassen's experiment) the sample must be discarded. By careful pulse-height or track-range discrimination, the scattering from such a two-component gas can be interpreted reliably to give the intensity of scattering due to each component (see also § 3.5).

A similar plant for handling ^{3}He using activated charcoal for purification has been described (Yarnell, Lovberg and Stratton, 1953). ^{4}He is usually present as an impurity to the extent of a few per cent, but since the chamber is unlikely to contain any other ^{4}He with which exchange could occur, the relative proportion of contamination will remain fixed over long periods.

It is advisable with precious targets to provide safeguards against failure of the thin window isolating the chamber from the main accelerator vacuum. This can be achieved by inserting a high gas impedance escape tube on the accelerator side of the chamber, an ion gauge or spark circuit to fire when the gas pressure commences to rise and a trip valve which closes off the chamber speedily before much gas escapes (Taschek *et al.* 1949).

While discussing the use of precious targets it is appropriate to refer also to experiment sin which ^{3}H and ^{3}He have served as bombarding ions (e.g. Hemmendinger, 1956; Holm and Argo, 1956). The gases are fed to the ion source in the usual way, recovered and purified by circulation through charcoal traps, then returned to the ion source. A closed system, which is subject to a gas loss of only 3 %, has been described (Allen and Almqvist, 1953).

We now proceed to a discussion of the purity of gas targets and measures—often quite elaborate—which have been adopted to

† Taschek *et al.* (1949) report that exchange proceeds about twice as fast while the beam is passing through the chamber.

free experiments of errors due to spurious scattering introduced by the presence of foreign atoms. The chief disturbance occurs at small angles and low energies where Coulomb scattering from atoms of air and organic vapours may grossly increase the apparent cross-section unless allowed for. If the degree and type of impurity is known (for instance from mass spectrometric sampling), the effect at small angles can be estimated from the well-known Rutherford formula:

$$\sigma(\Theta) = \left(\frac{Zze^2}{4E_0}\right)^2 \csc^4\left(\frac{\Theta}{2}\right), \qquad (3\cdot3)$$

where E_0 is the incident particle energy, z its atomic number and Z the atomic number of the nucleus. For protons,

$$\sigma(\Theta) = 0\cdot00129\left(\frac{Z}{E_0}\right)^2 \csc^4\left(\frac{\Theta}{2}\right)$$

in barns if E_0 is in MeV.

In p-p scattering the problem is at its worst in the range of the interference minimum (§7.7) which occurs at $\sim 380\,\mathrm{keV}$. Oxygen is a principal contaminant of commercial hydrogen and its cross-section at this energy is 1760 times the p-p cross-section to be measured, which means that the purity must be $\sim 99\cdot999\,\%$ if the oxygen scattering is to be kept below 1 % of the total. (Commercially available H_2 is rarely better than $99\cdot8\,\%$.) However, this is an extreme case. As the energy increases, the E^{-2} dependence rapidly diminishes the contributions from contaminants relative to the nuclear p-p yield. With reasonable care as to purity, corrections are appreciable only below about $\Theta = 25°$.

There are specifically two approaches to the problem of reducing background due to impurities, which either singly or jointly will generally cut the error from this source to $\sim 0\cdot1\,\%$, except at small angles where it may be several times larger. They are (a) by taking elaborate precautions to purify the target gas, and (b) by taking full advantage of the ability of modern detectors to discriminate between particles of different energies, in the present case between *bona fide* and spurious scattering. Contaminant atoms are chiefly O, N and C from small air leaks and outgasing of the chamber, or sometimes counter gas (A, CO_2, etc.) from slow leakage through thin separating windows.

In collision with heavier nuclei, incident particles lose less energy, giving bigger pulses in counters (assuming the full range to be absorbed in the counter) or longer tracks in photographic emulsions. For instance in 5 MeV. p-p scattering using emulsions (§ 3.5.5), the range of protons scattered by hydrogen at $\Theta = 21°$ is 139μ, while for heavy-atom scattering it is some 33μ longer. These ranges can be reliably resolved.

Fig. 12 (b) is a section drawing of a palladium leak for supplying pure hydrogen. Palladium dissolves H_2† which is re-evolved when the metal is heated in vacuum to more than 300° C. (Certain other metals, notably tantalum, behave as 'getters' in like manner.) In the experiment of Panofsky and Fillmore hydrogen was admitted to the leak, consisting of a $\frac{1}{4}$ in. palladium tube of wall thickness 0·006 in. at a pressure of 500 lb./in.². The tube was supported internally on ceramic insulators so that it could be heated directly from the 70 A. secondary of a step-down transformer supplied from the a.c. mains. Operating at this pressure, the leak supplied pure H_2 at ~ 200 cm.³/min. During the experiment the leak operated continuously, flushing H_2 into the chamber at a rate which changed the gas filling every 20 min.

Other workers have elected to fill the chamber initially and then either exclude any residual impurity scattering which builds up as the run proceeds, by pulse-height or track-length discrimination, or else monitor and subtract it. An example of the latter approach is afforded by the highly accurate p-p experiments (1·8–4·2 MeV.) of Worthington et al. (1953). In addition to the usual movable counter for hydrogen scattering, a fixed counter was set through the chamber lid at 90° to the beam. At this angle hydrogen scattering and all recoil particles have zero energy, whereas particles scattered from heavier atoms retain a large proportion of their energy. The 90° counter therefore responds specifically, and only, to contaminant scattering. The monitor rate was recorded at each experimental angle, and found to be appreciable, by comparison with H scattering, only below 25°.

At these small angles, and in the low energy range where the whole effect of gas purity is important, all scattering cross-sections

† According to Strong (1945) PdH_2 is formed, the metal dissolving ~ 0·006 % of H_2 by weight.

take the Rutherford form, equation (3.3). To apply corrections at small Θ based on the 90° monitor rate, it was necessary, therefore, to know only the ratio of the actual to the Rutherford cross-sections at 90°. Values of the ratio were inferred from a separate experiment in which the angular distribution of scattering from the gases which accumulated in the initially evacuated chamber was determined. Other experiments were made with traces of air deliberately admitted to the H-filled chamber and the results compared with the observed impurity scattering. The corrections at small angles in this work were uncertain by $\pm 30\%$, which allowed a maximum error in the p-p cross-sections of $\sim 0.3\%$ at about 10° and 4 MeV. With all these precautions however, the impurity correction remained the largest source of error at small scattering angles. It was negligible ($< 0.1\%$) beyond 25°.

A comparable procedure can be resorted to with photographic plate chambers (§ 3.5.5). At the higher angles impurity tracks, if present, are clearly apparent as a distinct range group; they may be counted separately so that the angular distribution of contamination scattering is determined during each run. Photographic chambers, it should be noted, are contaminated chiefly by water vapour evolved from the plates, which should be subjected to a long pre-exposure pumping period (~ 10 hr.) in order to dehydrate them. Nuclear research emulsions are now well-bonded to the glass and will usually withstand exposure to high vacuum (especially if prepared with extra plasticizer by the manufacturer), but if there is any tendency to peel, the edges may be painted with collodion.

3.3.2. Solid targets. There are two reasons for preferring solid to gas targets under certain conditions. In the first place, they are essential for scattering at energies of the order of 100 MeV., to obtain a reasonable yield, and secondly, they offer a convenient and elegant means (though this seems paradoxical at first sight) of avoiding the impurity scattering to which gas targets are subject (§ 3.3.1). They are used generally in experiments involving the coincidence technique, fig. 7, and mainly for scattering by hydrogen because of the ready availability of natural and synthetic hydrogenous solids.

Consider first the low energy applications. A thin foil, usually of the order of 5×10^{-4} cm. (the selected thickness depending, as with gas targets, on the degree of multiple scattering which can be tolerated) is set at the centre of an evacuated and continuously pumped chamber. The vacuum system is therefore much simpler than for gas targets, the only elaboration being a well-baffled diffusion pump and trap to prevent deposition of carbon on the target during operations.

The foil is bombarded by the beam and two counters set at conjugate angles (§ 2.4) to record coincidences between pulses from scattered and recoil particles. In the case of p-p scattering, if counters are set at $90°$, true coincidences can arise only from strictly conjugate particles corresponding to the collision of an incident proton with a proton in the target. The angle setting required for any other type of collision in which $m_1 \neq m_2$ would be quite different. In principle, therefore, the coincidence rate is independent of the presence of C, N and O in the foil and the method thereby avoids the problem of gas purity which is inherent in single counter or photographic plate detectors used with gas targets.

The method was first used by Wilson and Creutz (1940) for p-p experiments, and was extended by Karr, Bondelid and Mather (1951) to the p-d study by bombarding hydrogenous foils with deuterons (cf. § 3.5.2).

Unfortunately, the use of solid targets in this way brings its own difficulties, some of which, in practice, tend to be severe:

(a) Organic materials are often unstable under bombardment and the first problem is to find a suitable hydrogenous target. It must have negligible water absorption, should preferably have a well-established chemical formula and should lend itself to placement as a thin uniform smooth foil.

Nylon $(C_{12}H_{22}N_2O_2)_n$, H content 9·7 %, was used in early experiments but in order to obtain a smooth target it must be stretched taut and attached to a target holder, which is difficult to accomplish while retaining a foil of uniform thickness. Examination of the fringes in a Michelson interferometer gives a convenient check on uniformity, but the accuracy is not high. Similar difficulties have been reported for cellophane $(C_6H_{10}O_5)_n$, H content 6·2 %, but polyethylene-terephthalate $(C_{10}O_4H_8)_n$, H content 4·2 %, has been used successfully by observing certain precautions. From the point of view of structural stability, it seems desirable to avoid polymers involving the O-H radical. Polyethylene $(CH_2)_n$, H content 14.4%, and polystyrene $(CH)_n$, H content 7.8%, which have been used extensively in recent experiments, satisfy this requirement.

Some workers have noticed crinkling and discoloration of organic foils under bombardment, accompanied by a drift in the counting rate, presumably due either to continued dehydration of the material or actual molecular rearrangement. A long pre-pumping period is desirable and foils should be stored in vacuum and analysed immediately after completion of runs. Moveable target holders have been used in some experiments (Karr *et al.* 1951; Bondelid, Braden, Battat and Bohlman, 1952) so that a larger and hence more representative area of foil could be subjected to bombardment, no one part being used unduly.

(*b*) It is desirable to have a high H content in the target, and hydrocarbon foils are not as satisfactory as might be desired from this point of view. In principle the presence of other atoms does not matter but in practice they all contribute to the individual counting rates N_1 and N_2 of the two counters, which are therefore much higher than the coincidence rate. The chance coincidence rate is $N = 2N_1 N_2 \tau$, where τ is the resolving time of the coincidence circuit. Moreover, the r.m.s. angle of multiple scattering in the target (to the value of which the coincidence method is more sensitive than single counter arrangements) is approximately proportional to Z, the mean atomic number of the target. Z_{av} is about 4.5 for poly-ethylene-terephthalate, 3.5 for polystyrene and 2.7 for polyethylene.

(*c*) Use of the coincidence technique entails slit systems of considerable complexity. This will be discussed in §3.5.2. For the present it is sufficient to point out that uncertainties are introduced which tend to cancel the benefit of avoiding a gas handling and purifying system.

We return now to a general discussion of solid targets and the measurements which must be made on them in order to obtain absolute cross-sections. As with gas targets it is necessary to determine N_a, the number of H atoms/cm.²

$$N_a = \frac{f N_A}{A} \cdot \frac{m}{s} \quad \text{or} \quad \frac{f N_A}{A} \rho x, \qquad (3.4)$$

where m is the mass of the scattering foil of area s, density ρ and thickness x; f is its fractional H content by weight, N_A Avogadro's number ($6 \cdot 0228 \times 10^{23}$) and A the atomic weight of H.

In the case of certain synthetic targets the amount of unpolymerized material is negligible ($< 0 \cdot 1 \%$) and f may be obtained reliably from the known chemical formula, e.g. $(CH_2)_n$ and $(CH)_n$. In general f is determined by chemical analysis, preferably on the foil actually used in the experiment. The area s can usually be measured directly to $0 \cdot 1 \%$ and m to $\sim 0 \cdot 03 \%$ without undue difficulty or elaboration of equipment even in low-energy experiments when target foils are thin and light. f is likely to be the chief source of error in N_a. Checks should be made for water content and attention paid to any tendency for counting rates to drift during runs, which indicates a change occurring in foil structure or form. It is advisable, when determining an angular distribution, to refer back from time to time to a selected angle which serves to monitor the foil behaviour against systematic changes.

Chemical analysis on the whole foil, which is always much larger than the beam area, gives an average f value and it can only be assumed that the relatively small area of foil actually used as target is typical of the whole. Likewise the use of m/s determined from the whole foil has implicit in it the assumption that the foil is uniform in thickness and density. Procedures aimed at reducing errors from these sources and the contingency of foil change have been mentioned above (Karr *et al.* 1951; Bondelid *et al.* 1952).

In high-energy experiments, solid targets have been used with single counters,† with coincidence arrangements and even with photographic plates. The coincidence technique was used ingeniously at the University of Rochester (Oxley and Schamberger, 1952) for 240 MeV. p-p scattering. A target made of a number of polyethylene laminations, each 12 mg./cm.² thick, was located inside the vacuum tank of the cyclotron where it was struck by the expanding beam (fig. 14). Two scintillation counters (§ 3.5.4), also in the tank, were positioned above and below the beam in a vertical plane through the target. The upper counter could be set in three different positions to cover the angular range studied, while

† Used here in a broad sense which is intended to include telescopes of counters as distinct from two-particle coincidence systems.

both the target and the conjugate counter (lower counter in fig. 14) were continuously moveable. By using different numbers of laminations for the target its thickness was changed from 12 to 72 mg./cm.^{2} in accordance with variation in energy of the scattered protons with angle. This prevented excessive energy loss and multiple scattering in the target.

A few experiments using hydrogenous targets have been carried out with a single detector by the difference method, which becomes

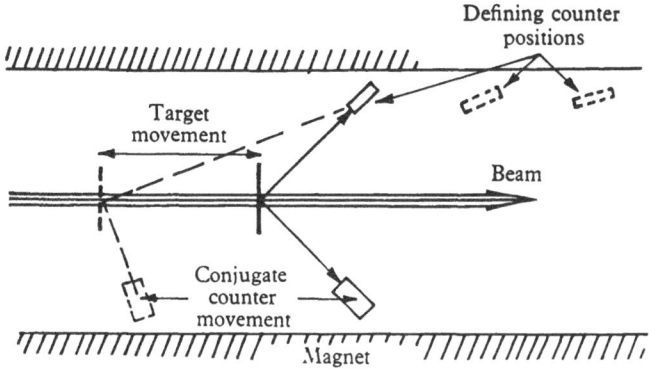

Fig. 14. Rochester arrangement for coincidence counting from an internal cyclotron target (Oxley and Schamberger, 1952).

more feasible at high energy where Rutherford scattering from carbon is reduced in proportion to the total scattering. In the Berkeley work at 340 MeV. (Chamberlain and Wiegand, 1950) the angular distribution of total scattering from polyethylene was determined first. A graphite target was then substituted, the thickness being chosen to have the same stopping power as the CH_2, and a similar distribution was taken. The scale of the latter was adjusted to correspond with the number of C atoms per cm.^{2} in the CH_2 target and subtracted at all angles from the total (cf. §3.5.4. and equation (3.7)).

So far we have referred exclusively to the use of hydrogenous targets, but there is no reason why 'heavy' or 'deuterated' targets should not be prepared, having some or all of their ^1H atoms replaced by ^2H (for the use of heavy paraffin and polyethylene, see Schamberger, 1952; Chamberlain and Clark, 1956).

3.3.3. *Miscellaneous targets.* Liquid hydrogen has replaced solid hydrocarbon in many recent experiments. Particularly is this an important improvement at small angles where a single detector working on difference measurements is subjected to a prohibitive intensity of Rutherford scattering from carbon, while the two-counter coincidence method is also in trouble because of the difficulty of detecting the low energy protons. The use of liquid targets enables the former system to function satisfactorily. However, scattering due to the liquid must still be determined by difference, since the containing vessel itself contributes some scattering (cf. §3.5.4).

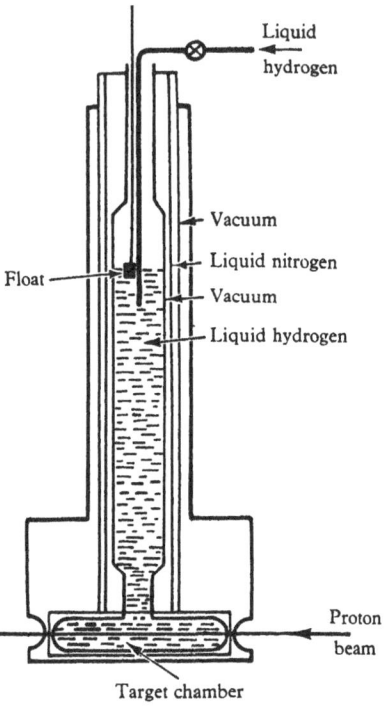

A typical assembly is shown in fig. 15 (Cook, 1951). Liquid hydrogen is contained in a stainless steel† tube 35 cm. long by 5 cm. diameter, closed at the ends by hemispherical foils 0·1 g./cm.² thick. Two identical hemispheres close off the vacuum jacket. A 10 cm. length of tubing above the vessel provides a reservoir for hydrogen,

Fig. 15. Schematic diagram of assembly for liquid hydrogen target (Cook, 1951).

a float serving to monitor the level. In the arrangement shown in the figure, the target vessel was entirely surrounded by a surface maintained at liquid nitrogen temperature. The evaporation loss rate of liquid hydrogen was found to be about 2 l/day. Other liquid ^{1}H, ^{2}H and ^{4}He target arrangements have been described in numerous reports (e.g. Whalin and Reitz, 1955; Marshall, Marshall, Nagle and Skolnik, 1953).

A less conventional technique which combines the target with

† Stainless steel does not catalyse conversion from ortho- to para-hydrogen. It also takes a highly polished surface which minimizes radiation loss.

the detector has been used with photographic emulsions and cloud chambers. When a beam of particles passes through either of these detectors, trajectories of scattered and recoil particles form conspicuous forked tracks which convey full information about the collision (§ 4.3.1). The method has historical interest because it was with cloud chambers that the first evidence of the nuclear contribution to p-p scattering was obtained (White, 1935; Wells, 1935). More recently, p-p scattering has been studied with emulsions in essentially the same way at 430 MeV. (Kao and Clark, 1955) and 3 BeV. (Kernan and Cester, 1955). However, the difficulty of achieving sufficient yield has prevented the method from reaching any real importance in charged-particle scattering. Diffusion cloud chambers (Snowden, 1953) and bubble chambers (Glaser and Rahm, 1955; Oswald, 1957) have been used with moderate success at very high energies. The latter have fast recycling times which enable them to be operated in synchronism with the pulses from a proton synchrotron (§ 3.2.6), and they also offer more concentrated targets. Pure liquid ^{1}H, ^{2}H and ^{4}He chambers have been operated successfully in this way.

An 18-in. diffusion chamber was used at 24 atmospheres with the Birmingham proton synchrotron for p-p scattering at 650 MeV. (Batson, Cooper and Riddiford, 1956; Batson and Riddiford, 1956). Particles scattered out of the vacuum box entered the chamber at the rate of 10–20 per pulse. From 25,000 photographs, some 300 two-pronged events were obtained. Provision of a 13 kG. magnetic field and measurement of track curvatures made possible a fairly positive identification of elastic events. Similar techniques were adopted with the Cosmotron at 900 MeV. (Melkanoff, Moszkowski, Nodvik and Saxon, 1955) and the Bevatron at 5·3 BeV. (Wright, Saphir, Powell, Maenchen and Fowler, 1955).

3.4. Measurement of beam current

Determination of an angular distribution on an arbitrary scale of intensity does not involve measuring the total beam traversing the chamber, but it does require that intensity measurements made at various angles be normalized. This can be achieved in two ways:

(a) By allowing the main beam to pass through a detector (whose efficiency need not be known) set at position (1), fig. 16.

5

Alternatively, the beam may be scattered from a second target so as to give a high counting rate in a *normalizing* or *monitoring* counter (position (3)).

(*b*) By referring the scattering at each angle back to that at one particular angle, e.g. position (2). A monitoring counter is set at a fixed angle inside the chamber.

Either method serves to normalize readings obtained with the detector over the entire range of the angular distribution. As much

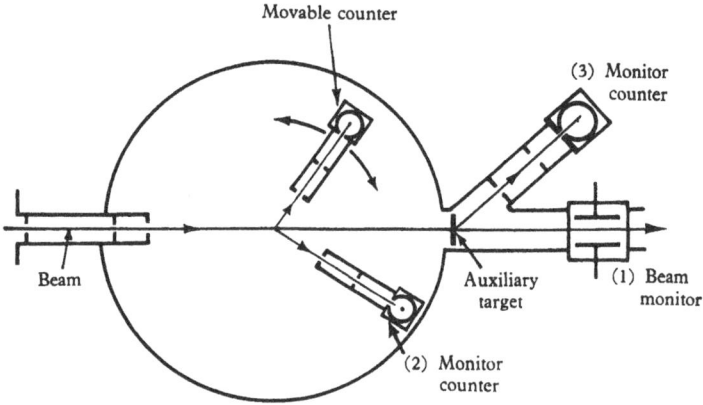

Fig. 16. Alternative methods of monitoring beam through scattering chamber. (1) ion chamber in main beam. (2) and (3) alternative positions for normalizing counters.

of the interest of scattering data centres on the angular dependence, relative measurements carried out in this way have proved valuable. (A large part of the neutron scattering has been confined to relative cross-sections (cf. chapter 4)). However, it is desirable when comparing experimental and theoretical results to have the most complete information possible, hence most experiments are designed to measure absolute values of differential cross-sections. This calls for a knowledge of the ion beam traversing the chamber.

Methods of beam measurement are among the more standardized features of scattering arrangements. At low energies the beam is usually stopped after passing through the chamber and the charge transported during a given length of run is measured. At high energies, the increase in range of particles in matter tends to make this method inconvenient, although there is no fundamental objection to it (see Kruse, Teem and Ramsey, 1954 for use of the Faraday cage at 40–95 MeV.). In general, less direct methods are used.

3.4.1. *Current collectors at low energy.* These have not departed in any important respect from the design employed by Herb *et al.* (1939) and shown in fig. 17. In their experiments a metal collecting cup which served as a Faraday cage was supported on an amber insulator free from the main body of the chamber. Its diameter

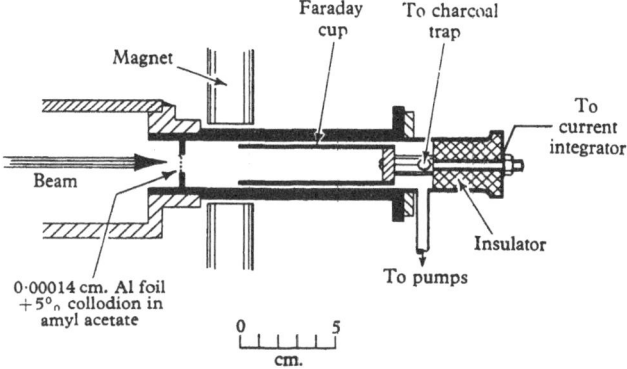

Fig. 17. Design of current collector for low energy experiments (Herb *et al.* 1939).

should be comfortably larger than the maximum beam spread allowed by the collimator slits and sufficient also to accommodate any further spread due to multiple scattering by the target.

In the case of thin foil targets multiple scattering takes place only in the target plane and its effects can be allowed for readily by computing the r.m.s. angle of scatter $\langle \phi^2 \rangle^{\frac{1}{2}}$ using equation (3.6). If a margin of several times $\Delta y = l \langle \phi^2 \rangle^{\frac{1}{2}}$, where l is the distance from target to collector, is allowed around the unscattered beam at the collector, there will be no significant loss of particles.

The calculation is more difficult for gas targets where multiple scattering occurs at all points x from collimator to collector. An approximate evaluation is obtained by following the procedure of Breit *et al.* (1939): Taking the whole path length as L and the r.m.s. angle of scatter due to thickness L of target as $\langle \phi^2 \rangle^{\frac{1}{2}}$, then $\langle \phi^2 \rangle^{\frac{1}{2}} \propto L^{\frac{1}{2}}$.

The effective r.m.s. lateral spread of beam is given, rather crudely,
by $\Delta y \sim \langle \phi^2 \rangle^{\frac{1}{2}} \int_0^L \left(\frac{x}{L}\right)^{\frac{1}{2}} dx = \frac{2}{3} L \langle \phi^2 \rangle^{\frac{1}{2}}$.

The ion beam is absorbed in the metal end of the cup, causing an accumulation of positive charge. Knowing the unit charge e ($1 \cdot 602 \times 10^{-19}$ C.), the number of particles N_i which cross the target is given by measuring the total charge. The current collector is isolated from the main vacuum by a thin foil and is evacuated to $\sim 10^{-6}$ mm. Hg pressure during runs by a separate pumping line. Unless this shielding is provided, ionization currents set up in the gas cause charge leakage and seriously vitiate beam measurement. It is also customary to provide a magnetic field across the region of the collector to prevent secondary electrons, which may be sprayed out of the foil, from reaching the cup. An appreciable yield of electrons is obtained from any surface struck by an ion beam. The most energetic ones probably result from direct ion-electron collisions and so possess energies up to ~ 10 keV. while the majority have much lower energy. Radii of curvature in a field of ~ 500 G. will not exceed ~ 6 mm., which means that all electrons from this source can be prevented effectively from reaching the cup merely by spacing it suitably behind the foil. Another method of achieving the same end is to introduce an electrostatic retarding field by applying a negative potential to the cup, say -150 V. relative to earth.

Electrons are also emitted from the end of the cup, their energies being of the order of 10 eV. The same magnetic field is sufficient to prevent their escape from the cup. Cork and Hartsough (1954a) reported that the application of a -20 V. bias served the same purpose but that no further change resulted from raising this to -300 V.

It has sometimes been found that intense γ-ray background causes appreciable ionization of air surrounding the chamber and leads to a measurable charge leakage between the outside of the collector and ground. One method of prevention is to surround the cup by an electrostatic shield held at the same potential as the cup. Simple check runs aimed at establishing the efficiency of precautions of this kind are highly important in absolute scattering measurements (see Herb *et al.* 1939; Yntema and White, 1954).

Several methods are available for measuring the positive charge collected on the cup, all capable of high accuracy if due instrumental precautions are taken. They have in common the measurement of the absolute potential developed on a condenser of accurately known capacity, but they differ considerably in procedure.

(a) The charge accumulated on a standard condenser may be determined by measuring its potential after completion of the run. The latter can be done with a quadrant electrometer (Karr et al. 1951), ballistic galvanometer (Herb et al. 1939; Blair, Freier, Lampi, Sleator and Williams, 1948) or vibrating reed electrometer (Kruse et al. 1954; Zimmerman, Kerman, Singer, Kruger and Jentschke, 1954), all of which may be calibrated for voltage measurement by means of a standard cell and potentiometer. The instruments and their calibration are adequately treated in texts on electrical measurements (Laws, 1938; Golding, 1942). A discussion of dynamic condenser electrometers, particularly relevant to measurement of very small ion currents, has been published by Palevsky, Swank and Grenchik (1947).

(b) The application of a known negative charge to a standard condenser which is then discharged by the beam current, the run being terminated when the condenser potential falls to zero. The same potential measuring devices may be used but their calibration is no longer important since they serve merely as null indicators. The negative charge may be applied by feeding a known current to the condenser for a known time. Worthington et al. (1953) introduced a balanced d.c. amplifier to indicate the null point, the counting of scattered particles being terminated automatically when this was reached.

(c) A strictly null method in which the Faraday cup is maintained at ground potential throughout has been used in several experiments (Cork, Johnston and Richman, 1950; Yntema and White, 1954). The current collector and circuit of one of these arrangements is shown in fig. 18. The cup is connected to a standard condenser and vibrating reed electrometer. The battery B supplies a comparatively low voltage which is regulated by two 10-turn helipots of 500 and 150 Ω so as to maintain the reading of the electrometer at zero. The total voltage required to do this is measured by the potentiometer at the completion of each run.

The condensers employed in all three methods are commonly of about $1\,\mu\mathrm{F}$. and in methods (a) and (b) they are charged to potentials of several volts. With beam currents of 10^{-8} to $10^{-9}\,\mathrm{A}$. each run then lasts $\sim 15\,\mathrm{min}$. It is sometimes convenient to use a network of condensers (Karr et al. 1951), which, by a simple switching of connexions, provides a wide range of capacities suitable for different beam intensities.

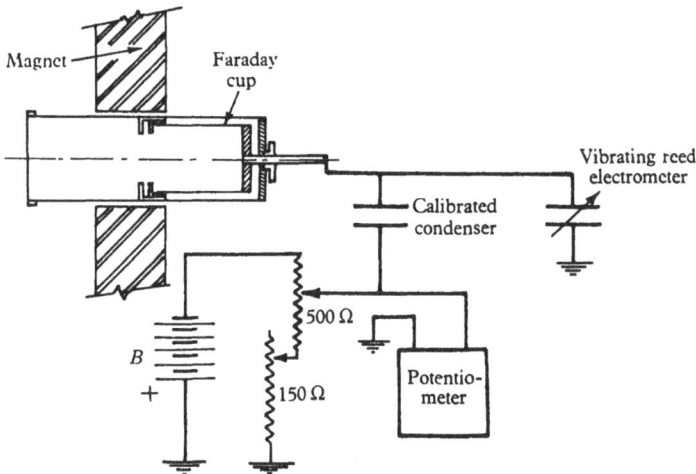

Fig. 18. Current collector, standard condenser and potential measuring circuit (Yntema and White, 1954).

(d) The same principle, viz. ascertaining the voltage rise of a standard condenser, can be incorporated into a fully automatic circuit. Meagher (1950) has described a 'Q-meter' based on a $33\,\mu\mathrm{F}$. condenser which was automatically shorted back to zero potential when it reached $3\,\mathrm{V}$., a charge of $10^{-10}\,\mathrm{C}$. being registered by a suitable scaling circuit. However, in view of the fact that the condenser was made up by the capacity of the Faraday cup to ground, the connecting leads and vacuum tubes, the Q-meter was not an absolute device but required calibration over the operating range of beam current. It was calibrated in situ by supplying to it currents established by measuring the voltage across standard resistances which varied from 5×10^{9} to $10^{11}\,\Omega$. An overall accuracy of about $1\cdot3\%$ was achieved which is rather poorer than generally applies to the other three methods. Reasonable precautions should confine beam current errors to $\pm 0 \cdot 5\%$ although errors as low as $0 \cdot 1\%$ have been claimed (Worthington *et al.* 1953).

Determination of the capacity of the condenser is far from being straightforward and has proved a troublesome source of small residual errors when high accuracy has been attempted (Yntema and White, 1954). In many cases the effective capacity includes as an appreciable part the capacity of a coaxial cable ($\sim 10 \mu\mu$F./ft.), which feeds the charge collected by the cup to the integrator, perhaps 100 ft. away.

Extreme care must also be exercised with some condensers against the effects of 'charge soakage' and leakage resistance, although the dielectric properties of modern polystyrene condensers are relatively good. Calibration of the condenser should be carried out by a procedure which simulates its conditions of use (rate of charge, maximum potential, temperature and relative humidity) as closely as possible. Both soakage and leakage, if present, will tend to increase the apparent capacity, and both are 'long-time' effects. Consequently, high frequency bridge measurements of capacity are likely to be inconclusive if the ultimate accuracy is required.† If a high-quality polystyrene condenser is used errors should not be in excess of $0 \cdot 1$ or $0 \cdot 2\%$ from these sources, for runs lasting ~ 30 min. The best method of calibration is to measure the time required to charge the condenser to the specified voltage, using known currents which cover the range pertaining to actual scattering runs. This procedure is applicable to currents as low as 10^{-11} A. (Royden and Caldwell, 1956).

3.4.2. *High energy devices for beam measurement.* There are two principal features which distinguish the technique of measuring beams of high energy ions. The first is the long range in matter of particles of several hundred MeV. energy (e.g. 38 cm. Al for a 400 MeV. proton); this tends to make a Faraday cage collector unwieldy. The second is the not-uncommon use of internal targets.

---

† We are referring here to effects which become important when the total error of the experiment must be kept to 1% or less, i.e. individual sources of error must be kept to $\sim 0 \cdot 1\%$. These are the conditions which prevail in low energy *p-p* and *p-d* scattering at the present time. It follows that unless new experiments achieve at least this level of precision they will be a waste of effort.

First consider devices which have been employed with external beams. In the Berkeley experiments at 340 MeV. (Chamberlain and Wiegand, 1950; Chamberlain *et al.* 1951) the beam was monitored with a thin-walled parallel-plate ionization chamber which had been calibrated against a conventional Faraday cup apparatus. The proton beam passed through the ion chamber before impinging on the Faraday cup which consisted of a brass cylinder 6 in. thick in an evacuated enclosure. During scattering runs, the current from the ion chamber was fed to a vacuum tube voltmeter, which served as an integrator by utilizing the capacity of a 50 ft. connecting cable plus that of the voltmeter itself. Beams of the order of 10^{-11} to 10^{-13} A were measured in this way.

Moreover, the Berkeley calibration at 340 MeV. could be extended semi-empirically to the intermediate range of energies where scattering experiments were carried out with absorbers in the beam. The ratio between the saturation current of the ion chamber and the Faraday cup current may be written as $l/w . (-dE/dx)$, where l is the length of path of beam through the chamber, w the energy required to produce an ion pair and dE/dx the specific energy loss. For argon, $dE/dx = 3 \cdot 08 \times 10^6$ eV. cm.^{2}/g. at 340 MeV. and w can be assumed to remain constant at $25 \cdot 5$ eV. Knowing the range-energy relation for argon (Aron, Hoffman and Williams, 1949), appropriate values of dE/dx may be inserted for different beam energies. Actually, if one knows the energy loss/proton/cm. and w, an *absolute* evaluation of the current output from the chamber in terms of protons/sec. is obtained. This was the basis of current integration in the Pittsburgh experiments (Mott, Sutton, Fox and Kane, 1953), using A for which the constants are known.

Several other absolute methods exist for calibrating an ion chamber. With feeble currents, it is possible to pass the beam through the chamber and a nuclear emulsion in tandem, then to count tracks to establish the absolute number of particles. Alternatively instead of an emulsion, a fast scintillation counter could be used (§ 3.5.4). The latter was employed in the beam measuring system at Chicago (Marshall *et al.* 1953). The beam was defined by two scintillation counters A and B (fig. 19) connected in coincidence so as to measure directly the number of protons falling on the target. However, at the beam intensity used, counting losses due to two particles traversing the counters within their resolving times amounted to 7 %. A monitor counter M was therefore located below the beam to record protons scattered from an auxiliary lead target placed in the beam several feet beyond the main target. In coincidence with B its counting rate was only 1/16th of the (A, B) rate. By measuring the correct value of the ratio $(M, B)/(A, B)$ at a beam intensity low enough for the (A, B) rate to be reliable, the correct value of (A, B) could be inferred from the

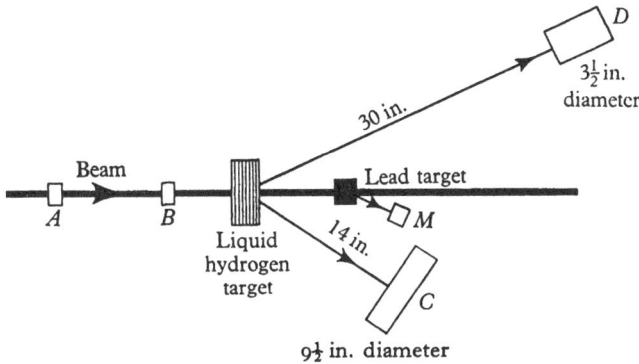

Fig. 19. Beam intensity measurement using auxiliary lead target and scintillation counters placed in the beam (Marshall, Marshall and Nedzel, 1953).

(M, B) rate at the higher beam intensities used in the actual runs. Scattered particles, in this experiment, were detected by two more counters C and D, and (A, B, C, D) were connected in 4-fold coincidence, which rendered the method of taking data almost independent of stray background radiation.

The Rochester experiments at high energies employed internal hydrogenous targets (fig. 14 and §3.3.2) which themselves served as beam integrators. Because the expansion of the beam per revolution in the synchrocyclotron is small, it is not practical to use an ion chamber or Faraday cup to collect the protons beyond the target in the usual way. However, the proton beam forms ^{11}C by the reaction ^{12}C $(p, pn)^{11}C$, the cross-section for which has been measured as a function of energy (Aamodt, Peterson and Phillips, 1949).† ^{11}C is β^+ active (0·95 MeV., 20·5 min.) and its

† Preliminary data at BeV. energies are also available (Cumming, Swartz and Friedlander, 1956; Horwitz, Murray and Crandall, 1956).

concentration in the target can be determined by an absolute count with a standard β-counter (Oxley and Schamberger, 1952). A different method has been used in experiments at 400–1000 MeV. with the Cosmotron (Smith, McReynolds and Snow, 1955). The target was located inside one of the straight sections of the machine and a signal was induced on a pair of pick-up electrodes which enclosed the beam path. The amplitude of the voltage pulse measured the density of the protons in the beam but, as this was not constant throughout the bunch, the shape of the pulse had to be known in order that the signal could be used as an absolute measure of the number of protons. Again it was found necessary to calibrate the signal against a Faraday collector and condenser arrangement. The device is sensitive to circulating beams containing $\sim 10^7$ protons.

The difficulty of accurately evaluating small beams of high energy particles has been emphasized by the discordance among published values of p-p cross-sections in the 100 to 450 MeV. range. Absolute values measured at Berkeley have been consistently lower than the Rochester values by $\sim 25\%$. Chicago and Pittsburgh values tended to support the former while those from Harvard and Harwell (which were also based on the ^{11}C method) agreed better with the latter. Although largely resolved by improved determinations of the $^{12}C(p, pn)^{11}$ C cross-section (Birnbaum, Crandall, Millburn and Pyle, 1955), the discrepancy serves to illustrate the need for more refined techniques in this part of the work. It also high lights the desirability of repeating experiments with basically different experimental techniques.

3.5. Charged-particle detectors and typical experimental arrangements

The function of the detector is to establish the intensity of scattering from the target, i.e. the number of particles scattered into a known solid angle. A detector of any one of several types may be placed behind a slit system (§ 2.4) which serves to define the solid angle through which scattered particles may enter.

Quite early in the development of experiments on scattering, detectors were adopted which responded to individual particles and there are now no exceptions to this procedure. The advantages of particle-by-particle detection are several: If the response of the detector is sensitive to energy, it is convenient to discriminate between particles of different energies, a case which arises frequently when scattered and recoil particles are not alike. In the same way it is possible to distinguish particles of a homogeneous energy group from the continuous spectrum due to γ-radiation and neutron recoils in the vicinity of an accelerator.

An equally cogent argument for single particle detection is that it offers scope for coincidence measurements, either between scattered and recoil particles or from the same particle using a telescope of counters in series.

In recent years virtually all charged particle scattering experiments have been carried out with proportional counters, scintillation counters or photographic plates. Ionization chambers and Geiger–Mueller counters are now only of historical interest in this field, but the cloud chamber (or perhaps some of its recent variants (§§ 3.3.3, 4.3.1., 4.3.2)) may return to favour for extremely high energy work.

The choice of detector for a particular scattering experiment, while to some extent a matter of personal preference, can be narrowed down by a consideration of the following points:

(1) The range of energy over which particles must be detected. The response of the detector at low energies is often the limiting factor that determines the angular range which can be covered.

(2) The energy discrimination required of the detector, e.g. so as to distinguish protons from deuterons in a p-d experiment, or genuine scattering from spurious background.

(3) Whether it is necessary to employ coincidence techniques to distinguish a particular kind of process amid competing events.

(4) The characteristics of the duty cycle of the accelerator (§ 3.2.4): if the particles arrive in short bursts the response of the detector and associated circuits must be appropriately fast.

In the following discussion the basic principle of operation of each detector has been assumed known. A number of comprehensive descriptions have been published and are quoted as general references after each section heading.

3.5.1. *Ionization pulse chambers* (Rossi and Staub, 1949; Wilkinson, 1950). These were employed with considerable success in early experiments. For energies up to a few MeV. it is possible completely to absorb the particle in the chamber gas and to amplify the electrode signal with a high gain linear amplifier. A parallel plate chamber used by Herb *et al.* (1939) for *p-p* experiments from 0·86 to 2·4 MeV. is shown in fig. 20. It was operated as an ion pulse device or 'slow chamber' by making the leak resistance big enough

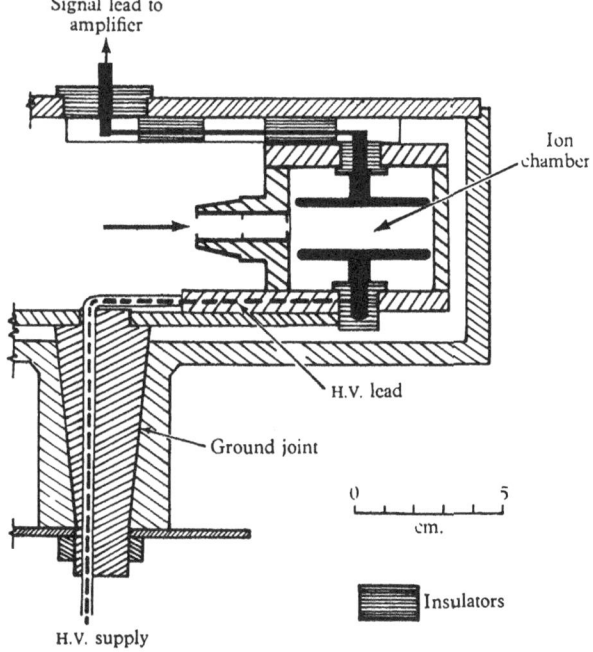

Fig. 20. Parallel plate ionization chamber used as detector in low energy scattering (Herb *et al.* 1939). The gas lead (not shown) to the ion chamber was also brought in through the cone at the bottom.

for the decay time constant (or clipping time) RC to be much larger than the ion collection time. The advantage of this system lay in the fact that the pulse height delivered was strictly proportional to the number of ions produced in the chamber, thus enabling it to be used for discriminating against scattering due to the heavier nuclei of contamination gases. A proton loses little energy by collision with C, N or O, and hence produces a pulse which can be distinguished readily from the smaller pulses due to *p-p* scattering. Visual monitoring with an oscilloscope was used to exclude large pulses. Similarly the chamber was found to operate satisfactorily in the presence of severe γ-radiation, which gave rise to comparatively small electron pulses.

The disadvantage of the ion pulse chamber is that the collection time is long, customarily of the order of msec., leading to slow counting rates and poorer statistics. The longest time constant in Herb's apparatus was 500 μsec. which, in order to guarantee that no counts were lost, kept the counting rate below 20 per sec. Another disadvantage is that the amplifier has to be capable of handling low frequency, of the order of kc./sec., which makes it vulnerable to a.c. pick-up and microphonics.

A superior arrangement is the 'fast chamber' in which the time constant is small compared with the ion collection time but still larger compared with the electron collection time ($\sim$ 1 μsec.). The grid-type fast chamber has definite advantages when rapid counting is necessary and, in addition, several different pulse heights must be distinguished. Pulse size is proportional to total ionization. Associated amplifiers may have a low frequency cut-off at 10–100 kc./sec., rendering them insensitive to microphonics. However, in medium energy experiments the difficulty of securing enough primary ionization in a small chamber to give pulses well above noise level† makes it advisable to operate the chamber with some gas amplification, i.e. in the proportional region. It is then practicable to generate a usable pulse from only a few electrons of primary ionization. Gas counters for scattering work are now invariably of this kind.

3.5.2. *Proportional counters* (Rossi and Staub, 1949; Wilkinson, 1950; Staub, 1953). In the strictly proportional region of gas counting the amplification is independent of, and the resulting pulse proportional to the initial ionization. Although the statistical nature of the ion multiplication process responsible for amplification introduces fluctuations in pulse size, in practice the energy resolution is not seriously impaired. Ability to discriminate between particles of different specific ionization is thus retained, while at

† The amplifier noise level is likely to be equivalent to 40 or 50 keV. of primary ionization in the chamber.

the same time a voltage pulse is delivered which is comfortably above background noise level in the ancillary electronic system. The distribution about the mean of pulse heights due to mono-energetic particles determines the energy resolution actually attainable. It depends on numerous factors, including the amount of primary ionization, the choice of gas filling (impurities which form negative ions tend to broaden the pulses), the counter construction and the stability of the high voltage supply. A typical figure for the width of the distribution at half maximum height is 5 % in the case of counters used in low energy scattering. At high energy the distribution becomes much broader (cf. §3.5.4). To some extent, the resolution may be worsened by the injudicious use of the counter. If its entrance window covers a large fraction of the counter diameter, the path lengths, and hence initial ionizations of particles traversing a cylindrical counter perpendicular to the wire, may differ consider-ably ($\sim$ 15 % in the case of the conjugate counter shown in fig. 22).

The total pulse length from a proportional counter is long (the diffusion time of positive ions to the cathode $\sim$ 10^{-4} sec.) but the rise is rapid, hence by using a short clipping time a constant fraction of the pulse can be used without serious loss of height. Resolving times of the order of 1 μsec. can then be realized. The counting rate is determined by the dead time of the counter, the safe upper limit being $\sim$ 10^4 per min.

The gas amplification can be as high as 10^3 or 10^4 but in scattering work it is rarely made larger than is required to boost the signal well above tube noise level, perhaps 10 or 100 times. For gas ampli-fications lower than about 10, the pulse shape becomes dependent on the location of the trajectory of the particle with respect to the centre wire of the counter, which is undesirable. Trajectories will lie at different distances from the wire, depending on the size of the counter window, so that, quite apart from the variation of primary ionization referred to above (which may be minimized by proper design of the counter), the pulse shape will depend on the spread in average time of arrival of electrons from different trajectories. The time constant of the attached circuit should, of course, be longer than the maximum time required for electrons to reach the wire.

The usual practice is to feed the anode signals from the counter to a single stage preamplifier with a gain of about 10, located alongside the scattering chamber. The space around an accelerator is subject to a radiation level well above human tolerance so that the counting circuits need to be situated a considerable distance away. The connecting cable attenuates the voltage pulse by an amount which depends on its capacity, usually $10-20\,\mu\mu$F./ft. The pre-amplifier assures that a pulse of ample size arrives at the input of the counting circuit. (An alternative is to use a cathode follower which supplies power to transmit the pulse, although it does not amplify the signal.) At the counting end of the cable the signal is received by a linear amplifier, pulse height selector (discriminator) and scaler.

Electronic methods of pulse sorting are highly desirable. Especially in p-d or p-t experiments where, for all angles smaller than $90°$, the two particles must be distinguished by their pulse heights, multi-channel analysers have an advantage both from the point of view of saving time and avoiding subjective bias. 30-channel analysers have been used in some scattering work, channel widths being set to about 1 V., stable to $\sim 1\,\%$. No attention will be given here to electronic circuit design which has been covered adequately elsewhere (e.g. Elmore and Sands, 1949; Van Rennes, 1952).

Proportional counters have, with a few exceptions, been confined to low and medium energy scattering work where accelerator beams are continuous and ultra-fast counting is not necessary (cf. § 3.5.4). They are, however, well suited to measurements at about 1 MeV. and lower, where scintillation counters (§ 3.5.4) and photographic plates (§ 3.5.5) are least satisfactory.

Typical experimental arrangements. Fig. 21 (*a*) illustrates the counter used by Sherr, Blair, Kratz, Bailey and Taschek (1947) in p-d scattering from $1\cdot5$ to $3\cdot5$ MeV. The outside surface was at ground potential and the central wire, which collected the signal, at high potential. When filled with butane at 5 cm. Hg pressure, it gave an amplification of 150. Tested with 5 MeV. α-particles, pulses were delivered having a half width of $5-7\,\%$. As the counter was required to operate inside the chamber, immersed in a deuterium atmosphere at 1 cm. Hg pressure, it was important to prevent corona and sparking to the high voltage lead supplying the centre wire. This was brought in through the rotating shaft which carried the counter and its slit unit, then fed to the counter through a gas-tight tube connected to the compartment above the counter proper. The lead was thus surrounded by air at atmospheric pressure over its full length. A second tube was brought in by the same method to supply the counter gas.

Outside the scattering chamber, the signal was separated from the high voltage by a 10 MΩ decoupler in the high voltage line, and a 100 $\mu\mu$F. coupling condenser delivered signals to a preamplifier,

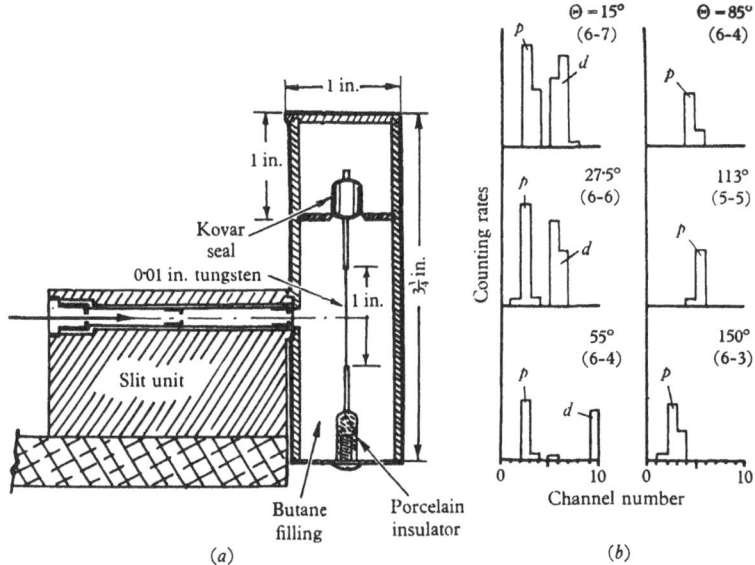

Fig. 21. (a) Proportional gas counter and defining slit unit. (b) Typical pulse height distributions obtained in p-d scattering at various angles Θ (Sherr *et al.* 1947). Numbers brackets in refer to amplifier gain settings, larger numbers implying higher gain.

linear amplifier and 10 channel pulse height discriminator. From 0 to 90°, the multi-channel analyser enabled protons and deuterons to be counted independently during a single run. The differential bias curves, fig. 21 (b), show the relative yield vs. channel number at various angles. The two groups were well resolved at most positions, with best separation at ∼ 55°.

In experiments at energies below 1 MeV. special attention must be paid to the energy loss caused by the counter window. Pulse heights are diminished and genuine counts tend to get lost among counts due to electrons. One unusual design for a fixed angle scattering chamber has been described (Cooper, Frisch and Zimmerman, 1954) in which the proportional counters are open to the main chamber space and therefore operate on the target gas, H_2 at 1·5 cm. Hg pressure. This arrangement avoided the stopping effect of windows altogether. The counter consisted of a 1·8-in. diameter cylinder directed axially at the target. The anode was a 12-in. rod of $\frac{1}{8}$-in. diameter cantilevered from a Kovar seal at the outside end of the counter and terminated in a $\frac{3}{8}$-in. round section at the other. Typical pulse heights were 15 times noise level. Another novel feature of this design was its reliance on a 'gas impedance collimator', with differential pumping, which avoided the usual foil separating scattering chamber from accelerator vacuum (cf. Worthington *et al.* 1953).

For work with solid targets (§ 3.3.2) the coincidence method is generally used. Two proportional counters are set at conjugate angles so that one records scattered and the other recoil particles when their signals are fed into a coincidence circuit. In non-relativistic *p-p* scattering the angle between the counters is fixed at 90°, but where the collision is between unlike particles the counters must be capable of independent movement.

A typical arrangement is shown in fig. 22 based on the 10 MeV. *d-p* experiment of Karr *et al.* (1951). The counters resembled that shown in fig. 21 (*a*) and were filled with a mixture of argon plus 5 % CO_2 at 20 cm. Hg pressure. Windows were made of polyethylene terephthalate stretched to 0·0001-in. and sealed to the counters between soft indium gaskets. The aperture in front of the defining counter (§ 2.4) was rectangular and, depending on the orientation employed, allowed a maximum angular spread of 1·8 or 0·8°. The solid angle subtended was approximately 10^{-4} steradians.

In this system, the counter walls were maintained at high potential (− 1000 V.) and the centre wires at signal level. The high voltage circuit was connected to the amplifier only *via* the small inter-electrode capacity of the counter so that disturbances produced by high voltage leakage in cables and the decoupling condenser were minimized. The centre wire was direct-coupled to the amplifier grid input.

High voltage and signal leads were brought into the chamber through three insulating vacuum seals in the bottom, and connected

6   M E S

to the counters by flexible insulated leads. Since the chamber was continuously evacuated no trouble arose from corona. The movement of the counters was provided by means of two concentric

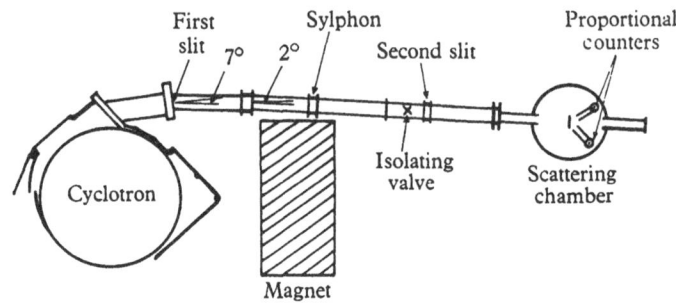

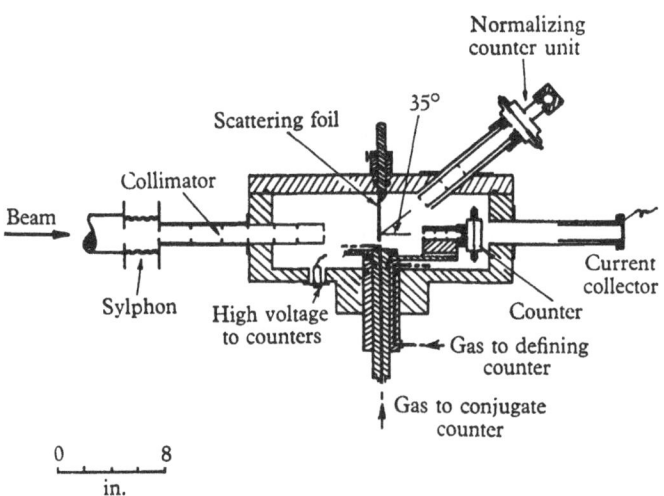

Fig. 22. Washington University coincidence scattering experiment (Karr *et al.* 1951). The upper diagram shows the general arrangement of the fringing-field analyser (not strictly to scale). The lower is a schematic elevation of the chamber. Only one of the two counters is shown inside the chamber.

shafts passing through a bushing at the bottom of the chamber, satisfactory vacuum seals being achieved by the use of O-rings (Kurie, 1948).

The usual operating procedure is to set the defining counter at the selected angle, then traverse across the expected conjugate angle (equation 2.1) with the second counter. Besides locating accurately the correct setting, this provides a valuable check on the size of the conjugate aperture. If its width is sufficient to accept all particles corresponding to those entering the defining counter, the coincidence rate should rise steeply, level off to a flat plateau and then decline again as the counter passes through the correct setting (cf. § 3.5.4).

The counting of coincidences has certain complicating features when, as in the d-p experiment described above, the two particles have different masses. Here the deuteron scattering angle Θ_1 increases from $0°$ to a maximum at $30°$ and then decreases again to $0°$ as the proton recoil angle Θ_2 increases steadily from 0 to $90°$ (fig. A. 2). There is an intermediate range of angles near $\Theta_1 \approx \Theta_2 \approx 30°$ where 'double counting' occurs. Besides *bona fide* coincidences from, say, deuterons entering the defining counter and protons the conjugate counter, genuine coincidences will also be recorded between protons entering the defining counter and deuterons the conjugate counter. Pulse height analysis must be relied upon to separate the two processes in this range. A similar situation occurs in the scattering of protons by deuterons between 50 and $55°$ (Schamberger, 1952).

In principle, either the scattered or recoil particle involved in a collision may be defined, but practical circumstances usually indicate which arrangement is superior over each angular range. The conjugate counter, which has the larger aperture, should be kept well away from the main beam or it may record an unduly high single counting rate from the carbon, etc., in the target, with the consequent risk of loss of true coincidences and increase in the chance coincidence rate. Multiple scattering effects, discussed later in the section, also influence the choice of arrangement.

Reference was made in § 3.3.2 to the geometrical difficulties introduced by the coincidence counting method and these will now be discussed in some detail. The following treatment applies also to coincidence counting with scintillators (§ 3.5.4).

(a) Consider first the relation between the heights of the defining and conjugate counter apertures, h_1 and h_2 respectively. Fig. 23 (a) represents a projection on a plane, perpendicular to the beam, of paths of scattered and recoil particles. Suppose a is the vertical height of the beam where it strikes the target, L_1 the distance between the centre of the target and defining aperture and L_2 the corresponding distance to the conjugate aperture. The angles at which the defining and conjugate counters are set to the beam are α and β respectively. Assume that the incident particles arrive at the target perfectly collimated. Then extreme paths of scattered

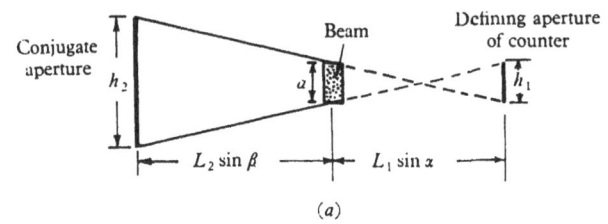

(a)

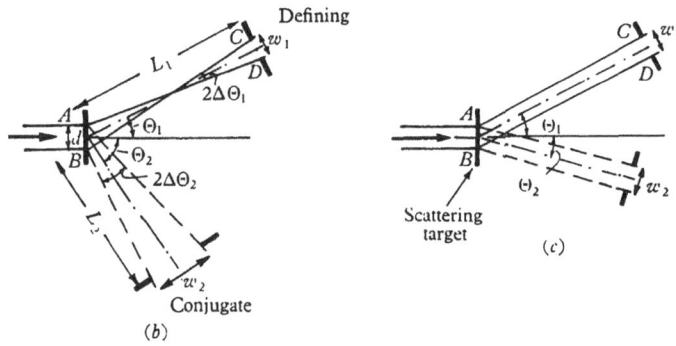

(b)

(c)

Fig. 23. Geometrical relationships between defining and conjugate counter apertures. (a) Projection on plane perpendicular to beam of extreme scattered and recoil rays. (b) and (c) Paths of extreme rays in the plane of scattering (not to scale) for d-p scattering when $\Theta_2 < 30°$ and $\Theta_2 > 30°$, respectively.

and recoil particles are as shown. These determine the height of the conjugate aperture required by geometrical considerations alone. It follows that

$$h_2 = a + (h_1 + a)\frac{L_2 \sin \beta}{L_1 \sin \alpha}. \qquad (3.5)$$

The ratio of the counter arms L_1/L_2 is often adjusted to keep h_2 within reasonable bounds. α and β are related by the equations (A.6), (A.38) of Appendix A. If the scattered particle is being defined, $\alpha = \Theta_1$, $\beta = \Theta_2$ or, if defining recoils, $\alpha = \Theta_2$, $\beta = \Theta_1$.

Fig. 23 (a) also emphasizes the importance of aligning the scattering chamber precisely. The two slits of the collimator and the two counter apertures must be coplanar. When $L_2 \sin \beta > L_1 \sin \alpha$, a small error in beam height becomes magnified at the conjugate aperture.

(b) The relation between the counter widths w_1 and w_2 can be followed through in like manner. Fig. 23 (b) and (c) show extreme paths of scattered and recoil particles in the plane of scattering. Either the diagonal rays AD and BC (b) or rays AC and BD (c) can determine w_2, depending on the mass ratio m_1/m_2, the angle setting of the defining counter, whether m_1 or m_2 is being defined and on the relative sizes of the defining aperture and beam width.

For any particular case the required width may be calculated readily enough. Fig. 23 (b) illustrates the case of d-p scattering (Karr et al. 1951) where the scattered deuteron is being defined and $\Theta_2 < 30°$. The diagonal rays determine w_2. d is the mean beam width and $2\Delta\Theta_1$ the angle between rays making angles $\Theta_1^* = \Theta_1 + \Delta\Theta_1$ and $\Theta_1^{**} = \Theta_1 - \Delta\Theta_1$. $2\Delta\Theta_2$ is the angle between corresponding recoil protons, i.e. $\Theta_2^* - \Theta_2^{**}$. Then w_2 can be computed by obtaining $\Delta\Theta_1 = (w_1 + d\cos\Theta_1)/2L_1$ and Θ_2^* and Θ_2^{**} from equation (A.6). Hence we get $2\Delta\Theta_2$ and then $w_2 = (d\cos\Theta_2 + L_2 . 2\Delta\Theta_2)$. The calculation is obviously similar for case (c), which illustrates the same collision but for $\Theta_2 > 30°$. Here the direct rays determine w_2. In practice it is found that the conjugate aperture needs to be considerably larger than the defining aperture at certain angle settings. For the apparatus shown in fig. 22, rectangular apertures were 0·26 × 0·10 and 1·0 × 0·85 cm.[2] respectively. By making collimator slits rectangular also, a useful flexibility was obtained. They were rotated through 90° to provide extra breadth or height as best suited the geometry.

The above method of computing h_2 and w_2 assumes that the incident beam is perfectly collimated and aligned. Small transverse components have the effect of increasing the size of aperture required. It has also been assumed that the target thickness is zero. If this were not the case, e.g. if gas targets were used with the coincidence method, demands on w_2 would become excessive. Nevertheless, experiments have been attempted in this way (Holm and Argo, 1956).

An effect which plays an important part in all experiments is multiple scattering. With single detectors, which will be discussed at this stage jointly with the coincidence technique, it imposes a 'compensation process' upon which the accuracy of the results very largely depends. Consider a beam of particles which has been scattered from the main target and is moving towards the detector. Particles which are multiply scattered out of this beam during passage through the gas in the chamber are replaced by other particles which are scattered into it. As much as 5 % of the beam

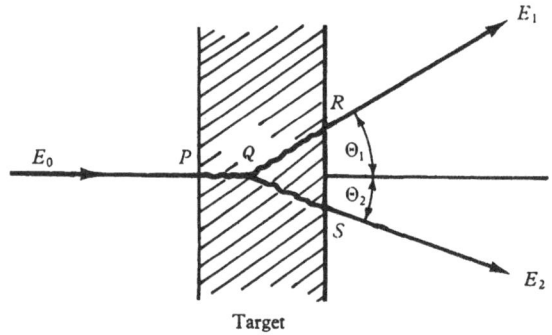

Fig. 24. Collision at point Q inside a solid target. Multiple scattering occurs along PQ, QR and QS.

may participate in this process at low energies (Worthington *et al.* 1953). If the density of scattered particles is constant over the range of Θ considered and in the absence of baffles in the slit unit, compensation is exact. In practice neither condition applies exactly, but at large angles, where the scattered intensity often varies slowly with angle, the net error is negligible if care is taken to see that the baffles do not interfere seriously with the process.

The single counter case has been considered in some detail (Worthington *et al.* 1953) and corrections have been calculated for non-compensation at small angles. The error at 1·8 MeV. and 7° was 0·15 %, which indicates the order of magnitude of the net error to be anticipated from breakdown of the compensation process under unfavourable conditions.

The problem is much more complex when the coincidence arrangement is used. Fig. 24 represents a collision inside a solid target. The incident particle enters at P, experiences a large angle scatter at Q and leaves the target along path QR. The struck particle recoils along QS. Multiple scattering occurs along PQ (where the particle has its full energy and hence is little affected), along QR where it has reduced energy E_1 and along QS where the recoil particle has energy E_2. The ultimate effect on a beam of incident particles is to spread the cones of both scattered and recoil particles by an amount which depends chiefly (for any given target) on the energy of the particles after collision.

Coincidences can be lost by either particle being multiply scattered so as to miss the aperture of its counter. At first sight, it might appear that if each particle has a similar chance of being lost, the coincidence loss rate might be something like twice as great. This is an over-simplification, but there is little doubt that multiple scattering will be more serious in coincidence arrangements and can lead to a significant loss of counts and low values for cross-sections.

A compensating process can function, however, if equipment is designed appropriately. Consider the case in which $E_1 \ll E_2$. The multiple scattering effect will then be dominated by what happens to the scattered particle, which has low energy. Some particles will be multiply scattered along QR (fig. 24) and fail to enter the defining aperture, causing a loss of coincidences even though the recoil particle is counted satisfactorily in the conjugate counter. The loss can only be balanced if a like number of scattered particles which should have just missed the defining aperture are multiply scattered into it *and if, at the same time, their recoils can enter the conjugate counter*. Effectiveness of compensation depends, therefore, on the conjugate aperture size. One can estimate by how much it needs to be increased beyond the geometrical size by calculating the conjugate angle $\delta\Theta_2$ corresponding to the r.m.s. multiple scattering angle $\delta\Theta_1$ for the path QR (see Karr *et al.* 1951).

The other case of interest occurs when $E_1 \gg E_2$. Here coincidences may be lost by recoil particles being multiply scattered so as to miss the conjugate aperture. In this case there can be no compensating process. The loss can only be minimized by making the conjugate aperture large enough to collect effectively all recoil particles, in spite of multiple scattering.

It appears, then, that both extreme conditions of energy disparity require an increase in the conjugate aperture to allow for multiple scattering effects—in the latter to catch all proper coincidences and in the former to catch the necessary compensating coincidences. The practical window size cannot be increased indefinitely on a counter and the final result of multiple scattering is to limit the range of angles which can be studied.

For the calculation of corrections due to multiple scattering the formula of Williams (1938) is generally used. The root mean square angle of scatter of particles after passing through a thickness of target is†

$$\langle \phi^2 \rangle^{\frac{1}{2}} \approx \left(\frac{\pi N_v x}{2} \right)^{\frac{1}{2}} \frac{Ze^2}{E} \left[\ln \left\{ \frac{\pi Z^{\frac{4}{3}} N_v x \hbar^2 m}{m_e^2 E} \right\} \right]^{\frac{1}{2}}. \tag{3.6}$$

Z is the mean atomic number of the target, E the energy of the particle and m its mass. m_e and e are the electronic mass and charge respectively. N_v is the number of nuclei/cm.^{3} and $\hbar$ is Planck's constant/2π. In equation (3.6) the particle is assumed to proceed perpendicularly across the target. In the case of a collision of the kind shown in fig. 24, Q may be located with equal probability at any point through the target and the distance travelled by the particle will lie between o and $x/\cos \Theta$ if it leaves at angle Θ. The effective value of the square root of the path length is obtained by integrating, giving $\frac{2}{3}(x/\cos \Theta)^{\frac{1}{2}}$, which should therefore be used in place of $x^{\frac{1}{2}}$ in equation (3.6).

This suggests that some improvement can be obtained at angles where the energy of one particle is small by rotating the target so as to place it more nearly perpendicular to the direction of exit of the particle. However, in the final analysis, it is probably safest to check the conjugate aperture by trial and error. Apertures of progressively increased size are placed in front of the counter until there is no further increase in the coincidence rate.

Equation (3.6) does not take account of energy loss, which, in the case of targets, windows of counters, etc., is usually negligible. In other instances, e.g. when particles are being slowed with foils to facilitate detection, it is obviously important. Dickinson and Dodder (1950) considered the energy loss and tabulated values of $\langle \phi^2 \rangle$ for

† The complete distribution of particles having deflexion ϕ is
$$P(\phi) \, d\phi = (2/\pi \cdot \langle \phi^2 \rangle)^{\frac{1}{2}} \exp \left[-\phi^2/2 \langle \phi^2 \rangle \right] d\phi,$$
where $\langle \phi^2 \rangle$ is the mean square angle of scatter. It is usually sufficient to calculate the r.m.s. angle of multiple scattering and the corresponding transverse spread $L \langle \phi^2 \rangle^{\frac{1}{2}}$ at distance L from the target.

protons, deuterons and α-particles in aluminium. Their paper also contains an interesting analysis of the effect of multiple scattering in various experimental situations, notably the loss of counts to be expected from slowing foils placed in front of counter windows.

3.5.3. *Geiger–Mueller counters.* The application of this kind of detector in scattering experiments is effectively denied by two features: (*a*) All pulses are equal in height regardless of the amount of primary ionization. (*b*) After the counter has been discharged by the passage of one particle there is a dead time lasting for several hundred μsec. during which a second particle entering the counter cannot be counted.

Of these, the former is the more serious in low energy scattering with continuous beam accelerators, because systematic errors due to spurious counts of a kind which could be excluded by pulse height analysis are generally larger than statistical errors likely to result from slower counting rates. On the other hand in the high energy range of pulsed accelerators, the short duration of each burst makes fast counting and short resolving times essential.

A Geiger–Mueller point counter was employed in early work in the 100–200 keV. range where particle energies were too low for ion chambers (Hafstad, Heydenburg and Tuve, 1938). Proportional counters had not then attained the general application which subsequently led to their replacing all other counters in low energy scattering.

3.5.4. *Scintillation counters* (Deutsch, 1948; *Scint. Counter Symp.* 1952; Curran, 1953; Swank, 1954; *Scint. Counter Symp.* 1954, 1956). Prior to 1948, the proportional counter was the principal electrical detector available for high energy nuclear experiments. That the scintillation counter has been developed since then to the point where it has almost universally replaced the gas counter is by no means accidental. During the same period, new accelerators have extended the energy range open to experiment to several BeV.

Attention was drawn in §§ 3.2.4. and 3.2.5 to the operating characteristics of high energy machines and the special counting problems which pulsed beams present. An excellent illustration is afforded by the first 340 MeV. *p-p* experiment of Chamberlain and

Wiegand (1950). Gas counters 3·8 cm. in diameter and 10 cm. long were used with conventional electronic circuits. The proton beam from the 184-in. cyclotron arrived in pulses probably as short as 0·1 μsec. with a repetition rate of 60 per sec. As the pulse length was shorter than the resolving time, each counter had to be limited to less than one count per pulse. Typical single counter rates were 1–10 per sec., while the coincidence rate was ∼0·5 per sec., which, though feasible, tended to render the gathering of data somewhat tedious.

In a subsequent experiment (Chamberlain *et al.* 1951), two major improvements were made: (i) A new method of extraction was used which lengthened the pulses to ∼25 μsec. (§3.2.4). (ii) The substitution of scintillation crystals, photomultipliers and circuits having a resolving time of 4×10^{-8} sec. was effected. Counting then averaged about 10 true counts/sec., an improvement which spells the difference between days and months of running time.

A short resolving time is the most essential requirement in high energy scattering technique using electrical counters. Apart from the pulsed nature of the beam, as the energy increases, competition from other nuclear processes grows in proportion, and coincidence techniques offer the best hope of separating the elastic process. Moreover, a high energy accelerator is surrounded by a very intense flux of protons, neutrons, deuterons, mesons, electrons and γ-radiation, which may cause the random counting rate of a single detector to be 100 times that for the process being investigated. The solution adopted to date has taken the form of multiple-coincidence telescopes of counters.

Although gas counters can be operated with rise times of ∼0·1 μsec., it is simpler to resort to scintillators. At high energies gas counters have definite disadvantages, associated chiefly with the low density of the ionizing medium. The great proportion of background particles have low energy, probably a few MeV.; these will be stopped in a large gas counter giving rise to big pulses. On the other hand, scattered protons of several hundred MeV. lose only ∼100 keV. in the gas and give very small pulses. Scintillators are used in a solid or liquid state, and this, even in convenient thicknesses, enables them to extract a considerable energy from a high speed particle.

A related problem concerns energy resolution. It is convenient to be able to measure the specific ionization as an aid to identification of a particle (see below). This calls for a counter of small stopping power and good proportionality, which advantages a gas counter might be thought to offer. Such is not the case when the energy which the particle can impart to an electron in a single collision is comparable with the average energy which the particle loses in the gas. Statistical fluctuations in ionization then become excessive, and the resolution in dE/dx deteriorates. This is the Landau–Symon effect (Landau, 1944; Symon, 1948), which has been investigated experimentally for 30 MeV. protons (Igo, Clark and Eisberg, 1953). At this energy, a counter filled with argon at 1 atmosphere pressure needs to be at least 20 cm. thick in order to have a resolution† of 25 %, while to realize 10 % would require a thickness of 1 m. Thin scintillators are much superior for this purpose.

Before proceeding to discuss the use of scintillation counters in scattering experiments, a brief reference will be made to the scintillation process itself. Fundamentally, this must be considered under two headings.

(a) In the transparent *phosphor*-type scintillator, in which an ionizing particle excites atoms to higher energy states, the excess energy is radiated subsequently as light when the atoms revert to their ground states. The resulting scintillation is detected by a photomultiplier tube which converts it to an electrical pulse. Among the phosphors which have been used as detectors of scattering, decay times for light emission are: naphthalene 6×10^{-8}, *trans*-stilbene 6×10^{-9}, anthracene 3×10^{-8}, liquid phosphors and plastics (typically) $3-5 \times 10^{-9}$ and sodium iodide, $2 \cdot 5 \times 10^{-7}$ sec.

(b) In the *Cerenkov* type (Jelly, 1953; Marshall, 1954; *Scint. Counter Symp.* 1954), scintillator particles passing through a dielectric medium with velocity greater than the phase velocity of light in the medium (c/n, where n is the refractive index) give rise to Cerenkov radiation. Photons are emitted instantaneously over the surface of a cone making an angle ξ to the particle direction, where $\cos \xi = (\beta n)^{-1}$ and β is the particle velocity relative to that of light

† Counter 'resolution' refers here to $\Delta E/E_a$, where E_a is the average energy loss of the particle in the counter and ΔE is the full width at half maximum height of the spectrum obtained from a monoenergetic beam (Wolfe, Silverman and De Wire, 1955).

in vacuum. The spectrum extends over all frequencies for which this condition is fulfilled, which implies a definite lower limit. In ordinary optical materials—glass, quartz, plastics and liquids—the visible spectrum is included, with most of the energy towards the violet end.

Used in conjunction with a photomultiplier, a Cerenkov scintillator offers an alternative method of fast counting. As the light pulse is generally shorter than 10^{-9} sec., extremely short resolving times are obtained. Such an arrangement also discriminates naturally against low energy background. So far, its chief application has been to the measurement of beam energy rather than as a scattered particle detector (cf. §5.5). The light output is smaller than that from phosphors and, until recently, energies available from accelerators have been little above the Cerenkov threshold.†

The properties of phosphor-type scintillators have proved adequate in most respects for high energy experiments. The intensity of the light pulse depends on the choice of crystal, its optical quantity and purity, transparency to its own radiation and on the operating temperature. NaI is rather slow, and is also hygroscopic, which is sometimes inconvenient. Organic phosphors have a high vapour pressure, requiring special attention when they are used in vacuum. However, their decay times are very short. Stilbene is exceptionally good, though its light output is rather weak. Liquid phosphors are easily prepared reproducible in quality and of large volume.

An important characteristic of detectors is the nature of their response to particles of different energy. The fundamental relationship between number of photons released and specific ionization has been reviewed elsewhere and will not be discussed here (e.g. see Curran, 1953; Swank, 1954). The chief point of interest is whether the output pulses are proportional to the particle energy absorbed in the phosphor. This has been investigated in the low energy region by Taylor, Jentschke, Remley, Eby and Kruger (1951), who found a significant difference between the behaviour of NaI on the one hand, and the organic phosphors on the other. For the former, pulse height vs. proton or deuteron energy is linear down to ~ 1 MeV. Response to α-particles becomes linear at ~ 10 MeV.

† If $n = 1.5$, the threshold is at $\beta = 0.67$, i.e. 325 MeV. for a proton.

Organic phosphors show non-linear response at low energies. Linearity is achieved at energies which differ for the various phosphors, but these are generally considerably higher than for NaI (e.g. for protons at ~ 4 MeV. in stilbene and 12 MeV. in anthracene).

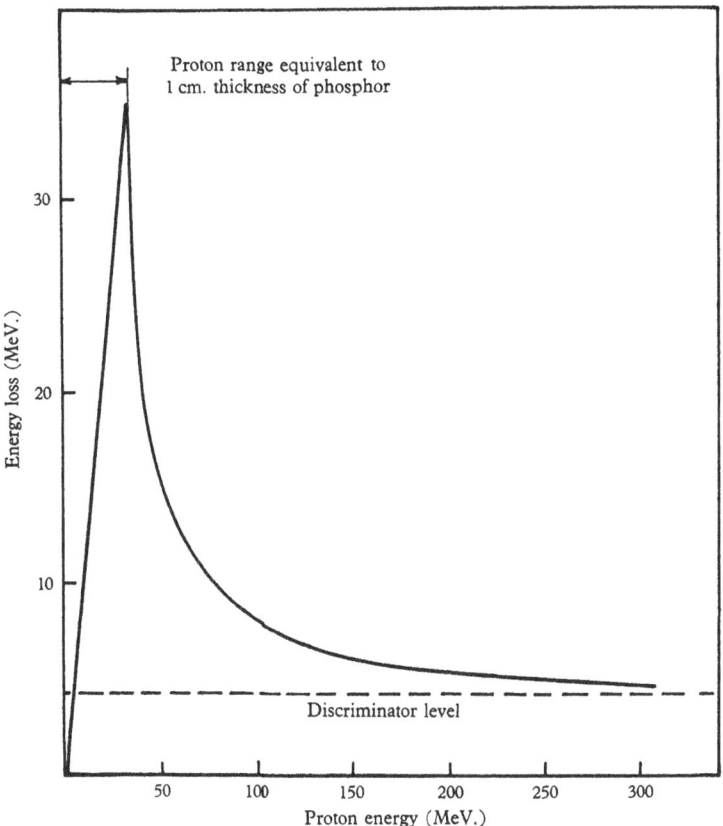

Fig. 25. Pulse height distribution due to protons traversing a 1 cm. thick phosphor (Wouters, *Scint. Counter Symp.* 1952).

NaI is therefore the only detector suitable for use in low energy scattering experiments. However, at high energies all scintillators show linear response, and the shorter decay times of organic types make them preferable.

The pulse height distribution for protons traversing a 1 cm. thick organic phosphor is shown in fig. 25, where it has been assumed that the light intensity is proportional to the energy lost by the particle. The dotted line represents a suitable discriminator setting, corresponding to an energy loss of about 4 MeV. Particles of all energies greater than this can then be detected, but it will be observed that, since the proton range is much greater than 1 cm., the distribution is double-valued. Protons of two energies yield the same pulse size. In practical arrangements, the scintillator would be shielded from the low energy protons, or a telescope of counters would be employed with an absorber between the counters so that low energy particles could not trigger more than the first.

The functions of electronic circuits used with scintillation counters parallel closely those referred to in connexion with proportional counters, although in millimicrosecond timing, design considerations are often very different. A number of reviews of the subject are available (Bell, 1954; Lewis and Wells, 1954; *Scint. Counter Symp.* 1956).

There are three approaches to the problem of counting scattered particles from pulsed accelerators: (1) The experiment may be designed so that commercially available circuits, some of which have dead times as short as 10^{-7} sec., may be used. Much scattering work has been done in this way, although it usually entails limiting the beam current and counting slowly. (2) In order to achieve really fast counting, special circuits may be designed which are capable of handling the transient pulses of particles. (3) Another line of approach, suggested by Wouters (*Scint. Counter Symp.* 1952), is to take advantage of the small duty cycle of high energy accelerators. Instead of concentrating on fast counting during the pulse, counts may be stored, together with appropriate time markers, in some form of memory device, and may subsequently be sorted and analysed during the relatively long dead time of the machine.

Typical experimental arrangements. Some applications of scintillators to charged-particle scattering will now be discussed in the following groupings:

(*a*) In conjunction with external targets, using a single telescope of several counters in coincidence, scintillators may be used to record either scattered or recoil particles (cf. §4.3.4).

(*b*) They may be used in conjunction with external targets, counting coincidences between scattered and recoil particles.

A combination of (*a*) and (*b*) is now commonly employed to give additional discrimination against background at high energy.

(*c*) Method (*b*) may be adapted for use with internal targets, though it is then attended by greater practical difficulties.

(*d*) They may be employed in double scattering experiments at high energy.

Examples of the first two applications are found in the Berkeley *p-p* experiments at 120–345 MeV. (Chamberlain *et al.* 1951). Method (*a*) was employed at small scattering angles (5–25°) where (*b*) was unreliable due to the low energy of one proton of the pair. Fig. 26 shows the arrangement. Two scintillators *A* and *B* were

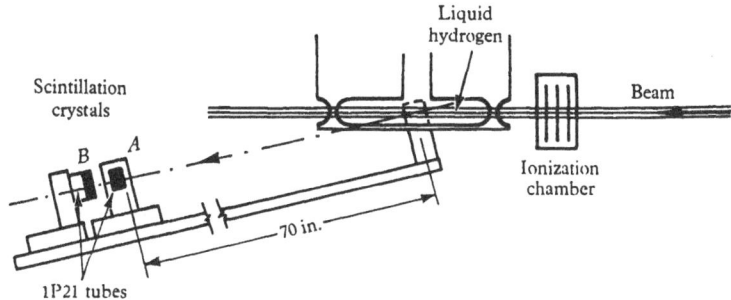

Fig. 26. Two-crystal telescope and liquid hydrogen target for high energy *p-p* scattering (Chamberlain *et al.* 1951).

placed adjacent to 1P21 photomultipliers operated in coincidence. *B* was made somewhat larger in order to minimize losses due to multiple scattering in *A*. The procedure followed was to determine the counting rate with the target empty, then full of liquid hydrogen, and attribute the difference to scattering by H. However, this is probably an over-simplification of the situation. The H-filled target has a higher stopping power, so that any low energy protons which have been produced in the collimator and which accompany the main beam will contribute to the coincidence rate of the telescope in the forward direction when the target is empty, but will be absorbed when it is full. If an absorber were inserted between *A* and *B* during the background run, the effect would be substantially corrected (Chamberlain and Garrison, 1954).

The greater part of an angular distribution can be covered satisfactorily by method (*b*), which carries numerous advantages at high energies. It has been widely employed in p-p experiments using hydrogenous targets and, to a lesser extent, in p-d experiments (e.g. Chamberlain and Stern, 1954; Chamberlain and Clark, 1956). Disposition of the detectors is the same as shown in fig. 22 except that scintillators replace gas counters; also the entire experiment is conducted in air. The overall arrangement used at Berkeley (Chamberlain et al. 1951) is shown in fig. 11.

The required size of conjugate counter is again determined by considerations of geometry and multiple scattering, as discussed in §3.5.2. Two trans-stilbene crystals were located 80 and 30 cm. respectively from the target, the defining one having dimensions 1·8 cm. high × 3·8 cm. wide, the conjugate 3·8 × 2·51 cm.². If the latter dimensions are sufficient, a plateau is obtained in the co-incidence counting rate as the conjugate counter traverses the angle Θ_2 corresponding to the scattering angle Θ_1 at which the defining counter is located. Fig. 27 illustrates this.† Then by re-setting the counters at 90° (the low energy conjugate condition) the new coincidence rate reveals the proportion of low energy proton con-tamination in the beam.

An important difference between low and high energy applica-tions of the coincidence technique is the relatively severe back-ground in the latter case, arising from many sources. Consider the events which can lead to coincidences between the two counters:

(i) True p-p elastic scattering.

(ii) At incident proton energies much greater than the binding energy of the nucleons in a carbon nucleus, a 'quasi-hydrogen' scattering may occur, involving an incident proton and a proton in the nucleus, the latter behaving essentially as free.

(iii) Inelastic events in carbon or other materials exposed to the beam, leading to two charged particles which trigger the counters within the coincidence resolving time, e.g. meson events, 'stars', etc.

(iv) True coincidences may also arise from stray particles of moderate energy passing through both counters.

(v) At small angles events in the class $p(p, \pi) X$ are possible.

---

† Since $(\Theta_1 + \Theta_2)$ is a function of kinetic energy at relativistic energies, an accurate measurement of the angle settings provides a useful check on the bombarding energy (cf. §5.5).

(vi) Accidental coincidences, i.e. particles from unrelated events triggering the counters within the resolving time.

As event (i) is the only one required, the remainder must be excluded from the observed coincidence rate. In practice this compels a return to a 'difference' experiment of the kind discussed

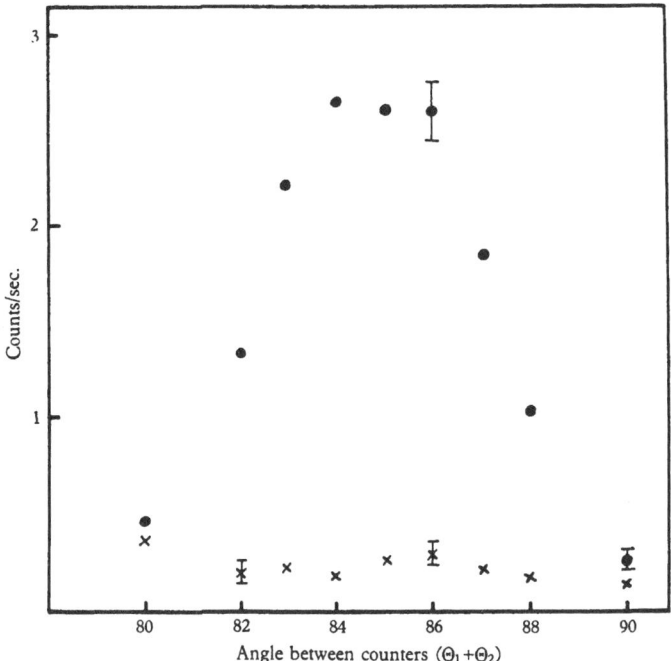

Fig. 27. Coincidence counting rate as a function of angle between counters in 345 MeV. p-p scattering (Chamberlain *et al.* 1951). ● polyethylene target, × carbon target.

earlier in connexion with single detectors. The procedure has been considered thoroughly by Sutton, Fields, Fox, Kane, Mott and Stallwood (1955). At each angle the coincidence rate should be measured with the hydrogenous target in place, then with a carbon target of the same stopping power and finally without any target. The accidental rate (vi) may be isolated by increasing the length of cable from either counter so that a pulse requires longer to reach the coincidence circuit. If the delay introduced is equal to the period of the cyclotron r.f. voltage, coincidences recorded can no longer originate from related physical events. (In Sutton's experiment, pulses from stilbene or plastic scintillators and RCA-5819 photo-tubes were clipped to 2×10^{-9} sec., the resolving time being $\sim 4 \times 10^{-9}$ sec.)

The true coincidence rate was taken as

$$N = (T - T_a) - (B - B_a) - R[(C - C_a) - (B - B_a)], \qquad (3.7)$$

where T, B and C are the observed coincidence rates for the hydrogenous target, no target and carbon target respectively. T_a, B_a and C_a are the accidental rates for the same target arrangements. R is the ratio of the number of C atoms per unit area in the hydrogenous target to that in the C target. Obviously, the beam current should be kept constant during the determination.

With these precautions some events of class (v) might still appear as true p-p events. Coincidences from the reaction $p + p \rightarrow \pi^+ + d$ are precluded by the geometry, but $p + p \rightarrow \pi^+ + p + n$ could interfere when the defining counter is at small angles, the meson then being recorded at a large angle. At this position the conjugate proton has low energy and can be stopped by placing a foil across the face of the scintillator whereas the mesons should be more energetic and continue to record. The difference indicates the general magnitude of the effect. However, when only a single telescope is used, for instance in p-p scattering at small angles, the $\pi^+ + d$ process also contributes to the spurious rate. Both processes can be estimated from published cross-sections (Rosenfeld, 1954) and indications are that they do not introduce net errors in excess of $\sim 1\%$ at 350 to 450 MeV.

Meson production increases sharply, both relative to elastic scattering and in absolute magnitude, from 400 to ~ 800 MeV. The total p-p cross-section rises from ~ 28 to 41 mb. between 460 and 660 MeV. (Meshchevyokov, Neganov, Soroko and Vzorov, 1954), while the elastic part remains fairly constant at 22 mb. Dzhelepov, Moskaliev and Miedviev (1955) break down the inelastic part at 650 MeV. into the following: $n + p + \pi^+$ (28%), $p + p + \pi^0$ (9%) and $d + \pi^+$ ($\sim 8\%$). Similar results have been obtained at Birmingham (Batson and Riddiford, 1956). Double meson production sets in at 600 MeV. but probably does not become important below 1 BeV. At the latter energy $\sigma_{el} \approx \sigma_{in}$ (Shapiro et al.

1954; Smith *et al.* 1955), while at 5 BeV. preliminary figures suggest that the elastic process accounts for only $\sim 15\%$ of the total p-p cross-section. At the latter energy $\sigma_{el} \approx 5$ mb. (Wright *et al.* 1955). The necessity for distinguishing between different particles and energies has placed a premium on discriminator techniques. Ideally, one wishes to measure the mass and charge of each particle negotiating the telescope. This requires a knowledge of at least two parameters such as velocity, energy, range, specific ionization, momentum or small angle scattering. At high (but not highly relativistic) energies, specific ionization (dE/dx) and energy (E) may be used conveniently. Fig. 28 illustrates the arrangement of a 340 MeV. p-d experiment (Chamberlain and Clark, 1956). Scintillators 1–4 form the defining telescope with scintillator 5 as the conjugate counter. No. 1 is the usual defining counter which operates in coincidence with 5. No. 2 is the 'dE/dx counter', whose thickness is small compared to the range of the deuterons. The pulse height is then approximately proportional to specific ioniza-tion.† No. 3 is a thick 'E counter', which stops all deuterons of correct energy. No. 4, a 'pass through' counter, registers a signal when an energetic background particle does not stop in the E counter. Aluminium absorber was used to decrease the deuteron energy at certain angles.

The outputs from the pulse height counters can be displayed, after amplification, in various ways. In the last mentioned experi-ment, coincidences (1, 5) triggered the horizontal sweep of an oscilloscope and, by the use of artificial delay lines, the pulses from 2, 3 and 4 appeared suitably spaced on the vertical trace. For each type of particle a fairly characteristic pattern was obtained. The range R of a particle depends on its mass m, charge z and energy E as follows:

$$R \propto \frac{E^{1\cdot8}}{m^{0\cdot8}z^2},$$

whence $$dE/dx = 1/(dR/dE) \propto \frac{m^{0\cdot8}z^2}{E^{0\cdot8}}.$$

A plot of dE/dx vs. E in terms of pulse heights gives a set of

---

† The same idea can be applied to gas counters at lower energies. In some work, gas and scintillation counters have been combined, the former to measure dE/dx, the latter to stop particles and measure E (Seagrave, 1955).

approximately parallel curves, the lowest corresponding to mass 1, the next to mass 2, etc. For doubly charged particles, dE/dx is very much higher.

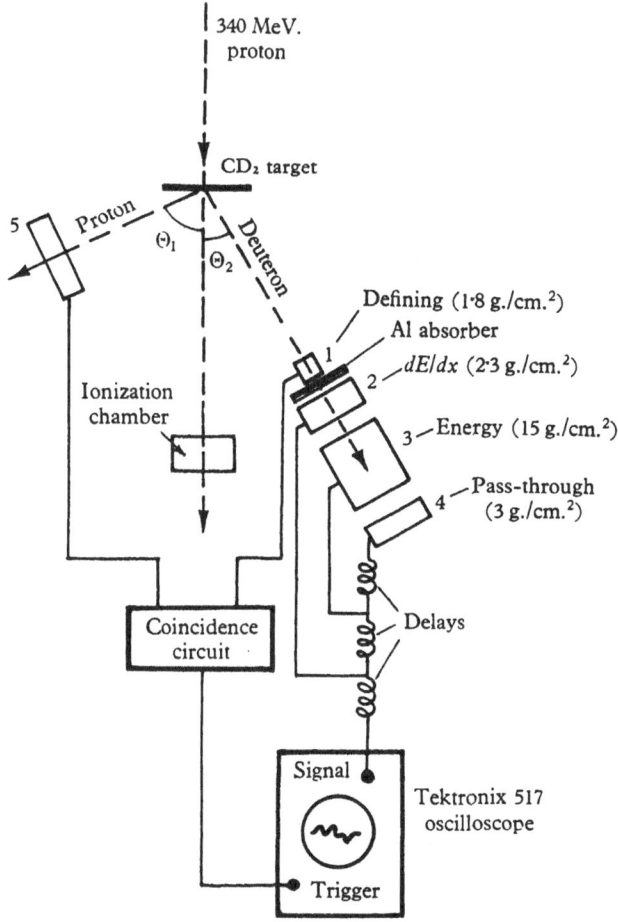

Fig. 28. Schematic diagram of 'dE/dx and E' telescope and block diagram of electronic circuits used in 340 MeV. p-d scattering (Chamberlain and Clark, 1956).

Another method of handling the information is to multiply the output pulses electronically. The product is essentially proportional to mass, but relatively insensitive to energy:

$$E\frac{dE}{dx} \propto m^{0.8}z^2E^{0.2}.$$

A factor of 2 in E affects the product by only $\sim 20\%$, hence a wide range of particle energies can be grouped together for mass and/or charge analysis. Details of the method, including the multiplying circuit, have been discussed by Wolfe et al. (1955).

The use of counters with internal targets (method (c) above), requires a relatively trivial adaptation of (b). The target intercepts the circulating beam inside the vacuum tank (cf. §3.3.2). Practical drawbacks are (i) Remote-controlled target and counter movements must be provided through vacuum seals. (ii) Counters are located in the magnetic field of the accelerator. (iii) Owing to the field, which curves the trajectories and gives rise to a focusing action, calculation of the effective solid angle is less straightforward. Nevertheless, these difficulties are not insurmountable and useful work has been done by this method at Rochester (Oxley and Schamberger, 1952; Schamberger, 1952; see also fig. 14) and at Harvard (Birge, Kruse and Ramsey, 1951).

Angular distributions of p-p scattering have also been measured from 440 to 1000 MeV. using a CH_2 target located inside one of the straight sections of the Cosmotron (Blewett, 1954). Scattered and recoil protons escaped through the thin aluminium walls of the vacuum vessel, to be detected in coincidence by two telescopes outside the machine. With a defining aperture of 0.25×0.25 in.^{2} and a conjugate aperture of 3×4 in.^{2}, the accidental coincidence rate varied from 3 to 9%. With a carbon target of the same stopping power inserted, the true coincidence rate (quasi-scattering, etc.) amounted to less than 4% of the elastic p-p rate.

Finally, we consider the application of scintillation counters to double scattering (method (d) above). Fig. 29 is a schematic diagram of the apparatus of Oxley et al. (1954) as used for p-p scattering. Here, and to an even greater extent in triple scattering, the principal difficulties are associated with intensity and background. The first target (polyethylene) was located inside the cyclotron in order to take advantage of the whole circulating beam. Protons scattered at $\sim 19°$ were collimated and entered the second scattering chamber through the aperture on the left. A 7.5 mm. stilbene crystal A defined the useful beam, which then impinged on a second polyethylene (or carbon) target. Hydrogen scattering was obtained by difference. The second scattering angle was $27°$.

Scattered and recoil protons were detected by the telescope (two crystals B and C separated by a Cu absorber which imposed a cut-off at 100 MeV.) and conjugate counter D respectively. The 'exploratory counter' E served only to establish initial alignment.

Long light pipes separated the scintillators from their photomultipliers, which were located inside a magnetically shielded region. Fourfold coincidences (A, B, C, D) were recorded with a circuit of resolving time $\sim 10^{-8}$ sec. The counting rate of A was

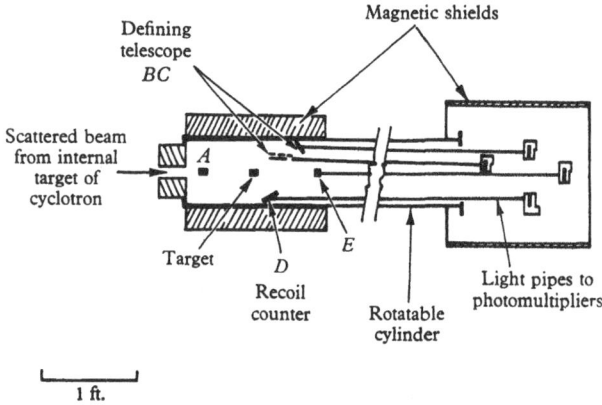

Fig. 29. Schematic illustration of scattering chamber used for 200 MeV. double scattering (Oxley et al. 1954).

$\sim 10^6$ per sec. and the coincidence rate ~ 40 per min., of which 5–15 % was due to background.

The entire unit shown in the figure could be rotated about its axis, and the experimental procedure was to measure the counting rate at various azimuthal settings and hence obtain the asymmetry ϵ of equation (2.16). Similar double scattering equipment has been described by various other workers (see especially Strauch, 1955; Donaldson and Bradner, 1955; Marshall et al. 1954; De Carvalho, Marshall and Marshall, 1954; Dickson and Salter, 1954).

Triple scattering calls for no new experimental methods but the 'signal to noise' level becomes distressingly small (Tripp, 1955). At the triple stage, counting rates may be ~ 1 per min., even with considerable sacrifice in angular resolution, and these few counts are likely to be lost among counts due to background from the first and second targets.

3.5.5. *Photographic plates* (Powell and Occhialini, 1947; Yagoda, 1949; *Fund. Mech. Phot. Sens. Symp.* 1951; Beiser, 1952; Goldschmidt-Clermont, 1953). A charged particle which passes through the AgBr grains of a photographic emulsion forms a latent image which, on development, reveals the track of the particle as a line of grains, clearly visible under a microscope. This fact has been known for some 30 years and employed occasionally for general nuclear and cosmic ray studies, but it was only during the post-war period that emulsions of sufficient sensitivity for precise investigations became available commercially. These are characterized by fine grain, high AgBr content and a high intrinsic sensitivity. However, their use as detectors in scattering experiments places few demands on their versatility. We list below the four properties which are important.†

(i) *Threshold sensitivity.* This is the upper limit of energy of a particle which registers as a just recognizable thread of grains. It is somewhat arbitrary, depending on the amount of fog background and also on what the observer is prepared to accept as a track. (Detection of a low-level signal among circuit noise is the analogous counter problem.) Ilford C2 emulsions, which have been extensively used in low and medium energy scattering, have a threshold sensitivity which is probably rather higher than 50 MeV. for protons, 100 MeV. for deuterons and 1500 MeV. for α-particles. It follows that these are the maximum energies for which it is profitable to employ C2 plates as detectors. Electrons of energy ~ 30 keV. record as a cluster of a few grains which add to the background.

Ilford G5 plates have an intrinsic sensitivity five or six times higher, and record singly-charged particles of all energies. The lowest grain density occurs at the minimum of specific ionization, viz. at ~ 2 BeV. for protons, 4 BeV. for deuterons, 8 BeV. for

† Recipes for chemical processing and descriptions of microscope technique, both of which are specific to the emulsion method, can be found in the general references cited.

α-particles or 1 MeV. for electrons. On a clear plate a grain density as low as 12 grains per 100 μ can just be recognized. The grain density depends, of course, on the development given to the plate. Moreover, plates exposed in the vicinity of accelerators receive high background fog making tracks more difficult to distinguish. Under these circumstances a grain density of 30 per 100 μ, or even higher, may be necessary (cf. Coates in *Fund. Mech. Phot. Sens. Symp.* (1951)).

The only other Ilford emulsion of interest in this connexion is the E1, which records protons relatively poorly and facilitates distinction between protons and α-particles. There is much to be gained by using the least sensitive emulsion capable of recording the required particles, in order to avoid fast electron tracks and reduce fog background.

(ii) *Stopping power.* The integral stopping power of an emulsion relative to air is $S = R_a/R_e$, where R_a and R_e are ranges of a charged particle of given energy in air and emulsion respectively. For a dry emulsion (density 3·92 g./cm.³), S for protons varies from ~ 1650 at 1 MeV. to ~ 2200 at 50 MeV., beyond which it increases very slowly. The stopping power is decidedly sensitive to the moisture content of the emulsion (Wilkins, 1951).

(iii) *Range-energy relations* (see also § 5.3). The use of emulsions as detectors of scattered particles rests chiefly on a knowledge of the range as a function of energy. In particular, observed ranges serve to identify specified groups of particles, or distinguish genuine scattering from spurious events such as slit scattering. Range-energy curves are reproduced in figs. 50 and 51. At low energies, ranges can usually be measured to ~ 1 μ and the error in energy determination confined to about 1 %.

(iv) *Range straggling.* Individual ranges of monoenergetic particles fluctuate appreciably from the mean, due to the statistical nature of the collision process responsible for slowing down in matter. If ΔR is the half-width at half the maximum height of the range distribution about a mean range R in the emulsion, then $\Delta R/R$ varies from 7 or 8 % (cf. 2·6 % for straggling in air) for 1 MeV. protons to rather less than 2 % above 5 MeV., beyond which it changes very slowly. Straggling far exceeds the error of range measurement of any one track and determines the effective discrimination between particles of different energy.

In the low and medium energy range, photographic plates are superior to other detectors in many respects. The lack of complication leads to a cheaper and smaller chamber. They can be cut to small dimensions or stood on edge for measuring cross-sections at positions close to the beam. The detectors themselves serve as particle discriminators and also, when track ranges have been accurately measured, fix the average beam energy that pertains during the actual run. Another important feature is that data may be collected at all angles simultaneously, which tends to make the relative values of cross-sections rather more reliable.

Finally, the economic advantage of gathering all data in one or at most several runs, each lasting perhaps an hour, should not be overlooked. By contrast, in counter experiments each angle must be studied in turn, which may involve weeks of running time. Considering the high cost of operating even a low energy cyclotron, a short experiment has much to commend it.

The limitations of plates are well known—in particular the tedium of scanning required to obtain reasonable statistics. Actually this aspect of plate work may have been over-emphasized in the past. Individual scattering experiments have been described which were based on a total of more than 10^5 tracks (e.g. Mather, 1952; Rosen and Allred, 1952). It is important to appreciate that a photographic plate chamber requires much less time and effort to get into operation than, say, a coincidence arrangement, so that one may legitimately reckon 'scanning time' against 'setting up time'. In some experiments, e.g. p-p and α-α, only tracks of a single range can occur at any one angle so that scanning involves a straight counting of tracks. At least 500 per hour can be counted under these conditions. However, under less favourable conditions it may be necessary to measure the length of each track and plot a histogram for each angle, in which case the rate is more commonly ~ 500 tracks/day.

At high energy, particle ranges become inconveniently long, also the emulsions tend to become flooded by background tracks which mask the elastic scattering and generally slow down the accumulation of data. Electrical coincidence methods appear to be preferable under these conditions.

Typical experimental arrangements. A number of plate chambers have been designed which provide for a single large plate, or perhaps two symmetrically disposed above and below the beam, to record particles scattered by a gas target. Details of experimental arrangements of this kind were described in early papers (e.g. Chamberlain and Garrison, 1954) and by Dearnley, Oxley and Perry (1948). The latter achieved an accuracy of about 2·5 % in p-p scattering at 7 MeV. A one-to-one correspondence between track position and scattering angle reduced scanning to the simple operation of counting. The chief drawbacks of single plate cameras are the more complex geometries and an uncertainty due to slit edge penetration which confuses the solid angle calculation.

These difficulties may be avoided by designing plate chambers after the fashion of electrical counter chambers, i.e. using a double-slit unit to define a beam of scattered particles. If the slit jaws are machined flush with the axis of the beam, no appreciable edge penetration can occur among particles which succeed in entering the detector placed behind the slit unit. Several photographic chambers have followed this principle, with similar slit units at every angle so that all data can be recorded simultaneously.

A multiple plate camera having a set of small photographic plates disposed at various radii about a target was described as early as 1940 (Wilkins, 1940), although the emulsions available at that time must have seriously limited its application. The idea was revived in a modified form when nuclear emulsions became available. Fig. 30 shows the plate chamber used in conjunction with the St Louis cyclotron by Mather (1951). A set of radial slit units defined scattering angles from 7 to 165° in ~ 5° intervals, and a plate was placed behind each so that the particles entered emulsions at angles α ranging from 14 to 25°. The smallest angle of incidence occurred in the forward direction in order to accommodate the full range of the higher energy particles within the emulsion.

The cross-section $\sigma(\Theta)$ is related to the yield and constants of the apparatus in a simple way if second order corrections for non-ideal geometry are neglected:

$$\sigma(\Theta) = n_a R_0 h \sin \Theta / N_i N_v 2b \sin \alpha, \qquad (3.8)$$

where n_a is the number of tracks entering unit area of plate located as shown in fig. 30. N_v and N_i were defined in connexion with equation (2.13) and α is the angle of incidence on the plane of the emulsion. Appropriate values of R_0 and h must be used for each area of plate scanned.

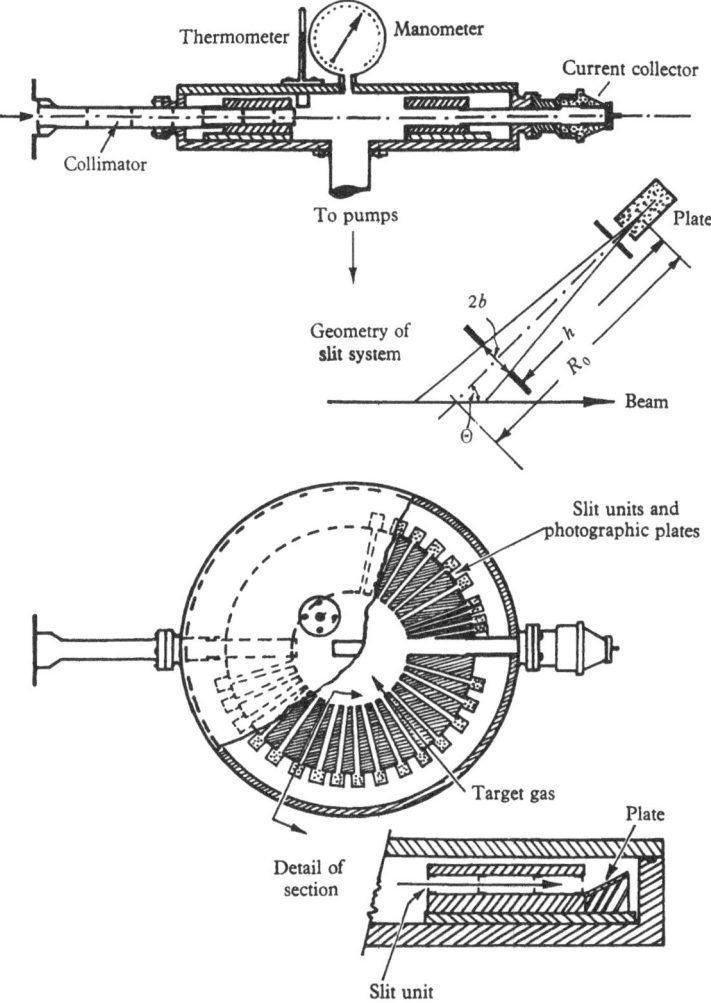

Fig. 30. Multi-plate photographic chamber for low energy scattering (Mather, 1951). The inset shows the geometry of the slit system.

The energy resolution given by plate chambers is good enough to discriminate against most spurious particles. Slit scattering, for instance, tends to become appreciable at small angles in spite of careful collimator design. Fig. 31 shows a histogram of track ranges obtained in 5 MeV. *p-p* scattering at $\Theta = 7°$ (Mather, 1951). The shaded block, which represents slit scattering, amounted to 14 %

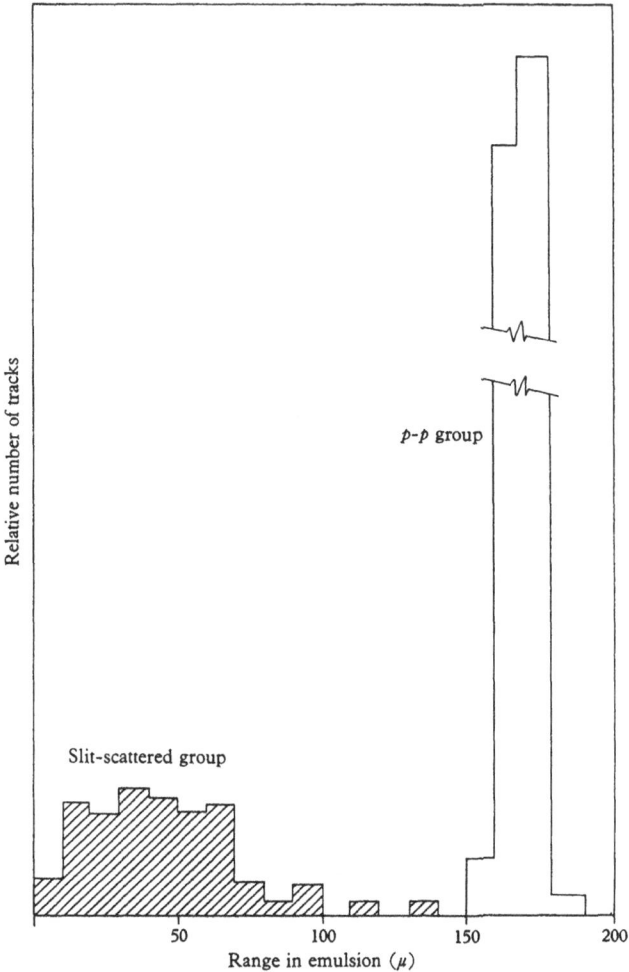

Fig. 31. Histogram of track ranges observed at 70° in 5 MeV. *p-p* scattering using a photographic plate chamber (Mather, 1951).

of the whole yield at this angle, but the ranges being so conspicuously shorter, it could be subtracted reliably.

It is less easy to cope with scattering in the forward direction due to gas impurities. Over most of the angular distribution tracks from this source are obviously longer than the others and can be counted separately, but this breaks down at $\Theta \gtrsim 20°$. In the experiment referred to above, p-p tracks had a range of 154μ at $15°$, those from heavy atom scattering being only some 18μ longer, a difference which could not be resolved. At small angles, therefore, a correction had to be applied to n_a values based on control runs designed to determine the impurity scattering when no target gas was admitted. On the assumption of Rutherford scattering (equation (3.3)), one can also extrapolate to small angles the measured impurity scattering at intermediate angles which is observed in an actual run. †

Sources of small errors such as these are numerous in scattering work and it becomes highly important to obtain as many independent checks as possible at the one energy. Several laboratories, recognizing this, have run parallel projects to study the same interaction by different techniques. An example is afforded by the St Louis d-p work which included both coincidence (fig. 22) and photographic (fig. 30) techniques. Moreover, the availability of either 10 MeV. deuterons or 5 MeV. protons from the same cyclotron made it possible to conduct d-p and p-d experiments with the same relative velocity of particles (collision energy 3·3 MeV. in the CM system). The collected data from all three experiments should be consistent, in the absence of systematic errors, when referred to the CM.

Tables I and II illustrate the difference in range between scattered and recoil particles in 5 MeV. p-d and d-p scattering. As discussed previously, both particles are confined to the 0–90° quadrant in the latter case. Below 30°, three distinct groups appear on each plate due to the fact that there are two kinds of collision (either a 'grazing' or a 'hard' one) which can scatter the deuteron at the same angle (cf. fig. A. 2). Nevertheless, the ranges always differ sufficiently to be resolved and the three groups can be counted separately. In fact one advantage associated with this feature of d-p work is that each plate below 30° yields three points which are well dispersed on the CM angular distribution. As they derive from observations with the same slit unit, their relative values should be particularly reliable.

† This is only valid at low energies where non-Rutherford scattering is known to be negligible.

TABLE I. *Ranges of the proton (p) and deuteron (d) tracks observed at various Lab angles Θ in 5 MeV. p-d scattering. θ is the corresponding CM angle*

Θ	θ	Kind of track	Range in μ
11°	16·4°	p	168
	158·0	d	89
16	23·9	p	162
	148·0	d	83
30	44·5	p	138
	120·0	d	61
50	73·5	p	92
	80·0	d	26
70	98·0	p	54
	40·0	d	5
105	133·8	p	19
125	141·9	p	12
145	161·6	p	8
165	172·4	p	7

TABLE II. *Ranges of the protons (p) and two deuteron tracks observed at various Lab angles Θ in 10 MeV. d-p scattering. sd denotes a short deuteron track, ld a long deuteron track. θ is the corresponding CM angle*

Θ	θ	Kind of track	Range in μ
7°	21·1°	ld	325
	166·0	p	440
	172·9	sd	13
16	48·4	ld	260
	148·0	p	396
	162·5	sd	16
25	82·7	ld	196
	130·0	p	324
	147·3	sd	25
35	110·0	p	230
55	70·0	p	70
75	30·0	p	7

In a photographic plate chamber constructed at Los Alamos (Allred, Rosen, Tallmadge and Williams, 1951) the plates were set on edge, instead of lying flat as shown in fig. 30. This allowed spacing at 2·5° intervals, giving more points on the angular distribution—a helpful feature where the distribution varies rapidly or oscillates as in α-α scattering (e.g. Steigert and Sampson, 1953). Typical range distributions obtained with the chamber appear in fig. 32. Both histograms show some evidence of disintegration

products although the continuous background is more severe in
(a), as expected. Tracks of α-particles from the ^{3}H(d, n)^{4}He
reaction are too short to cause confusion. The intermediate range
group at 90 μ was contributed by about 5·5 % ^{1}H (by number of
atoms) in the tritium, which latter is not easily obtained in high
purity. The d-t study is, of course, a difficult one because of gas impurity and competing reactions. However, even when the continuous

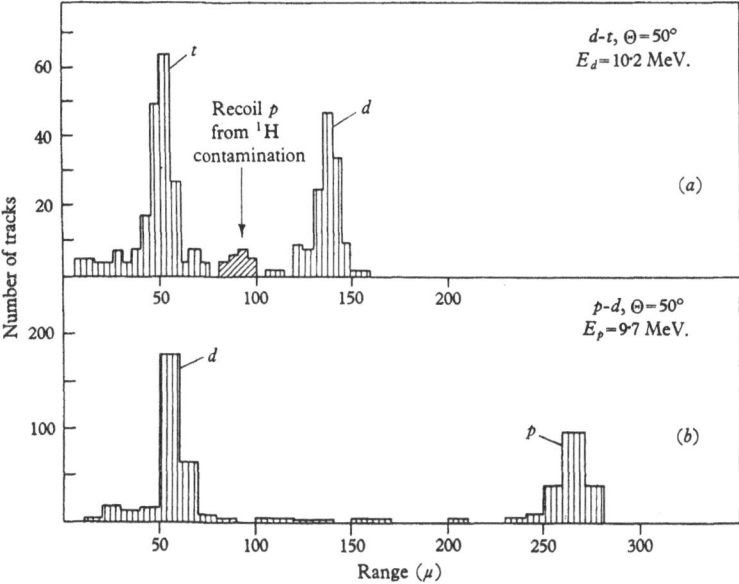

Fig. 32. Range distribution obtained with Los Alamos multi-plate chamber for
(a) d-t and (b) p-d scattering (Allred, Armstrong, Hudson, Potter, Robinson,
Rosen and Stovall, 1952a; Allred, Armstrong, Bondelid and Rosen, 1952b).

background is worse than shown in (a), for instance in d-^{3}He
scattering, reasonable values for the elastic process can still be
secured by subtracting a level of background based on that on either
side of the peak. The final error in cross-sections, under such
unfavourable conditions, is likely to be ~ 10 %, contributed chiefly
by the correction for continuous background.

  Consider now the influence of random radiation upon plates
exposed in the vicinity of an accelerator. Selectivity of tracks in
emulsions rests upon certain criteria which must be satisfied by a
bona fide track: (i) The track must begin at the emulsion surface.

This requirement alone excludes most of the knock-on protons†
from fast neutron flux. (ii) It must have the correct range (within
limits of straggling) corresponding to the particular collision and
angle of observation involved. (iii) It must enter the plate at the
correct angle defined by the slit unit. (iv) The angle of dip α must
be at least approximately correct. (v) The grain density along the
track must be compatible with that of its fellows of the same
particle group. In practice the first three conditions are more important, the last two providing useful confirmation in some cases;
e.g. in p-α scattering the high grain density of α-recoils helps to distinguish them from short slit-scattered proton tracks. Discrimina-
tion achieved by application of these conditions during scanning is,
in most cases, remarkably efficient.

A field of application to which the photographic plate is better
suited than any other detector is that of double and triple scattering
at low energies, although little has yet been done in this direction.
The design of the Minnesota chamber (Heusinkveld and Freier,
1952), which first detected polarization of protons by helium,‡ is
shown in fig. 33. The two scatterings were coplanar and selected
to occur at $\Theta_1 = 76°$, corresponding to $\theta = 90°$ in the CM system
(equations 2.1, 2.23). The first collimator defined a beam of
several μA., which was centred on the quartz plate to give correct
alignment. The second collimator defined a beam of single-
scattered (polarized) protons which entered the large chamber to
the right.

Particles which experienced a second scatter could enter photographic plates inclined at 60° above and below the beam. The whole
unit thus comprised a polarizer and analyser combination, the latter
portion depending upon a measurement of the asymmetry of
scattering in the usual way (§2.4). The angle of second scatter
($\sim 76°$) was defined by accepting only tracks in either plate which
entered the emulsion with angle of dip between specified limits.
Notice also that particles scattered from the slit edges of the second
collimator could not reach the emulsions.

† Nevertheless, a high background even of non-confusable tracks makes
scanning more tedious and calls for rather more concentration. It is a wise
move to surround the chamber with several feet of concrete shielding, especially
when bombarding with deuterons.
‡ The experiment is discussed in §8.7.

The He gas pressure was ∼ 1 atmosphere, which allowed excessive multiple scattering by the usual criteria of single scattering, but was dictated by the difficulty of obtaining a measurable yield.

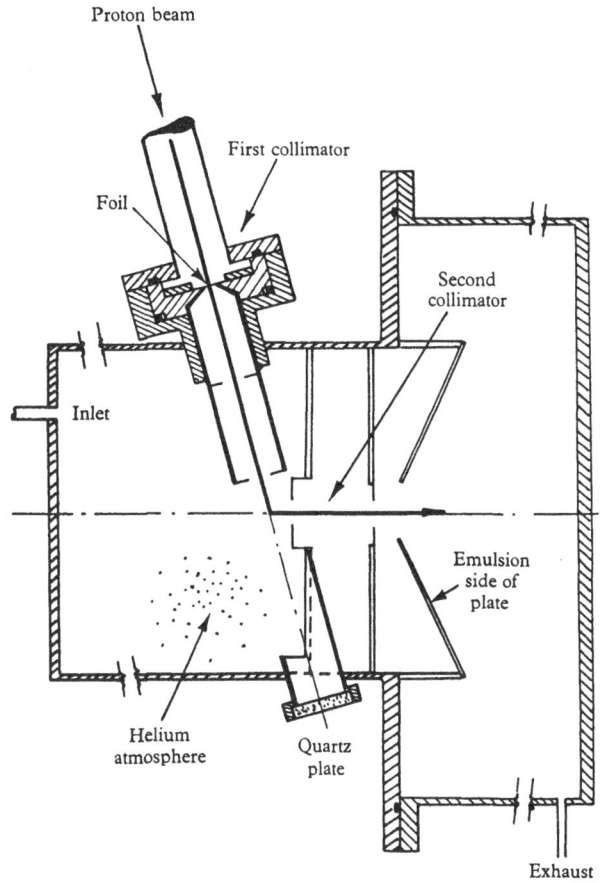

Fig. 33. Apparatus for double p-α scattering at approximately 3 MeV. (Heusinkveld and Freier, 1952).

With solid angles enlarged to ∼ 0·05 sterad. the yield was still only a few hundred tracks after a 24 hr. exposure.

Photographic plate chambers have not been employed for scattering at energies of the order of 100 MeV., although this is by no means out of the question. (Little has been attempted so far in the way of using magnetic fields to analyse and discriminate between

8   M E S

scattered particles, before reaching the detector.) However, at intermediate energies useful plate experiments have been conducted by Panofsky and Fillmore (1950). No defining slits were used, the history of each scattered particle being deduced from measurements made on its track in the emulsion. The beam passed through the centre of a symmetrical array of plates as shown in fig. 34. Scattering occurred at all points through the gas and some of the particles entered the emulsions. In these experiments the plates were rather more exposed to slit scattering from the collimator than was the case in the multi-plate chambers discussed above.

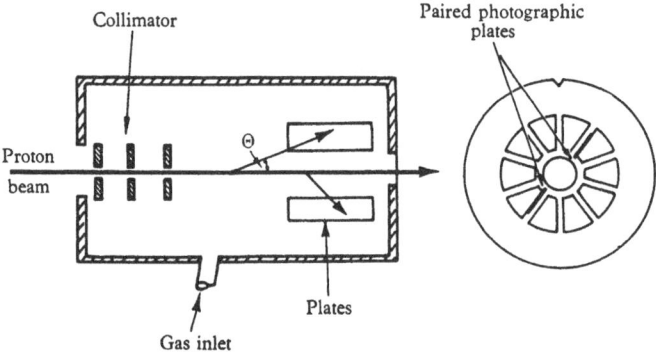

Fig. 34. Photographic plate chamber used at 30 MeV. for p-p scattering (Panofsky and Fillmore, 1950).

Two measurements had to be made on each track, its angle of entry into the emulsion and the distance of the point of entry from the beam centre. An obvious condition that these should give a reliable picture of the scattering event is that multiple scattering both in the gas and the emulsion be negligible. By using a number of plates, as shown, and basing cross-sections on the sum of the track counts in paired plates, any error due to misalignment of beam is adequately corrected. Sources of error are discussed fully in the original paper.

Attempts have been made at the University of Rochester to study p-p scattering at 240 MeV. by inserting photographic plates inside the cyclotron tank (Towler, 1952). Protons scattered from the circulating beam by a hydrocarbon target described helical paths and entered plates located near the lower pole tip of the magnet. The most striking feature of the experiment was the simplicity of

the equipment compared with that of other high energy arrangements, but the price paid for this one virtue appears to have been
high. As in other internal target experiments (§3.3.2), the calculation of solid angle was complicated because of effects due to the
magnetic field. Moreover, although in principle it is possible to
separate H from C scattering, the high C yield made scanning rather
unrewarding. Out of 21,000 tracks (from CH_2 and C) only 6600
were attributable to p-p events.

CHAPTER 4

NEUTRON SCATTERING TECHNIQUES

4.1. Differences between charged-particle and neutron scattering

The study of neutron interactions with light nuclei is significantly different from the charged-particle case, both in theory and experiment. The most important difference is the absence of any Coulomb barrier, which enables even thermal† neutrons to interact with nuclei of any atomic number. The range of energy available to neutron physics is therefore much broader. Actually, some of the most interesting effects in neutron scattering—para- vs. ortho-hydrogen scattering, diffraction by hydrogenous crystals, polarization by ferromagnetic materials, etc.—are all encountered in the thermal range.

Much of the experimental method described in the present chapter arises from this circumstance and has no counterpart in charged-particle scattering. For instance, thermal neutrons by their very nature can only be produced with a wide velocity distribution and various kinds of monochromator have had to be developed which isolate narrow bands from the distribution.

However, within the familiar range of charged-particle scattering (0·1 MeV. to $\sim$ 1 BeV.), neutron techniques do bear a fairly close resemblance to those described in chapter 3. The chief differences may be listed under five headings:

(a) There is no prime source of neutrons. A universal handicap in neutron scattering is the lack of adequate intensity, imposed by the fact that neutrons must be produced from nuclear reactions, then recollimated into a beam, which is a two-stage process.

(b) Because neutron targets are not available, less flexibility exists is the design of experiments. The maximum thermal neutron flux has been realized in the core of a reactor (§4.2.3). Taking this as $10^{14} n/\text{cm.}^2/\text{sec.}$, the corresponding neutron density is $\sim 10^9 n/\text{cm.}^3$, which is still lower by a factor of 10^{10} than the usual

---

† That is, neutrons in thermal equilibrium with the medium. Thermal energy $E = Tk = 0·0252$ eV. at 293°K. (see §4.2.7).

gas target density employed in scattering. Neutrons must therefore form the bombarding particles, and this rules out the possibility of doing complementary *n-d* and *d-n* experiments, say, at the same CM energy (cf. § 3.5.5).

(*c*) Except in the low energy range ($\sim$ 10 MeV.), no methods of generating highly monoenergetic neutrons are known and, in fact, with available sources the spread is generally considerable (cf. § 4.2.6). Even at low energies many sources are heterogeneous, e.g. fission and *bremsstrahlung* (γ, n) sources, and the sources provided by numerous reactions. Hence the onus of selecting particles of specified energy is often transferred to auxiliary equipment (e.g. velocity selectors) or to the detector itself (e.g. threshold detectors).

(*d*) Neutrons are difficult to detect with uniform efficiency over the range of energy corresponding to scattering angles 0–180°. One is generally restricted to detecting the charged recoil particle. Hence there is no counterpart to the valuable coincidence technique (§§ 2.4, 3.5.2).

(*e*) For the same reason, absolute differential cross-sections are more difficult to determine. The usual practice is to combine a relative measurement of the angular distribution with a measurement of total cross-section obtained by transmission (§§ 2.4, 4.3.5).

(*f*) Neutron scattering possesses at least one instrumental advantage. The freedom from ionization loss means that fast neutron beams may be sent through air for considerable distances without degradation, hence experiments may often be conducted in air instead of in vacuum.

In the following treatment we shall adopt the organization of chapter 3, discussing first the available neutron sources and then some typical experiments, grouped according to the apparatus employed.

4.2. Neutron sources

Practical sources of neutrons are always based upon reactions between nuclear particles or the interactions of photons with nuclei, e.g. (α, n), (p, n), (d, n), (t, n), (γ, n) and fission. These all give rise to fast neutrons. Slow neutrons are produced by reducing the energy of fast neutrons in a succession of elastic collisions with light nuclei (§ 4.2.7).

A few of the sources mentioned, notably those involving naturally radioactive substances, although too weak for useful scattering work, are valuable as standards of neutron flux for calibration of detectors. They have also been employed extensively for exploring the neutron density and energy distribution inside moderating media.

4.2.1. *Natural sources.* The earliest sources of neutrons were based on the $^9\mathrm{Be}(\alpha, n)\,^{12}\mathrm{C}$ reaction, using natural α-emitters such as radium, radon and polonium. The reaction has a Q value of $5 \cdot 65\,\mathrm{MeV}$. but for several reasons (dependence of neutron energy on direction of emission, residual excitation of $^{12}\mathrm{C}$, thick target effects and neutron moderation within the Be) the emergent neutrons from an actual source possess a complex spectrum which spreads over a wide energy band. That from a Po-α-Be source extends from $10 \cdot 8$ to $\sim 1\,\mathrm{MeV}$. (Whitmore and Baker, 1950).

The yield of an (α, n) source comprising $M(\alpha)\,\mathrm{g}$. of α-emitting substance and $M(n)\,\mathrm{g}$. of neutron-emitting target varies with the proportions of constituents approximately as follows:

$$N = N_0 \frac{M(n)}{M(n) + M(\alpha)}, \tag{4.1}$$

where N_0 is the yield if $M(n)$ is infinitely greater than $M(\alpha)$. Quoting a comprehensive review of the subject by Anderson (1948), typical values for N_0 are Ra-α-Be 17, Rn-α-Be 15, Po-α-Be 3, in units of 10^6 fast neutrons/curie/sec. The most satisfactory source is probably Ra-α-Be. Radium in the form of $\mathrm{RaBr_2}$ can be compressed with beryllium powder to give a small source ($\sim 6\,\mathrm{cm.}^3$) which generates $\sim 10^7\,n/\mathrm{sec.}$, i.e. the flux near the surface is $\sim 6 \times 10^5\,n/\mathrm{cm.}^2/\mathrm{sec.}$ The maximum neutron energy is about $13 \cdot 2\,\mathrm{MeV}$.

Rn-α-Be sources have been widely used because of the easy availability of radon, but the half-life is too short (3.8 d.). Polonium provides a source fairly free from γ-rays, but preparation is more difficult and its half-life is also rather short for convenience (140 d.).

Pile-produced plutonium, though not strictly a natural α-emitter, has been employed in the same way, apparently with certain advantages (Steward, 1955). Its long half-life ($2 \cdot 2 \times 10^4\,\mathrm{yr.}$)

renders it suitable for use in neutron standards, particularly as its γ-ray emission is smaller than that of Ra. The inter-metallic compound $PuBe_{13}$ is more homogeneous and stable than the usual mixtures. However, the available yield is rather smaller, e.g. $1\cdot2 \times 10^6\,n/\text{sec}$. from a source of 13 g. Pu and 7 g. Be.

Other (α, n) studies using artificially-accelerated particles are referred to in § 4.2.4.

During the 1930's, (γ, n) reactions on ^{9}Be and ^{2}H using γ-radiation from ThC″ and RaC were commonly employed as sources for scattering in the 0·1–1 MeV. interval. They are now preferred for standard neutron sources of constant intensity—especially the ^{9}Be$(\gamma, n)\,^8$Be reaction based on γ-radiation from Ra. As the need for physical mixing of the components is removed, the risk of a gradual alteration in properties is much reduced. However, a Ra-γ-Be source is generally only ~ 10 % of the strength of a Ra-α-Be source and the average neutron energy is much lower. Moreover, the spectrum is still heterogeneous. In the design evolved by the U.S. Bureau of Standards (Curtis and Carson, 1949), a capsule of 1 g. of Ra is located at the centre of a 4 cm. Be sphere. Since α-particles from the Ra cannot reach the Be, the source strength is simply that due to the Ra-γ-Be yield, viz. ~ $1\cdot1 \times 10^6\,n/\text{sec}$.

Apart from the natural sources, (γ, n) reactions can be produced by the radiation from pile-produced isotopes or the *bremsstrahlung* from electron accelerators, both of which are dealt with in the following section.

4.2.2. *Photoneutron sources.* Deuterium and beryllium are the only possible target substances for (γ, n) sources based on either natural or artificial γ-emitters. Their neutron binding energies are 2·226 and 1·666 MeV. respectively compared with more than 4 MeV. for all other stable nuclei, and more than 6 MeV. for most.

From the conservation of energy and momentum, the neutron energy E_n can be expressed in terms of the γ-ray energy E_γ by

$$E_n = \frac{A-1}{A}\left[E_\gamma - Q - \frac{E_\gamma^2}{1862(A-1)}\right] + \delta\cos\Phi, \qquad (4.2)$$

where Q is the binding energy of a neutron in the nucleus of mass number A. The term involving δ determines the energy spread of the neutrons, which depends on the angle Φ between the γ-ray and the direction of emission of the neutron,

$$\delta \approx E_\gamma \left[\frac{2(A-1)(E_\gamma - Q)}{931 A^3} \right]^{\frac{1}{2}}. \qquad (4.3)$$

The maximum spread is $\Delta E_n = 2\delta$, which becomes proportionately smaller at higher neutron energies. It is also smaller for ^{9}Be than for ^{2}H sources. The 220 keV. neutrons emitted from a ^{24}Na-D$_2$O source can be expected to show a 14 % spread, and those of energy 830 keV. from ^{24}Na-Be about 3·5 %. These figures refer, of course, to thin target yields. In the interests of intensity, actual sources are made by surrounding the γ-emitter with several cm. of D$_2$O or Be, both of which happen to be efficient neutron moderators (cf. § 4.2.7). Compton scattering also reduces the average γ-energy, hence the observed neutron yield has a lower mean energy and a wider energy spread than calculated by equation (4.3). The full width at half maximum height is $\sim 30\%$ for the two sources referred to above (Hughes and Eggler, 1947).

In addition to the natural photoneutron sources (§4.2.1), the principal (γ, n) sources are based on *pile-produced radio-isotopes*, and are actually much cheaper than natural sources of equivalent strength. To be useful, the isotope must emit copious mono-energetic radiation, $E_\gamma > Q$, and its half-life should be at least one hour. Not many satisfy these requirements. The 'artificial' sources listed in Table III were recommended by Wattenberg (1947) and a number of them have been employed for low energy trans-mission measurements (e.g. Fields, Russell, Sachs and Wattenberg, 1947).

The saturation activity attainable in a given mass of substance exposed to a thermal flux ϕ in a reactor is $A_\infty = \Sigma_{act}\phi$, where Σ_{act} is the macroscopic cross-section (§2.3) for thermal activation. In the case of an ^{124}Sb-γ-Be source, if $\phi \sim 10^{12} n/\text{cm.}^2/\text{sec.}$, a few cm.^{3} of Sb ($\sigma_{act} = 2\cdot5$ b.) can be given an activity of several curies. If we assume a 4 c. source located in a Be cylinder of the design due to Wattenberg, the m.f.p. for γ-radiation being $\sim 0\cdot7$ cm., the neutron yield will be $N_i N_v \sigma_{\gamma n} x \sim 10^7 n/\text{sec}$. Strengths of about 10^6–10^8 can be realized with all the sources listed in Table III when the radio-isotopes are prepared in reactor fluxes of $\sim 10^{12}$. With reactors now operating considerably above this, the possibility of raising neutron yields by an order of magnitude seems reasonable. Of growing importance for time-of-flight measurements (see below) is the *bremsstrahlung-neutron* source, which requires a high energy, high current electron accelerator (Feld, 1951). Cockcroft, Duckworth and Merrison (1949) described the Harwell arrangement consisting of a 3.2 MeV. linear accelerator bombarding a lead target with an average current of $40 \mu A$. By absorbing the γ-ray output in D_2O or Be, an average yield of $\sim 10^9$ n/sec. resulted. Pulse lengths imposed by the accelerator were $\sim 2 \mu$sec., the repetition rate was 200 per sec. and the yield at the peak of each pulse $\sim 2 \times 10^{12}$ n/sec.

TABLE III. *Photoneutron sources using pile-produced radio-isotopes*

(After Wattenburg, 1947)

Source	Half-life	E_γ (MeV.)	E_n(observed) (keV.)	Approximate yield
^{24}Na $+ D_2O$	14·8 hr.	2·76	220	27
^{24}Na $+$ Be	14·8 hr.	2·76	830	13
^{56}Mn $+ D_2O$	2·6 hr.	2·7	220	0·31
^{56}Mn $+$ Be	2·6 hr.	1·81, 2·13, 2·7	~ 150,† 300	2·9
^{72}Ga $+ D_2O$	14 hr.	2·51	130	~ 6
^{72}Ga $+$ Be	14 hr.	1·87, 2·21, 2·51	~ 700‡	~ 5
^{116}In $+$ Be	54 min.	1·8, 2·1	~ 150, 300§	0·82
^{124}Sb $+$ Be	60 d.	1·7	24	19
^{140}La $+ D_2O$	40 hr.	2·50	151	$\sim 0·8$
^{140}La $+$ Be	40 hr.	2·50	620	$\sim 0·3$

† 90 % of the neutrons in this group.   ‡ Estimated.
§ ~ 60 % in this group.
Yields are in units of $10^4 n$/sec./curie (1 g. of target at 1 cm.).
Sources of information, Wattenberg (1947); Russell, Sachs, Wattenberg and Fields (1948); Hanson (1949); Hughes (1953).

Very high yields can also be obtained with Van de Graaff electron accelerators. The characteristics of these machines were discussed in §3.2.2 in connexion with positive ion sources, and apart from the positions of cathode and anode being interchanged, the electron machine is essentially the same. The commercially available High Voltage Corporation Type-FD 3 MeV. machine delivers 4 mA. (Foster *et al.* 1953) which, if used with D_2O or Be, should generate $\sim 10^{11} n$/sec. The electron source may be pulsed for time-of-flight measurements (§§4.3.4, 4.4.1).

At 20 MeV. and higher energies it is advantageous to employ targets of medium or heavy elements in order to benefit from their larger cross-sections for (γ, n) processes.† Consider the case of a 20 MeV. linear accelerator bombarding a thick Pb target which serves both as *bremsstrahlung* target and (γ, n) converter. The Pb threshold is about 7 MeV., hence neutrons of all energies up to about 13 MeV. will be produced. The effective (γ, n) cross-section is 0·3 b., derived from $\int_{7}^{20} \dfrac{n(E)\,\sigma(E)}{nE} \, dE$ by integrating numerically over the shape of the *bremsstrahlung* spectrum (Katz and Cameron, 1951) and the variation of (γ, n) cross-section with energy (Montalbetti, Katz and Goldenberg, 1953). It follows that an average beam of several μA. of electrons absorbed into a lump of Pb target would yield about $10^{11}\,n/$sec. Similar values have been estimated for 40 MeV. electrons on thick uranium targets (Biram, 1954).

4.2.3. Nuclear reactors. These have been reviewed thoroughly from the point of view of neutron sources by Hughes (1953), while much of the same ground has been covered by treatments of the fission process and the design of reactors (e.g. Soodak and Campbell, 1950).

It has been found empirically (Bonner, Ferrell and Rinehart, 1952; Hibdon and Muehlhause, 1951; Watt, 1952) that the energy distribution of neutrons emitted in the fission process can be represented to within experimental errors by a hyperbolic function of the form

$$n(E) = \left(\frac{2}{\pi e}\right)^{\frac{1}{2}} \sinh (2E)^{\frac{1}{2}} e^{-E}, \qquad (4.4)$$

which, for large E, becomes

$$n(E) = \left(\frac{1}{2\pi e}\right)^{\frac{1}{2}} e^{-E + (2E)^{\frac{1}{2}}}, \qquad (4.5)$$

where $n(E)$ is the number of neutrons per unit energy, normalized to one fission neutron, and E is the neutron energy in MeV. Neutrons of all energies from thermal to ~ 17 MeV. are released, the maximum of the distribution being at about 1·5 MeV.

† Cross-sections have been listed by various authors, e.g. Mott *et al.* (1953); Nathans and Halpern (1954). The cross-section for Pb at the peak of the giant resonance $(E = 13\cdot7$ MeV.) is $\sim 0\cdot8$ b., compared with $\sim 0\cdot001$ b. for Be at ~ 3 MeV.

Intense sources of neutrons having this distribution can be obtained by putting a uranium 'converter' in the external beam of a pile (Hughes, Spatz and Goldstein, 1949). In this arrangement, the normal neutron spectrum from the pile itself is considerably degraded in energy by the moderator (§ 4.2.7). It is customary to classify reactors according to the neutron energy upon which they depend for fissioning of the fuel elements. Research piles, with which we are chiefly concerned here, viz. those using natural uranium or uranium slightly enriched in ^{235}U and moderated with graphite or D_2O, depend on fission by *thermal neutrons* (~ 0.03 eV.). Other types can be designed to operate on intermediate energy neutrons, say ~ 10 keV., or the fast flux above 100 keV. (see catalogues, Isben, 1952; Kowarski, 1955; *Proc. Int. Conf. Atom. Energy*, 1956).

For the purposes of the present treatment, the significant questions concern availability of external beams of neutrons, their energy distribution and intensity. Fig. 35 is a schematic drawing of the Harwell reactor E443, illustrating the facilities for internal irradiation and extraction of beams from various quarters. Near the pile centre and among the fuel elements, the energy is higher than in the reflector where many neutrons have experienced a sufficient number of collisions to become thermalized.

The density of fast neutrons in the pile is actually quite small, but in terms of the flux ϕ ($= nv$, where n is the number per cm.^{3} and v is the neutron velocity) the fast flux may not be much smaller than the slow. (The ratio fast/slow depends chiefly on the 'slowing down power' $\xi N_v \sigma_s$ referred to in § 4.2.7). The main difficulty in the way of utilizing the fast fission flux for, say, cross-section measurements by transmission, arises from confusion caused by *resonance* neutrons†—partly moderated neutrons of energies between ~ 1 eV. and 1 keV. for which the cross-sections are much larger. However, the resonance spectrum is of great importance itself; it determines the availability of intermediate energy neutrons for use with fast choppers or crystal monochromators (§ 4.4), instruments whose development was made possible by the intense fluxes available from reactors.

---

† So-named because of the large number of distinct absorption resonances (cf. chapter 12) occurring in this energy range among medium weight and heavy nuclei.

The spectrum of neutrons in process of slowing down from a monoenergetic source takes the form (§4.2.7)

$$\phi_E \, dE = K \frac{dE}{E}, \qquad (4.6)$$

where ϕ_E is the flux slowing down through energy E and K is a constant which includes the primary source strength and character-

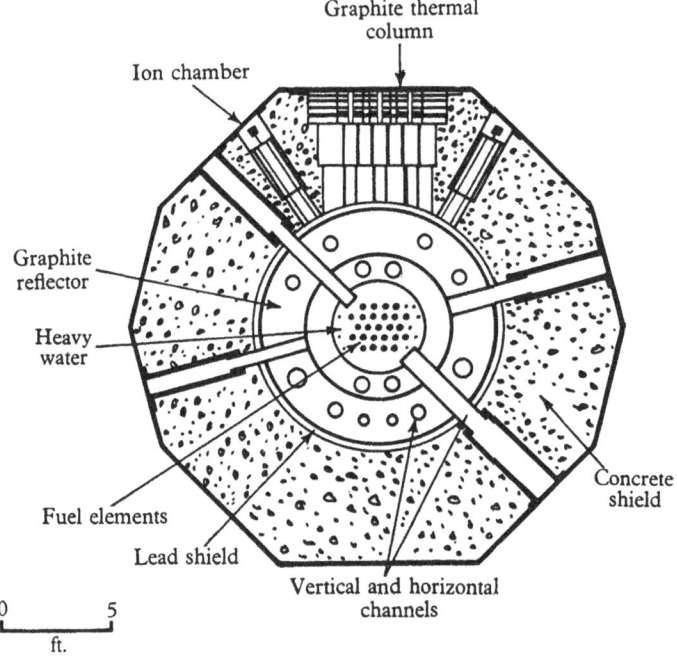

Fig. 35. Schematic section through Harwell reactor E443 showing facilities for irradiation and beam extraction. Collimators inserted into the large horizontal ports provide beams for monochromators, etc.

istics of the moderator. Since a fission spectrum is heterogeneous, the actual slowing down spectrum from a pile will be a sum over many such dE/E spectra, each having a different upper limit of energy.

This slowing down spectrum merges into a thermal neutron distribution† which approximates to the Maxwellian form

$$n(v) \, dv = \frac{4n}{v_0^3 \pi^{\frac{1}{2}}} v^2 \, e^{-v^2/v_0^2} \, dv, \qquad (4.7)$$

† A common method of determining the ratio of resonance and thermal fluxes makes use of the Cd absorption below 0·04 eV. (see Hughes, 1953).

where $n(v)\,dv$ is the number of neutrons per cm.[3] with velocities between v and $v+dv$, n is the total number per cm.[3] and v_0 the velocity at which $n(v)$ is a maximum. The corresponding temperature of the distribution is determined by $kT=\frac{1}{2}mv_0^2$, where m is the neutron mass and k the Boltzmann constant. If the measured velocity spectrum (e.g. using a velocity selector, §4.4) fits an equation of this form, T can be obtained directly. However, instrumental difficulties arise (Hughes, 1953) and less direct methods have to be employed. One can assume that the distribution is Maxwellian and compare the theoretical and experimental transmission cross-sections for a material whose cross-section is known accurately as a function of v.† It is generally found that thermal neutrons extracted from inside a reactor have an equivalent temperature $\sim 100°$ K. higher than the actual moderator temperature (e.g. Anderson, Fermi, Wattenberg, Weil and Zinn, 1947), implying that true thermal equilibrium has not been achieved. The discrepancy presumably reflects the presence of the high energy tail of resonance neutrons.

For obtaining thoroughly thermalized neutrons all research piles are provided with a large *thermal column* formed by stacking graphite so that it extends through the shield as shown in fig. 35. As neutrons diffuse through the column, the thermal flux falls off as $e^{-x/L}$, where the relaxation length L is ~ 30 cm., rather shorter than the true diffusion length in graphite (~ 50 cm., §4.2.7) because of leakage from the sides of the stack. The resonance flux decreases much faster with $L\sim 13$ cm. (Hughes, 1953).

Consider now the absolute beam intensities. It is usual to rate reactors by their central (lattice) flux when operating at full power. In existing types, this ranges from 10^{10} to several times $10^{14}\,n/$cm.[2]/sec. However, for scattering experiments it is the external flux which is more relevant and this is necessarily much smaller. From a channel let in a distance l it will be $\phi_r A/4\pi l^2$, where ϕ_r is the

---

† For boron, $\sigma_a = 1.64 \times 10^6/v$. The average absorption cross-section $\bar{\sigma}_a$ is

$$1.64 \times 10^6 \frac{\int n(v)/v \cdot dv}{\int n(v) \cdot dv} = 1.64 \times 10^6 \left(\frac{2}{\pi^{\frac{1}{2}}v_0}\right) = 1.128\sigma_{v_0},$$

where v is in m. sec.$^{-1}$. σ_{v_0} is the 'thermal cross-section' generally listed (see §4.2.7).

prevailing flux at the reactor end of the channel. In the case of E443, a 6 in. diameter channel let into the centre ($\phi_r \sim 10^{14}$) would yield $\sim 10^{10} n/\text{cm}^2/\text{sec}$. outside the concrete shield. The intensity of the resonance neutron flux in any decade interval (1–10, 10–100 eV., etc.) is, typically, $\sim 15\%$ of the integrated thermal flux inside the pile lattice, say $\sim 10^{13} n/\text{cm}^2/\text{sec}$. or $\sim 10^9$ outside the shield.

Although ϕ_r is a maximum near the pile centre, the high resonance flux there tends to interfere with measurements, hence it is often more convenient to draw off a beam from the reflector or (for pure thermal neutrons) from some point in the thermal column. External beam fluxes are then correspondingly weaker. Taking ϕ_r as 10^{12} at the outer face of the reflector, at the end of a 4-in. channel the flux will be $\sim 3 \times 10^8$. However, many existing piles give much smaller fluxes than this, perhaps 1–10 % of the figures quoted above.

4.2.4. *Reactions produced by positive ions.* By bombarding targets of various elements with ions which have been accelerated to high energy, various reactions can be initiated which release neutrons. Accelerators best suited to this task are high current H.V. sets, Van de Graaff generators and fixed-frequency cyclotrons (§§ 3.2.1–3.2.3). External currents of $\sim 100\,\mu\text{A}$. and even ~ 1 mA. are not uncommon, while E. O. Lawrence, in a review of high current accelerators, refers to a source producing ~ 1 A. of particles at ~ 100 keV. (Lawrence, 1955). The possibility of using such unprecedented ion currents, which could yield $\sim 10^{14} n/\text{sec}$., rests on the question of dissipating ~ 100 kW. of power in the target!

Advantages of neutron sources based on charged-particle reactions are threefold: (i) The energy depends on the incident ion energy and the angle of neutron emission (see below), which allows a useful range of variation to be obtained in many cases. Utilizing this variation with the reactions of different Q-value listed below, it is possible to cover most of the neutron energy spectrum from a few keV. to ~ 20 MeV. (ii) Monoenergetic neutrons are made available, providing the ion bean is monoenergetic.† (iii) Certain

---

† This assumes that neutron and recoil nucleus share the available energy. If a three-particle break-up occurs, the neutron spectrum is continuous. It is also assumed that a single energy state of the nucleus is involved.

reactions, in particular the *t-d* and *d-d*, have large cross-sections in the low energy range accessible to H.V. sets.

The fundamental energy equation of two-particle reactions is

$$E_3^{\frac{1}{2}} = \frac{(m_1 m_3 E_1)^{\frac{1}{2}} \cos \Theta}{m_3 + m_4} \pm \left[\left\{ \frac{m_4 - m_1}{m_3 + m_4} + \frac{m_1 m_3 \cos^2 \Theta}{(m_3 + m_4)^2} \right\} E_1 + \frac{m_4}{m_3 + m_4} Q \right]^{\frac{1}{2}},$$

$$(4.8)$$

where m_1, m_2, m_3 and m_4 are masses of the incident particle, target, neutron and recoil nucleus, respectively. E_1 is the incident energy, E_3 that of the neutron emitted at angle Θ to the incident direction and Q the mass-energy $[(m_1 + m_2) - (m_3 + m_4)] c^2$.

When Q is positive (*exothermic* reaction), the radical in square brackets always exceeds the first term in magnitude. Only one real value of E_3 exists, i.e. the neutron energy is single-valued. However, when Q is negative (*endothermic* reaction), E_3 is single-valued only if

$$E_1 > E_c = -\left(\frac{m_4}{m_4 - m_1} \right) Q. \qquad (4.9)$$

For $E_1 < E_c$, E_3 is double-valued, i.e. two neutron energies occur at any angle of emission. Moreover, a definite *energy threshold* below which a reaction cannot proceed governs the endothermic case. This occurs at

$$E_{th} = -\left(\frac{m_1 + m_2}{m_2} \right) Q, \qquad (4.10)$$

which is obtained by equating the energy available for excitation to Q at the threshold (see equation (2.19)). At the threshold, neutrons appear with zero velocity in the CM system and hence their velocity in the Lab system is that of the CM, viz. $v_0(m_1/m_1 + m_2)$ confined to the forward direction (cf. Appendix A). The corresponding neutron energy is then

$$E_3 = \frac{m_1 m_3}{(m_1 + m_2)^2} E_{th}. \qquad (4.11)$$

As E_1 rises above the threshold, the direction of neutron emission fans out into a forward cone whose half-angle Θ_m is found by equating the radical of equation (4.8) to zero. This gives

$$\Theta_m = \arcsin \left[\frac{m_2 m_4 (E_1 - E_{th})}{m_1 m_3 E_1} \right]^{\frac{1}{2}}. \qquad (4.12)$$

For $\Theta < \Theta_m$, two distinct neutron energies are associated with each value of Θ. When E_1 reaches the value E_c the cone is spread out over the whole forward hemisphere, while for still higher energies neutrons are emitted in all directions. At all incident energies, highest energy neutrons occur at $0°$, lowest at $180°$.

The characteristics of some of the most useful sources are listed below (based partly on the review by Hanson, Taschek and Williams (1949)).†

Exothermic reactions

(i) ^{2}H(d, n) ^{3}He $+ 3.265$ MeV. Because of its high yield at low bombarding energies this has been the most extensively used reaction. Total cross-sections are shown in fig. 36. By employing incident deuterons of several MeV. energy, the d-d reaction also becomes an important source of neutrons of 4–7 MeV. However, at higher energies there is considerable continuous background (Laughlin and Kruger, 1948). Yield curves from a thick heavy ice target indicate that $\sim 7 \times 10^7$ n/sec. are emitted per μA. of 1 MeV. deuterons. High current accelerators of several MeV. can achieve $\sim 10^{11}$ n/sec. (cf. §§3.2.1.–3.2.3). Angular distribution studies reveal a concentration of neutrons in the forward direction (Hanson et al. 1949). Another important feature is that the companion reaction ^{2}H(d, p) ^{3}H has a comparable cross-section (Bretscher, French and Seidel, 1948; Hanson et al. 1949; Arnold, Phillips, Sawyer, Stovall and Tuck, 1954) and may be used for monitoring the neutron emission.

(ii) ^{3}H(d, n) ^{4}He or ^{2}H(t, n) ^{4}He $+ 17.58$ MeV. The highest Q-value associated with any reaction yielding monoenergetic neutrons occurs in the t-d interaction. Deuterium gas or various 'deuteride' targets (heavy ice, etc.) may be bombarded with tritons. Alternatively, tritium, either gaseous or absorbed into zirconium or other getters (cf. §3.3.1), can be bombarded with deuterons (Peck and Eubank, 1955; Johnson and Banta, 1956). The reaction has a high yield at low energies due to the prominent resonance ($\sigma_{\max} \approx 5$ b.) which occurs at $E_d \approx 100$ keV. or $E_t \approx 150$ keV. The neutron energy is $E_3 = 4/(4+1). 17.6 = 14.1$ MeV. when $E_1 \approx 0$. As the α-particles

† Neutron energy as a function of incident energy and angle of emission can be represented conveniently by nomographs (McKibben, 1946).

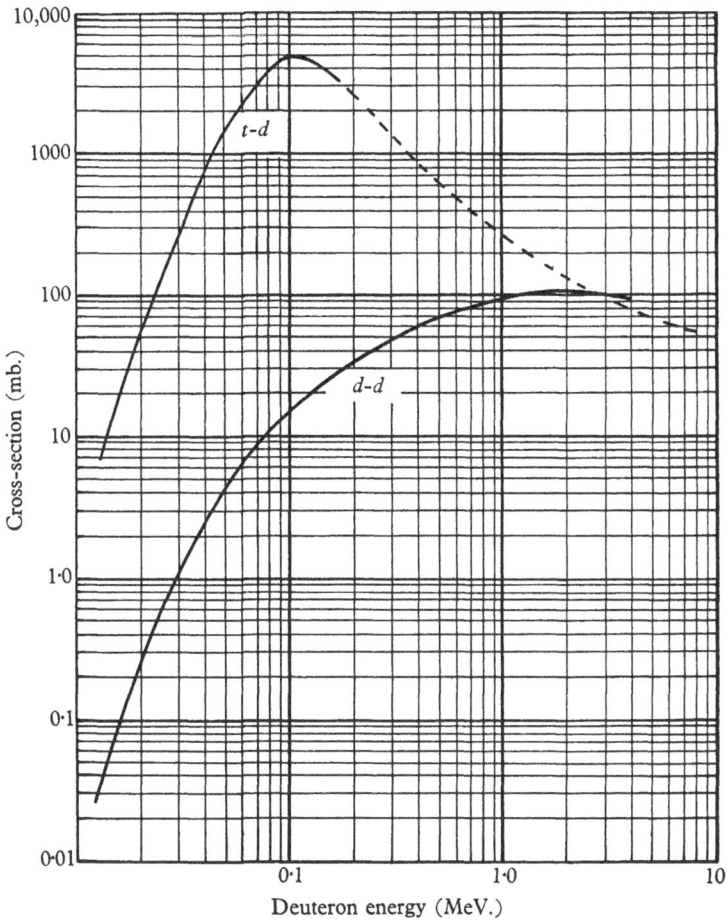

Cross-section (mb.)

Deuteron energy (MeV.)

Fig. 36. Total cross-sections for ^{2}H(d, n) ^{3}He and ^{3}H(d, n) ^{4}He reactions. T-d data beyond resonance are rather uncertain. Chief sources of information: Hanson *et al.* (1949); Brolley, Fowler and Stovall (1951); Argo, Taschek, Agnew, Hemmendinger and Leland (1952); Arnold *et al.* (1954); Johnson and Galonskey (1955); Hemmendinger and Argo (1955).

gain appreciable energy, they can be counted without undue difficulty to monitor the neutron flux (Allred, Armstrong and Rosen, 1953). By using deuterons of ~ 3 MeV., the reaction covers the range of neutron energy from about 12 to 20 MeV. Cross-sections are shown in fig. 36. Typical thick target yields are $\sim 5 \times 10^8\,n$/sec./μA. for $E_d = 600$ keV. Peck and Eubank (1955) obtained $2 \times 10^9\,n$/sec./mA. at $E_d = 200$ keV. using a target containing 1 atom of ^{3}H per 4 atoms

9   MES

of Zr. (For a detailed discussion of the mechanics and yield of this reaction see Benvenisti and Zenger, 1954.)

(iii) ^{7}Li(d, n) ^{8}Be $+ 15 \cdot 02$ MeV. Prior to the advent of t-d sources, this reaction was useful because of its high Q. It was widely employed for n-p scattering between about 11 and 25 MeV. (Salant and Ramsay, 1940; Ageno, Amaldi, Bocciarelli and Trabacchi, 1943; Shapiro *et al.* 1954; Ageno *et al.* 1947; Sleator, 1947), generally with an energy-selective detector (§ 4.3.5). The neutron spectrum is complex owing to the excitation of higher states of ^{8}Be and the $2\,^4$He $+ n$ break-up. At $\sim$ 100 keV. the yield follows the Gamow function, reaching $17 \times 10^6\,n$/sec./μA. on thick targets at $E_d = 600$ keV. (Ralph and Dunnan, 1954). For $E_d \gtrsim 1$ MeV., the yield exceeds that from d-d sources.

(iv) ^{9}Be(d, n) ^{10}B $+ 4 \cdot 36$ MeV. This also gives a heterogeneous neutron spectrum corresponding to the many levels of ^{10}B (Ajzenberg, 1951). It, and also (d, n) reactions on $^{10, 11}$B, was used as source in several early scattering experiments (Powell, Heitler and Champion, 1940; Champion and Powell, 1944; Ageno *et al.* 1947). Thick target yields from Be range from $4 \times 10^6\,n$/sec./μA. at 400 keV. to $\sim 10^8$ at 800 keV., but above about 1 MeV. the yield from Li is more intense. Deuterons of 10 MeV. on thick Be yield $\sim 3 \times 10^{10}\,n$/sec./μA. (Smith and Kruger, 1951).

Endothermic reactions

(i) ^{3}H(p, n) ^{3}He $- 0 \cdot 764$ MeV. A high yield of monoenergetic neutrons is obtainable for proton energies up to at least $2 \cdot 5$ MeV. (Jarvis, Hemmendinger, Argo and Taschek, 1950). T-Zr targets are often used as mentioned above (Adair, 1952). The Coulomb barrier is low, hence yields are considerable even close to the threshold ($1 \cdot 019$ MeV.), at which the neutron energy is ~ 60 keV. Cross-sections are:

E_p (MeV.)	$1 \cdot 1$	$1 \cdot 4$	$1 \cdot 7$	$2 \cdot 0$	$2 \cdot 3$
σ (barns)	$0 \cdot 11$	$0 \cdot 25$	$0 \cdot 35$	$0 \cdot 44$	$0 \cdot 53$

At higher energies the reaction may be monitored by counting the associated ^{3}He particles. The competing process ^{3}H(p, γ), which has a cross-section near the (p, n) threshold of $\sim 0 \cdot 03$ mb. (Perry and Bame, 1955), gives rise to 20 MeV. γ-rays that tend to cause confusion when counting neutrons.

(ii) $^7Li(p, n)$ $^7Be - 1.645$ MeV. Until 1949, neutrons from this source were believed to be monoenergetic (Grosskreutz and Mather, 1949; Johnson, Laubenstein and Richards, 1950). It is now known that for $E_p > 2.4$ MeV. ($E_{th} = 1.88$ MeV.) there are two neutron groups† corresponding to the ground state of ^7Be and the first excited state at 431 keV. (Ajzenberg and Lauritsen, 1955). However, the reaction is still commonly employed as providing a source of keV. neutrons. The cross-section is characterized by a wide resonance which rises to 0.5 b. at ~ 2.2 MeV. The angular distribution is strongly peaked in the forward direction. At the resonance, the forward yield is $\sim 10^7 n$/sterad./sec./μA. (See also Hanson et al. 1949).

A similar reaction $^9Be(p, n)$ $^9B - 1.851$ MeV. has also been used to a limited extent. A strong continuous spectrum robs it of much of its value as a neutron source (Marion, Brugger and Bonner, 1955).

(iii) $^{51}V(p, n)$ $^{51}Cr - 1.53$ MeV. When heavy targets are used, the neutron energy is almost independent of angle of emission, which is convenient in certain types of experiment. Also, because the CM velocity is low, neutrons emitted at threshold have energies of the order of keV. However, the Coulomb barrier (~ 6 MeV. for V) is high, and the yield therefore rather low near the threshold. The ^{51}V reaction has been studied in some detail as a source of 5–120 keV. neutrons (Gibbons, Macklin and Schmitt, 1955). At an incident energy of 2 MeV., the thin target yield is $6 \times 10^3 n$/sterad./sec./μA. in the forward direction. As the first excited state of ^{51}Cr is at 780 keV., monoenergetic neutrons should be available for all proton energies up to 2.34 MeV., the threshold for excited state production. Thick target yields are heterogeneous in energy but of about 10 times the above intensity (Hanson et al. 1949).

(iv) $^{45}Sc(p, n)$ $^{45}Ti - 2.84$ MeV. Higher incident energies are required in this case to reach the threshold at 2.91 MeV. However, proton penetration through the barrier is more probable, the yield exceeding that from ^{51}V targets‡ by about 40 times at the threshold. At 3.5 MeV. the thick target forward yield has been quoted as $2.5 \times 10^6 n$/sterad./μA. (Hanson et al. 1949). Absolute

† The intensity of the second group is about 10 % of that of the ground state group (see Feld, 1953).

‡ Vanadium and scandium are both isotopically pure, ^{51}V and ^{45}Sc, respectively.

cross-sections are also available (Brugger, Bonner and Marion, 1955). No evidence has been found indicating excited states in ^{45}Ti. Other (p, n) reactions which have been suggested as sources of keV. neutrons are ^{63}Cu(p, n) ^{63}Zn $- 4 \cdot 15$ MeV. (Brugger *et al.* 1955) and ^{55}Mn(p, n) ^{55}Fe $- 1 \cdot 0$ MeV. (Stelson and Preston, 1951; McCue and Preston, 1951).

(v) ^{12}C(d, n) ^{13}N $- 0 \cdot 281$ MeV. This is an exception to the general class of (d, n) reactions which are usually exothermic. As the laboratory threshold is $0 \cdot 328$ MeV., this is a convenient reaction for low voltage machines unable to reach the ^{3}H(p, n) or ^{7}Li(p, n) thresholds. The first excited state of ^{13}N is at $2 \cdot 4$ MeV., hence neutrons should be monoenergetic up to this energy, i.e. corresponding to $E_d \approx 2 \cdot 7$ MeV. Drawbacks of the reaction are: (a) The unavoidable yield of d-d neutrons due to deuterons absorbed in the target and collimating slits. (b) The ^{13}C(d, n) ^{14}N $+ 5 \cdot 32$ MeV. process must also be expected (^{13}C abundance $1 \cdot 1 \%$). Probably the chief significance of the carbon reactions lies in the difficulty of avoiding them, since carbon tends to deposit from pump oil vapours on whatever target is being employed.

In addition to the yields quoted above for various reactions, most of which apply to fairly low bombarding energies, a substantial body of data now exists on high energy (cyclotron) yields. We shall not deal with this here. An introduction to the literature can be obtained from the following references: Ridenour and Henderson (1937); Allen, Nechaj, Sun and Jennings (1951); Feld (1953); Blosser and Handley (1955); Millburn, Moyer, Tai and Kaplan (1955). For deuterons and protons of 10–30 MeV., neutron yields are very considerable, e.g. $\sim 10^{10}$ to 10^{11} n/sec./μA.

4.2.5. *Deuteron stripping reactions at high energy.* The (d, n) process at high energies was historically the first source of neutrons having energies of the order of 100 MeV. When a beam of deuterons of energy much greater than the binding energy of the deuteron impinges on a target of any substance, neutrons which are fairly homogeneous in energy and strongly collimated in the forward direction are 'stripped off' in considerable intensity (Helmholtz, McMillan and Sewell, 1947). At 190 MeV. the neutron yield from a $0 \cdot 5$-in. Be target is about 2 % (Serber, 1947) and the mean energy is

~ 90 MeV. The half-width of the angular spread is a function of the atomic number of the target. Expressed in radians, it is $(0.155 + 0.0006Z)$, i.e. about $10°$ for 190 MeV. deuterons. The theory of stripping reactions, leading to estimates of energy and angular spread, is dealt with in § 11.2.

4.2.6. *Charge exchange by very high energy protons.*† Another technique for generating very high energy neutrons from accelerated charged particles makes use of charge exchange in nucleon-nucleon collisions. This is the source of the highest energy neutrons available. Miller, Sewell and Wright (1951), using 330 MeV. protons, observed a forward collimation of neutrons, although the distribution was less sharp than obtained by (d, n) stripping (§ 4.2.5). The full angular width at half maximum intensity was $54–59°$ for Be, Al, Cu and U targets, when neutrons of all energies higher than 20 MeV. were included.

Fig. 37 (a) shows the neutron energy spectra obtained at several angles when bombarding LiH (0.44 g./cm.²) with 95 MeV. protons (Hofmann and Strauch, 1953). A marked forward collimation occurs in the case of the highest energy neutrons. At $0°$ the peak yield appears at ~ 85 MeV. with a half-width of rather more than 20 MeV., but below this the yield declines to a plateau where the intensity is about half that at the peak. Similar results were obtained for LiD and Be, whereas C, Al, Cu and Pb showed no peak in the energy spectrum at $0°$. Hofmann and Strauch suggest that the chief contribution to the width of the forward peak arises from energy spread in the initiating proton beam, due to radial oscillations, energy loss and scattering in the target during one or more traversals. Typical distributions taken in the forward direction for protons of 300–500 MeV. are shown in fig. 37 (b). (Methods of measuring these high energies are discussed in chapter 5.)

The collimation of high energy neutrons into a beam of suitable size is effected by setting several feet of steel or lead tubing into the concrete shield round the accelerator (see fig. 44). Working fluxes range from 10^4 to ~ 5×10^5 n/cm.²/sec., i.e. an order of magnitude lower than typical proton fluxes, which renders neutron scattering experiments difficult, though still feasible.

† For a general review and bibliography see Hofmann and Strauch (1953).

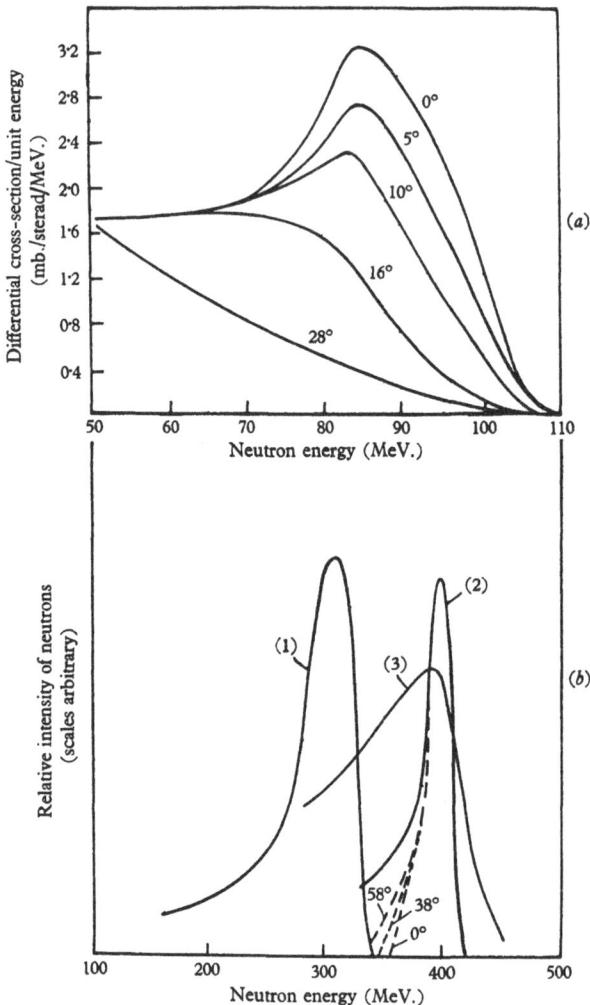

Fig. 37. (a) Neutron energy distribution observed at various angles from LiH bombarded with 95 MeV. protons (Hofmann and Strauch, 1953). (b) Relative intensity of neutrons emitted at 0° from various targets bombarded by protons. Curve (1) 1¾-in. LiD target, 340 MeV. protons (De Pangher, 1955). Curve (2) 1-in. Be target, 427 MeV. protons (Hartzler and Siegel, 1954). The dotted curves indicate the effective spectrum for *n-p* scattering, obtained by inserting a Cu absorber between counters of the recording telescope (see fig. 44). Curve (3) Be target, 480 MeV. protons (Dzhelepov and Kazarinov, 1954). Intensity scales of the three curves are unrelated. Experimental points are omitted in all cases.

Neutrons of ~ 1 Bev. energy have been generated by the same process. $2 \cdot 2$ Bev. protons bombarding a Be target produce a neutron beam which lies chiefly within a forward cone of half-angle $\sim 6°$ (see Coor, Hill, Hornyak, Smith and Snow, 1955; Fowler, Shutt, Thorndike and Whittemore, 1954).

4.2.7. *Production of slow neutrons.* Slow neutron scattering forms an important and considerable part of neutron scattering generally and warrants separate discussion of its sources. Whereas the methods of production treated in earlier sections are capable of generating neutrons ranging in energy from ~ 1 keV. to many MeV., there is no primary source of unadulterated slow neutrons. These must be produced in a secondary stage by slowing down fast neutrons. In the case of a thermal reactor, the fission flux is slowed within the body of the reactor, although even here a comparable fast flux exists and additional slowing has to be provided for experimental purposes in the form of the thermal column (§ 4.2.3). In all other cases, the procedure is to surround the source with a sufficient thickness of moderating substance, whose physical function is to effect a transfer of energy from the neutrons to itself by means of elastic collisions.

The moderation of neutrons has been discussed extensively elsewhere (Soodak and Campbell, 1950; Glasstone and Edlund, 1952; Hughes, 1953) and the present treatment will be confined to a summary.

Three properties are desirable in the moderator. They are:

(i) *A low mass number A.* The nearer A approaches to unity, the more energy can be transferred from the neutron per collision (cf. equation (A. 12)). If the angular distribution of scattering is assumed to be the same for all nuclei, viz. isotropic, a given number of collisions will transfer most energy, on the average, to the moderator of smallest atomic mass. A convenient measure is the *logarithmic decrement* $\xi = \overline{\ln (E_i/E_{i+1})}$, which is the amount by which $\ln E$ is reduced, on the average, in each collision. This possesses a characteristic value for every material,

$$\xi = 1 + \frac{(A-1)^2}{2A} \ln \left(\frac{A-1}{A+1}\right)^{\dagger}.$$ (4.13)

† A close approximation is $2/(A+\tfrac{2}{3})$ for $A > 1$.

If E_0 were the initial neutron energy and E_i that remaining after i collisions,

$$i = \frac{\ln E_0 - \ln \bar{E}_i}{\xi} \tag{4.14}$$

is the number of collisions required to degrade neutrons to a mean energy E_i, where $\ln \bar{E}_i \approx \overline{\ln E_i}.$†

(ii) *A large scattering cross-section σ_s.* A large ξ can only be effective if the probability of collision is also large. Since the probability that a collision will occur per cm. path is $N_v \sigma_s$, a better measure of the quality of a moderator is the *slowing down power* $\xi N_v \sigma_s$ or $\xi \Sigma_s$ (see § 2.3).

(iii) *A small absorption cross-section $(\sigma_a \ll \sigma_s)$.* Certain light elements, boron and lithium, have favourable slowing down powers but are useless as moderators because neutron absorption is far more probable than elastic scattering. The *moderating ratio* takes account of this, $\xi \Sigma_s / \Sigma_a$ or $\xi \sigma_s / \sigma_a$. The larger the moderating ratio, the more satisfactory the material as a medium for slowing neutrons. The moderating properties of several commonly used materials are shown in Table IV.

Neutrons of all intermediate energies are made available by the slowing down process and considerable importance attaches to the actual distribution. Assume that q_0 neutrons are produced per cm.³ throughout a large volume of moderator. (This describes fairly closely the conditions in a pile.) For neutrons of velocity v there are $v N_v \sigma_s$ collisions per sec., where σ_s is the elastic scattering cross-section—assumed constant during the slowing down. The energy loss per collision is ξE, hence if $n(E)$ is the density of neutrons per unit energy, the number slowing down through energy E per cm.³ per sec. (called the *slowing down density*) is

$$q(E) = n(E) v N_v \sigma_s \xi E.$$

If σ_a is negligible compared with σ_s, the neutrons passing per sec. through any energy interval must equal their production rate, $q(E) = q_0$, whence the slowing down flux is

$$\phi_E \, dE = (nv)_E \, dE \approx \frac{q_0}{\xi N_v \sigma_s} \frac{dE}{E}. \tag{4.15}$$

This is the 'dE/E spectrum' referred to in § 4.2.3.

† This is exact if $\bar{E}_i$ is taken as the geometric mean.

When σ_a is not negligible, but still small compared with σ_s,

$$\phi_E \, dE = \frac{q_0}{\xi N_v \sigma_s} \exp\left[-\frac{1}{\xi \sigma_s} \int_E^{E_0} \sigma_a \, dE/E \right]. \qquad (4.16)$$

The exponential involving the integral over the slowing down range is the probability that a neutron will survive down to energy E and is called the *resonance escape probability*.

TABLE IV. *Slowing down properties of various moderators*

Moderator	ξ	i†	$\xi\Sigma_s$	$\xi\dfrac{\Sigma_s}{\Sigma_a}$
^{1}H	1·0	19	1·2 (H_2O)	72
^{2}H (D)	0·725	26	0·17 (D_2O)	12,000
Be	0·209	89	0·18	160
C	0·158	118	0·064	170

† For moderation of 3 MeV. neutrons to thermal energy (0·025 eV.).

More complex problems are presented by the following important cases, in which the distribution of neutrons in space must be considered also. In the absence of absorption, approximate solutions are gaussians (Soodak and Campbell, 1950).

(*a*) A plane source (*y*, *z* plane) emitting Q fast neutrons/cm.^{2}/sec. in an infinite medium. The slowing down density along the x direction is

$$q(x, E) = \frac{Q}{(4\pi\tau)^{\frac{1}{2}}} e^{-x^2/4\tau}. \qquad (4.17)$$

Neutrons traversing the reflector of a pile are represented fairly well by this equation.

(*b*) A point source emitting Q fast neutrons/sec., e.g. an accelerator target embedded in a moderator effectively infinite in extent:

$$q(r, E) = \frac{Q}{(4\pi\tau)^{\frac{3}{2}}} e^{-r^2/4\tau}. \qquad (4.18)$$

The initial energy is E_0 in both cases. τ is the neutron 'age' (dimensions cm.^{2}),

$$\tau = \int_E^{E_0} \frac{\lambda_{\mathrm{tr}} \lambda_s \, dE}{3\xi E} = \int_t^{t_0} \frac{\lambda_{\mathrm{tr}} v}{3} \, dt. \qquad (4.19)$$

Here λ_s is the ordinary m.f.p. for scattering, equation (2.9). If the scattering is not isotropic in the Lab system,‡ the quantity which enters into the diffusion equation is the *transport mean free path* λ_{tr}, which is the average distance a neutron moves between collisions

‡ At the energies being considered, all scattering is isotropic in the CM system (chapter 6), but in the Lab. system the neutron retains a predominantly forward velocity component unless $A \gg 1$.

in a direction away from the source. It can be shown from the kinematics of collisions that

$$\lambda_{tr} = \frac{\lambda_s}{\left(1 - \frac{2}{3A}\right)}, \qquad (4.20)$$

which implies a transport cross-section

$$\sigma_{tr} = \sigma_s \left(1 - \frac{2}{3A}\right). \qquad (4.21)$$

For large A, $\lambda_{tr} \rightarrow \lambda_s$, but for hydrogen (the extreme case), $\lambda_{tr} = 3\lambda_s$. The integral over time in equation (4.19) is taken from the instant of neutron emission, t_0, to the time t required to slow to energy E.

The thickness of moderator needed to slow down neutrons can be specified by the root mean square distance moved from the source, $\langle x^2 \rangle^{\frac{1}{2}}$ for a plane source and $\langle r^2 \rangle^{\frac{1}{2}}$ for a point source. These are related to τ by $\langle x^2 \rangle = 2\tau$ and $\langle r^2 \rangle = 6\tau$ respectively, where

$$\tau = L_s^2 = \left(\frac{\lambda_{tr}\lambda_s i}{3}\right) \qquad (4.22)$$

and i is the number of collisions required to slow down from E_0 to E, i.e. to age τ. This also brings out the physical significance of τ as a measure of the width of the spatial distribution of slowing down density. $q(r, E)$ falls to $1/e$ of its maximum value in a distance $r = 2(\tau)^{\frac{1}{2}}$. A large τ means that neutrons have slowed down considerably, hence have diffused far from the source so that the q-distribution is broad.

Neutrons which have been slowed down to thermal equilibrium with the medium, exhibit a velocity distribution described by equation (4.7) and a flux distribution obtained by multiplying the right-hand side by v. The *most probable velocity* v_0 is related to the absolute temperature T by $\frac{1}{2}mv_0^2 = kT$ (§ 4.2.3). The *average velocity* found by integrating over the Maxwell distribution is $v_a = 2v_0/\pi^{\frac{1}{2}}$, i.e. about 13 % larger. The *most probable energy* is found by transforming to an energy distribution and differentiating, giving $\frac{1}{2}kT$. The *average energy* is $\frac{3}{2}kT$. These distinctions are important because in considering the diffusion of thermal neutrons it is convenient to treat them as monoenergetic, which is legitimate provided the appropriate velocity, v_0 or v_a, is used. Cross-sections for 'thermal neutrons' are always quoted for 2200 m./sec., which is the value of v_0 at 293° K. The corresponding energy is 0·0252 eV. but the average energy at this temperature is 0·0378 eV. The spatial distribution of thermal neutrons about a source can be obtained by solving the following diffusion equation. The neutrons are assumed to have a single (average) energy and the absorption cross-section σ_a is taken to be small compared with the scattering cross-section σ_s.

$$D\nabla^2 n \; - \; \frac{vn}{\Lambda} \; + \; q \; = \; \frac{\partial n}{\partial t}. \qquad (4.23)$$

$$\text{leakage} \quad \text{absorption} \quad \text{production}$$

This expresses the net rate of change of neutron density n at any point as the sum of three contributing rates. In the steady state, $\partial n/\partial t = 0$ and (leakage + absorption) = production. q is the production rate in n/cm.^{3}/sec. Λ ($= 1/\Sigma_a$) is the m.f.p. for absorption, hence vn/Λ is the number of neutrons absorbed per cm.^{3} per sec. ∇^2 is the Laplacian and $(D\nabla^2 n)$ expresses the rate of change of density due to neutrons diffusing into and out of an elementary volume. The diffusion coefficient D takes the usual form $\lambda_{tr} v/3$, where λ_{tr} is now the transport m.f.p. for thermal neutrons and is related to the thermal scattering m.f.p. λ_s and cross-section σ_s by equations analogous to equations (4.19) and (4.20).

The diffusion properties of the medium can be combined in a single constant, the diffusion length L defined by $L^2 = D\bar{\tau}$, where $\bar{\tau}$ is the mean life of a neutron against capture ($= \Lambda/v$). Hence

$$L^2 = \left(\frac{\lambda_{tr}\Lambda}{3}\right) = \left(\frac{1}{3N_v^2 \sigma_{tr}\sigma_a}\right). \qquad (4.24)$$

Diffusion properties of a number of materials are shown in Table V. Note that for a $1/v$ absorber, the correct absorption cross-section to insert here is not that usually listed, which corresponds to the most probable velocity (2200 m./sec.), but should correspond to the average velocity of the Maxwell distribution. The latter cross-section is smaller by a factor $\pi^{\frac{1}{2}}/2$ or 0·8865.

Equation (4.23) can be rewritten as

$$\nabla^2 n - \frac{n}{L^2} + \frac{q\bar{\tau}}{L^2} = 0. \qquad (4.25)$$

The solution can be found for any particular geometry of source and medium by applying appropriate boundary conditions (Soodak and Campbell, 1950). Special cases have been listed and their solutions displayed graphically by Wallace (1949).† Two cases, corresponding to those referred to earlier, are of special interest in neutron research:

(a) The density distribution along the x direction when neutrons diffuse away from a plane source emitting Q thermal neutrons/cm.^{2}/ sec. approximates to the behaviour of neutrons traversing the thermal column of a reactor (§ 4.2.3). If $x \gg \lambda_{tr}$,

$$n(x) = \frac{Q\bar{\tau}}{2L} e^{-|x|/L}. \tag{4.26}$$

L is here a 'relaxation length' which reduces the density to $1/e$. As in the slowing down problem, the mean square distance diffused in the x direction before capture is $\langle x^2 \rangle = 2L^2$.

TABLE V. *Thermal diffusion properties of some moderators*

Substance	Density ρ (g./cm.^{3})	L (cm.)	λ_{tr} (cm.)	Λ (cm.)
H_2O	1	2·8	0·43	55
Paraffin (CH_2)	0·895	2·4	0·40	43
D_2O	1·1	170	2·4	36,000
Be	1·84	25	1·7	1,100
C	1·6	51	2·5	3,100

(b) For a point source in an infinite medium, emitting Q n/sec., the radial distribution is

$$n(r) = \frac{Q\bar{\tau}}{4\pi L^2 r} e^{-r/L}, \tag{4.27}$$

where $\langle r^2 \rangle = 6L^2$. If $L \gg r \gg \lambda_{tr}$, $n(r) \propto 1/r$ instead of $1/r^2$ as in free space.

The total range covered by neutrons during slowing down and subsequent diffusion can be represented by the *migration length*,

$$M = (\tau + L^2)^{\frac{1}{2}}. \tag{4.28}$$

For distances greater than a few migration lengths from a fast source, the density distribution is virtually identical with that of a thermal source of the same strength.

4.3. Neutron scattering from ~1 MeV. to 1 BeV.

Techniques described in this section are applicable, with appropriate modifications of scale and detail, over a very wide energy

† Including also 'slowing down' distributions from fast sources.

range. Basic equipment (detectors, targets, etc.) is little different from that described in chapter 3 and will be treated here only when significant departures are made to meet the requirements of neutron cattering. For a review of cross-sections and techniques for neutron measurements in the 10- to 30-MeV. range, see Fowler and Brolley (1956).

When a well-collimated neutron beam traverses a thin target, elastic scattering causes target nuclei to recoil with an energy E_2 given by equation (2.3). E_2 depends on the recoil angle Θ_2, which is related to the angle of neutron scatter Θ_1 by equation (2.1). It follows that a particular elastic event is completely specified by measurable properties of the recoil particle.

If both the incident neutron energy E_0 and its direction are known, it is only necessary to record the yield of recoil particles from o to 90° in the Lab system in order to determine the relative angular distribution of neutron scattering. When the incident beam is heterogeneous in energy, which considerations of § 4.2 showed to be not infrequent, a determination of E_2 at each recoil angle fixes the relevant value of E_0 and the method remains practicable. Conversely, if the energy E_0 is single-valued, measurement of E_2 values by track length in a cloud chamber, or photographic plate or pulse height from a counter, provide sufficient information to determine the angular distribution even though the neutron flux is uncollimated. However, the latter is a situation which presents itself less often in practice.

These possibilities cover the basic principles of a variety of experimental arrangements employed in fast neutron scattering. Moreover, the observed yield of the recoil particles per unit solid angle (§ 2.4) may be transformed to relative scattering cross-sections in the CM system by equation (2.22).

However, such procedures will not determine absolute differential cross-sections without further information about either of the following: (a) Knowledge of the intensity of the neutron beam (analogous to the current collector information of §§ 3.4.1, 3.4.2). (b) The total cross-section for scattering measured in a separate experiment (see 'transmission experiments' § 2.4). In the latter case if $2\pi \displaystyle\int_\theta Kn(\theta)\sin\theta\,d\theta$ is equated to σ, where $n(\theta)$ is the relative yield at CM angle θ, the normalizing constant K is obtained.

Both procedures are used in neutron scattering, the latter rather more often, especially at high energies. (a) is available only in the special case where the source strength of neutrons has previously been measured by comparison with a standard, or can be monitored reliably. The measurement of fast neutron fluxes has been discussed in detail by Barschall, Rosen, Taschek and Williams (1952). For instance, with d-d sources, one can use the competing ^{2}H(d,p)^{3}H yield and with t-d sources the α-particle emission.

4.3.1. *Cloud chamber experiments.* As was the case for charged-particle scattering, the Wilson cloud chamber (Wilson, 1951) provided a useful tool in the early days of neutron scattering, in addition to which it has retained a significant place much longer. The theoretical prediction of spherical symmetry in the n-p distribution at energies below about 10 MeV. (cf. § 6.3) was examined by many workers, the earliest of whom all employed cloud chambers (Auger and Monod-Herzen, 1933; Kurie, 1933; Meitner and Philipp, 1934; Harkins, Kamen, Newson and Gans, 1936). However, there was marked disagreement among the results, attributable (at least in retrospect) to shortcomings of equipment and sources then available. Dee and Gilbert (1937) described the first reliable chamber experiments in this field and set the pattern for much future work.

They observed recoil proton tracks in a Wilson cloud chamber containing a mixture of 60 % methane and 40 % argon at a pressure of $\sim 3 \cdot 5$ atmospheres. Under these conditions the maximum proton range, corresponding to a head-on collision, was $3 \cdot 5$ cm.— a length which could be contained conveniently in the chamber. A stereoscopic pair of photographs was made of every track, which, when re-projected enabled the recoil angle Θ_2 and range to be measured. Knowing the recoil energy E_2 from the range in the gas, the incident neutron energy E_0 could be calculated from

$$E_0 = E_2/\cos^2 \Theta_2$$

(equation (2.3) with $m_1 = m_2$). Unless this conformed within reasonable tolerances with the known beam energy, the track was rejected as spurious.

Careful attention to the neutron source and beam purity is necessary in this kind of experiment when, in contradistinction to charged-particle scattering, only a portion of the whole event is available on the record. Spurious recoil tracks may arise from a low energy component in the beam or from the superposition of a semi-isotropic background of neutrons due to scattering in the chamber walls and surrounding equipment.† A major advantage of Dee and Gilbert's technique lay in the use of homogeneous d-d neutrons instead of the $Be(\alpha, n)$ neutrons which, if obtained from natural α-emitters, have an energy spread from o to $\sim$ 10 MeV. (§ 4.2.1). Since $E_2 \propto \cos^2\Theta_2$ and, also, tracks shorter than a certain range escape detection, there must be a tendency to lose the larger-angle recoils. The angle of cut-off will be smaller for smaller E_0 and, in the case of a spectrum with a long tail on the low energy side, this will almost certainly distort the angular distribution. Even in the case of d-d neutrons, short, steeply inclined tracks making a small projection in the plane of the chamber may not be recorded. The effect in both cases would be to deplete the apparent yield at large angles, which in the CM system would lead to a low value of $\sigma(o°)$, making $\sigma(9o°)/\sigma(o°) > 1$, even though the distribution were really isotropic. More recently, some workers have introduced a third camera, located vertically over the chamber, to provide better accuracy when measuring short tracks which are directed towards any one camera. This assures at least two good views in all cases (Laughlin and Kruger, 1948; Catala and Gibson, 1951).

Similar trouble may be encountered from high energy forward recoils due to nearly head-on collisions. Tracks lying in the plane of the chamber record satisfactorily, but those which dip appreciably may leave the illuminated volume and fail to record altogether, or record with shortened range. This tends to deplete the $\sigma(18o°)$ value in the CM distribution. Certain of the early experiments which appeared to favour strong forward scattering may have been in error from this cause.

It is essential in such experiments to determine the azimuthal distribution of tracks in every angle interval. Fig. 38 illustrates a track in the cloud chamber space, making Θ_2 with the neutron

† The back-scattered intensity of neutrons due to the surroundings of a laboratory can be estimated simply (Biram and Tait, 1950).

direction. The angle between its projection on the plane through the point of collision normal to the neutron direction and the horizontal plane is the azimuthal angle ϕ. An element of solid angle is $d\Omega = \sin\Theta_2 \, d\Theta_2 \, d\phi$. For any given value of Θ_2 and interval $\Delta\Theta_2$ there should be axial symmetry, i.e. all elements $\Delta\phi$ should contain the same number of tracks.† A polar plot of Θ_2 vs. ϕ gives the so-called 'clock diagram' fig. 39. In the case illustrated it is immediately apparent that tracks were missed in the normal direction,

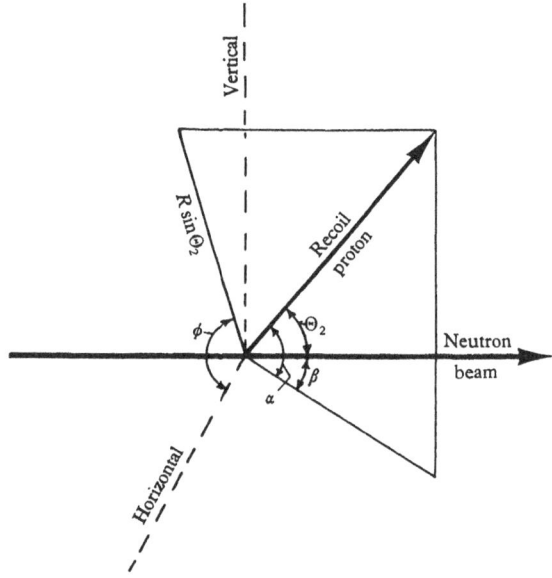

Fig. 38. Definition of angles made by proton recoil track. Θ_2 is the recoil angle, ϕ the azimuthal angle, α the dip angle and β the beam angle.

and it is not difficult to apply corrections (see Kruger, Shoupp and Stallman, 1937; Laughlin and Kruger, 1948).

An estimate of the losses to be expected can be made from geometrical considerations following the treatment of Laughlin and Kruger (1948). Suppose b is the radius of the neutron beam and $2d$ the height of the illuminated volume. Fig. 40 represents the azimuthal plane, the beam being indicated by the small circle. A recoil track originating at P in the N.E. quadrant remains in the

† This assumes the absence of any polarization effects and is therefore generally applicable to all first scattering processes where neither beam nor target particles have any preferred alignment.

region of illumination if its azimuthal projection lies between extremes PA and PB, where A and B are the points where the circle of radius $r = R \sin \Theta_2$ intersects the boundaries of the light beam. R is the track range in the chamber gas. For the quadrant shown, the chance of the track being fully illuminated is given by

$$P = \left[\int_0^b \int_0^{(b^2 - y^2)^{\frac{1}{2}}} \frac{2}{\pi} \arcsin \frac{d - y}{r} \, dx \, dy \right.$$
$$\left. + \int_0^b \int_0^{(b^2 - y^2)^{\frac{1}{2}}} \frac{2}{\pi} \arcsin \frac{d + y}{r} \, dx \, dy \right] \frac{1}{dx \, dy}. \quad (4.29)$$

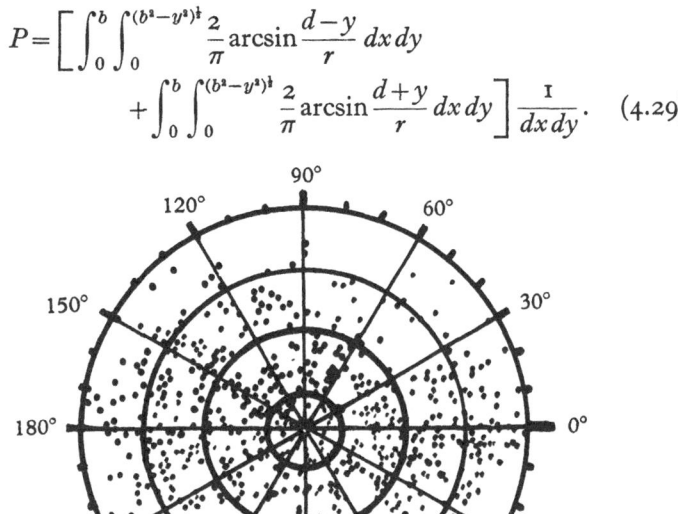

Fig. 39. 'Clock diagram' showing distribution in azimuth ϕ (angle variable 0–360°) for 635 recoil protons scattered at angle Θ_2 (radial variable). Circles correspond to $\Theta_2 = 10$, 30, 50 and 70°. Correction for loss near the normal plane increased the number of tracks to 1163 (Kruger *et al.* 1937).

P can be determined by numerical integration by dividing the quadrant into a number of segments and the azimuthal angle into a number of intervals. Two values of r must be chosen corresponding to upper and lower limits of R imposed by beam inhomogeneity, straggling, variation in chamber pressure, etc. Then the actual azimuthal distribution observed should lie between extremes calculated from the two r values.

Later studies have improved and extended the application of cloud chamber techniques to scattering, especially along the following lines: (*a*) High chamber pressures have enabled high energy interactions to be investigated. Laughlin and Kruger (1948) employed a methane-filled chamber operating at 25 atmospheres for 12 MeV. *n-p* scattering. The yield was about one track per two expansions, which occupied about 5 min. (*b*) Statistics have been improved. The labour of photographing large numbers of expansions, and measuring and analysing the tracks militates against any very generous yield of data. Whereas early experiments commonly reported entire angular distributions based on ∼ 100 tracks, it is now common to use several thousand.

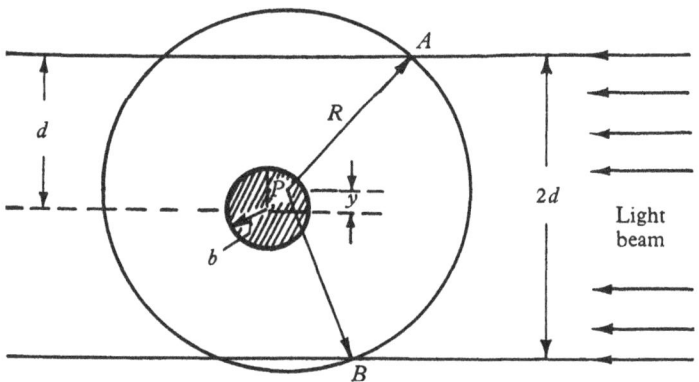

Fig. 40. Plane perpendicular to neutron direction showing azimuthal angles made by track projections. Shaded area represents the neutron beam.

n-d scattering, which has also been attempted with cloud chambers, presents special problems, viz. provision of a deuterated gas and correction for spurious tracks. Heavy methane (CD_4) has been used by most workers (Darby and Swan, 1948; Griffith, Remley and Kruger, 1950; Caplehorn and Rundle, 1951), and deuterium, or a mixture of gases including deuterium, by others (Kruger, Shoupp, Watson and Stallman, 1938; Thorndike and Wotring, 1951). Even when D_2O is used to provide the vapour, it is difficult to prevent some ^{1}H contamination which will give rise to *n-p* recoils. Mass spectrographic analysis of the gas used in one experiment showed it to comprise 85·4 % CD_4 and 9·2 % CD_3H, the balance being N_2 and O_2 (Griffith *et al.* 1950). It is highly important to have such a mass analysis available (preferably carried out on a sample taken from the actual chamber filling) and also to know the energy distribution of the neutron beam in order that the yield of possible spurious tracks may be estimated. Apart from proton recoils, inelastic n-d processes become possible at bombarding energies greater than $\sim 3 \cdot 3$ MeV. Protons from the ^{2}H$(n, 2n)p$ break-up will have a continuous energy distribution and contribute an appreciable background of spurious tracks at bombarding energies of about 10 MeV. (see also §§ 4.3.3, 8.3). Finally, there is the possibility of (n, p) reactions on heavier elements and vapours present in the chamber, in particular C, N and O. At low and intermediate energies these are not very significant as sources of spurious tracks. ^{12}C(n, p) ^{12}B and ^{16}O(n, p) ^{16}N are both highly endothermic ($Q = -12 \cdot 6$ and $-9 \cdot 6$ MeV. respectively).

Two high energy n-p experiments have been conducted with cloud chambers at Berkeley, using neutron beams produced by deuteron stripping and proton charge exchange respectively (Brueckner, Hartsough, Hayward and Powell, 1949; De Pangher, 1955). In the former of these, 90 MeV. neutrons were collimated into a $\frac{5}{8}$-in. diameter beam by a 10 ft. paraffin tube. They then entered a 16-in. cloud chamber which was operated at 110 cm. Hg pressure, using hydrogen saturated with an alcohol-water mixture. Knock-on protons were photographed by twin cameras in the usual way. The chamber was surrounded by Helmholz coils which were pulsed with 4000 A. every 2 min. The field developed to $\sim 14{,}000$ G. in 2 sec. and remained steady for $0 \cdot 15$ sec. before being cut off. During this interval the chamber was expanded and the cyclotron pulsed toward the end of the sensitive time. Very sharp tracks resulted, the energy of each proton being given by the radius of curvature of its track measured in the horizontal plane:

$$E_p = \left(\frac{e^2}{2m}\right) \left(\frac{B\rho}{\cos \alpha}\right)^2, \qquad (4.30)$$

where α is the angle of dip of the beginning of the track (fig. 38), B the flux density, m the proton mass and e its charge (e.m.u.). The slant radius of curvature actually measured, ρ_s, is related to ρ by $\rho^2 = \rho_s^2 \cos^4 \alpha$ and the dip angle is related to other angles specifying the recoil by

$$\cos \Theta_2 = \cos \alpha \cos \beta \qquad (4.31)$$

and $\qquad\qquad\qquad \tan \alpha = \tan \phi \sin \beta. \qquad (4.32)$

The neutron energy follows from

$$E_n = \left(\frac{e^2}{2m}\right) \left(\frac{B\rho}{\cos^2\alpha}\right)^2 \frac{1}{\cos^2\beta}. \tag{4.33}$$

Relativistic corrections are small at 90 MeV. and equations (4.30) and (4.31) are sufficient for most purposes (for rigorous formulae, see De Pangher, 1955).

Similar methods were used by De Pangher at 300 MeV. The important addition in both experiments is the magnetic field. This provides an alternative means of measuring particle energies without recourse to the extremely high pressures capable of stopping the particles within the chamber—clearly impracticable at these energies.

None of the cloud chamber experiments described above yields absolute cross-sections. All angular distributions have an arbitrary scale of intensity.

Diffusion chambers (Snowden, 1953) and bubble chambers (Glaser and Rahm, 1955; Oswald, 1957), which have come into use more recently than Wilson chambers, could be employed for neutron scattering in the same way, although little has been attempted so far. The bubble chamber presents a liquid target at least 100 times denser than that attainable with a high pressure vapour chamber (cf. § 3.3.3).

4.3.2. Photographic plate experiments.† Lampson, Mueller and Barton (1937) measured about 100 recoil proton tracks obtained by passing neutrons through plates. Powell *et al.* (1940) employed Ilford 'half-tone' emulsions to study the angular distribution of recoil protons due to neutrons of ~9 MeV. from a B(d, n) source. Considering the limitations of plates then available the results were surprisingly good.

Subsequent work with emulsions on neutron scattering falls into two categories: (a) When the emulsion itself serves as target as well as detector (cf. § 3.3.3). The neutron beam is passed directly through an emulsion, the hydrogen content of which becomes the target for *n-p* scattering. Recoiling protons leave tracks, the length

† For a description of the properties of photographic emulsions affecting their use in scattering, see § 3.5.5.

and orientation of which provide enough information to determine an angular distribution. As in the case of the cloud chamber, precautions must be taken against loss of tracks which make unfavourable azimuthal angles. The technique is applicable to n-p studies, and possibly n-d by loading the emulsion with heavy water. (*b*) Plates are disposed about a target in much the same way as in the multi-plate chambers described in § 3.5.5. This arrangement possesses the advantage of being applicable to any process yielding recoil particles of suitable range.

Method (*a*) has not been used appreciably, although when only weak neutron fluxes are available, the large effective solid angle presented by an emulsion is helpful. However, the method is beset by serious difficulties, of which more tedious and exacting scanning is the most discouraging. The normal content of Ilford nuclear emulsions is 3×10^{22} H atoms/cm.^{3}. A 200 μ plate exposed to a flux of 10^5 n/cm.^{2}/sec. at 10 MeV. yields $\sim$ 600 recoil tracks/ cm.^{2}/sec. An exposure of a few minutes provides suitable scanning conditions. Complications may arise from: (1) plate fogging by γ-radiation accompanying the neutron beam, (2) the difficulty of determining spatial orientation of tracks after allowance is made for emulsion shrinkage during processing, (3) inelastic processes occurring on Ag, Br, C, N, O, etc., present in the emulsion.

In very weak fluxes the effective H content may be augmented by $\sim 2 \times 10^{23}$ atoms/cm.^{3} of dry emulsion by soaking the plate in water. (The exact uptake depends on the soaking time, temperature and pH of the plate.) Loading with D_2O results likewise in a very considerable target thickness of ^{2}H atoms. However, in both cases the recoil tracks will be more tenuous and corrections for shrinkage more uncertain.

A more conventional procedure for neutron scattering is illustrated by the 14 MeV. Los Alamos experiments (Allred *et al.* 1953). The apparatus is shown in fig. 41. Neutrons from a *t*-*d* source were collimated by a $\frac{3}{4}$-in. channel through an 18-in. iron slab followed by a $1\frac{1}{4}$-in. hole through 6 in. of paraffin. Iron was chosen because of its relatively large cross-section for inelastic scattering, on account of which it degrades the neutron energy in a single collision. The paraffin further moderated the uncollimated neutrons.

The beam entered and left the vacuum space of the scattering chamber through 0·003 in. Pt windows. The target consisted of 0·007 in. Pt, on both sides of which paraffin was evaporated (ordinary in the case of n-p, deuterated for n-d work),† giving surface densities of 3–10 mg./cm.[2]. This provided essentially a double target which, if set at 45° to the beam, allowed all angles,

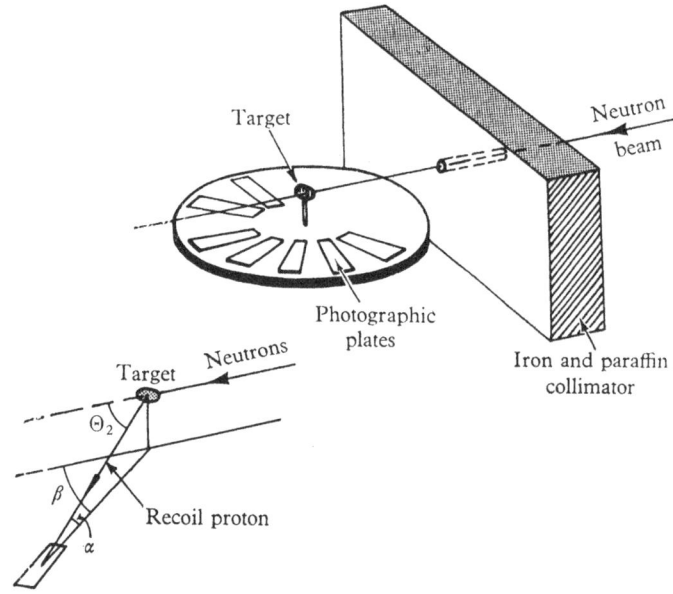

Fig. 41. Essential components of multi-plate neutron scattering chamber used by Los Alamos group at 14 MeV. (Allred *et al.* 1953).

forward and backward, to be covered in a single run. The integrated neutron flux at the target was several times 10[9] per cm.[2]. Photographic plates, 200–400 μ thick, were laid flat on the chamber floor surrounding the target. The differential cross-section for recoil can be expressed as

$$\sigma(\Theta_2) = \frac{n(A)\,R^2}{N_a N_i A \sin \alpha},\qquad (4.34)$$

where $n(A)$ is the number of acceptable tracks entering area A of plate at angle α, R is the mean distance of A from the target, N_a is

† Preparation and analysis of paraffin targets capable of remaining stable under vacuum is described in the paper.

the number of target atoms/cm.^{2} and N_i the total number of neutrons which traverse the target in the course of the run. The recoil angle Θ_2 is related to α and β by equation (4.31). N_i was calculated from the number of neutrons generated per sec. by the t-d source, which must equal the number of α-particles emitted per sec. A known fraction of the latter was counted with a proportional counter. Absolute cross-sections could therefore be obtained in this experiment.

Methods of scanning and criteria for track selection run parallel to those treated in §3.5.5. The range, dip α† and orientation β in the horizontal plane were measured for every track found on a prescribed area of each plate. The information was used to reconstruct each scattering event, whence tracks not aligning with the target could be eliminated. These might arise from (i) knock-on protons due to stray neutrons, (ii) (n,p) reactions in the chamber lining and in vapour traces remaining in the chamber, (iii) protons produced in the collimator with sufficient energy to enter the chamber. 'Blank' runs (i.e. no target) isolate some of these contributions.

By range analysis on the remaining tracks, the groups due to elastic scattering can be distinguished without difficulty. However, a fairly continuous spectrum of other ranges occurs, amounting to a few per cent of the main proton group in the case of n-p scattering. This is believed to be due chiefly to neutrons degraded in the collimator and to stray neutrons.

In the n-d case, spurious tracks arise from various factors and may amount to 10% at the worst angles. Deuterated targets generally contain some residual ^{1}H atoms which give rise to proton tracks. These are always much longer than deuteron recoils and present no difficulties. Proton tracks also originate in the deuteron break-up ^{2}H$(n, 2n)$^{1}H, which has a threshold at 3·34 MeV. The maximum energy of the proton can be calculated by assuming the two neutrons to be emitted in the same direction with equal velocities. Recoil n-p protons are not always distinguishable from these (see fig. 42). However, if the ^{1}H content of the target is established, for instance by determining the proton recoils at small angles where they can be resolved confidently (or by mass

† α is the dip angle after correction for emulsion shrinkage.

spectrometric analysis), the numbers at other angles can be calculated by assuming isotropic n-p scattering in the CM system.

Degraded neutrons, giving rise to short deuteron tracks indistinguishable from break-up protons, are a serious trouble in n-d experiments. Complete information must be obtained on the

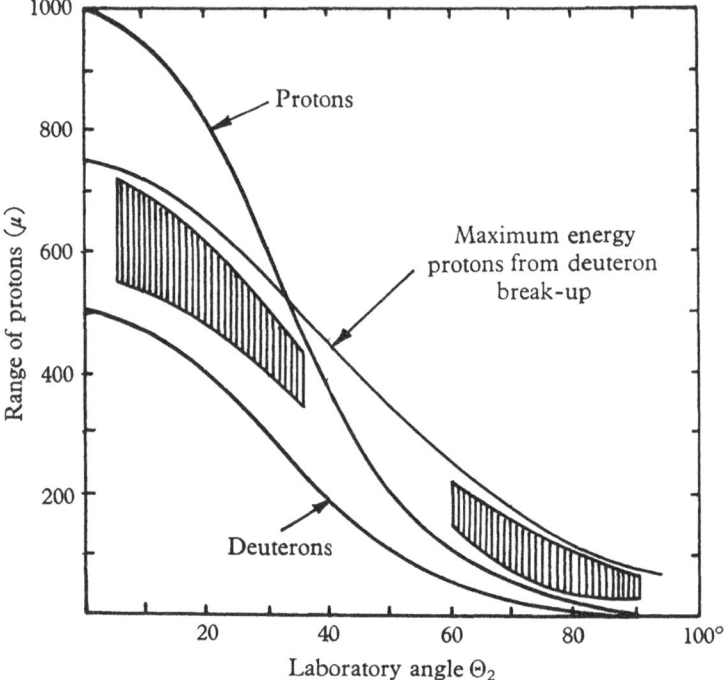

Fig. 42. Range in photographic emulsion of particles which can result from n-d scattering at 14 MeV. in the presence of some hydrogen contamination. Protons from deuteron break-up can be distinguished in the region shown shaded (Allred *et al.* 1953).

primary neutron spectrum, for example, by carrying out an n-p experiment first. Knowing the approximate form of the n-d angular distribution over the range of neutron energies present in the primary spectrum, the number of recoil deuterons corresponding to each energy interval can be estimated for each scattering angle. Since the yield of recoil protons from ^{1}H impurity can be determined fairly readily (as above), the remainder of the short tracks encountered can be attributed to deuteron break-up.

The disagreement between total cross-sections for n-d scattering as obtained by integrating over the angular distribution and by transmission is probably largely due to deuteron break-up (Seagrave, 1955; Poss, Salant, Snow and Yuan, 1952). Available information, both experimental and theoretical, is displayed in Table XI. Inelastic n-d processes are discussed further in § 8.3.

At 14 MeV. and higher energies, the 4-body case presents analogous disintegration problems: (a) $n - {}^3H$: Thresholds for $(n, 2n)$ and $(n, 3n)$ processes occur at 8·5 and 11·5 MeV., respectively. (b) $n - {}^3He$: The (n, p) reaction is exothermic. The threshold for (n, d) ${}^2H$ is 4·4 MeV., for (n, pn) ${}^2H$ 7·3 MeV. and for complete disintegration $\sim$ 10 MeV. Inelastic n-α thresholds range from about 22 to 35 MeV. Table XXI shows cross-sections for the possible processes at 90 MeV. where inelastic contributions are considerable.

Photographic chambers have not been used very widely for high energy studies, although Wallace (1951) employed a radial disposition of plates for 90 MeV. n-p scattering, resembling in general the arrangement shown in fig. 34. An exposure of 7 hr. in a flux of $\sim 10^5$ neutrons/cm.²/sec. gave sufficient track density for scanning.†

4.3.3. *Electrical counter experiments.* In general, these offer better counting statistics, their chief drawback being poorer discrimination against spurious and competing yields. Some of the earliest applications of electrical counting to neutron scattering combined target gas with detector after the fashion of cloud chamber experiments (Barschall and Kanner, 1940; Coon, Davis and Barschall, 1946; Coon and Barschall, 1946; Hall and Koontz, 1947). Instead of individual tracks being made visible, the ions or electrons were collected and recoil particle energies deduced from pulse heights after linear amplification. Both ion and electron pulse devices were used in early work, although the latter are now generally preferred. In spite of obvious disadvantages, scintillators have been applied to n-p scattering in the same way (Cross, 1952). On the whole, this technique is better suited to n-p and n-α measurements than say n-d, n-t or n-${}^3He$, the latter being subject to competing disintegration reactions which electrical discrimination alone

† An interesting by-product of this experiment was the severe fogging action of hydrogen at a pressure of two atmospheres, presumably due to chemical reduction. It was found that the blackening over a period of 8 hr. could be diminished sufficiently by reducing the temperature to about $-15°C$.

cannot be expected to handle. In such cases, it is preferable to keep the detector clear of the neutron beam by adopting the more conventional arrangements described later.

Fig. 43 shows a counter used for n-α scattering at various energies up to 2·7 MeV. (Adair, 1952). When this contained 1·5 atmospheres of He, the same pressure of A and 1·4 cm. of CO_2,† the most energetic particles had ranges of 7 mm. Gas multiplication was ~ 15 with 2000 V. on the centre wire. Guard rings were held at an intermediate potential to minimize the loss of gas amplification that occurs near the ends of the wire when terminating sections have

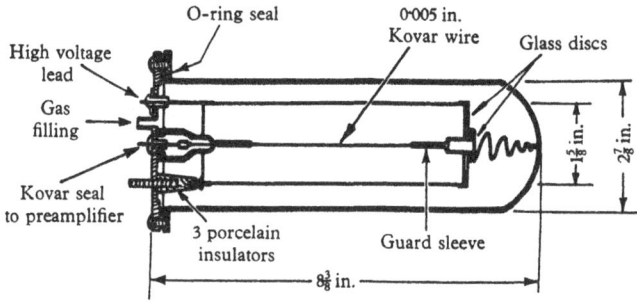

Fig. 43. Proportional counter for determining angular distribution of α-recoils by pulse height measurement. (Low energy n-α scattering, (Adair, 1952.))

the same potential as the wire (see Cockroft and Curran, 1951). The output was fed to a preamplifier and amplifier of rise time 0·5 μsec., clipping time 32 μsec. The resulting pulses were analysed by a single channel differential discriminator of channel width 1·5 V., the position of which was variable from 0 to 50 V.

Suppose $P(E_\alpha)$ is the experimental distribution of recoil energies, i.e. the number of α-particles per unit energy interval. If $\sigma(\theta)$ is the differential scattering cross-section in the CM system, the number of α-particles recoiling between θ and $\theta + d\theta$ is proportional to $\sigma(\theta) 2\pi \sin\theta \, d\theta$, or $\sigma(\theta) \, d(\cos\theta)$. By equation (A. 12), an α-particle of mass m_α, recoiling at Θ_2 (Lab.) after collision with a neutron of mass m_n and energy E_0, has an energy $E_\alpha = E_{\max} \cos^2 \Theta_2$, where $E_{\max}$ is the maximum possible recoil energy:

$$E_{\max} = 4m_n m_\alpha E_0/(m_n + m_\alpha)^2.$$

† CO_2 reduced rise times from 50 to ~ 5 μsec. With a clipping time of 32 μsec., the maximum pulse height was insensitive to variations in rise time due to small differences in electron collection time from different tracks.

Also from equation (2.23), $\Theta_2 = (\pi - \theta)/2$, hence

$$\cos\theta = (E_{max} - E_\alpha)/E_{max} \quad \text{and} \quad d(\cos\theta) = -dE_\alpha/E_{max}.$$

Thus the number of α-particles becomes proportional to $\sigma(\theta)\, dE_\alpha/E_{max}$. This is proportional to $P(E_\alpha)\, dE_\alpha$. The measured energy distribution of recoils in the Lab. system (number of pulses vs. pulse height) has the same shape as the curve of $\sigma(\theta)$ vs. $\cos\theta$ ($\theta = 0$, $\cos\theta = 1$ at zero recoil energy; $\theta = \pi$, $\cos\theta = -1$ at maximum recoil energy (see Barschall and Kanner, 1940)).

However, small edge corrections are necessary to the pulse height distribution because some of the tracks will lie partly inside and partly outside the active volume. At the ends of the electrode the electric field strength, and hence also the gas multiplication, varies somewhat unless correctly designed guard rings are used. (For the calculation of corrections for these effects, see Seagrave, 1953; Wilkinson, 1950.) At small neutron scattering angles, the α-particle recoil energy is small and pulses tend to become confused with electron and heavy atom recoil pulses. Data are not very satisfactory at small angles when this method is employed.

The energy resolution of the counter can be determined by supplying monoenergetic particles of known energy. Adair (1952) used proton pulses from the (n, p) reaction on nitrogen admitted to the counter, and also α-pulses from a Po source. In each case the width at half maximum height was $0.05E$.

A number of similar investigations have been carried out. A parallel plate ion chamber containing 16 atmospheres of He (maximum α-particle range 7 mm.) was used by Huber for n-α scattering from 3 to 4.1 MeV. (Huber and Baldinger, 1952). Seagrave (1953) succeeded in extending the method to 14 MeV. by the addition of 5.4 atmospheres of heavy 'stopping' gas, krypton. Without this, Huber's design would call for He at ~ 100 atmospheres to reduce the recoil range sufficiently. Seagrave's counter was calibrated by reference pulses provided by a ^{239}Pu deposit on the centre wire.

A superior method of investigating neutron scattering employs the conventional assembly shown in fig. 6. As used in early n-p and n-d work, the detector consisted of coincidence sets of counters or ion chambers (Tatel, 1942; Ageno et al. 1947). The technique closely resembles that of charged-particle scattering discussed in chapter 3, and will be treated here very briefly.

Fig. 44 illustrates the experimental equipment employed by the Pittsburgh group for n-p scattering at 400 MeV. (Hartzler and Siegel, 1954). The monitor telescope served to normalize the yields measured at each angle with the 3-counter telescope. The second crystal of the latter defined the acceptance angle. The copper

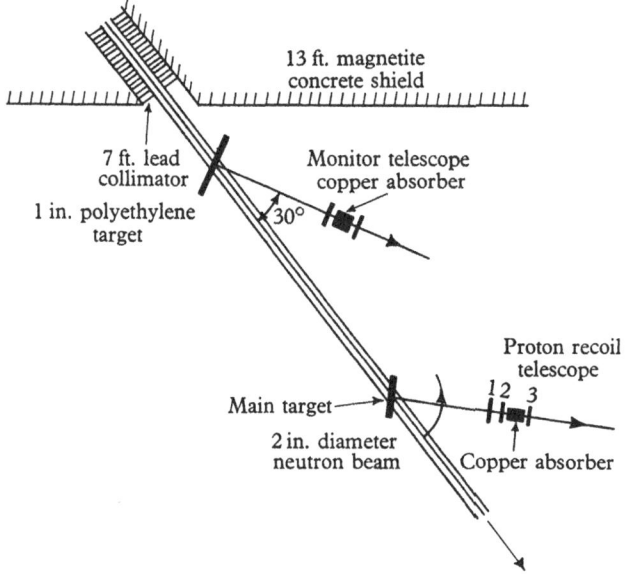

13 ft. magnetite
concrete shield

7 ft. lead
collimator

Monitor telescope
copper absorber

1 in. polyethylene
target

30°

Proton recoil
telescope

1 2 3

Main target

2 in. diameter
neutron beam

Copper absorber

Fig. 44. Experimental arrangement for 400 MeV. neutron scattering, using proton recoil telescope. 1, 2 and 3 are scintillation crystals (Hartzler and Siegel, 1954).

absorber between second and third crystals was adjusted at each angle so that only recoil protons corresponding to an incident energy above 350 MeV. would be recorded. This is a familar practice in neutron scattering, which renders the detector insensitive to the low energy tail of the spectrum (cf. fig. 37). However, as the absorber was required to vary in thickness from 0·03 to 4·45 in. at this incident energy, a large variation in efficiency occurred due to proton loss by absorption and scattering. Hence the efficiency had to be calibrated at each angle. The external proton beam from the cyclotron was scattered into the same telescope from a similar polythene target and with exactly the same geometry. Then the ratio $(1, 2, 3)/(1, 2)$ of coincidences (fig. 44) gave the efficiency under operating conditions for protons of one energy. The ratio was determined at each angle and for a set of proton energies (357–428 MeV.) which straddled the peak of the neutron spectrum. From the known spectrum the net efficiency at each angle could be calculated for the neutron experiment. In this case, it ranged from 0.35 at $0°$ to ~ 0.6 at $60°$ (Lab.).

The use of hydrocarbon targets involves subtraction at all angles of the scattering from an equivalent carbon target. Quasi-n-p scattering is the principal source of particles from carbon (cf. §3.5.4). Allowance is also necessary for inelastic n-p events, chief of which at ~ 400 MeV. is $n+p \rightarrow \pi^0 + d$. This has essentially the same angular distribution as $p+p \rightarrow \pi^+ + d$, i.e. $(0.2 + \cos^2 \theta)$ in the CM system (Hildebrand, 1953) but, according to charge independence, only half the cross-section—say of the order of 0.5 mb. at 400 MeV. At this energy σ_{np}(elastic) is ~ 34 mb., hence only small corrections are introduced. $n+p \rightarrow \pi^- + 2p$ reactions are several times less probable (Wright and Schluter, 1954). Complications from competing meson processes of course increase manyfold, at energies of the order of 1 BeV. (see Fowler *et al.* 1954).

In the Pittsburgh experiment described above, a check was made on the identity of the particles causing threefold coincidences by measuring the specific ionization dE/dx in the first crystal. Using the range-energy relation for stilbene and the $\cos^2 \Theta$ dependence of proton energy, the variation of pulse height with angle can be calculated and compared with the observed behaviour. Combined E, dE/dx telescopes can also be applied to distinguish charged products from neutron scattering (cf. §3.5.4).

Many other experiments covering a wide energy range have been carried out by essentially the same method. For further technical details see (a) n-p: Barschall and Taschek (1949); Hadley, Kelly, Leith, Segrè, Wiegand and York (1949); Brolley, Coon and Fowler (1951); Guernsey, Mott and Nelson (1952); Selove, Strauch and Titus (1953); Dzhelepov and Kazarinov (1954); Seagrave (1955); (b) n-d: Coon and Taschek (1949); Wallace (1951); Seagrave (1955); (c) n-t: Coon and Barschall (1946).

4.3.4. *Time-of-flight experiments.* These depend on the accurate timing of neutron transits over a known flight path. Application of the method to MeV. neutrons is an outgrowth of thermal velocity selectors (§4.4), made possible as techniques for measuring very short time intervals ($\sim 10^{-8}$ sec.) improved.

The source of neutrons may either be pulsed (e.g. by modulating the ion source of an accelerator producing d-d or d-t neutrons) so as to obtain short bursts of neutrons about 1 mμsec. long, or an associated charged particle (e.g. the α-particle accompanying the d-t reaction) may be detected to provide a zero-time marker. The latter arrangement avoids the necessity of pulsing the source, wastes almost none of the ion beam and yields better resolution, since this is limited only by the electronics of the associated particle counter. However it is restricted to simple neutron reactions which are accompanied by energetic recoil particles.

Scattered neutrons are detected in a counter located a known distance (~ 1 m.) from the source. The delay between zero-time signals (ion source or associated particle) and signals from the scattered particle detector measures the neutron velocity. Using a fixed flight path, the pulses may be displayed on the same linear sweep of an oscilloscope and photographed, or fed to a direct-reading chronotron. Alternatively, by using a fast coincidence circuit, the coincidence rate can be determined as a function of the length of flight path. Peaks in the coincidence curve correspond to elastic-ally and inelastically scattered groups. The overall time resolution in these experiments is typically a few mμsec. which, at 3 MeV. and over a 1 m. flight path, corresponds to $\sim 15 \%$ in energy.

A complete account of the method as applied to inelastic scattering of neutrons from a number of elements has been given by O'Neill (1954). A beam of 0·1 μA. 100 keV. deuterons bombarded a thin tritium-zirconium target (§4.2.4), yielding $\sim 10^5$ neutrons per sec. with 14·8 MeV. energy. The 3 MeV. α-particles accompanying the reaction were recorded in a scintillation counter subtending a solid angle of 1 % at the target. This provided the zero-time markers for all neutrons emitted within a narrow cone, thereby defining the neutron beam also, without additional collimation. The α-particle counting rate was about 1500 per sec. Scatterers were placed close to the target within the neutron cone and scattered neutrons detected in an organic liquid scintillator at a distance of 50–100 cm. in a direction at 90° to the incident beam. Neutron counting rates were ~ 1000 per hr. with the scatterers employed. Flight times, ranging from 20–65 mμsec., were measured with a chronotron circuit having nine 4·7 mμsec. channels, each of which registered directly.

Sources of error in this experiment were chiefly associated with (a) *lengths*: finite size of scatterer and counter; (b) *time*: finite resolution of electronics (especially transit times of photoelectrons), flight time of 14·8 MeV. neutrons inside scatterer, variation in flight time of α-particle (~ 0·3 mμsec.); (c) *primary neutrons*: energy spread (~ 90 keV.) due to spread in deuteron energies (~ 2 %) and finite size of cone of emitted neutrons. At energies greater than 1 MeV., time errors considerably exceeded all others. The full width at half maximum of the overall error function was ~ 8 mμsec. Energy resolution was therefore rather crude, particularly above several MeV., but the technique appears amenable to significant improvement. About 10 % energy resolution should be attainable at 12 MeV. with a 3 m. flight path (O'Neill, 1954). Rapid developments are also occurring in mμsec. pulse techniques for neutron spectrometry at keV. to MeV. energies (see, for instance, Various authors, 1956).

4.3.5. *Measurement of total cross-section by transmission.* Angular distribution experiments with neutrons do not in general lead directly to absolute cross-sections, but are normalized to the total cross-section which is measured in a transmission experiment (see also § 2.4). Two arrangements, typical of low and high energy experiments respectively, are shown in fig. 45. That shown in (a) was used to measure the attenuation by C, H, D, O and N at 14·1 MeV. (Poss *et al.* 1952). The procedure, which is a quite general one, was to determine the exponential decrease in intensity of a highly collimated beam as a function of sample thickness. In some cases, e.g. for H, the result can be obtained most conveniently from the difference in transmission of two substances. Fig. 46 shows the data for pure graphite and polyethylene $(CH_2)_n$. By taking $-\ln T / \sigma N_v x$, one obtains σ_C and $(\sigma_C + 2\sigma_H)$ and hence σ_H.

The neutron beam must be monitored in order to normalize each set of data and this was accomplished by counting recoil

α-particles from the t-d source. The overall accuracy of transmission experiments is often high. Good counting statistics ($\sim$0·3 % in the present case) result from the fact that transmitted rather than scattered intensity is being measured. A usual source of error which warrants a small correction even in arrangements

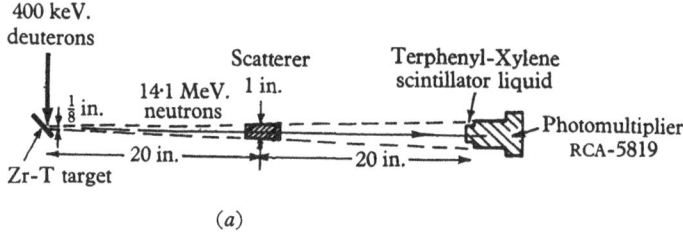

(a)

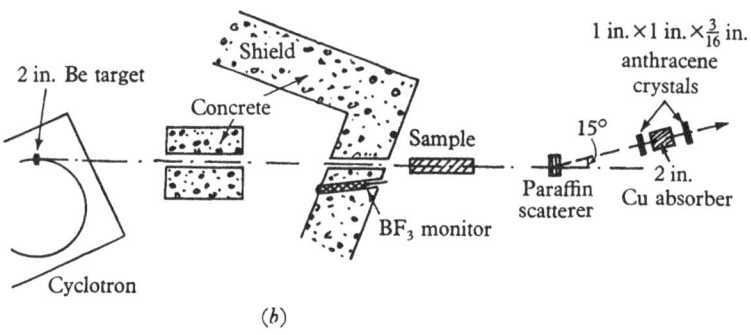

(b)

Fig. 45. Typical arrangements for measuring total neutron cross-sections by transmission. (a) at 14·1 MeV. (Poss *et al.* 1952), (b) at 280 MeV. (Fox, Leith, Wouters and MacKenzie, 1950).

having 'good geometry', arises from the so-called 'scattering-in' effect. The apparent transmission through a sample may be written as

$$T = T_0 \left(1 + \frac{\Delta T_1}{T_0} + \frac{\Delta T_2}{T_0} + \ldots \right), \qquad (4.35)$$

where T_0 is the true transmission, ΔT_1 the distortion due to singly scattered neutrons, ΔT_2 that due to double scattering, etc. All the corrective terms have the effect of increasing the apparent transmission, i.e. decreasing the measured cross-section. In 'good geometry' only the first order term introduces an appreciable

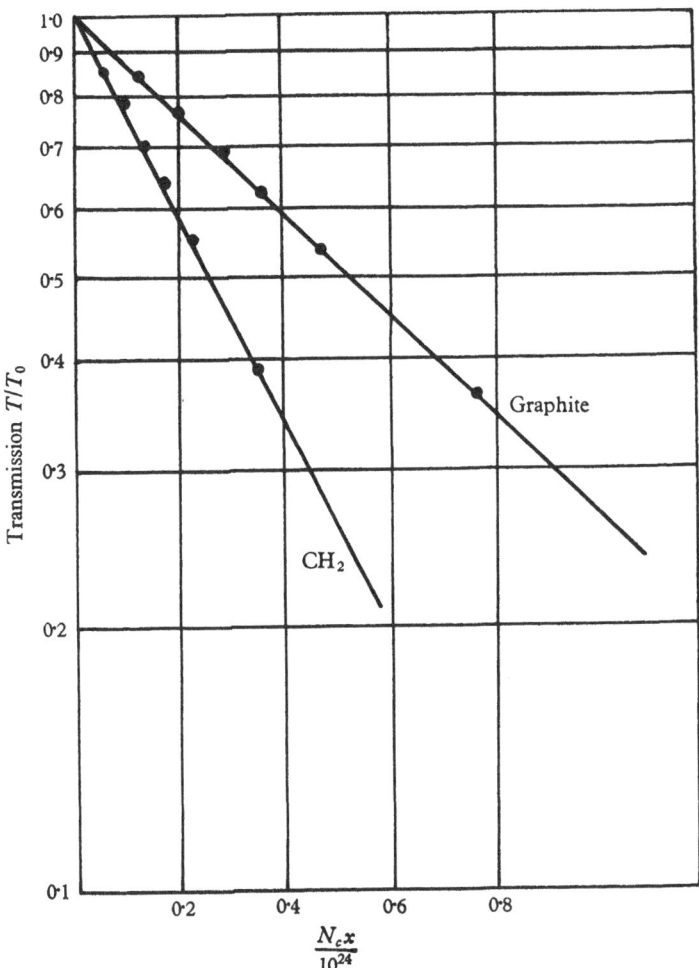

Fig. 46. Transmission of 14·1 MeV. neutrons through graphite and poly-ethylene vs. thickness in units of number of carbon atoms/cm.². σ_H is obtained by difference (Poss *et al.* 1952).

correction. If L_1 is the source to scatterer distance, L_2 that from scatterer to detector and A the area of the scatterer, then

$$\frac{\Delta T_1}{T_0} \approx \sigma(0) N_v x A \left(\frac{L_1+L_2}{L_1 L_2}\right)^2. \qquad (4.36)$$

This expression has a minimum when $L_1 = L_2$, which is the geometry generally adopted. From equation (2.4), $d\sigma/\sigma = dT/T \ln T$.

MES

Applying this to equation (4.36), the first order correction to the cross-section becomes

$$\frac{\Delta\sigma}{\sigma} \approx \frac{\sigma(0)}{\sigma} A \left(\frac{L_1 + L_2}{L_1 L_2} \right)^2. \qquad (4.37)$$

$\sigma(0)$ is the differential cross-section for forward scattering, which is usually evaluated (Bethe and Placzek, 1940) by treating it as a diffraction or 'shadow' effect (see also § 12.1):

$$\sigma(\Theta) = R^2 \left[\frac{J_1(R\Theta/\lambda)}{\Theta} \right]^2, \qquad (4.38)$$

where R is the nuclear collision radius, λ the neutron wavelength/2π and J_1 a Bessel function of the first kind. With 'good geometry' it is almost invariably justifiable to approximate J_1 by the first term of the expansion and write $J_1 \approx R\Theta/2\lambda$. Poss *et al.* (1952) quote the correction to the transmission as varying from 0·04 to 0·2% over the range of sample thickness used. As shown in fig. 45 (*a*), each piece was machined accurately to a truncated cone which matched the divergence of the beam, thereby minimizing the 'scattering-in'. Further details of the sources of error and corrections involved in transmission experiments are described in numerous papers, (Amaldi, Bocciarelli, Cacciapuoti and Trabacchi, 1946; Sleator, 1947; Poss *et al.*, 1952; Hafner, Hornyak, Falk, Snow and Coor, 1953).

The steadily improving accuracy of transmission experiments is due to the use of more efficient detectors and the availability of higher fluxes, both of which make for better counting rates and geometry. Detectors of many types have been tried, including gas counters, activation foils, fission chambers and crystal and liquid scintillators. For high energy monitoring, the Bi fission chamber (threshold 60 MeV.) has been extensively used (Wiegand, 1948). When suitable monoenergetic sources are not available, a lower limit cut-off can be obtained by using threshold detectors, based generally on $(n, 2n)$ and, to a lesser extent, (n, p) processes (see tabulations, Cohen, 1951; Feld, 1953). A non-homogeneous source of neutrons of maximum energy E_m may be combined with a detector which responds only to neutrons of energy greater than E_{thr}. If the latter is somewhat less than E_m, the response of the detector occurs only in the interval $E_m - E_{thr}$. Salant and Ramsay (1940) used the

Li(d, n) continuum, for which $E_m \approx 15$ MeV., in combination with the ^{63}Cu($n, 2n$) ^{62}Cu activation reaction which has a threshold at $\sim 11\cdot4$ MeV. ^{62}Cu decays by β^+ emission with a convenient 10 min. half-life. The $(E - E_{\mathrm{thr}})^2$ dependence of the ($n, 2n$) reaction throws emphasis on the higher energies in the interval (Fowler and Slye, 1950). Other convenient detectors are ^{107}Ag($n, 2n$), $E_{\mathrm{thr}} = 9\cdot6$ MeV., half-life $24\cdot5$ min.; ^{12}C($n, 2n$), $E_{\mathrm{thr}} = 20\cdot3$ MeV., half-life 21 min. (Sherr, 1945).

Fig. 45 (b) illustrates the apparatus of Fox et al. (1950) for the determination of total cross-sections at 280 MeV. Here the detector is a 2-crystal scintillation telescope directed towards a paraffin converter and making an angle of 15° to the neutron beam. The beam monitor consists of a BF_3 slow neutron counter set into the concrete shield, its rate being proportional to the fast neutron intensity. The solid angle was 4×10^{-4} sterad., which was sufficiently 'good' to avoid scattering-in corrections in the case of light elements but required an addition of $\sim 1\%$ for heavy elements.

One of the problems of high energy neutron work arises from the lack of homogeneity of the neutrons (§§ 4.2.5, 4.2.6). In a transmission experiment one measures $\int n(E) p(E) \, dE$ with the sample out and $\int n(E) p(E) \, e^{-\sigma(E) N_v x} \, dE$ with the sample in, where $n(E) \, dE$ is the number of neutrons with energies between E and $E + dE$, $p(E)$ is the probability of detection and $\sigma(E)$ is the total cross-section at energy E. If $n(E)$ is single-valued, the ratio of these two integrals yields σ immediately. The difficulty presents itself in the case of a heterogeneous spectrum, which becomes distorted as it traverses the sample, higher energy neutrons having a smaller cross-section for scattering. The same phenomenon is troublesome at thermal energies where it is referred to as 'hardening of the beam'. The function of absorber A in the telescope of fig. 45 (b) is to establish a minimum energy below which recoil protons cannot reach the second counter. This serves the same purpose as an ($n, 2n$) threshold detector. Thus, the telescope can be set to respond only to a selected upper edge of the proton (and hence neutron) spectrum, e.g. 240–280 MeV. in Fox's experiment. The same principle has been used for neutron transmission studies in the BeV. range (Coor et al. 1955).

4.4. Scattering at very low energies

Interactions of neutrons with nuclei have been most intensively investigated from 0 to 10 eV. and, with greater technical difficulty, to ~ 10 keV. Machine sources using endothermic reactions (§ 4.2.4) will generate approximately monoenergetic neutrons down to keV. energies. Moreover time-of-flight techniques discussed below in connexion with slow neutron measurements involving μsec. timing, are rapidly being extended to MeV. energies by mμsec. timing devices. There is thus a continuous coverage of the spectrum from thermal to BeV. energies. However, resolution is still poorest and detectors least satisfactory in the 100 eV.–100 keV. range. The following discussion applies chiefly to neutrons below ~ 100 eV., a region which is notable for the numerous sharp resonances observed in the absorption spectra of medium weight and heavy nuclei. Fundamental data concerning the free n-p cross-section have also been obtained with slow neutrons.

4.4.1. Velocity selectors. Devices which single out neutrons of a particular energy from a continuous spectrum of energies are known as 'velocity selectors', or, if they make available a beam of monoenergetic neutrons, 'monochromators'. The first of these was built in 1935 but inadequate source strength (Ra-Be) limited its application to thermal neutrons (Dunning, Pegram, Fink, Mitchell and Segrè, 1935). Post-war developments in this field have been prodigious, due chiefly to the intense thermal and slowing down (dE/E) fluxes made available by reactors (§§ 4.2.3, 4.2.7). Partly moderated neutrons produced by high current accelerators also compare favourably in intensity at keV. energies and are widely used.

Although selectors of many types are in use, the details depending largely on the source, only two distinct principles are involved.

Time-of-flight selector.† These isolate neutrons of a particular velocity by recording only those that traverse a known distance in a prescribed time. A pulse of neutrons emitted during time δt(~ 1 μsec.) is recorded by a detector, usually an enriched ^{10}BF$_3$ ionization chamber or proportional counter,‡ which is placed at

† See also § 4.3.4 for time-of-flight at MeV. energies.
‡ ^{10}B(n, α) ^{7}Li has a $1/v$ dependence with a large cross-section, 750 b. at 2200 m./sec.

a distance d from the source and modulated at the same frequency as the source but with a time lag t. The detector is sensitive for a short time δt (ideally $\delta t \ll t$) which allows it to register only neutrons having velocity d/t. Since path lengths are usually 5–50 m., the flight time† for 100 eV. neutrons is $\sim 100\,\mu\text{sec}$., which is easily measured electronically.

In practice, neutrons from a single burst are counted in successive time 'channels', $t + \delta t$, $t + 2\delta t$, ..., which enables a wide range of energies to be covered simultaneously. If the channel time is made the same as the pulse duration, the maximum counting rate for any particular resolution of the system is obtained. δt is adjustable from a few μsec. to any desired larger value and the number of channels may be 100 or even 1000. After the slowest channel has counted, another neutron burst is emitted and so the whole process continues cycling. Transmission measurements can be made for each channel by interposing the sample in the flight path.

The usefulness of a selector is determined by its energy resolution and counting rate per channel, the design constants requiring a compromise between these two features. The resolution in energy is

$$\frac{\Delta E}{E} = \frac{2\Delta v}{v} = -\frac{2\Delta t}{t}. \qquad (4.39)$$

The uncertainty in flight time Δt includes the length of neutron pulse (part of which is the slowing down time in the case of machine sources) and the channel time, both δt. Actual resolution depends on the shape of the counting-rate vs. flight-time function, which usually resembles an isosceles triangle of base Δt. $\frac{1}{2}\Delta t$ would therefore be more consistent with usual definitions of resolution.

Energy resolution must clearly depend on the flight path d, but choice of the latter is bound up with the counting rate per channel, which depends on $\delta t/d^2$. Hence Δt is commonly reckoned per unit flight path and adopted as a convenient measure of the inherent resolution 'quality' of a selector. If Δt is expressed in μsec./m. and energy in eV., then $\Delta E/E = -0.028E^{\frac{1}{2}}\Delta t$. Typical figures are 0.05–1 μsec./m. (see tables, Hughes, 1953; Feld, 1953).

† For neutrons of energy E (eV.), the flight time per unit path length is $71.5/E^{\frac{1}{2}}$ μsec./m.

Techniques for producing pulsed beams can be classified according to the neutron source:

(a) *Pile neutrons.* Here a fast, rotating shutter or 'chopper' which is alternately transparent and opaque to neutrons, is placed in the beam. Thermal (slow) choppers make use of the high cadmium absorption (Fermi, Marshall and Marshall, 1947) and resonance (fast) choppers use a large thickness of steel, e.g. 16 in. in the Argonne selector (cf. Seidl, Hughes, Palevsky, Levin, Kato and Sjostrand, 1954).

(b) *Machine sources.* These include positive ion accelerators, betatrons (Yeater, Gaerttner and Baldwin, 1953) and linear accelerators. In the case of the positive ion machines the usual practice is to modulate the ion source† (Baker and Bacher, 1941; Rainwater and Havens, 1946). Good resolution has been obtained with *bremmstrahlung*-neutron sources (§ 4.2.2) based on linear accelerators, which give short pulses of very high peak current (Merrison and Wiblin, 1951; Feld, 1951; Hodgson, Gallagher and Bowey, 1952). (For a general discussion of pulsed sources, see Taylor and Havens, 1949; Feld, 1953).

Crystal selector. In principle, this resembles an X-ray spectrometer. It makes use of the fact that neutrons of velocity v show wave properties corresponding to a wavelength $\lambda = h/mv$, where h is Planck's constant and m the neutron mass. In terms of energy, $\lambda = 0.287/E^{\frac{1}{2}}$ if E is in eV. and λ in Å. A beam of resonance neutrons from the pile falls upon a large single crystal (LiF, Be, etc.) at a small glancing angle and a counter is placed so as to detect the reflected beam (Zinn, 1946; Borst, Ulrich, Osborne and Hasbrouck, 1946; Sawyer, Wollan, Bernstein and Peterson, 1947; Sailor, Foote, Landon and Wood, 1956).‡ Diffraction maxima occur when the Bragg relation is satisfied:

$$n\lambda = 2d\sin\alpha, \qquad (4.40)$$

where d is the distance between atomic planes involved in the reflexion and α is the glancing angle for the nth order. For example, $d = 2.32$ Å. for the (1, 1, 1) planes of LiF. For $n = 1$, $\alpha = 13°$ at 0.1 eV., $3.5°$ at 1 eV. and $1.1°$ at 10 eV.

---

† For example, by pulsing the arc of a cyclotron. However, much faster modulation can be effected (e.g. Van der Spuy, 1956).

‡ Sailor *et al.* (1956) also gives a useful bibliography of selectors in general.

The resolution is determined by the quality of the collimation. Differentiating equation (4.40) leads to $\Delta\lambda/\lambda = \cot\alpha \cdot \Delta\alpha \approx \Delta\alpha/\alpha$ when α is small. In terms of energy,

$$\frac{\Delta E}{E} = -\frac{2\Delta\lambda}{\lambda} \approx -2\frac{\Delta\alpha}{\alpha}. \qquad (4.41)$$

Since $\Delta\alpha$ is, at best, about 0·1°, the energy resolution for the case quoted would be 1·5 % at 0·1 eV. and 18 % at 10 eV. Better results have been obtained using Be which has closer lattice packing. Borst and Sailor (1953) quote 2% at 1 eV. and 16% at 50 eV. for the Brookhaven instrument. Notice that $\Delta E/E \propto E^{\frac{1}{2}}\Delta\alpha$, which is analogous to the energy dependence of the time-of-flight selector's resolution.

Both types of selector are used primarily for transmission measurements of total cross-section (§ 4.3.5). Direct measurements of scattering are difficult because of insufficient intensity, although the higher flux reactors now coming into use will improve the position considerably. Activation cross-sections can be measured with the crystal type only, since this is also a monochromator. The crystal instrument, however, has a more restricted range $\sim$ 0·01–100 eV. At sub-thermal energies, higher order reflexions of more energetic neutrons which are present in greater abundance tend to interfere. Some attempts to remove these have been made with neutron filters or mechanical monochromators in conjunction with crystal selectors. This would extend the range to below 0·001 eV. At the upper limit, fast choppers and pulsed sources yield better resolution beyond a few eV. Moreover the reflected intensity from a crystal varies as $1/E$.

Fast choppers are restricted by practical speeds of rotation to an upper limit of a few keV. Pulsed machine sources can be used from 0·001 eV. up to MeV. energies (§ 4.2.4). At high energy the chief difficulty is to produce sufficiently short ($\sim m\mu sec.$) pulses and to measure the short flight times involved. At thermal energies, the diffusion time of neutrons within the moderator is the main contribution to Δt (and may exceed 20 $\mu sec.$). Slowing down times from, say, a few MeV. to $\sim$ 100 eV. are less than 1 $\mu sec.$ and therefore relatively unimportant.

4.4.2. *Sub-thermal energies.* 'Cold' neutrons having a Maxwellian distribution corresponding to a subnormal temperature can be obtained by moderating neutrons down to equilibrium with a medium which is maintained at low temperature (cf. §6.10). A simpler method is to filter out the high energy component of ordinary thermal neutrons.

From the Bragg law, equation (4.40), it is evident that for neutrons of wavelength exceeding $\lambda_c = 2d$, no possible angle of scattering exists except $\alpha = 0$. Such neutrons, having an energy less than $E_c = h^2/2m\lambda_c^2$, will be transmitted by a crystal, providing also that nuclear capture is negligible at sub-thermal energies (Wick, 1937; Anderson, Fermi and Marshall, 1946). Typical values of E_c range from 0·0052 eV. for Be to 0·00052 eV. for Mn. The graphite cut-off occurs at 0·00183 eV. When a neutron beam enters a long column of the filter (e.g. ∼40 cm. of Be or BeO), randomly oriented single crystals cause many Bragg reflexions which attenuate all energies above E_c. Although some attenuation of the lower energy neutrons also occurs, due to crystal imperfections, thermal motions and capture, the emergent beam will consist chiefly of a useful band of sub-thermal energies. Such neutrons may be used for transmission measurements in the usual way. Cold neutrons can also be produced by total reflexion from certain mirrors (Fermi and Marshall, 1947). Transmission of a neutron beam through a scattering medium can be described in terms of a refractive index,

$$n = 1 - \frac{\lambda^2 N_v a}{2\pi}, \qquad (4.42)$$

where N_v is the number of atoms per unit volume of the medium, λ the neutron wavelength and a the scattering length (§6.6). n is negative if a is positive and total reflexion occurs. The limiting angle for a neutron beam incident from air is

$$\theta_c \approx [2(1-n)]^{\frac{1}{2}} = \lambda \left(\frac{N_v a}{\pi}\right)^{\frac{1}{2}}, \qquad (4.43)$$

which is usually about 10 min. of arc. For glancing angles less than this, total reflexion occurs.

Important applications of sub-thermal neutrons are discussed in §6.9 *et seq.*†

† For a complete discussion of experimental technique see Hughes (1953).

169

CHAPTER 5

DETERMINATION OF BEAM ENERGY

5.1. Introduction

In all scattering experiments it is necessary to determine the energy of the incident particles at the target. However, it is not generally convenient to make the measurement exactly at the target position, hence in charged-particle scattering small corrections have to be applied for absorption by entrance windows, target gas, etc.† Moreover, as the target itself has finite thickness, a mean collision energy must be found. Each of these effects is likely to contribute a correction of $\sim$ 10 keV. in the region of low energy scattering where highest precision has been attained so far. The problem of reproducibility of an energy calibration is equally important, for unless the energy is measured during the course of the actual run, the accelerator or analyser must be relied upon to maintain its calibration. In some experiments, these aspects of the scattering problem have been treated rather summarily.

The degree of precision required depends on the particular interaction being studied and also on the range of energy and angle involved. In low energy charged-particle scattering, $\sigma(\theta)$ varies as E^{-2} at small angles (equation (3.3)). On the other hand, in the region of specifically nuclear scattering, p-p cross-sections vary approximately as E^{-1}. Jackson and Blatt (1950) have pointed out that over the region of Coulomb-nuclear interference (§ 7.6), the energy in p-p scattering should be known to 0·1 % even though the yield is measured to 0·5–1 %, otherwise spurious indications of p- or d-phase-shifts result. It is difficult to stipulate any general rule for the accuracy needed in other cases without recourse to a similar detailed analysis. However, from the experimental point of view, there is much to be said for determining $d\sigma(\theta)/dE$ at all θ in order that an appropriate tolerance (corresponding, say, to 1 % of $\sigma(\theta)$) can be applied to E at each angle investigated.

† The material of this chapter applies in the main to precision charged-particle scattering. Measurement of neutron energy, except in a few special cases noted in the text, reduces to a measurement of charged-particles, e.g. recoil proton energy.

When the effects of resonance are superposed upon potential scattering, $\sigma(\theta)$ tends to become unduly sensitive to changes in E. For instance in p-α scattering, the broad resonance due to ^{5}Li causes $\sigma(\theta)$ to change rapidly in the low energy region both at large and small CM angles (Freier, Lampi, Sleator and Williams, 1949). (90° scattering is fairly insensitive to energy.) At 170°,

$$d\sigma(\theta)/dE = 0\cdot4\text{--}0\cdot5 \text{ b./sterad./MeV.} \quad \text{at} \quad 1\cdot7 \text{ MeV. and } 0\cdot3 \text{ b.}$$

At 25°, $d\sigma(\theta)/dE$ is $0\cdot8$–$0\cdot9$ at $2\cdot25$ MeV. and $0\cdot67$ b. In both cases E must be controlled to ~ 7 keV. if $\sigma(\theta)$ is to be correct to 1 %, but in practice ΔE has often been allowed to rise to ~ 20 keV.

In the following sections we discuss the methods available for determining primary beam energies and tabulate useful calibration data.

5.2. Calibration at low energies

In the case of direct-type accelerators (H.V. sets and Van de Graaff generators) operating below about 1 MeV., absolute electrometers and voltmeters of several types have been utilized, some of which are capable of measuring voltage to $0\cdot1$ % (see review by Craggs and Meek, 1954, ch. 7). Although useful for an approximate measure of beam energy and for linear interpolation between reference voltages, or voltage control (Herb *et al.* 1939; Cooper *et al.* 1954), they are inadequate for absolute energy measurements. Strictly speaking, they measure the potential developed by the generator rather than the kinetic energy of ions actually accelerated down the tube and extracted from the machine. Owing to several factors, such as the use of an auxiliary focusing potential, or a potential to extract the beam from the ion source, the particle energy may not be the same as that inferred from the current through a resistance chain, or from a generating voltmeter. The usual practice now is to deflect the beam after it emerges, using either an electrostatic or a magnetic analyser of known dimensions and field strength. The former is the more limited in the energy range which it can handle without the application of excessive potentials between the quadrant plates (cf. §3.2.2).

A primary calibration of the electrostatic analyser requires that the voltage between the deflecting plates be determined with reference to a standard cell. The procedure has been described fully by Herb, Snowdon and Sala (1949) for the apparatus shown in fig. 10, and has been employed successfully up to about 4 MeV. (Worthington *et al.* 1953). The field strength of a magnetic analyser can be found from the e.m.f. generated by a flip coil in the field, or in terms of the magnetic moment of the proton (which is known to 0·01 % (Gardner and Purcell, 1949; Poss, 1949)) by the magnetic resonance method (Bloembergen, Purcell and Pound, 1948). The procedure and the important sources of error have been discussed by Collins, McKenzie and Ramm (1953).

If the (non-relativistic) velocity of a particle of charge e and mass m is v, then the radius of curvature ρ of its path perpendicular to a field of strength B is given by

$$Bev = \frac{mv^2}{\rho}, \qquad (5.1)$$

whence
$$(B\rho)e = p, \qquad (5.2)$$

where p is the momentum. If e is in e.m.u., B in gauss and v and c (the velocity of light) both in cm./sec., then ρ is given in cm. $B\rho$ is the so-called 'magnetic rigidity' and is a measure of the momentum per unit charge of the ion. The kinetic energy T (relativistic) can be derived from $B\rho$ as follows:

Momentum is related to total energy mc^2 and rest energy $m_0 c^2$ by the familiar relation (Appendix A, equation (A. 26))

$$(pc)^2 = (mc^2)^2 - (m_0 c^2)^2, \qquad (5.3)$$

which can be written in the form

$$\frac{T}{m_0 c^2} + 1 = \left[\left(\frac{p}{m_0 c} \right)^2 + 1 \right]^{\frac{1}{2}}, \qquad (5.4)$$

where $T = (mc^2 - m_0 c^2)$. Substituting for p from equation (5.2) gives

$$\frac{T}{m_0 c^2} + 1 = \left[\left(\frac{B\rho e}{m_0 c} \right)^2 + 1 \right]^{\frac{1}{2}}. \qquad (5.5)$$

T is in ergs if the other quantities are expressed in the units quoted above. For low energy particles, $(B\rho e/m_0 c)^2 \ll 1$ and the term in square brackets may be expanded thus:

$$\left. \begin{aligned} \frac{T}{m_0 c^2} + 1 &= \left[1 + \frac{1}{2} \left(\frac{B\rho e}{m_0 c} \right)^2 - \frac{1}{8} \left(\frac{B\rho e}{m_0 c} \right)^4 + \dots \right], \\ T &= \frac{(B\rho e)^2}{2m_0} \left[1 - \frac{1}{4} \left(\frac{B\rho e}{m_0 c} \right)^2 + \dots \right]. \end{aligned} \right\} \qquad (5.6)$$

The square brackets contain the relativity correction, which amounts to about 0·1 % for 2 MeV. protons. Writing E for the classical kinetic energy, then $E = (B\rho e)^2/2m$. For $B\rho$ in G-cm., the following expression gives E in MeV.: $E = (B\rho/10^5 k)^2$, where $k = 1·445$ for protons, 2·043 for deuterons and 1·440 for α-particles.

The great advantage of magnetic analysers is their wide range of applicability, not only in the low energy field but at the highest energies of interest in scattering (§ 5.5).

Besides electrostatic and magnetic deflexion methods, a third absolute method capable of a precision of 0·1 % is known. Here the absolute velocity of ions is measured by timing their transit between two radio frequency gaps, or through a resonant cavity. The kinetic energy of the particle is obtained in terms of its known mass and readily measurable parameters of length and frequency (Altar and Garbuny, 1949; Shoupp, Jennings and Jones, 1949).

The energies of a number of reaction thresholds and sharp γ-ray resonances have been fixed by one or more of the above methods. Some of these are collected in Table VI. The most reliable and convenient are probably the ^{7}Li(p, n) threshold and the ^{27}Al(p, γ) and ^{19}F$(p, \alpha'\gamma)$ resonances. These, or other values believed to be sufficiently reliable, then serve as fixed points on the high voltage scale and may be used in turn for calibration.

A valuable extension of the calibration of a magnetic analyser can be obtained by taking advantage of the multiplicity of beams available from H.V. sets or Van de Graaff generators. Using hydrogen, deuterium, tritium or a combination of these produced by feeding a mixture of the gases to the ion source, 16 different molecules can be formed, if we include the triatomic species (see footnote, p. 42). All masses from 1 (H$^+$) to 9 (TTT$^+$) occur, some in different ways. Since $\rho = (2Em)^{\frac{1}{2}}/Be$ (see above), a mass 9 mole-cule accelerated to 1 MeV. has the same ρ as a 9 MeV.† proton. Hence a calibration at very low energies (e.g. by using established resonances) can be extrapolated to considerably higher energy.

Two other sources of fixed points on an energy scale are avail-able: (a) energies of α-particles from natural emitters; (b) known Q-values of nuclear reactions.

† Based on atomic mass numbers. For precise extrapolation exact ionic masses must be used.

The α-particle energies cover a range up to $\sim 8\cdot 8$ MeV. (ThC') and in some cases are known with great precision. The subject has been reviewed by Briggs (1954), whose best values for several sources appear in Table VII. Protons and α-particles of the same energy and deuterons of half the energy all have the same $B\rho$, hence a proton calibration is available to the ThC' energy.

TABLE VI. *Energy values of well-established calibration points from reactions among light nuclei*

(Chief sources: Fowler and Lauritsen, 1949; Richards, Smith and Browne, 1950; Bonner and Butler, 1951; Jones, Douglas, McEllistrem and Richards, 1954; Ajzenberg and Lauritsen, 1955; Marion *et al.* 1955.)

Reaction	Energy (keV.)
$^{19}F(p, \alpha'\gamma)$	$340\cdot 0 \pm 2$
$^{7}Li(p, \gamma)$	$441\cdot 4 \pm 0\cdot 5$
$^{12}C(p, \gamma)$	$456\cdot 0 \pm 2$
$^{13}C(p, \gamma)$	$554\cdot 0 \pm 2$
$^{19}F(p, \alpha'\gamma)$	$873\cdot 5 \pm 1$
$^{9}Be(d, n)$	$920\cdot 0 \pm 2$
$^{19}F(p, \alpha'\gamma)$	$933\cdot 0 \pm 1$
$^{9}Be(p, \gamma)$	$998\cdot 0 \pm 4$
$^{3}He(p, n)$	$1020\cdot 3 \pm 1\cdot 5$
$^{7}Li(p, p'\gamma)$	$1030\cdot 0 \pm 5$
$^{9}Be(p, \gamma)$	$1087\cdot 0 \pm 2$
$^{19}F(p, \alpha'\gamma)$	$1092\cdot 0 \pm 2$
$^{16}O(d, n)$	$1833\cdot 0 \pm 3$
$^{7}Li(p, n)$	$1881\cdot 4 \pm 1\cdot 1†$
$^{9}Be(d, n)$	$1916\cdot 0 \pm 4$
$^{14}N(d, n)$	$1967\cdot 0 \pm 4$
$^{9}Be(p, n)$	$2059\cdot 0 \pm 2$
$^{16}O(d, n)$	$2393\cdot 0 \pm 4$
$^{11}B(p, n)$	$3015\cdot 0 \pm 3$
$^{13}C(p, n)$	$3236\cdot 0 \pm 3$

† This threshold, which has provided the most widely accepted secondary standard, is quoted here from Jones *et al.* (1954). A weighted mean of various determinations is $1881\cdot 1 \pm 0\cdot 5$.

Reactions which yield clearly resolved groups of particles can be employed in much the same way. The important quantity is the Q-value,

$$Q = \left(\frac{m_1 - m_4}{m_4}\right) E_1 + \left(\frac{m_3 + m_4}{m_4}\right) E_3 + \left[\frac{E_1^2 + E_3^2 - E_4^2}{2m_4 c^2}\right]$$

$$- 2(m_1 m_3 E_1 E_3)^{\frac{1}{2}} \cos \Theta \left[\left(1 + \frac{E_1}{2m_1 c^2}\right)\left(1 + \frac{E_3}{2m_3 c^2}\right)\right]^{\frac{1}{2}} \bigg/ m_4. \quad (5.7)$$

The quantities are the same as defined in §4.2.4. Depending on the precision required, the terms set in square brackets may often be omitted if the kinetic energies of particles are small compared with their rest energies. Taking the non-relativistic version, and recasting into a more convenient form for calculating the energy of the emitted particle E_3 when Q is known, equation (5.7) leads to equation (4.8). A considerable number of disintegration energies are known to better than 1 in 1000. Some of these appear in Table VIII. They are particularly useful for calibrating a scale at higher energies when the bombarding energy E_1 of a low energy machine is known, e.g. from the data of Table VI. A detailed discussion of sources of error affecting energy determinations of reaction products with electrostatic and magnetic analysers is contained in the appendix of a paper by Brown, Snyder, Fowler and Lauritsen (1951).

TABLE VII. *α-particle Bρ and energy values useful for calibration* (Briggs, 1954)

(Errors quoted are standard deviations.)

Isotope	$B\rho$ (10^5 G-cm.)	α-energy (MeV.)
^{210}Po	$3\cdot31649\pm0\cdot0008$	$5\cdot3007\pm0\cdot0026$
^{212}Bi(ThC)		
α_1 (70 %)	$3\cdot55389\pm0\cdot0007$	$6\cdot0861\pm0\cdot0024$
α_0 (27 %)	$3\cdot54232\pm0\cdot0008$	$6\cdot0466\pm0\cdot0027$
^{214}Po(RaC')	$3\cdot99274\pm0\cdot00022$	$7\cdot6804\pm0\cdot0009$
^{212}Po(ThC')	$4\cdot26934\pm0\cdot0009$	$8\cdot7801\pm0\cdot004$

5.3. Range methods

Determination of beam energy by range measurements in air, aluminium, or some other material of known stopping power is widely employed at all energies. The method is not absolute but requires a range-energy relation to be established for the material, e.g. by using particles of known energy (§5.2). It is also inherently less accurate than deflexion methods, due chiefly to range straggling which occurs when particles are stopped in a medium.

Numerous accounts of typical procedures have been published (e.g. Meagher, 1950; Cork and Hartsough, 1954a; Yntema and White, 1954). The simplest technique is merely to replace the current collector of the scattering chamber with a thin window and allow the beam to pass through into air where its range can be measured. Range-energy values for air are generally quoted for a pressure of 76 cm. Hg and a temperature of $15°$ C., hence the observed range must be corrected:

$$R = R_{\text{obs}} \frac{P}{76} \frac{288 \cdot 2}{T}, \tag{5.8}$$

where P and T are the prevailing pressure and temperature.

TABLE VIII. *Q-values of nuclear reactions useful in calibration*

(For source references, see Van Patter and Whaling, 1954.)

Reaction	Q (MeV.)
^{27}Al(p, α) ^{24}Mg	$1 \cdot 595 \pm 0 \cdot 002$
^{16}O(d, p) ^{17}O	$1 \cdot 919 \pm 0 \cdot 004$
^{9}Be(p, α) ^{6}Li	$2 \cdot 126 \pm 0 \cdot 002$
^{23}Na(p, α) ^{20}Ne	$2 \cdot 378 \pm 0 \cdot 003$
^{12}C(d, p) ^{13}C	$2 \cdot 722 \pm 0 \cdot 003$
^{16}O(d, α) ^{14}N	$3 \cdot 115 \pm 0 \cdot 0025$
^{6}Li(p, α) ^{3}He	$4 \cdot 023 \pm 0 \cdot 002$
^{2}H(d, p) ^{3}H	$4 \cdot 031 \pm 0 \cdot 006$
^{9}Be(d, p) ^{10}Be	$4 \cdot 588 \pm 0 \cdot 006$
^{6}Li(d, p) ^{7}Li	$5 \cdot 027 \pm 0 \cdot 003$
^{24}Mg(d, p) ^{25}Mg	$5 \cdot 097 \pm 0 \cdot 007$
^{13}C(d, p) ^{14}C	$5 \cdot 942 \pm 0 \cdot 004$
^{9}Be(d, α) ^{7}Li	$7 \cdot 153 \pm 0 \cdot 003$
^{11}B(p, α) ^{8}Be	$8 \cdot 585 \pm 0 \cdot 006$
^{14}N(d, p) ^{15}N	$8 \cdot 614 \pm 0 \cdot 007$
^{10}B(d, p) ^{11}B	$9 \cdot 235 \pm 0 \cdot 011$
^{19}F(d, α) ^{17}O	$10 \cdot 039 \pm 0 \cdot 010$
^{7}Li(p, α) ^{4}He	$17 \cdot 346 \pm 0 \cdot 010$
^{6}Li(d, α) ^{4}He	$22 \cdot 386 \pm 0 \cdot 011$

Alternatively, the particles may be scattered by a target located at the centre of the chamber and their ranges measured in some convenient absorber. The incident energy E_0 is related to the measured energy E_1 at angle Θ_1 by equation (2.2). In the $9 \cdot 7$ MeV. experiment of Cork and Hartsough (1954 a), the range was measured at $30°$, after scattering in the helium-filled chamber. These workers connected the first two units of a triple counter in coincidence, and the third in anti-coincidence, and determined the thickness of Al which had to be placed in front of the triple counter to stop protons in a $1 \cdot 9$ mg./cm.^{2} foil between the second and third units. The accuracy claimed was $0 \cdot 5$ %.

For the conversion of particle ranges to energies, comprehensive compilations are available for many materials (Aron *et al.* 1949;

Rich and Madey, 1954). Ranges in air, carbon, aluminium and copper, all of which are commonly used in this connexion, appear in figs. 47, 48 and 49.

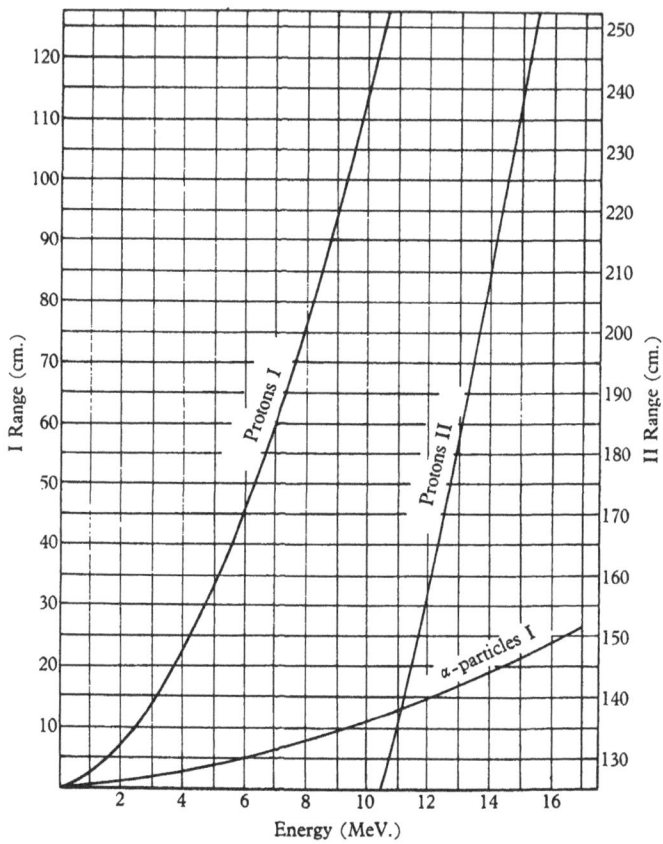

Fig. 47. Ranges of protons and α-particles in air at 76 cm. Hg pressure and $15°$C. (Data based on Aron *et al.* 1949.) The ranges of deuterons and tritons follow from the usual relations:

$$R_d(E) = 2R_p(E/2) \quad \text{and} \quad R_t(E) = 3R_p(E/3).$$

Photographic scattering chambers (§3.5.5) present a special case. Here each photographic plate serves both as scattering detector and for range measurement, and thus the mean beam energy which applies during the actual run is obtained. The procedure for measuring range with a microscope is straightforward.

The horizontal range H can be obtained using either a calibrated eyepiece micrometer, or a stage micrometer in the case of long tracks. In most cameras, the plates are positively positioned, so that

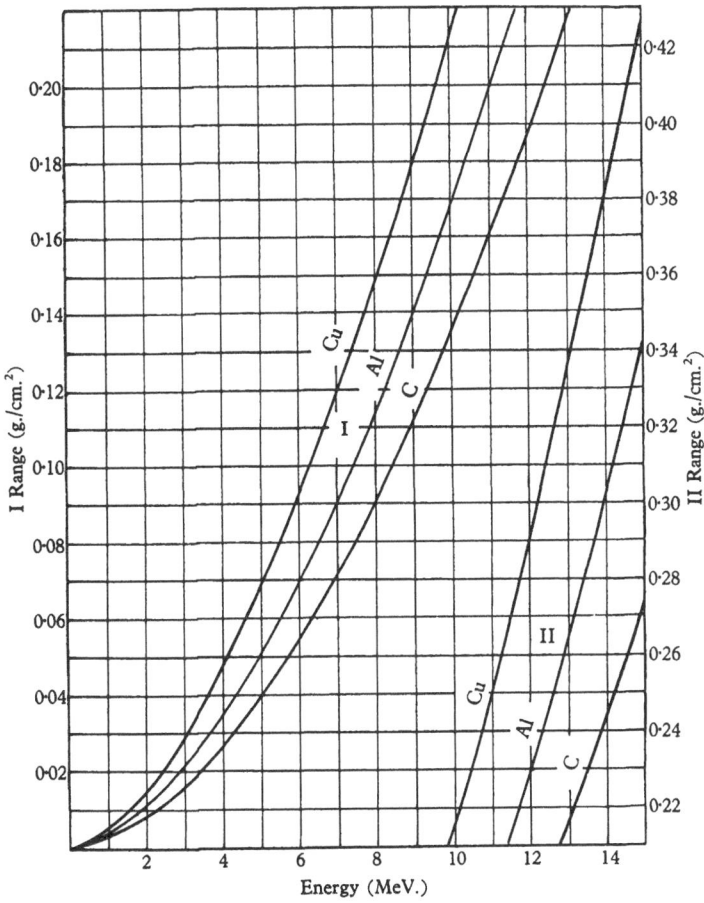

Fig. 48. Ranges of protons in C, Al and Cu at low energies. Ranges are conveniently expressed in g./cm.² (Data based on Rich and Madey, 1954.) For deuterons and tritons see caption to fig. 47.

the angle of incidence α of the particles on the emulsion surface is fixed. Then $R = H/\cos \alpha$ if it is assumed that no *lateral* shrinkage of the emulsion has occurred during processing. (The latter appears to be true if the region near the plate edges is avoided.)

MES

The ranges of different types of particle in emulsions are shown in figs. 50 and 51.

It is important to know the degree of accuracy warranted in the measurement of angle when E_0 is inferred from E_1 or E_2, measured at Θ_1 or Θ_2 respectively (equations (2.2) and (2.3)). In p-p scattering, if $\Theta = 45°$, an error $\Delta\Theta = 0\cdot1°$ would correspond to $\Delta E \approx 0\cdot3\%$,

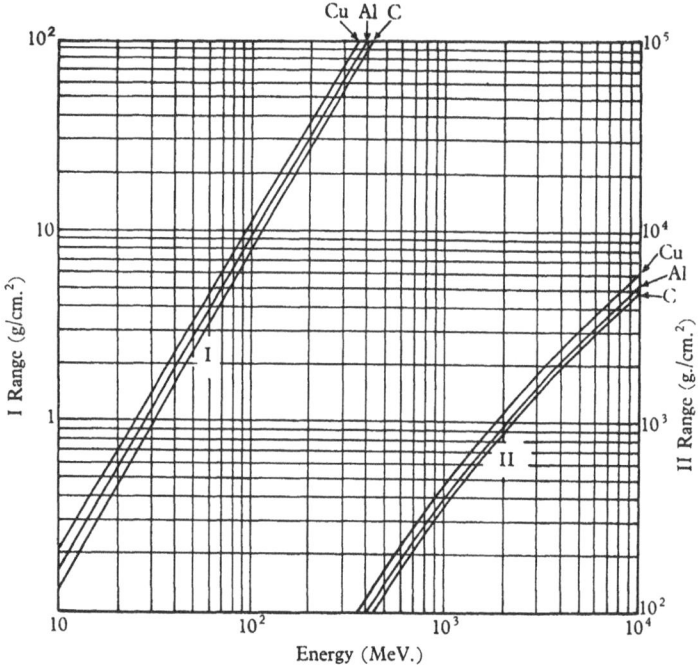

Fig. 49. Ranges of high energy protons in C, Al and Cu.
(Data based on Rich and Madey, 1954.)

whereas in p-α scattering at the same angle ΔE would be only $\sim 0\cdot03\%$. The same point arises when the energy being measured is deliberately scaled down, so that a reliable portion of the range-energy curve may be used, by making Θ large. For example, in 30 MeV. p-p scattering, if proton ranges are measured at about 55°, E_1 is near 10 MeV., at which energy ranges are well known. However, E is then rather sensitive to Θ, $0\cdot1°$ corresponding to $\Delta E \sim 0\cdot5\%$. For any given uncertainty in angle, the error introduced in E depends on both the mass ratio involved and the particular angular setting.

Variation in stopping power of emulsion with atmospheric conditions is often the principal source of error in energy measurements

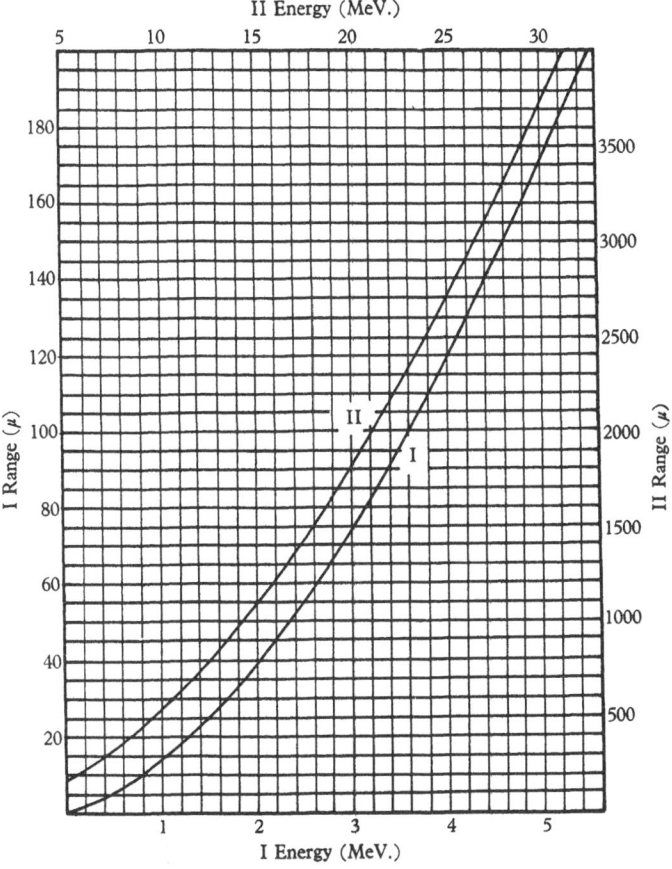

Fig. 50. Ranges of protons in dry photographic emulsion. (Data taken chiefly from Catala and Gibson, 1951; Rotblat, 1951; Wilkins, 1951; Gibson, Prowse and Rotblat, 1954). For derivation of deuteron and triton ranges, see caption to fig. 47.

by this method. Actually, modern emulsions are remarkably standard products which vary in composition very little, even from batch to batch (Rotblat, 1951). However, gelatin does take up moisture readily from the atmosphere and changes amounting to several per cent may occur in the stopping power as atmospheric humidity and temperature vary over their accustomed range. Figs. 50 and 51 apply to vacuum dried Ilford C 2 emulsion of density

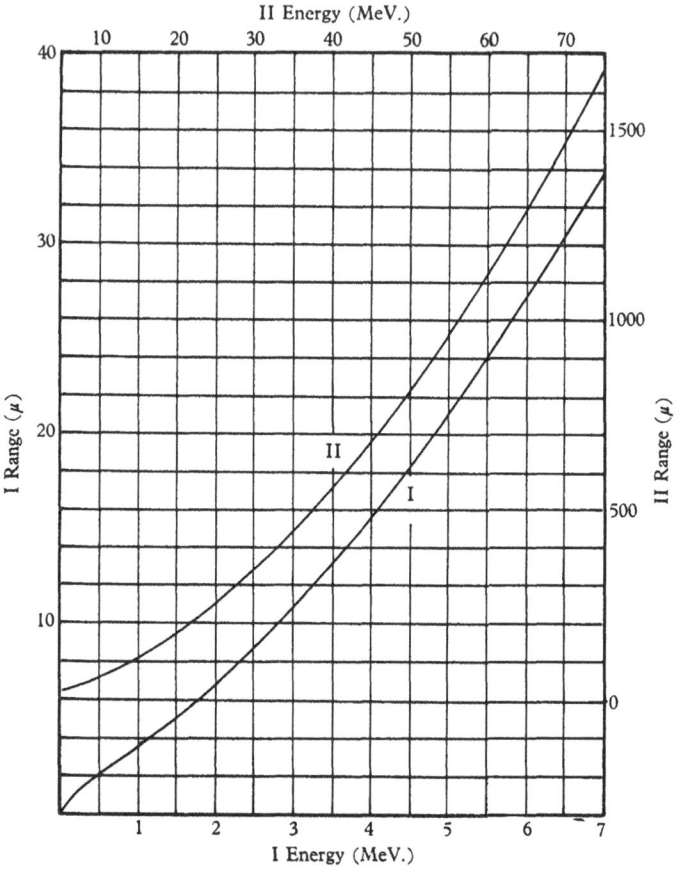

Fig. 51. Ranges of α-particles in dry photographic emulsion.
(For sources of information, see caption to fig. 50.)

about 3·94 g./cm.³,† which probably represents the state of the plates as used in most scattering experiments. The loaded chamber should always be pumped for at least several hours before exposure. If the emulsion density is brought to within 0·5 % of 3·94 g./cm.³,

† According to information supplied by Ilford Ltd., this corresponds to a residual water content of about 0·08 g./cm.³ of emulsion.

the residual effect on integral stopping power and hence on range will be no more than 0·25 % for a 10 MeV. proton. Further details are given by Wilkins (1951).

5.4. Time-of-flight methods

The principle of these has already been discussed in relation to low energy neutron scattering (§§ 4.3.4, 4.4.1). An interesting application, in a slightly modified form, to the measurement of beam energy from a fixed-frequency cyclotron has also been described (Manley and Jakobson, 1954). A 10 MeV. proton beam was passed through two scintillators, the first of which was fixed, while the second could be moved along the direction of the beam. Since a cyclotron beam is modulated (§ 3.2.3), each scintillator delivered periodic signals, the separation in phase of which depended on the proton velocity and the distance between the scintillators.

The fixed scintillator was used solely to provide a reference phase, the signal being displayed in a convenient manner on a fast oscilloscope. A pickup coil situated near the deflector provided the horizontal sweep and the two output signals were mixed and fed to the vertical scale. The procedure was then to determine the distance along the beam between two successive positions of the movable counter which differed in phase by 2π. Only length and frequency measurements are involved, both of which can be made simply. Potentially, the method seems capable of higher precision than actually obtained (~ 2 %).

5.5. Energy measurement above 100 MeV.

At high energy, other sources of error are usually more important than uncertainties in beam energy. Particularly in the case of neutrons, it is probably more difficult to generate an approximately monoenergetic beam than to measure its energy (§§ 4.2.5, 4.2.6). Moreover, at very high energies two new principles for measuring energy become available (numbers 5 and 6 below).

Altogether, six different approaches are possible:

(1) Estimates may be based on a knowledge of a cyclotron field strength and orbit radius.

(2) An external magnetic field may be used, the procedure being similar to that at low energies.

(3) Range measurements may be made in materials of known stopping power (§ 5.3).

(4) The time-of-flight method may be applied. With present timing techniques this is insufficiently accurate at high energy. It must, in any case, fail when the particle velocity approaches that of light, because velocity (the quantity actually determined) is then insensitive to large changes in energy. For protons or neutrons and for a 1 % error in E_0, v must be measured to 0·5 % at low energy, ~ 0·25 % at 500 MeV. and ~ 0·16 % at 1 BeV.

(5) When coincidences are counted between scattered and recoil particles (§ 2.4), a check on the energy is available from the angles Θ_1 and Θ_2 at which counters must be placed, because the relationship between these angles is energy-dependent in the relativistic region (equations (A. 39), (A. 40), (A, 41)). Fig. 27 illustrates a case in point.

(6) The angle at which Cerenkov radiation (cf. § 3.5.4) is emitted by a fast charged particle traversing a transparent medium may be observed. This method becomes available only when $\beta > n^{-1}$, where $\beta = v/c$ and n is the refractive index of the medium. For ordinary transparent materials the threshold is at 300 MeV. or higher, but by employing special lead glasses ($n > 2$) proton energies above 150 MeV. can be measured. Because the Cerenkov counter yields extremely short pulses, it is also the best choice for time-of-flight determinations at high energy.

The most accurate experiments above 100 MeV. have incorporated several of the above techniques to provide a cross check on the energy. (See Sutton *et al.* 1955, for detailed discussion of methods 3, 5 and 6. For general procedures of beam measurements, see Chamberlain *et al.* 1951; Kruse *et al.* 1954; Bogachev and Vzorov, 1954; Smith *et al.* 1955.)

The spectrum of high energy neutrons may be investigated by determining the differential range curve of recoil protons ejected from a hydrogenous target (Hadley *et al.* 1949; Hartzler and Siegel, 1954). Copper, the absorption properties of which are convenient and well known, has been commonly used for this purpose. The procedure is essentially the same as that described in § 5.3 for 9·7 MeV. protons.

TABLE IX. *Energy loss data for protons in various materials*

(From Aron et al. 1949; Rich and Madey, 1954.)

Energy loss (MeV./g./cm.²)

Energy (MeV.)	Hydrogen	Deuterium	Helium	Air	(CH)	(CH₂)	Carbon	Aluminium	Copper
1	691	346	282	—	276	306	242	—	—
2	396	199	166	—	169	185	150	115	—
3	283	142	119	102	123	135	110	86·2	—
4	224	112	95·7	81·7	98·0	107	87·6	69·6	46·1
5	185	92·7	79·9	68·7	82·0	89·4	73·4	58·8	40·5
6	159	79·5	68·9	59·4	70·8	77·1	63·5	51·2	36·2
7	140	69·9	60·7	51·7	62·5	68·0	56·1	45·5	32·8
8	125	62·4	54·5	46·0	56·1	61·0	50·3	41·0	30·0
9	113	56·5	49·4	42·1	50·9	55·4	45·8	37·5	27·8
10	103	51·7	45·3	38·8	46·7	50·8	42·0	34·5	24·2
12	88·5	44·3	39·0	33·9	40·2	43·7	36·2	29·9	19·5
16	69·4	34·7	30·7	27·5	31·9	34·4	28·6	27·8	16·4
20	57·5	28·8	25·6	23·1	26·4	28·7	23·9	20·0	12·0
30	40·9	20·5	18·3	16·6	19·0	20·5	17·1	14·6	9·63
40	32·2	16·1	14·4	13·2	15·0	16·2	13·6	11·6	8·12
50	26·8	13·4	12·1	11·0	12·5	13·6	11·4	9·74	7·07
60	23·1	11·6	10·4	9·57	10·9	11·7	9·83	8·46	5·71
80	18·4	9·19	8·32	7·66	8·67	9·36	7·86	6·79	4·85
100	15·5	7·74	7·01	6·48	7·32	7·90	6·64	5·76	3·66
150	11·5	5·74	5·23	4·85	5·46	5·89	4·96	4·31	3·04
200	9·41	4·71	4·31	4·00	4·50	4·85	4·09	3·58	2·40
300	7·34	3·67	3·36	3·13	3·52	3·79	3·20	2·81	2·08
400	6·30	3·15	2·89	2·70	3·03	3·26	2·76	2·43	1·89
500	5·68	2·84	2·62	2·44	2·74	2·95	2·49	2·20	1·68
700	5·00	2·50	2·31	2·16	2·42	2·60	2·21	1·95	1·55
1,000	4·54	2·27	2·10	1·97	2·20	2·37	2·01	1·79	1·45
2,000	4·16	2·08	1·94	1·83	2·04	2·19	1·86	1·66	1·47
3,000	4·15	2·08	1·94	1·84	2·04	2·19	1·87	1·68	1·50
4,000	4·21	2·10	1·97	1·87	2·08	2·23	1·90	1·71	1·54
5,000	4·27	2·14	2·00	1·90	2·12	2·27	1·94	1·75	1·69
10,000	4·57	2·29	2·16	2·06	2·28	2·45	2·09	1·90	

5.6. Corrections for energy loss

As pointed out in § 5.1, corrections often have to be applied to the measured energy for small losses sustained in the equipment itself. Where these originate depends on how and where the energy is determined and on the general arrangement of the equipment. In most cases they can be recognized very easily. The energy required is that obtaining at the target centre during the run.

Corrections for energy losses are generally small and can be calculated to sufficient accuracy if specific ionization data are available for the materials involved. Table IX lists energy-loss data for some common target, window and absorber materials (Aron *et al.* 1949; Rich and Madey, 1954).

CHAPTER 6

LOW ENERGY NEUTRON-PROTON
SCATTERING

6.1. Introduction

The previous four chapters have dealt with experimental methods. We now proceed to consider the theory of scattering processes and the interpretation of experimental results in terms of the nuclear force and nuclear models. The following sections relate to the scattering of low energy (0–10 MeV.) neutrons by protons, assuming the two-body interaction to involve central forces only.

In principle, the cross-section for elastic collisions depends on the way in which the protons are bound in molecules. However, it is legitimate to treat the protons as free if the neutron energy $E \gg 1$ eV., which is the order of magnitude of chemical binding energies. For slow neutrons (in particular, thermal and sub-thermal neutrons) the cross-sections differ according to whether the scatterer consists of hydrogen gas, heavy molecules such as paraffin, or a crystal lattice. Experiments along these lines all lead directly to information about the characteristics of the two-body interaction.

The Schrödinger equation describing two interacting particles distance r apart and of equal mass m is given by

$$\nabla^2 \psi + \frac{2\mu}{\hbar^2} [E - V(r)] \psi = 0, \qquad (6.1)$$

where $\psi(r, \theta, \phi)$ is the de Broglie wave function of the system in terms of spherical polar co-ordinates. E is the relative energy of the particles in the CM system, equal to half the energy in the Lab. system, and $\mu = \frac{1}{2}m$ is the reduced mass (equation (A.15)). $V(r)$ is a short range potential function of r only.

In the scattering experiment we shall assume an incident flux of one neutron per second per unit normal area passing along a z-axis and being scattered along a direction r at an angle θ to it. Because $V(r)$ is a central force, the scattering is axially symmetric so that $\psi = \psi(r, \theta)$. Thus ψ may be expanded at very large distances from the scattering centre as the sum of a plane wave e^{ikz} along the z-axis and a scattered wave along r such that its probability density obeys an inverse square law:

$$\psi(r) \sim e^{ikz} + f(\theta) e^{ikr}/r \quad \text{(for } r \to \infty), \qquad (6.2)$$

where $f(\theta)$ is the amplitude of the scattered wave in the direction θ. The number of particles scattered between angles θ and $\theta + d\theta$ is therefore

$$d\sigma = |f(\theta)|^2 d\omega, \qquad (6.3)$$

where $d\omega = 2\pi \sin \theta d\theta$ is the solid angle into which the particles are scattered.

To solve equation (6.1), we shall reduce it to a set of ordinary differential equations by making an expansion of ψ in terms of Legendre polynomials:

$$\psi(r, \theta) = \frac{1}{r} \sum_{l=0}^{\infty} A_l f_l(r) P_l(\cos \theta). \qquad (6.4)$$

Substituting equation (6.4) in equation (6.1) we find:

$$\frac{d^2}{dr^2} f_l + \left[k^2 - U(r) - \frac{l(l+1)}{r^2} \right] f_l = 0, \qquad (6.5)$$

where $k^2 = (m/\hbar^2)/E$, $U(r) = (m/\hbar^2) V(r)$, and l is called the orbital angular momentum quantum number. Denoting the incident and scattered waves by u_l and v_l respectively, then u_l is a solution of equation (6.5) and v_l a solution for $V(r) = 0$, both satisfying the initial condition $u_l(0) = v_l(0) = 0$. It is easily found that the asymptotic forms for $r \to \infty$ are

$$u_l \sim C \sin(kr + \tfrac{1}{2}l\pi), \quad v_l \sim C \sin(kr + \tfrac{1}{2}l\pi + \delta_l), \qquad (6.6)$$

where δ_l is the phase-shift suffered by an incident wave of orbital angular momentum l.

It is shown by Mott and Massey (1949) that the scattered amplitude $f(\theta)$ of equation (6.2) is

$$f(\theta) = \frac{1}{2ik} \sum_{l=0}^{\infty} (2l+1)(e^{2i\delta_l} - 1) P_l(\cos \theta). \qquad (6.7)$$

From equation (6.3) the angular distribution or differential cross-section in barns per sterad. in the CM system is

$$\frac{d\sigma}{d\omega} = \frac{1}{4k^2} \left| \sum_{l=0}^{\infty} (2l+1) P_l(\cos \theta) (e^{2i\delta_l} - 1) \right|^2, \qquad (6\cdot8)$$

where the relation between CM and Lab. angles is $\theta = 2\Theta$, Θ being the scattering angle in the Lab. system (Appendix A). The total probability of scattering, or total cross-section, follows by integrating over θ:

$$\sigma = \frac{4\pi}{k^2} \sum_{l=0}^{\infty} (2l+1) \sin^2 \delta_l, \qquad (6.9)$$

which is interpreted as the effective area of interception offered by one proton to a single neutron impinging normally anywhere on a plane circular area of one square centimetre surrounding the proton.

Assuming a short range potential $V(r)$, it may be shown that only a small finite number of phases δ_l contribute to the cross-section at a given energy, while for incidence energies of 10 MeV. Lab. or less only δ_0 is appreciable.

6.2. Energy dependence of phase shifts

We require the qualitative dependence of the phases δ_l on the energy and to find this both classical and wave-mechanical methods may be used, the first being of only doubtful validity.

(a) *Classical approach.* As the orbital angular momentum is conserved, we have
$$\mathbf{r} \times \mathbf{p} = dp = l\hbar,$$

where $\mathbf{p}$ is the momentum of the incident neutron, distance r from the scattering centre, and d is the distance of closest approach. Then $l = (P/\hbar)d = d/\lambda$, where $\lambda = 2\pi\lambdabar$ is the de Broglie wavelength of the neutron. The distance of closest approach d must be less than the range b of nuclear force in order that interaction take place and so we require $l < b/\lambda$. Hence $E(\text{CM}) > l^2(\hbar^2/mb^2)$ or

$$E(\text{Lab.}) > l^2 \frac{2\hbar^2}{mb^2}. \qquad (6.10)$$

On taking a rough value $b = 2\cdot8 \times 10^{-13}$ cm.† and converting to eV., one finds δ_0 to be important but that δ_l $(l = 1, 2, \ldots)$ can be neglected for $E(\text{Lab.}) \lesssim 10$ MeV., which is confirmed by the wave-mechanical treatment below. However, the application of the formula to higher values of l indicates that δ_2 is negligible for $E(\text{Lab.}) < 40$ MeV. and

---

† Recent values for triplet and singlet ranges are $b = 2\cdot05$ and $2\cdot58 \times 10^{-13}$ cm. (cf. § 10.10), leading to δ_1 negligible for $E(\text{Lab.}) < 22$ MeV. and 14 MeV. respectively. However, very small p and d phases may have an important effect on $d\sigma/d\omega$, even though σ is practically unaltered (cf. § 10.1).

δ_3 for $E(\text{Lab}) < 90$ MeV. These values are much too large, illustrating the danger of extending classical ideas to quantum mechanics.

(b) *Wave mechanical approach.* This is based on a perturbation calculation of δ_l which, for simplicity, is carried out for p-waves only. An integral equation for the phase shift δ_l may be found (Mott and Massey, 1949) of the form

$$\sin \delta_l = -\frac{1}{k} \int_0^\infty U(r)\, u_l(r)\, v_l(r)\, dr. \qquad (6.11)$$

If δ_l is small, the scattered wave $v_l(r)$ does not differ very much from the incident plane wave $u_l(r)$, so $v_l(r) \approx u_l(r)$ and equation (6.11) reduces to the Born approximation for the phase:

$$\delta_1 \approx -\frac{1}{k} \int_0^\infty U(r)\, u_1^2(r)\, dr. \qquad (6.12)$$

For p-waves, $u_1 = \sin kr/kr - \cos kr$ and as $U(r)$ is short range in (6.12), we are only interested in $u_1(r)$ for kr small, where $u_1 \approx \frac{1}{3}(kr)^2$.
Thus $\delta_1 \approx -\frac{1}{9}k^3 \int_0^b U(r)\, r^4\, dr$ for a well of range b. Also

$$\int_0^b U(r)\, r^4\, dr = \frac{1}{5}b^5 U$$

for the spherical well and from equation (1.7),

$$b^2 U \approx \frac{1}{4}\pi^2 \approx \frac{5}{2}.$$

Thus in general, $\displaystyle\int_0^b U(r)\, r^4\, dr = \frac{1}{2}\nu b^3$, where ν is a constant of order of magnitude 1, which for the spherical well turns out on numerical integration to be 0·89. This gives $\delta_1 \approx -\frac{1}{18}\nu(kb)^3$, which is comparatively small for kb and hence E small compared to $U(r)$. This is the condition for the validity of (6.12), viz. δ_1 small. As $V(r) \sim 20$ MeV. for a rectangular well of range $2 \cdot 8 \times 10^{-13}$ cm., it follows that $E(\text{Lab.}) = 10$ MeV. is roughly the limit for ignoring δ_1 in comparison with δ_0, thus checking with the classical argument.

The corresponding formula for arbitrary l is

$$\delta_l \approx \frac{5}{2} \frac{\nu}{(2l+1)^2} \cdot \frac{(kb)^{2l+1}}{(2l+3)}, \qquad (6.13)$$

so that for $kb < 1$, δ_l decreases very rapidly for increasing l.

6.3. Spherical symmetry of scattering for $E < 10$ MeV.

As all phase shifts except δ_0 are negligible for $E \lesssim 10$ MeV., the total cross-section reduces to

$$\sigma = \frac{4\pi}{k^2} \sin^2 \delta_0, \tag{6.14}$$

which is independent of angle and therefore spherically symmetric. This property has been verified experimentally by a number of workers to within a few per cent, confirming the fundamental assumption of short range nuclear forces. However, the influence of the $P_1(\cos \theta)$ term for p-wave scattering should cause a departure from spherical symmetry at energies higher than 10 MeV.

6.4. The scattering of low energy neutrons by protons

For the case of incident neutrons of only a few MeV., equation (6.14) for the cross-section may be simplified somewhat. The wave equation is

$$\frac{d^2 u}{dr^2} + \kappa'^2(r)\, u = 0, \tag{6.15}$$

where $\kappa'(r) = [(m/\hbar^2)\{E - V(r)\}]^{\frac{1}{2}}$. Defining $k = [(m/\hbar^2)\,E]^{\frac{1}{2}}$, the boundary conditions are

$$u(0) = 0, \quad u(r) \sim \sin(kr + \delta_0) \quad \text{for} \quad r \to \infty.$$

We replace $V(r)$ by some average value $-V_0$ for $r < b$, and by zero for $r > b$, so that for this rectangular well approximation,

$$\kappa'(r) \to \kappa' = [(m/\hbar^2)\,(E + V_0)]^{\frac{1}{2}}.$$

The solution of equation (6.15) is then

$$u(r) = A \sin \kappa' r, \qquad r < b$$
$$= B \sin(kr + \delta_0), \quad r > b.$$

As the logarithmic derivative $u'(r)/u(r)$ is continuous at $r = b$,

$$k \cot(kb + \delta_0) = \kappa' \cot \kappa' b. \tag{6.16}$$

Also the ground state of the deuteron, binding energy $-W$, is described by equation (1.6):

$$\kappa \cot \kappa b = -\alpha, \quad \text{where} \quad \kappa = [(m/\hbar^2)\,(V_0 - W)]^{\frac{1}{2}}, \quad \alpha = (mW/\hbar^2)^{\frac{1}{2}}.$$

If E is only a few MeV., $E \ll V_0$ and also $W \ll V_0$, and hence $\kappa' \approx \kappa$, which leads to

$$\kappa' \cot \kappa' b \approx -\alpha. \qquad (6.17)$$

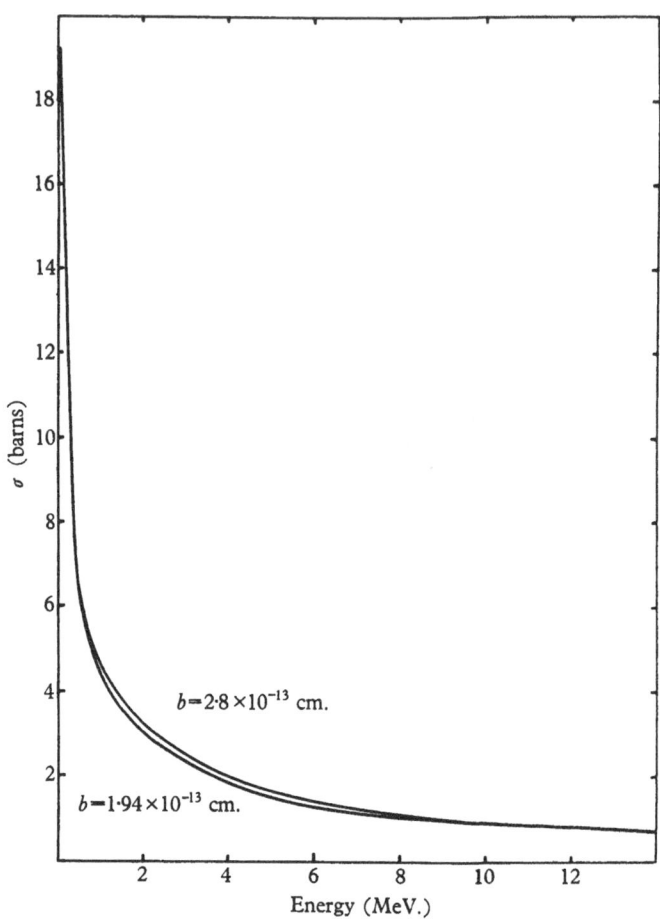

Fig. 52. Scattering cross-section σ for n-p collisions as a function of neutron energy (Rosenfeld, 1948). Two different values of the range b for the spherical well are assumed.

On taking a zero range first approximation $b = 0$ and combining equations (6.16) and (6.17), we find

$$k \cot \delta_0 = -\alpha. \qquad (6.18)$$

Hence from equation (6.14),

$$\sigma \approx \frac{4\pi}{k^2 + \alpha^2} = \frac{4\pi\hbar^2}{m} \cdot \frac{1}{E + W}. \tag{6.19}$$

A second approximation leads to the range correction factor $(1 + \alpha b)$ as a multiplying factor in equation (6.19), which, however, is not very sensitive to range. A range factor of the order $b \sim 2 \times 10^{-13}$ cm. gives $\alpha b \sim 0.5$ so that a 30 % change in b does not produce a large change in the cross-section (fig. 52).

For thermal neutrons $(E \to 0)$, equation (6.19) predicts a cross-section of $\sigma = 2.33$ b. The experimental value (Melkonian, 1949) is (20.36 ± 0.10) b., which is much too large to be explained by range correction factors as these can, at most, double the result of equation (6.19). However, at several MeV. the discrepancy is only 20–25 %. It is clear, therefore, that a scattering resonance, which is not predicted by equation (6.19), occurs near zero energy. A possible inference is that a second state of the deuteron having a binding energy W very close to zero contributes to the scattering.

6.5. The singlet state of the deuteron

Wigner postulated in 1935 that the deuteron ground state is a triplet corresponding to the three possible orientations in a magnetic field of a deuteron with parallel neutron and proton spins. Denoting eigenstates of spin $+\frac{1}{2}$ and $-\frac{1}{2}$ along an arbitrary z-axis by α and β respectively, and taking $S =$ total spin, $M = z$-component of spin, then the appropriate spin functions comprise a symmetric triplet:

$$S = 1, \text{ triplet} \begin{cases} \alpha(1)\,\alpha(2), & M = 1, \\ 2^{-\frac{1}{2}}[\alpha(1)\,\beta(2) + \alpha(2)\,\beta(1)], & M = 1, \\ \beta(1)\,\beta(2), & M = 1. \end{cases}$$

Wigner's suggestion was that there is also an antisymmetric spin state of the deuteron with neutron and proton spins anti-parallel:

$$S = 0, \text{ singlet}; \quad 2^{-\frac{1}{2}}[\alpha(1)\,\beta(2) - \alpha(2)\,\beta(1)], \quad M = 0.$$

Inserting statistical weights of $\frac{3}{4}$ and $\frac{1}{4}$ respectively in front of the triplet and singlet cross-sections σ_T and σ_S, the neutron-proton cross-section becomes

$$\sigma = \tfrac{3}{4}\sigma_T + \tfrac{1}{4}\sigma_S, \tag{6.20}$$

or approximately

$$\sigma = \frac{\pi \hbar^2}{m} \left(\frac{3}{E + W_T} + \frac{1}{E + W_S} \right) (1 + \alpha b), \qquad (6.21)$$

where W_T and W_S are the triplet and singlet binding energies of the deuteron.

Taking the value $\sigma = 20 \cdot 36$ b. for the thermal neutron cross-section in the energy range 0–100 eV. and also $W_T = (2 \cdot 226 \pm 0 \cdot 003)$ MeV., substitution in (6.21) gives $W_S \approx 60$ keV. However, from other experiments on the scattering of sub-thermal neutrons by ortho- and para-hydrogen (§ 6.10), it turns out that W_S is positive so that the singlet state of the deuteron is not bound. The expression for the singlet cross-section in equation (6.21) is thus purely formal, and $W_S \approx 60$ keV. is a virtual level corresponding to a scattering resonance near zero energy, explaining why the measured thermal neutron cross-section is so much larger than the predicted triplet cross-section.

6.6. The scattering length a

For a well of range b, denote the wave function for $r < b$ by $u(r)$ and for $r > b$ by $v(r)$. At zero neutron energy E, the radial wave functions of (6.6) satisfy

$$u'' - \frac{m}{\hbar^2} V(r) u(r) = 0, \quad v''(r) = 0. \qquad (6.22)$$

If we normalize $u(r)$ such that

$$u(r) \to v(r) = \sin(kr + \delta)/\sin \delta \quad \text{for} \quad r \to \infty, \qquad (6.23)$$

then $u(0) = 1$.

Also the solution of equation (6.22) is $u(r) \to v(r) = (1 - r/a)$, where a is a constant giving the first zero of the wave function (figs. 53, 54). We have

$$\left. \begin{aligned} \lim_{k \to 0} \frac{\sin(kr + \delta)}{\sin \delta} &= \lim_{k \to 0} (1 + kr \cot \delta_0) = 1 - \frac{r}{a}, \\ a &= \lim_{k \to 0} \left(-\frac{\tan \delta}{k} \right), \end{aligned} \right\} \qquad (6.24)$$

where a is called the zero energy scattering length. For triplet scattering, the wave function is 'pulled in' by the strong attractive potential, so that we might expect a zero energy phase shift of π, which is confirmed by equation (6.18). On the other hand, for singlet scattering the wave function is not pulled in and a phase at zero energy of o is obtained. In fact, it is shown in Appendix B

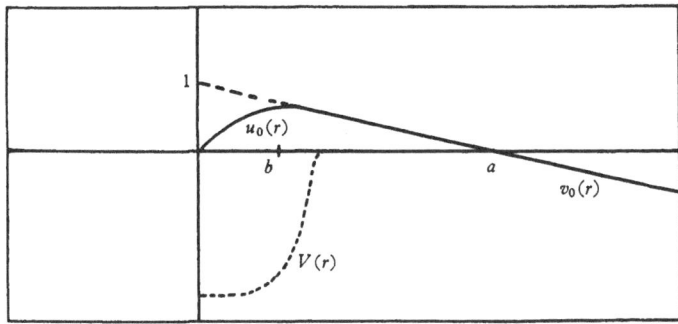

Fig. 53. Zero energy wave functions $u_0(r)$ ($r < b$) and $v_0(r)$ ($r > b$) plotted against r in a potential field $V(r)$ for the triplet scattering of neutrons by protons. The zero energy scattering length a is positive.

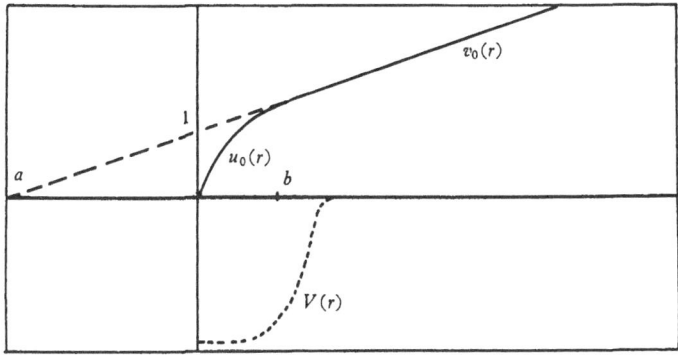

Fig. 54. Zero energy wave functions $u_0(r)$ ($r < b$) and $v_0(r)$ ($r > b$) plotted against r in a potential field $V(r)$ for the singlet scattering of neutrons by protons. The zero energy scattering length a is negative.

that if $\delta(k)$ is defined as a function of k, and also $\delta(\infty) = 0$, then $\delta(0) = n\pi$, where n is the number of composite bound states of the two body system. Thus for triplet scattering, a is positive, but for singlet scattering a is negative, which is illustrated in figs. 53 and 54 respectively.

At zero energy, $(\sin\delta)/k = \pm (\tan\delta)/k$, so that the cross-section σ of equation (6.15) becomes

$$\sigma = 4\pi a^2, \qquad (6.25)$$

which is the same as the scattering by an impenetrable sphere of radius a.

6.7. The effective range and shape-independent approximation formula

It is clear from the expression (6.21) that low energy neutron-proton scattering experiments do not provide detailed information about the shape of the nuclear potential, but at best only values of the range and strength of the potential. This was pointed out by Landau and Smorodinsky (1944), who produced a formula showing that the phase shift for scattering below about 10 MeV. energy is independent, to a close approximation, of the shape of the potential well and depends only on two parameters common to all wells. Their formula was later proved by Schwinger using a variational method, and applied to n-p and p-p scattering by Blatt (1948) and Blatt and Jackson (1949). A simpler non-variational proof based directly on the properties of the Schrödinger equation has been given by Bethe (1949) and is followed here.

Let $u(r)$ be the 'inside' wave function satisfying the Schrödinger equation for an energy k^2:

$$u''(r) + \left[k^2 - \frac{m}{\hbar^2} V(r) \right] u(r) = 0. \qquad (6.26)$$

Similarly at zero energy,

$$u_0''(r) - \frac{m}{\hbar^2} V(r) u_0(r) = 0. \qquad (6.27)$$

Multiplying equation (6.26) by u_0 and equation (6.27) by u and subtracting, we have

$$\frac{d}{dr}(uu_0' - u_0 u') = k^2 uu_0, \qquad (6.28)$$

subject to the boundary condition $u(0) = 0$.

Similarly the 'outside' wave function $v(r)$ satisfies the equation

$$v'' + k^2 v(r) = 0, \qquad (6.29)$$

leading to
$$\frac{d}{dr}(vv_0' - v_0 v') = k^2 vv_0. \qquad (6.30)$$

We shall normalize $v(r)$ by the factor $1/\sin \delta$ so that

$$v(r) = \sin(kr + \delta)/\sin \delta.$$

Then $\qquad v(0) = 1, \quad v'(0) = k \cot \delta, \quad v_0'(0) = -a.$

Subtracting equations (6.28) and (6.30) and integrating between the limits 0 and ∞, we find the exact relation

$$k \cot \delta = -1/a + k^2 \int_0^\infty (v v_0 - u u_0) \, dr. \qquad (6.31)$$

The main contribution to the integral comes from the region where $u(r)$ and $u_0(r)$ differ appreciably from their asymptotic forms $v(r)$ and $v_0(r)$, namely the 'inside' region of the nuclear well. The potential is here of the order of 20 MeV. or more so that provided the incident energy $E \ll |V(r)|$, say $E < 10$ MeV., then the wave number $\kappa(r)$ is nearly independent of E so that we may replace $v(r)$ by $v_0(r)$ and $u(r)$ by $u_0(r)$ in equation (6.31). The result is the *shape-independent approximation*,

$$k \cot \delta = -1/a + \tfrac{1}{2} r_0 k^2 + o(k^4), \qquad (6.32)$$

where the 'effective range' r_0 is defined by

$$r_0 = 2 \int_0^\infty (v_0^2 - u_0^2) \, dr. \qquad (6.33)$$

r_0 depends on both the width and the depth of the potential well. It is called the *effective range*, as the integral in equation (6.33) is of the order 1×10^{-13} cm. so that the factor 2 makes r_0 of the order of size of the well range. Thus low-energy neutron-proton scattering is, to a close approximation, independent of the detailed shape of the interaction, depending only on the parameters a and r_0 for $E < 10$ MeV.

For bound states, we replace E by $-W$ so that

$$k^2 = -\frac{m}{\hbar^2} W = -\frac{1}{R^2}, \quad v(r) = e^{-r/R},$$

where R is the size of the deuteron. The boundary conditions for $v(r)$ are

$$v(0) = 1, \quad v'(0) = -1/R, \quad v_0'(0) = 0.$$

From equations (6.28) and (6.30) there follows the relation

$$1/R \approx 1/a + \tfrac{1}{2} r_0/R^2. \qquad (6.34)$$

Thus in the zero range approximation, the zero energy scattering length a is equal to the size R of the deuteron.

The cross-section for the triplet or singlet states is found from equations (6.14), (6.32) and (6.34):

$$\sigma_{S,T} = \frac{4\pi}{k^2} \sin^2 \delta$$

$$= 4\pi \Big/ \left[\left(k^2 + \frac{1}{R^2} \right) \left\{ 1 - \frac{r_0}{R} + \frac{r_0^2}{4} \left(k^2 + \frac{1}{R^2} \right) \right\} \right], \qquad (6.35)$$

where r_0 may of course have different values for triplet and singlet scattering. As the total cross-section given by equation (6.20) has both singlet and triplet contributions, the values of r_0 should in principle follow from equation (6.35) but the experimental accuracy is unfortunately too low to lead to a unique pair of values of r_0. These can however be found by other methods (§ 6.10).

If a higher approximation than equation (6.32) is aimed at and the energy dependence of r_0 is taken into account, then an infinite series in k^2 is obtained:

$$k \cot \delta = -1/a + \tfrac{1}{2} r_0 k^2 - Pr_0^3 k^4 + Qr_0^5 k^6 - \dots \qquad (6.36)$$

Blatt and Jackson (1949) examined a number of potentials of different shape and found that in most cases the shape parameter P was less than 0·05 and never more than 0·14, an example being the rectangular well for the triplet neutron-proton interaction, $P = 0·0327$. Thus for $kr_0 < 1$, the k^4 term may be safely neglected as its effect is less than the accuracy of the neutron-proton scattering experiments.

6.8. The intrinsic range b

The effective range r_0 was defined by equation (6.33), with $v_0(r)$ given as $v_0(r) = 1 - r/a$. Hence

$$r_0 = 2 \int_0^\infty \left[(1 - r/a)^2 - u_0^2 \right] dr. \qquad (6.37)$$

We define the intrinsic range b as the range which corresponds to a resonance at zero energy, provided the well-depth is the smallest one which will give this zero energy resonance. Then

$$1/a = 0, \qquad b = 2 \int_0^\infty \left[1 - u_0^R(r)^2 \right] dr. \qquad (6.38)$$

For the rectangular well, the intrinsic range is equal to the ordinary range or width w of the well, as $u_0^R(r) = \sin[(\pi/2w)r]$. Next a well-depth parameter s is defined such that

$$V(r) = sV^{(R)}(r). \tag{6.39}$$

Then $s = 1$ implies a resonance at zero energy, $s < 1$ implies a virtual level and $s > 1$ a real level.

To treat the dependence of b on r_0 and a, we consider the case $s = 1$, and increase s to obtain a bound level, which requires a to be positive. Putting $u_0 = 1 - r/a - g(r)$, equation (6.37) becomes

$$r_0 = 2\int_0^\infty \left[2g(r) - g^2(r) - \frac{2r}{a}g(r) \right] dr. \tag{6.40}$$

Substitution of equation (6.38) in equation (6.40) gives the effective range as

$$r_0 = b - 4\int_0^\infty \frac{r}{a}g^{(R)}(r)\,dr, \tag{6.41}$$

where $g(r)$ is assumed approximately unaltered from its value at resonance, $g^R(r)$. For the rectangular well, $u_0 = \sin[(\pi/2b)r]$ and hence $g^R(r) = 1 - \sin[(\pi/2b)r]$, so we find

$$r_0 \approx b\left(1 - 0.38\frac{b}{a}\right). \tag{6.42}$$

The coefficient of b/a differs from well to well, a long-tailed well (exponential or Yukawa) extending further out than $r = b$, thus giving larger values of $\int_0^\infty rg^{(R)}(r)\,dr$.

The intrinsic range b is not an experimental quantity but must always be inferred from the experimental effective range using equation (6.41). Hence a given experimental r_0 leads to different intrinsic ranges for different assumed well shapes. On the other hand, the effective range is independent of well shape.

Fig. 55 shows a plot (Blatt and Jackson, 1949) of b/r_0 against r_0/a for four different well shapes, the rectangular, gaussian, exponential and Yukawa wells. Thus if the scattering length a is determined at zero energy, the intrinsic range b may be found from the effective range r_0 for each well shape chosen.

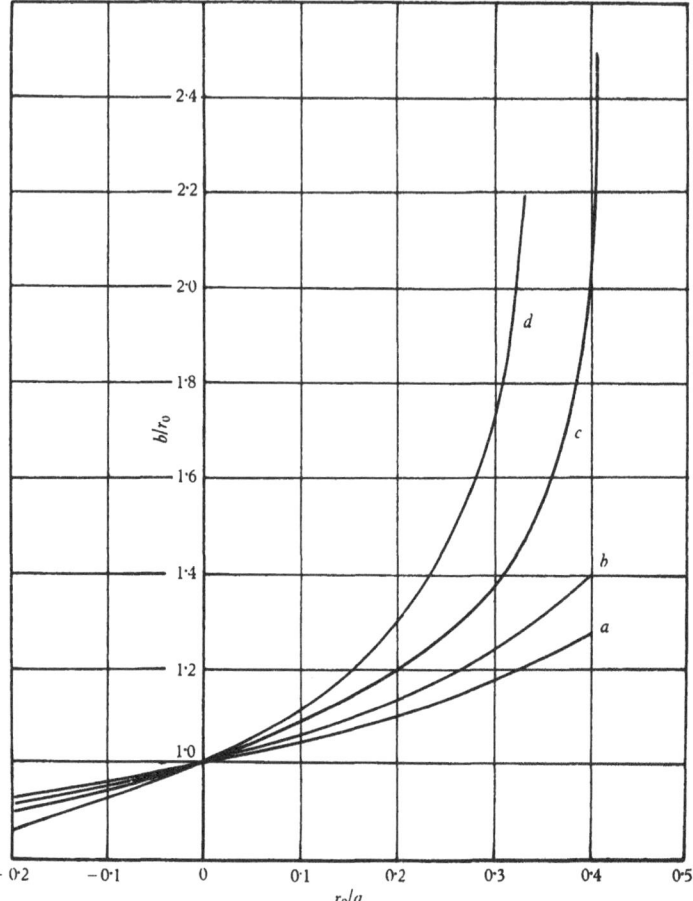

Fig. 55. Ratio of intrinsic to effective range b/r_0 given as a function of r_0/a for four different well shapes: (a) spherical well, (b) gaussian well, (c) exponential well, (d) Yukawa well (Blatt and Jackson, 1949).

6.9. Thermal neutron scattering by protons bound in molecules

In all actual cases, the proton involved in neutron-proton scattering is not free as previously assumed, but is bound in some molecule such as paraffin or water. It is necessary to take into account the effect of the chemical bond for neutron energies of the order of chemical binding energies, i.e. a few eV.

The reduced mass μ of the system varies according to whether the proton is free or bound, so that in terms of the nucleon mass we have:

(a) proton free, $\mu = \frac{1}{2}m$;

(b) proton bound in a heavy molecule, $\mu = m$.

We would like to be able to use the Born approximation expression for the scattering by a potential $V(r)$:

$$d\sigma = \frac{1}{4\pi^2\hbar^4}\mu^2 \left| \int \psi_f^* \, V(r) \, \psi_0 \, d\tau \right|^2 d\omega, \qquad (6.43)$$

where ψ_f and ψ_0 are the final and initial wave functions respectively and $d\omega$ is an element of solid angle.

The condition for the validity of equation (6.43), viz. that the scattering potential $V(r)$ be small compared to the incident energy E, is clearly not satisfied for $E \sim 1$ eV. However, the approximate condition (1.7) for constant binding energy of the deuteron, using a well of range b, viz. $V_0 b^2 \approx \pi^2\hbar^2/4m$, can also be seen from equation (6.17) to be the condition for a constant scattering cross-section near zero energy. Hence using equation (1.7), we can construct an artificial potential of negligible magnitude but great range such that the same scattering is obtained and equation (6.43) is valid. Therefore, the cross-section is proportional to the square of the reduced mass. Hence

$$\sigma(\text{bound}) = 4\sigma(\text{free}). \qquad (6.44)$$

The proton is regarded as bound if the incident energy is very much less than the chemical bond, which for the CH bond in paraffin is $0\cdot4$ eV. Thus $E \ll h\nu$, where $h\nu$ is the energy of the proton in the sub-group of the molecule. For $E > h\nu$, one quantum of energy is given to the vibration and there is a rise in the cross-section; similar but smaller effects occur at $2h\nu$, $3h\nu$, etc., until for $E \gg h\nu$, the curve flattens out at $\sigma(\text{free})$. In the case of faster neutrons (say $E \gtrsim 50$ eV.), the chemical binding effect is quite negligible and can be ignored, the protons in the molecule being treated effectively as free.

6.10. The scattering of thermal neutrons by ortho- and para-hydrogen

Just as there are parallel and anti-parallel spin states of the deuteron, so the hydrogen molecule can exist in states for which the two proton spins are respectively parallel and anti-parallel. An ortho-hydrogen molecule has a total spin of 1 so that the former has a symmetric spin function and the latter an anti-symmetric spin function; hence ortho- and para-hydrogen have statistical weights of 3 and 1 respectively.

The primary object of thermal neutron scattering experiments is to measure the total cross-section near zero energy and hence to show the spin dependence of the n-p interaction; furthermore accurate values of the zero energy scattering lengths and the range of nuclear force are obtained and the singlet deuteron state is shown to be virtual.

The most direct procedure is to compare the transmissions (§§ 2.3, 2.4) of samples of hydrogen gas containing different proportions of ortho- and para-hydrogen for slow neutrons. At energies of the order of eV. the cross-section varies very little, hence the values obtained can be regarded as the limiting cross-section at zero energy.

The earliest experiments in this field were not very accurate and as the nuclear range of force is very sensitive to the ortho-hydrogen cross-section, the rather ridiculous result of a negative range was obtained (Rosenfeld, 1948). However, later experiments at Los Alamos (Sutton, Hall, Anderson, Bridge, De Wire, Long, Snyder and Williams, 1947) gave reasonable results. Neutrons were moderated to sub-thermal energies by paraffin soaked in liquid air† and a narrow band of energies between 0·0008 and 0·0025 eV. (10–30° K.) was selected by the time-of-flight technique (§ 4.4.1). The scattering material was hydrogen gas enclosed in a tube at the temperature of liquid hydrogen boiling at 60 cm. pressure. Two samples of gas were used successively, one consisting of the normal hydrogen mixture of 75 % ortho-hydrogen and 25 % para-hydrogen, and the other of a pure sample of para-hydrogen.

Para-hydrogen was prepared by condensing liquid hydrogen over activated charcoal for two days, 99 % purity being obtained. The charcoal acts as a catalyst, speeding up the normally slow low temperature spin transition. The reason for the transition follows from the exclusion principle, which states that the total wave function, as

† Temperatures above 260° K. induce transitions from para- to ortho-hydrogen.

given by the product of the spin and spatial wave functions, must be anti-symmetric. As para-hydrogen has an anti-symmetric spin function, it must have a symmetric orbital wave function, and hence can exist only in the states of even orbital angular momentum, $l = 0, 2, 4, \ldots$ Similarly ortho-hydrogen can exist only in states of odd l, $l = 1, 3, 5, \ldots$ At very low temperatures, almost all the H_2 molecules are in the two ground states $l = 0$ and $l = 1$, the $l = 0$ state having the lower energy of the two. Hence by cooling sufficiently and using a catalyst, the ortho-hydrogen can undergo a spin transition and pass over into the lower energy $l = 0$ ground state of para-hydrogen.

The scattering cross-section for ortho-hydrogen should be larger than for para-hydrogen because for parallel proton spins the scattered waves from the two protons interfere constructively, whereas for anti-parallel spins interference is destructive. The total cross-section is

$$\sigma_t = f_o \sigma_o + f_p \sigma_p + \sigma_c, \qquad (6.45)$$

where f_o and f_p are the fractions of hydrogen molecules in the ortho- and para-states respectively and σ_o and σ_p are the corresponding scattering cross-sections. σ_c is the capture cross-section per molecule.

For normal hydrogen, equation (6.45) reduces to

$$\sigma_{t_1} = \tfrac{3}{4} \sigma_o + \tfrac{1}{4} \sigma_p + \sigma_c,$$

For pure para-hydrogen, equation (6.45) becomes

$$\sigma_{t_2} = \sigma_p + \sigma_c.$$

σ_c is obtained by extrapolation from the measured value at $0 \cdot 025$ eV. by assuming that it varies inversely with the neutron velocity; thus if the transmission is measured for the two cases, σ_o and σ_p follow immediately.

The cross-section for para- and ortho-hydrogen may be derived in terms of the zero energy scattering amplitudes a_0 and a_1, where the latter are the amplitudes of the scattered neutron wave for singlet and triplet states respectively for scattering by a free proton:

$$a_{0,1} = \lim_{k \to 0} \left(-\frac{\tan \delta^{0,1}}{k} \right). \qquad (6.46)$$

The total nuclear spin operator S of a system containing a neutron and a proton is given in terms of neutron and proton spin operators S_n, S_p:

$$S = S_n + S_p, \qquad (6.47)$$

where S, S_n and S_p are half the Pauli spin operators σ, σ_n, σ_p respectively. Squaring equation (6.47),

$$S^2 = S_n^2 + S_p^2 + 2S_n \cdot S_p.$$

S^2, S_n^2, S_p^2 are constants of motion with eigenvalues $S(S+1)$, $S_n(S_n+1)$ and $S_p(S_p+1)$ respectively, with $S_n = S_p = \frac{1}{2}$. Hence

$$S_n \cdot S_p = \frac{1}{2}S(S+1) - \frac{3}{4}.$$

Writing $\sigma = 2S$ in units of $\hbar$, then

$$\sigma_n \cdot \sigma_p = 2S(S+1) - 3$$
$$= 1 \text{ for triplet } (S=1),$$
$$= -3 \text{ for singlet } (S=0).$$

The scattered amplitude for a single proton is

$$A = \tfrac{1}{4}(1 - \sigma_n \cdot \sigma_p) a_0 + \tfrac{1}{4}(3 + \sigma_n \cdot \sigma_p) a_1$$
$$= \tfrac{1}{4}(a_0 + 3a_1) + \tfrac{1}{4}(a_1 - a_0)\sigma_n \cdot \sigma_p. \qquad (6.48)$$

Equation (6.48) reduces to a_1 and a_0 for triplet and singlet states respectively and hence is taken as the basis of the theory. As the molecular separation of the protons is assumed small compared to the incident neutron wavelength, then to a first approximation the small phase difference in scattering may be neglected and the amplitudes may be added directly. The scattered amplitude A for a hydrogen molecule becomes

$$A = \tfrac{1}{2}(a_0 + 3a_1) + \tfrac{1}{2}(a_1 - a_0)\sigma_n \cdot S_H,$$

where $S_H = \frac{1}{2}(\sigma_{p_1} + \sigma_{p_2})$ is the total spin of the hydrogen molecule.

The scattered intensity $d\sigma/d\omega$ is proportional to $|A|^2$ and must be averaged over all polarizations of the neutron beam. We have

$$A^2 = \tfrac{1}{4}(a_0 + 3a_1)^2 + \tfrac{1}{4}(a_1 - a_0)^2(\sigma_n \cdot S_H)^2$$
$$+ \tfrac{1}{2}(a_0 + 3a_1)(a_1 - a_0)(\sigma_n \cdot S_H),$$

where the average of $\sigma_n \cdot S_H$ is 0 (obtained by expansion in Cartesian co-ordinates) and the average of $(\sigma_n \cdot S_H)^2$ is

$$S_H^2 = S_H(S_H + 1). \qquad (6.49)$$

Here S_H has the value 1 and 0 for ortho- and para-hydrogen respectively so that unless $a_1 = a_0$ the ortho-hydrogen cross-section must be larger than the para-hydrogen cross-section.

However we have neglected the chemical bond effect, the slight phase shift due to the finite proton separation, and also the effect

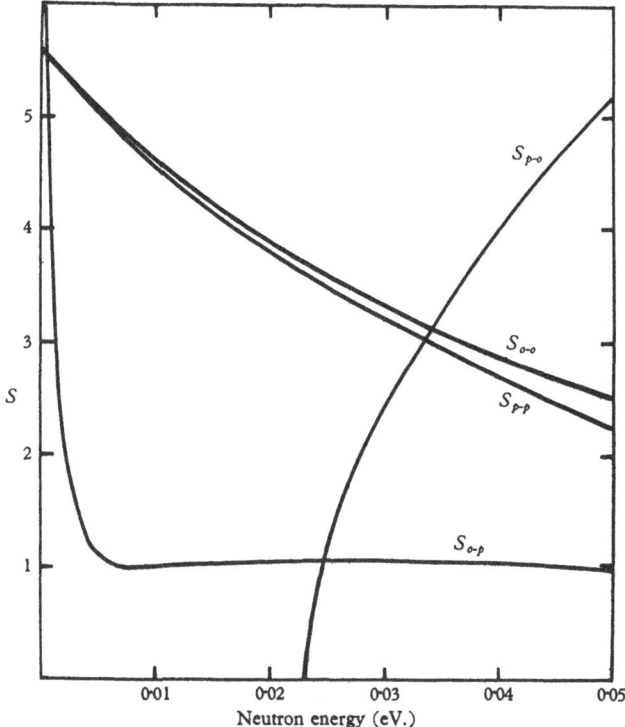

Fig. 56. Functions $S_{p \to p}$, $S_{o \to o}$, $S_{o \to p}$ and $S_{p \to o}$ plotted vs. neutron energy E. $S_{p \to p}$ and $S_{o \to o}$ refer to elastic scattering and $S_{o \to p}$ and $S_{p \to o}$ to inelastic transitions (Rosenfeld, 1948).

of molecular motion. Investigation of these effects (Schwinger and Teller, 1937; Schwinger, 1940) leads to the result

$$\sigma(\text{para}) = S_{p \to p}(3a_1 + a_0)^2 + S_{p \to o}(a_1 - a_0)^2,$$

$$\sigma(\text{ortho}) = S_{o \to o}(3a_1 + a_0)^2 + (2S_{o \to o} + S_{o \to p})(a_1 - a_0)^2, \quad (6.50)$$

where the four S's are functions of energy (fig. 56). $S_{p \to p}$ and $S_{o \to o}$ refer to elastic scattering in the lowest para- and ortho-states, whereas $S_{o \to p}$ and $S_{p \to o}$ describe inelastic scattering processes (spin transitions). In the very low energy region, $S_{o \to o} \approx S_{p \to p}$ and $S_{p \to o} = 0$. The experiments of Sutton *et al.* (1947) give σ(ortho) and σ(para) in the form shown in fig. 57 as a function of neutron energy, the ortho-hydrogen cross-section being about 30 times the para-hydrogen cross-section, which overwhelmingly proves the *n-p*

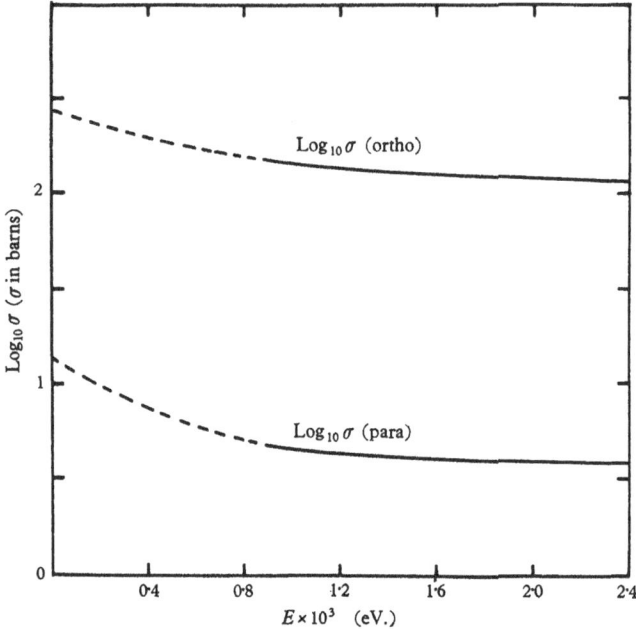

Fig. 57. The logarithms of the cross-sections σ(ortho) and σ(para) as functions of neutron energy E (Sutton *et al.* 1947). Values below 0.8×10^{-3} eV. are interpolated using the deduced zero energy *n-p* scattering lengths

$$a_1 = 0.522 \times 10^{-12} \text{ cm.}, \quad a_0 = -2.34 \times 10^{12} \text{ cm.}$$

force to be spin-dependent. From (6.50), two equations are obtained, leading to

$$a_1 = \pm 0.522 \times 10^{-12} \text{ cm.}, \quad a_0 = \pm 2.34 \times 10^{-12} \text{ cm.}$$

As the triplet state is known to be bound, a_1 must be positive, i.e. $a_1 = +0.522 \times 10^{-12}$ cm. To find the sign of a_0, we observe that opposite signs for a_1 and a_0 lead to the value 30 for the ratio σ(ortho)/σ(para) in agreement with experiment, whereas if a_1 and

a_0 are both taken with the same sign, the ratio is approximately 1·4. Hence

$$a_1 = +0\cdot522 \times 10^{-12}\,\text{cm.}, \quad a_0 = -2\cdot34 \times 10^{-12}\,\text{cm.}$$

From §6.6, the negative sign of a_0 means that the singlet state of the deuteron is virtual.

Again, if we try to see whether a neutron spin other than $\frac{1}{2}$ is possible and use the value $\frac{3}{2}$ say, the scattering amplitude then becomes

$$A = \tfrac{1}{8}(3a_0 + 5a_1) + \tfrac{1}{8}(a_1 - a_0)\,\boldsymbol{\sigma}_n \cdot \boldsymbol{\sigma}_p,$$

leading to the result for scattering by ortho- or para-hydrogen:

$$d\sigma/d\omega = \tfrac{1}{16}(3a_0 + 5a_1)^2 + \tfrac{1}{16}(a_1 - a_0)^2\, S_H(S_H + 1).$$

This gives the ratio $\sigma(\text{ortho})/\sigma(\text{para}) \approx 2$, which is much below the correct value of 30. Thus the neutron spin is $\frac{1}{2}$ and not $\frac{3}{2}$ or some other such half-integral value.

From equation (6.25), the cross-section σ for scattering of slow neutrons by a free proton is $\sigma = \frac{3}{4}\sigma_1 + \frac{1}{4}\sigma_0$, where $\sigma_1 = 4\pi a_1^2$ and $\sigma_0 = 4\pi a_0^2$ are the triplet and singlet n-p cross-sections respectively. The value thus obtained for σ is 19·7 b., compared to $(20\cdot36 \pm 0\cdot10)$ b., measured directly (Jones, 1948; Melkonian, 1949) by using some hydrogenous substance such as paraffin for the scatterer; thus the ortho- and para-hydrogen experiments check well with ordinary n-p scattering.

To find the effective range r_0, we use the shape-independent approximation (6.34), obtaining

$$r_0 = (1\cdot6 \pm 0\cdot2) \times 10^{-13}\,\text{cm.} \qquad (6.51)$$

The intrinsic range can be found for the various well-shapes from equation (6.41). For the rectangular well, the intrinsic range b is equal to the ordinary range so that equation (6.16) may be used:

$$\delta = \arctan\left(\frac{k}{\kappa'}\tan\kappa' b\right) - kb$$

$$\rightarrow k\left(\frac{\tan\kappa' b}{\kappa'} - b\right), \quad (k \rightarrow 0).$$

Hence for zero energy,

$$a_1 = b - \frac{\tan\kappa' b}{\kappa'}, \qquad (6.52)$$

where $\kappa'^2 = (m/\hbar^2) V_0$. We can also use the relation for the binding energy of the deuteron:

$$\cot \kappa b = -\alpha/\kappa, \qquad (6.53)$$

where $\qquad \kappa^2 = \dfrac{m}{\hbar^2}(V_0 - W), \quad \alpha^2 = \dfrac{m}{\hbar^2} W.$

The nuclear potential V_0 can be eliminated from equation (6.52) and equation (6.53), giving the infinite series

$$\alpha^2 a_1^2 = 1 + \alpha b + 0.3447 \alpha^2 b^2 + 0.0246 \alpha^3 b^3 + \dots \qquad (6.54)$$

The best fit for the experimental a_1 is given by the triplet range value $b = (1.54 \pm 0.4) \times 10^{-13}$ cm.

The singlet state calculation depends too much on the exact value of the virtual energy of the deuteron to obtain a definite range value. However, the singlet amplitude is in rough agreement with a range value $2.6 - 2.8 \times 10^{-13}$ cm., checking with the value deduced from p-p scattering experiments (§ 7.9).

Summary. The experiments on the scattering of sub-thermal neutrons by ortho- and para-hydrogen prove the following points:

(a) The n-p force is spin-dependent.

(b) The singlet deuteron level is virtual.

(c) The zero energy n-p cross-section is confirmed to be about 20 barns.

(d) The effective range of the triplet n-p potential is

$$(1.6 \pm 0.2) \times 10^{-13} \text{ cm.}$$

The intrinsic range for the rectangular well is

$$(1.54 \pm 0.4) \times 10^{-13} \text{ cm.,}$$

but the singlet range can only be said to be compatible with the p-p singlet range of $2.6 - 2.8 \times 10^{-13}$ cm.

(e) The neutron spin is $\tfrac{1}{2}$.

6.11. The diffraction of thermal neutrons by crystalline hydrides

A different approach to the same problem has been developed by Shull, Wollan, Morton and Davidson, (1948) which confirms the results of the ortho- and para-hydrogen experiments. Shull's method uses the diffraction of neutrons by some crystalline hydride, sodium hydride being chosen because of its ease of preparation and low absorption cross-section.

An intense source of mono-energetic thermal neutrons obtained from a pile is directed at a powdered sample of NaH and the

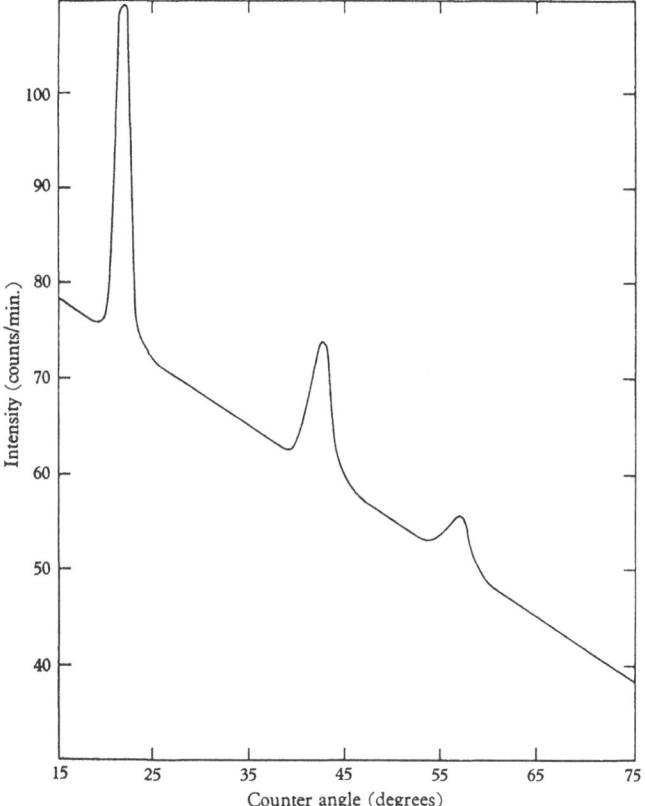

Fig. 58. Scattered intensity as a function of angle for an intense source of thermal neutrons diffracted by NaH crystals (Shull *et al.* 1948).

scattered intensity is measured at different angles. The intensity-angle graph (fig. 58) has several sharp resonances, of which only the first two peaks can be determined accurately. The intensities of the diffraction peaks can be used to determine the coherent scattering cross-section of hydrogen, employing crystal structure theory, and this permits a determination of the scattering amplitude

in the usual manner. The coherent scattering amplitude f_H is given (Shull *et al.* 1948) by
$$f_H = 2(\tfrac{3}{4}a_1 + \tfrac{1}{4}a_0). \qquad (6.55)$$

The scattering cross-section for the free proton is
$$\sigma = 4\pi(\tfrac{3}{4}a_1^2 + \tfrac{1}{4}a_0^2). \qquad (6.56)$$

On solving equations (6.42) and (6.43), $a_1 = +0.520 \times 10^{-13}$ cm., leading to a rectangular well range for the triplet $n\text{-}p$ interaction of $r_0 = (1.6 \pm 0.2) \times 10^{-13}$ cm., in excellent agreement with Sutton and Hall's value of $(1.54 \pm 0.4) \times 10^{-13}$ cm. (§ 6.10).

6.12. The liquid-mirror reflexion of neutrons

Another more accurate method of finding the effective range is by total reflexion of thermal neutrons from a liquid mirror (§ 4.4.2) containing protons. The refractive index of the liquid chosen must be less than unity so that neutrons impinging on the surface at glancing angles are totally reflected. Hughes and co-workers (Hughes, Burgy and Ringo, 1950; Ringo, Burgy and Hughes, 1951) reflected thermal neutrons at the critical angle from tri-ethyl benzene ($C_{12}H_{18}$), a liquid with a H/C ratio of 1.5. The critical angle θ_c is related to the de Broglie wavelength λ_c of the neutrons and the C and H scattering lengths f_C, f_H by the formula
$$\theta_c = \lambda_c [N_v(f_C + 1.5f_H)/\pi]^{\frac{1}{2}}, \qquad (6.57)$$

where N_v is the number of C atoms/cm.^{3}. The measured value is $f_H = -(3.78 \pm 0.02) \times 10^{-13}$ cm., which leads to the parameters for triplet and singlet states respectively (Ringo *et al.* 1951; Salpeter, 1951):

$$\left.\begin{array}{l} a_1 = (0.538 \pm 0.002) \times 10^{-12} \text{ cm.,} \\ r_0 = (1.70 \pm 0.03) \times 10^{-13} \text{ cm.,} \end{array}\right\} \text{triplet,}$$

$$\left.\begin{array}{l} a_0 = (-2.369 \pm 0.006) \times 10^{-12} \text{ cm.,} \\ r_0 = (2.7 \pm 0.5) \times 10^{-13} \text{ cm.} \end{array}\right\} \text{singlet.} \qquad (6.58)$$

This is within the error of the methods of § 6.10 and § 6.11 but is far more accurate.

6.13. The capture of neutrons by protons

At very low energies the capture of neutrons by protons with the emission of γ-rays can take place:
$$n + p \rightarrow {}^2H + h\nu. \qquad (6.59)$$

This is the inverse process to the photodisintegration of the deuteron so that the capture cross-section may be obtained from the latter using the principle of detailed balance. If a box contains an initial state A and a final state B of a reversible interaction, then at equilibrium the number of nuclei in the energy range between E and $E + dE$ is proportional to the density of states $\rho(E)$ and to the Boltzmann factor $e^{-E/kT}$, where k is the Boltzmann constant and T the absolute temperature. The conservation of energy relation for the whole system leads to the cancellation of the Boltzmann factors and to the equilibrium relation

$$\rho_A W_{A \to B} = \rho_B W_{B \to A}, \tag{6.60}$$

where ρ_A and ρ_B are the densities per unit energy of states A and B respectively and the W's represent probabilities for the direct and inverse processes. The ρ's are given in each case by

$$gh^{-3}\, d\mathbf{p}/dE = gh^{-3} 4\pi p^2\, dp/dE \tag{6.61}$$

per unit volume of the box, where g is the statistical weight of the states and the probabilities are related to the corresponding cross-sections by

$$W_{A \to B} = v_A\, \sigma_{A \to B}, \quad W_{B \to A} = v_B\, \sigma_{B \to A}. \tag{6.62}$$

Using the relativistic formulae (Appendix A)

$$E^2 = c^2(p^2 + m^2 c^2), \quad Ev = c^2 p, \tag{6.63}$$

we find $\qquad\qquad dp/dE = 1/v. \tag{6.64}$

Equation (6.61) thus results in a general relation, often called the reciprocity theorem:

$$g_A\, p_A^2\, \sigma_{A \to B} = g_B\, p_B^2\, \sigma_{B \to A}. \tag{6.65}$$

g_A is taken as the product of the neutron and proton weighting factors, both of which are 2, corresponding to the two directions of spin, giving $g_A = 4$. Similarly, g_B is the product of the deuteron and γ-ray weighting factors, these being 3 and 2 respectively, corresponding to the triplet deuteron ground state and the two possible polarization directions for a photon. Thus $g_B = 6$. Also $p_A = \mu v$, $p_B = h\nu/c$. Hence the capture cross-section is given in terms of the photodisintegration cross-section as follows:

$$\sigma_{\text{cap}} = \frac{3h^2\nu^2}{2\mu^2 v^2 c^2}\, \sigma_{\text{dis}}. \tag{6.66}$$

The cross-section for photodisintegration is (Heitler, 1954)

$$\sigma_{\text{dis}} = \frac{4\pi\nu\mu^2 v}{\hbar^3 c} \mid H \mid^2, \tag{6.67}$$

where H is the matrix element for the transition of the electric or magnetic dipole moment, ν is the frequency of the radiation and $\mu = \frac{1}{2}m$ is the reduced mass of the system.

The electric interaction involves the matrix element

$$H_{\text{el}} = \frac{1}{2}e\boldsymbol{\epsilon} \int \psi_f^* \mathbf{r} \psi_0 \, d\mathbf{r}, \tag{6.68}$$

where $\boldsymbol{\epsilon}$ is the polarization vector for the incident photon. This vanishes unless the final state is a p-state. However, at low energies we can neglect the interaction between neutron and proton in the p-state and take instead the wave function for a free particle of angular momentum 1 for ψ_f. The initial function ψ_0 is approximately equal to

$$\psi_0 = (\alpha/2\pi)^{\frac{1}{2}}(1 + \frac{1}{2}\alpha b) e^{-\alpha r}, \tag{6.69}$$

where b is the force range. Thus

$$H_{\text{el}} = \frac{1}{2}e(\alpha/2\pi)^{\frac{1}{2}}(1 + \frac{1}{2}\alpha b) \int e^{-i\mathbf{k} \cdot \mathbf{r}} \, \mathbf{r} \frac{e^{-\alpha r}}{r} \, d\mathbf{r}. \tag{6.70}$$

This leads to the cross-section

$$\sigma_{\text{el}} = \frac{8\pi}{3} \left(\frac{e^2}{\hbar c} \right) \frac{1}{\alpha^2} \frac{(W_1 E)^{\frac{3}{2}}}{(W_1 + E)^3}(1 + \alpha b) = \sigma_0(1 + \alpha b), \tag{6.71}$$

where W_1 is the energy of the triplet deuteron state and $E = h\nu - W_1$. $(1 + \alpha b)$ is the range correction to the zero range approximation σ_0.

The corresponding calculation using non-central forces leads to an almost identical result (Feshbach and Schwinger, 1951) but replacing $(1 + \alpha b)$ by $1/(1 - \alpha r_0)$, where r_0 is the triplet effective range:

$$\sigma_{\text{el}} = \sigma_0/(1 - \alpha r_0). \tag{6.72}$$

Photomagnetic disintegration uses the magnetic dipole moment

$$(e\hbar/2mc)(\mu_p \boldsymbol{\sigma}_p + \mu_n \boldsymbol{\sigma}_n), \tag{6.73}$$

the matrix element for the transition becoming

$$H = (e\hbar/2mc) \sum_{\text{spin}} \chi_1^*(\mu_p \boldsymbol{\sigma}_p + \mu_n \boldsymbol{\sigma}_n) \chi_0 \int \psi_f^* \psi_0 \, d\mathbf{r}, \tag{6.74}$$

where χ_1 and χ_0 are the triplet and singlet spin states respectively. The final state must have $l=0$ to obtain a non-vanishing matrix element and must be ^1S as the excited ^3S states are orthogonal to the ground state. We thus find for the photomagnetic cross-section:

$$\sigma_m = \tfrac{2}{3}\pi \left(\frac{e^2}{\hbar c}\right)\left(\frac{\hbar}{mc}\right)^2 (\mu_p - \mu_n)^2 \frac{(W_1 E)^{\frac{1}{2}}(W_1^{\frac{1}{2}} + W_0^{\frac{1}{2}})^2}{(E+W_1)(E+W_0)}, \quad (6.75)$$

where W_0 is the energy of the virtual ^1S state of the deuteron. The introduction of non-central forces leads to a rather complicated singlet and triplet range correction to equation (6.75) (see Feshbach and Schwinger, 1951; Bethe and Longmire, 1954).

At high energies the electric contribution is much greater than the magnetic, but for $E \ll W_1 \approx 2.2$ MeV., the magnetic cross-section is greater by a factor $0.013 W_1^2/[E(W_0+E)]$. At thermal energies neutron capture is important, hence σ_{el} may be neglected in evaluating the capture cross-section. Thus

$$\sigma_{\text{cap}} = \frac{3}{2}\frac{(E+W_1)^2}{mc^2 E}\sigma_m$$

$$= \pi \frac{e^2}{(mc^2)^2}\frac{\hbar}{mc}(\mu_p - \mu_n)^2 \left(\frac{W_1}{E}\right)^{\frac{1}{2}}\frac{(W_1^{\frac{1}{2}} + W_0^{\frac{1}{2}})^2}{(W_0+E)}, \quad (6.76)$$

where E is the CM energy of the neutron or proton. For $E \to 0$, σ_{cap} is proportional to $E^{-\frac{1}{2}}$, i.e. to $1/v$, so that for thermal neutrons, $E \approx 0.025$ eV., $\sigma_{\text{cap}} \sim 0.30$ b., in good agreement with experiment.

If the correction for non-zero triplet and singlet effective range is made in equation (6.76), the singlet range may be deduced by equating it to experimental values. The result is between 2 and 3×10^{-13} cm., in qualitative agreement with scattering data.

212

CHAPTER 7

LOW ENERGY PROTON-PROTON
SCATTERING

7.1. Introduction

Compared with neutron-proton scattering, proton-proton scattering appears complicated by the addition of Coulomb repulsion
to the potential. However, a number of factors combine to offset
this:

(a) On the experimental side, p-p data are decidedly more precise,
for reasons brought out in chapters 3 and 4.

(b) As p-p scattering involves identical particles, Fermi statistics
apply and the Pauli principle excludes the totally symmetric states
($^3S, ^1P, ^3D, ...$). Removal of half the spin states greatly simplifies
interpretation of the experiments, especially below about 15 MeV.
where only ^1S scattering is appreciable (cf. §§ 6.2, 10.1).

(c) Interference effects between Coulomb and nuclear potentials
can be used to provide information regarding the sign of the phase
shifts.

7.2. Wave mechanical treatment of charged-particle collisions neglecting the nuclear potential

For the case of particles of equal mass m, without spin, impinging
on each other in two streams, the wave equation for an interaction
$V(r)$ is

$$\nabla^2\psi + \frac{2\mu}{\hbar^2}[E - V(r)]\psi = 0, \qquad (7.1)$$

where the reduced mass μ is half the proton mass m. Using the
convention of taking the impinging particles travelling along the
z-axis and being scattered in a direction (r, θ), the solution has the
asymptotic form for large r

$$\psi \sim e^{ikz} + f(\theta)\, e^{ikr}/r. \qquad (7.2)$$

The particles obey Bose–Einstein statistics, and the wave function
ψ, given as a function $\psi(\mathbf{r}_1, \mathbf{r}_2)$ of the positions of the two particles,
must be symmetric as the latter are indistinguishable experi-
mentally. It has the form $\psi(\mathbf{r}_1, \mathbf{r}_2) + \psi(\mathbf{r}_2, \mathbf{r}_1)$, where the effect of

interchanging r_1 and r_2 is to replace $\mathbf{r}$ by $-\mathbf{r}$. This is equivalent to keeping r unaltered and replacing θ by $(\pi - \theta)$. Hence

$$\psi \sim e^{ikz} + e^{-ikz} + [f(\theta) + f(\pi - \theta)] e^{ikr}/r$$

$$\sim 2 \cos kz + [f(\theta) + f(\pi - \theta)] e^{ikr}/r, \quad r \to \infty. \quad (7\cdot3)$$

As the scattering amplitude is $[f(\theta) + f(\pi - \theta)]$, the effective cross-section for scattering of identical particles becomes

$$| f(\theta) + f(\pi - \theta) |^2 \, d\omega, \quad (7.4)$$

where $d\omega$ is an element of solid angle into which the particles are scattered at an angle θ with their incident direction. An example of particles without spin to which equation (7.4) applies involves the scattering of α-particles by helium.

However, for particles with spin $\frac{1}{2}$ (neutrons or protons), Fermi statistics apply so that the total wave function, given as a product of spin and spatial functions, must be anti-symmetric. If the spins are parallel, the spin wave function is symmetric and the spatial wave function anti-symmetric. From equation (7.2), the cross-section for scattering into solid angle $d\omega$ is

$$| f(\theta) - f(\pi - \theta) |^2 \, d\omega. \quad (7.5)$$

On the other hand if the particles have anti-parallel spins, their spin function is anti-symmetric and their spatial wave function symmetric, so that the cross-section is

$$| f(\theta) + f(\pi - \theta) |^2 \, d\omega. \quad (7.6)$$

Assigning statistical weights 3 and 1 (the number of spin states in each case) to equations (7.5) and (7.6) respectively and taking the sum, the differential cross-section in the CM system is

$$\frac{d\sigma}{d\omega} = \tfrac{3}{4} | f(\theta) - f(\pi - \theta) |^2 + \tfrac{1}{4} | f(\theta) + f(\pi - \theta) |^2$$

$$= | f(\theta) |^2 + | f(\pi - \theta) |^2 - | f(\theta) f(\pi - \theta) |. \quad (7.7)$$

Equation (7.7) contains ordinary, exchange and interference terms respectively. $f(\theta)$ was first derived for a Coulomb field by Mott (cf. Mott and Massey, 1949, p. 34).

Writing $k^2 = (m/\hbar^2) E$, $V(r) = e^2/r$, $\beta = (m/\hbar^2) e^2$, the wave equation (7.1) becomes $\nabla^2 \psi + (k^2 - \beta/r) \psi = 0. \quad (7.8)$

Substituting $\psi = e^{ikz} F$, the transformed equation

$$\nabla^2 F + 2ik \frac{\partial F}{\partial z} - \beta \frac{F}{r} = 0 \quad (7.9)$$

has a solution of type $F(r-z)$ from equation (7.2). Substitution of $F(r-z)$ into equation (7.9) leads to

$$2\left(1-\frac{z}{r}\right)\frac{d^2F}{dr^2}+\frac{2}{r}\frac{dF}{dr}+2ik\left(\frac{z}{r}-1\right)\frac{dF}{dr}-\frac{\beta}{r}F=0. \qquad (7.10)$$

or

$$\zeta\frac{d^2F}{d\zeta^2}+(1-ik\zeta)\frac{dF}{d\zeta}-\tfrac{1}{2}\beta F=0, \qquad (7.11)$$

where $\zeta=r-z$. The solution to equation (7.11) is the hypergeometric function

$$F=F_1(-i\alpha,1,ik\zeta)=\sum_{n=0}^{\infty}(ik)^{n+1}\zeta^n\prod_{s=0}^{n}(s+\tfrac{1}{2}\beta/ik)/(s+1)^2, \qquad (7.12)$$

where $\alpha=\tfrac{1}{2}\beta/k=e^2/\hbar v$. Also $F=W_1+W_2$, where for r large it can be shown that

$$\left.\begin{aligned} W_1 &\sim \frac{e^{\frac{1}{2}\pi\alpha}}{\Gamma(1+i\alpha)}\left(1-\frac{\alpha^2}{ik\zeta}\right)\exp(i\alpha\ln k\zeta),\\ W_2 &\sim -\frac{i\,e^{-\frac{1}{2}\pi\alpha}}{\Gamma(-i\alpha)}\cdot\frac{e^{ik\zeta}}{k\zeta}\exp(-i\alpha\ln k\zeta), \end{aligned}\right\} \qquad (7.13)$$

where $W_1 e^{ikz}$ and $W_2 e^{ikz}$ are incident and scattered waves respectively. As an incident wave of unit amplitude is required, we take the total wave function as

$$\psi(r,\theta)=e^{-\frac{1}{2}\pi\alpha}\Gamma(1+i\alpha)\,e^{ikz}F_1(-i\alpha,1,ik\zeta). \qquad (7.14)$$

This has the asymptotic form for a pure Coulomb field:

$$\psi^c\sim I+Sf^c(\theta), \qquad (7.15)$$

where $\quad I=[1+\alpha^2/ik(r-z)]\exp[ikz+i\alpha\ln k(r-z)]$,

$$\left.\begin{aligned} S &=\frac{1}{r}\exp[ikr-i\alpha\ln kr], \quad \sigma_l=\arg\Gamma(i\alpha+l+1),\\ f^c(\theta) &=\frac{e^2}{mv^2}\operatorname{cosec}^2\frac{\theta}{2}\cdot\exp\left[-i\alpha\ln\left(\sin^2\frac{\theta}{2}+\right)i\pi+2i\sigma_0\right]. \end{aligned}\right\} \qquad (7.16)$$

Substituting the scattered amplitude $f^c(\theta)$ in equation (7.7), we obtain the Mott scattering formula (CM system):

$$\frac{d\sigma}{d\omega}=\left(\frac{e^2}{mv^2}\right)^2\left[\frac{1}{\sin^4\frac{1}{2}\theta}+\frac{1}{\cos^4\frac{1}{2}\theta}-\frac{\Phi}{\sin^2\frac{1}{2}\theta\cos^2\frac{1}{2}\theta}\right], \qquad (7.17)$$

where $\quad\quad\quad\quad \Phi=\cos(\alpha\ln\tan^2\tfrac{1}{2}\theta)$.

For an incident energy $E\gtrsim 1$ MeV., $\alpha<\tfrac{1}{7}$, so $\Phi\approx 1$ except near $\theta=0$ and $\tfrac{1}{2}\pi$. Equation (7.17) compares with the Rutherford formula obtained by classical methods:

$$\frac{d\sigma}{d\omega}=\left(\frac{e^2}{mv^2}\right)^2\left[\frac{1}{\sin^4\frac{1}{2}\theta}+\frac{1}{\cos^4\frac{1}{2}\theta}\right], \qquad (7.18)$$

the two components allowing for a scattered particle at an angle θ and an identical recoil particle at an angle $(\pi - \theta)$ in the CM system. The extra term $\Phi \cosec^2 \tfrac{1}{2}\theta \sec^2 \tfrac{1}{2}\theta$ in the Mott formula is

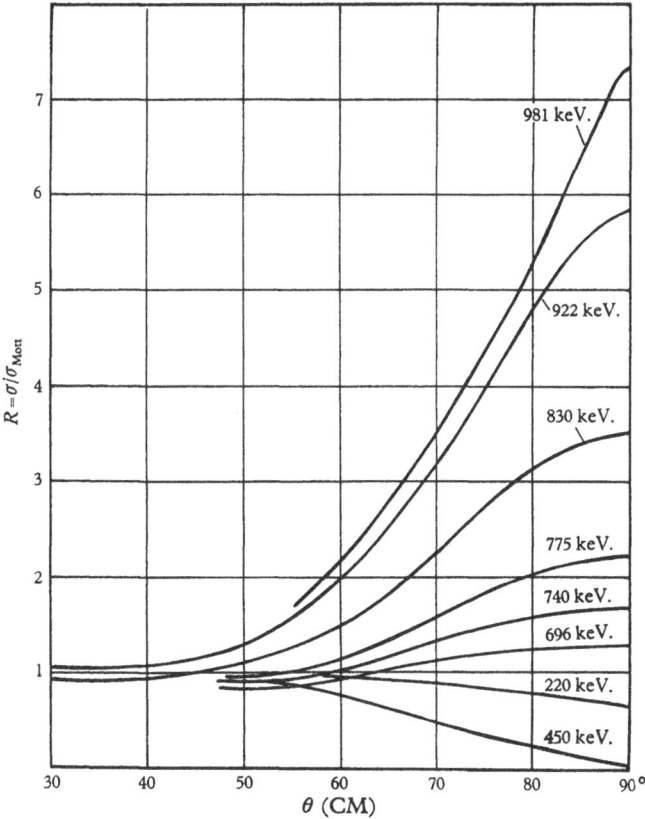

Fig. 59. The ratio R of measured p-p scattering to Mott scattering as a function of scattering angle in the CM system (Tuve *et al.* 1936; Hafstad *et al.* 1938).

due to interference between the wave functions of the incident and recoil protons, and has no classical analogy.

However early experiments (Tuve, Heydenburg and Hafstad, 1936) at 0·6–1 MeV. showed a systematic departure from Mott's formula (7.17), which increased with incident energy. If the ratio R of actual scattering to Mott scattering is plotted against the angular distribution of the scattered protons, R at $\theta = 90°$ (CM) increases from about 1·08 to 696 keV. to about 7· at 981 keV. (fig. 59).

This indicates the presence of a potential other than Coulomb. The rapid increase of R with energy is due to the incident particle coming close enough to the short range nuclear well to interact strongly with the target proton.

7.3. Effect of the nuclear potential

Assuming that the nuclear potential between two protons has a short range b similar to that of the n-p potential, p-p scattering at low energies should involve only s-wave scattering. The problem of a mixed Coulomb and nuclear field was first solved for the anomalous scattering of α-particles by hydrogen (Taylor, 1932) and later extended to p-p scattering (Breit, Condon and Present, 1936).

In a pure Coulomb field and in the CM system, the asymptotic solution at large distances of the wave equation is found from equations (7.15) and (7.16) to be $\psi^c \sim I + Sf^c(\theta)$. I is the incident wave, being almost a plane wave but with a small space-dependent phase shift due to the long range Coulomb potential. $Sf(\theta)$ is the spherical scattered wave and $|f(\theta)|^2$ the cross-section per unit solid angle.

First we consider the effect of the nuclear force, ignoring the identity of the two protons, and correcting only the $l = 0$ component of ψ^c. An expansion of $\psi^c(r)$ is made in terms of Legendre polynomials:

$$\psi^c(r) = \sum_{l=0}^{\infty} A_l P_l(\cos\theta) L_l^c(r).\qquad(7.19)$$

The wave equation for a nuclear field is

$$\nabla^2 \psi + \frac{m}{\hbar^2}[E - V(r)]\,\psi = 0,\qquad(7.20)$$

where
$$V(r) = f(r),\quad r < b,$$
$$= e^2/r,\quad r > b.$$

The general solution of the wave equation which possesses axial symmetry can be written

$$\psi = \sum_{l=0}^{\infty} A_l P_l(\cos\theta) L_l(r),\qquad(7.21)$$

where the A_l are arbitrary constants and the L_l are solutions of

$$\frac{d^2 L_l}{dr^2} + \frac{2}{r}\frac{dL_l}{dr} + \left[\frac{m}{\hbar^2}(E - V) - \frac{l(l+1)}{r^2}\right]L_l = 0.\qquad(7.22)$$

We now define A_l^c and L_l^c as the values of A_l and $L_l(r)$ giving the solution ψ for a pure Coulomb field, and A_l^s and L_l^s as the corresponding values for the mixed Coulomb and nuclear field.

Gordon (1928) showed that the required functions for the Coulomb field are

$$A_l^c = \frac{1}{k} i^l (2l+1) e^{i\sigma_l},$$

$$L_l^c \sim \frac{1}{r} \cos\left\{kr - (l+1)\frac{\pi}{2} - \alpha \ln 2kr + \sigma_l\right\}, \quad r \to \infty. \right\} \quad (7.23)$$

In the region outside the nuclear well, $r > b$, L^s and L^c both satisfy the same differential equation and hence are merely different combinations of the two independent solutions of this equation, given asymptotically by

$$\frac{1}{r} \exp\left[\pm i\left\{kr - (l+1)\frac{\pi}{2} - \alpha \ln 2kr + \sigma_l\right\}\right]. \quad (7.24)$$

Hence
$$L_l^s(r) \sim \frac{1}{r} \cos\left\{kr - (l+1)\frac{\pi}{2} - \alpha \ln 2kr + \sigma_l + K_l\right\}, \quad (7.25)$$

where K_l is an arbitrary phase factor.

As both ψ^s and ψ^c represent the same incident wave plus the appropriate scattered wave, their difference must be a scattered wave only so that $A^s L^s(r) - A^c L^c(r)$ is a function of e^{ikr} alone. Also

$$A_l^s = A_l^c e^{iK_l} = \frac{1}{k} i^l (2l+1) e^{i(K_l + \sigma_l)}.$$

Hence
$$A_l^s L_l^s(r) - A_l^c L_l^c(r) \sim (e^{2iK_l} - 1) A_l^c L_l^c. \quad (7.26)$$

Now ψ^c is given by equations (7.15) and (7.16), so that from equations (7.21) and (7.26), assuming pure s-wave scattering,

$$f^s(\theta) = f^c(\theta) + \frac{i}{2k}(e^{2iK_0} - 1). \quad (7.27)$$

From equations (7.27), (7.26) and (7.7), the angular distribution in the CM system is

$$\frac{d\sigma}{d\omega} = \frac{1}{4}\frac{e^4}{E_0^2}\left[\operatorname{cosec}^4\frac{\theta}{2} + \sec^4\frac{\theta}{2} - \Phi \operatorname{cosec}^2\frac{\theta}{2}\sec^2\frac{\theta}{2}\right.$$

$$\left. - \frac{2}{\alpha}\sin K_0 \cos K_0 \operatorname{cosec}^2\frac{\theta}{2}\sec^2\frac{\theta}{2} + \frac{4}{\alpha^2}\sin^2 K_0\right], \quad (7.28)$$

where $\Phi = \cos(\alpha \ln \tan^2 \tfrac{1}{2}\theta) \approx 1$ for $E > 1$ MeV. (α small), and $E_0 = \tfrac{1}{2}mv^2$ is the incident proton energy in the Lab system. Equation (7.28) reduces to the Mott formula (7.17) when $K_0 = 0$. The fourth term is due to interference between Coulomb and nuclear scattering. Its linear dependence on K_0 allows determination of whether the nuclear potential is attractive or repulsive, as attractive and repulsive potentials cause positive and negative δ_0 respectively. Actually experiments show the S potential is attractive.

For large energies, the $\sin^2 K_0$ term, which represents the linear increase of nuclear scattering with proton energy, becomes predominant.

7.4. Extension of theory to arbitrary angular momenta

For energy $E > 10$ MeV., the interactions involving angular momenta $l > 0$ become appreciable; in fact even below 10 MeV., p-wave scattering may impose a small correction on equation (7.28). The expression for the cross-section must therefore be reconsidered. We now have

$$\psi^s - \psi^c \sim \sum_l P_l(\cos\theta)\,[A_l^s L_l^s(r) - A_l^c L_l^c(r)].$$

Also
$$\psi^s - \psi^c \sim \sum_l [f^s(\theta) - f^c(\theta)]\,S.$$

Hence
$$f^s(\theta) = -\left(\frac{e^2}{mv^2}\right)\operatorname{cosec}^2\frac{\theta}{2}\exp\left(-i\alpha\ln\sin^2\frac{\theta}{2} + i\pi + 2i\sigma_0\right)$$

$$+ \frac{1}{2k}\sum_l P_l(\cos\theta)\,i^l(2l+1)\,e^{2i\sigma_l}(e^{2iK_l} - 1)\,e^{-(l-1)\frac{1}{2}\pi i}.$$

$$\tag{7.29}$$

The full expression for the angular distribution becomes

$$P = \frac{d\sigma}{d\omega} = \frac{1}{4}\frac{e^4}{E_0^2}\left[\operatorname{cosec}^4\frac{\theta}{2} + \sec^4\frac{\theta}{2} - \Phi\operatorname{cosec}^2\frac{\theta}{2}\sec^2\frac{\theta}{2}\right.$$

$$- \frac{2}{\alpha}\sum_l g_l(2l+1)\left\{\cos\phi_l^s\operatorname{cosec}^2\frac{\theta}{2} + (-1)^l\cos\phi_l^c\sec^2\frac{\theta}{2}\right\}$$

$$\times P_l(\cos\theta)\sin K_l + \frac{4}{\alpha^2}\sum_l g_l(2l+1)\{P(\cos\theta)\}^2\sin^2 K_l$$

$$+ \frac{8}{\alpha^2}\left\{\sum_{l,\,l'\,\text{even}} + 3\sum_{l,\,l'\,\text{odd}}\right\}(2l+1)(2l'+1)P_l(\cos\theta)P_{l'}(\cos\theta)$$

$$\left. \times \sin K_l \sin K_{l'}\cos(\theta_l - \phi_{l'})\right], \quad (l' < l) \tag{7.30}$$

where $g_l = 1$ and 3 for even and odd l respectively,

$$\phi_l = K_l + 2(\sigma_l - \sigma_0), \quad \phi_l^s = \phi_l + \alpha \ln \left(\sin^2 \frac{\theta}{2} \right), \quad \phi_l^c = \phi_l + \alpha \ln \left(\cos^2 \frac{\theta}{2} \right),$$

$$e^{i\sigma_l} = \Gamma(i\alpha + l + 1)/|\Gamma(i\alpha + l + 1)|,$$

$$\sigma_l = \alpha \left(-\gamma + \sum_{n=1}^{l} \frac{1}{n} \right) + \sum_{\nu=1}^{\infty} \left\{ \frac{\alpha}{l+\nu} - \arctan \frac{\alpha}{l+\nu} \right\},$$

$$\gamma = \text{Euler's constant} = 0 \cdot 577215 \ldots, \quad \sigma_l - \sigma_{l-1} = \arctan \frac{\alpha}{l+\nu}.$$

On grouping the cross-section terms in a series dependent on l,

$$P = P_M + (\Delta P)_0 + (\Delta P)_1 + (\Delta P)_2 + (\Delta P)_3 + \ldots.$$

Writing $s = \sin \frac{1}{2}\theta$, $c = \cos \frac{1}{2}\theta$, $t = \tan \frac{1}{2}\theta$:

$$\left. \begin{aligned} 4(E_0^2/e^4) P_M &= s^{-4} + c^{-4} - s^{-2}c^{-2} \cos (\alpha \ln t^2), \\[6pt] 4(E_0^2/e^4) (\Delta P)_0 &= -\frac{2}{\alpha}(s^{-2} \cos \phi_0^s + c^{-2} \cos \phi_0^c) \sin K_0 + \frac{4}{\alpha^2} \sin^2 K_0, \\[6pt] 4(E_0^2/e^4) (\Delta P)_1 &= -\frac{18}{\alpha}(s^{-2} \cos \phi_1^s - c^{-2} \cos \phi_1^c) P_1(\cos \theta) \sin K_1 \\[4pt] &\quad + \frac{108}{\alpha^2}[P_1(\cos \theta)]^2 \sin^2 K_1, \\[6pt] 4(E_0^2/e^4) (\Delta P)_2 &= -\frac{10}{\alpha}(s^{-2} \cos \phi_2^s + c^{-2} \cos \phi_2^c) P_2(\cos \theta) \sin K_2 \\[4pt] &\quad + \frac{100}{\alpha^2}[P_2(\cos \theta)]^2 \sin^2 K_2 \\[4pt] &\quad + \frac{40}{\alpha^2} \sin K_0 \sin K_2 \cos (\phi_2 - \phi_0) P_2(\cos \theta), \\[6pt] 4(E_0^2/e^4) (\Delta P)_3 &= -\frac{42}{\alpha}(s^{-2} \cos \phi_3^s - c^{-2} \cos \phi_3^c) P_3(\cos \theta) \sin K_3 \\[4pt] &\quad + \frac{588}{\alpha^2}[P_3(\cos \theta)]^2 \sin^2 K_3 \\[4pt] &\quad + \frac{504}{\alpha^2} \sin K_1 \sin K_3 \cos (\phi_3 - \phi_1) P_3(\cos \theta). \end{aligned} \right\}$$

$$(7.31)$$

P_M is simply the Coulomb scattering and $(\Delta P)_l$ denotes the scattering of protons with angular momentum l. A curious feature is that whereas s- and p-scattering is discrete, waves of higher angular

momentum undergo scattering which includes interference terms
between the different states, e.g. the $\sin K_0 \cos K_0$ terms in $(\Delta P)_2$.
The total cross-section is infinite due to the long range nature of
the Coulomb potential, the scattering at large collision distances
(small angle scattering) giving an infinite contribution.

7.5. Calculation of phase shifts for arbitrary potentials

We assume there is some interaction which may vary for different
l. Because of their spins the two protons may be in either the singlet
or triplet state. If the state is a singlet, then the orbital wave
function is symmetric and contains terms with even l only, but if
it is a triplet, the orbital wave function is anti-symmetric and
therefore contains terms with odd l only. Thus the even l occur only
in singlets and the odd l only in triplets.

Let F_l and G_l be the regular and irregular solutions respectively
of the wave equation for a pure Coulomb field, having the asymptotic form for large r

$$\left.\begin{array}{l} F_l(\alpha,\rho) \sim \sin\left(\rho - \tfrac{1}{2}l\pi - \alpha\ln 2\rho + \sigma_l\right), \\ G_l(\alpha,\rho) \sim \cos\left(\rho - \tfrac{1}{2}l\pi - \alpha\ln 2\rho + \sigma_l\right), \end{array}\right\} \quad (7.32)$$

where $\rho = kr$. Let F_{li} and G_{li} be the corresponding regular and
irregular solutions of the wave equation for the combined Coulomb
and nuclear fields, having asymptotic forms

$$\left.\begin{array}{l} F_{li}(\alpha,\rho) \sim \sin\left(\rho - \tfrac{1}{2}l\pi - \alpha\ln 2\rho + \sigma_l + K_l\right), \\ G_{li}(\alpha,\rho) \sim \cos\left(\rho - \tfrac{1}{2}l\pi - \alpha\ln 2\rho + \sigma_l + K_l\right), \end{array}\right\} \quad (7.33)$$

where the p-p phase shift is written as K_l to distinguish it from the
n-p phase shift δ_l. Then for a value of r such that the nuclear
potential is negligible (very small compared to the Coulomb field)
we may write

$$\left.\begin{array}{l} F_{li} = F_l \cos K_l + G_l \sin K_l, \\ G_{li} = G_l \cos K_l - F_l \sin K_l. \end{array}\right\} \quad (7.34)$$

Hence $\qquad \left.\begin{array}{l} F_l'F_{li} - F_lF_{li}' = (F_l'G_l - F_lG_l')\sin K_l, \\ G_lF_{li}' - G_l'F_{li} = (F_l'G_l - F_lG_l')\cos K_l, \\ \tan K_l = (F_l'F_{li} - F_lF_{li}')/(G_lF_{li}' - G_l'F_{li}), \end{array}\right\} \quad (7.35)$

where the prime denotes $d/d\rho$. For a potential other than the
spherical, F_{li} and F_{li}' must be found by numerical integration. The

quantities F_l, F_l', G_l, G_l' are found in terms of Coulomb wave functions (see §7.6) at a distance such that the nuclear potential is negligible compared to the Coulomb potential. This distance is about three times the nuclear range for long tailed wells like the exponential or Yukawa. These quantities can be tested using the Wronskian relations

$$F_l'G_l - F_lG_l' = 1, \quad F_{l-1}G_l - F_lG_{l-1} = l(l^2+\alpha^2)^{-\frac{1}{2}}. \quad (7\cdot36)$$

7.6. Coulomb wave functions

The radial wave equation for a repulsive Coulomb field is

$$\frac{d^2F_l}{d\rho^2} + \left\{1 - \frac{2\alpha}{\rho} - \frac{l(l+1)}{\rho^2}\right\}F_l = 0, \quad (7\cdot37)$$

where $\alpha = e^2/\hbar v = 1/ka$, $a = \hbar^2/\mu e^2$. Taking $z = 2i\rho$, $\kappa = i\alpha$, $m = l+\frac{1}{2}$, equation (7.37) becomes

$$\frac{d^2F_l}{d\rho^2} + \left\{-\frac{1}{4} + \frac{\kappa}{z} + \frac{\frac{1}{4} - m^2}{z^2}\right\}F_l = 0. \quad (7\cdot38)$$

The solution of equation (7.38) is Whittaker's function $M_{\kappa,m}(z)$ (Whittaker and Watson, 1953), a hypergeometric function which can be expressed as a linear combination of the confluent hypergeometric functions $W_{\kappa,m}(z)$ and $W_{-\kappa,m}(z)$:

$$M_{\kappa,m}(z) = e^{-\pi i \kappa} \frac{\Gamma(2m+1)}{\Gamma(\frac{1}{2}+m-\kappa)} W_{-\kappa,m}(-z)$$

$$- e^{\pi i(m-\kappa-\frac{1}{2})} \frac{\Gamma(2m+1)}{\Gamma(\frac{1}{2}+m+\kappa)} W_{\kappa,m}(z), \quad (7\cdot39)$$

where $M_{\kappa,m}(z)$ is a regular solution for m real and positive. The asymptotic form of $M_{\kappa,m}(z)$ determines the factor by which it must be multiplied to give F as defined by the asymptotic form (7.32). Define

$$Y = [\Gamma(\tfrac{1}{2}+m-\kappa)/\Gamma(\tfrac{1}{2}+m+\kappa)]^{\frac{1}{2}} e^{\frac{1}{2}\pi i(m+\frac{1}{2}-\kappa)} W_{\kappa,m}(z), \\
Y^* = [\Gamma(\tfrac{1}{2}+m+\kappa)/\Gamma(\tfrac{1}{2}+m-\kappa)]^{\frac{1}{2}} e^{-\frac{1}{2}\pi i(m+\frac{1}{2}+\kappa)} W_{-\kappa,m}(z). \quad (7\cdot40)$$

Then

$$F = \tfrac{1}{2}(Y+Y^*) = \tfrac{1}{2}\,|\,e^{\pi i\kappa/2}\,\Gamma(\tfrac{1}{2}+m-\kappa)/\Gamma(2m+1)\,|\\
\times e^{-\frac{1}{2}\pi i(m+\frac{1}{2})} M_{\kappa,m}(z),\\
G = \tfrac{1}{2}(Y-Y^*) = \tfrac{1}{2}\,|\,e^{\pi i\kappa/2}\,\Gamma(\tfrac{1}{2}+m-\kappa)/\Gamma(2m+1)\,|\\
\times e^{-\frac{1}{2}\pi i(m+\frac{3}{2})} \overline{M}_{\kappa,m}(z). \quad (7\cdot41)$$

where if m is not a positive integer,

$$\overline{M}_{\kappa,m}(z) = \frac{i[\cos 2\pi m + e^{-2\pi i\kappa}]}{\sin 2\pi m} M_{\kappa,m}(z)$$

$$- \frac{2i\Gamma(2m+1)\Gamma(2m)}{\Gamma(\tfrac{1}{2}+m+\kappa)\Gamma(\tfrac{1}{2}+m-\kappa)} e^{-\pi i(m-\kappa)} M_{\kappa,-m}(z). \quad (7.42)$$

If m is a positive integer, M is obtained by taking the limit.

Complex series expansions for equation (7.41) are available (Yost, Wheeler and Breit, 1936). However, these are inconvenient for computation, so equivalent real series are found by substituting back into the wave equation a power series with the same form, thereby obtaining recurrence formulae between the series coefficients:

$$F_l = C_l\rho^{l+1}\Phi_l, \quad G_l = D_l\rho^{-l}\Theta_l,$$

where

$$\Phi_l = \sum_{l+1}^{\infty} A_j\rho^{j-l-1}, \quad \Theta_l = \Psi_l + \rho^{2l+1}(p\ln 2\rho + q)\Phi_l,$$

$$\Psi_l = \sum_{j=-l}^{\infty} a_j\rho^{j+l}, \quad A_{l+1} = 1, \quad A_j = (2\alpha A_{j-1} - A_{j-2})/(j+l)(j-l-1),$$

$$a_{-l} = 1, \quad a_{l+1} = 0, \quad a_j = \{2\alpha a_{j-1} - a_{j-2}$$
$$- p(2j-1)A_j\}/(j+l)(j-l-1),$$

$$p = (2l+1)(e^{2\pi\alpha} - 1)C_l^2/\pi,$$

$$q = p\left\{\sum_{s=1}^{l} \frac{s}{s^2+\alpha^2} - \sum_{s=1}^{2l+1} \frac{1}{s} + \text{Re}\,\frac{\Gamma'(-i\alpha)}{\Gamma(-i\alpha)}\right.$$
$$\left. + (-1)^l \frac{2^l}{(2l)!} \sum_{-l}^{l} \text{Im}\, \frac{2^n(i\alpha+n-1)\dots(i\alpha-l)}{(l+n)!(l-n+1)} + 2\gamma\right\},$$

$$(2l+1)C_l D_l = 1,$$

$$C_l = \frac{2^l}{(2l+1)!}(l^2+\alpha^2)^{\frac{1}{2}}[(l-1)^2+\alpha^2]^{\frac{1}{2}}\dots(1+\alpha^2)(2\pi\alpha)^{\frac{1}{2}}/(e^{2\pi\alpha}-1)^{\frac{1}{2}},$$

$$\text{Re}\quad \Gamma'(-i\alpha)/\Gamma(-i\alpha) = -\gamma + \alpha^2 \sum_{n=1}^{\infty} \frac{1}{n(n^2+\alpha^2)},$$

$$\gamma = \text{Euler's constant} = 0.57722\dots\dots \quad (7.43)$$

In particular, for s-wave scattering,

$$C_0^2 = 2\pi\alpha/(e^{2\pi\alpha} - 1), \quad (7.44)$$

and for r small, $r \ll R$,

$$\left.\begin{aligned} F_0(r) &= C_0 kr(1 + r/2R + \dots), \\ G_0(r) &= \frac{1}{C_0}\left[1 + \frac{r}{R}\left(\ln\frac{r}{R} + 2\gamma - 1 + h(\alpha)\dots\right)\right], \end{aligned}\right\} \quad (7.45)$$

where $\qquad h(\alpha) = \mathrm{Re}\,\{\Gamma'(-i\alpha)/\Gamma(-i\alpha)\} - \ln\alpha$

and $\qquad R = \hbar^2/me^2 = 2 \cdot 88(15) \times 10^{-12}\,\mathrm{cm}.$

is the Bohr radius of a proton bound to a fixed unit charge. C_0^2 is the Coulomb penetration factor, interpreted as the probability of finding two protons together as compared to that of two uncharged particles.

Tables of Coulomb wave functions exist (Nat. Bur. Standards, 1952) which enable $F_l(\alpha,\rho)$ and $F_l'(\alpha,\rho)$ to be found by double interpolation. However $G_l(\alpha,\rho)$ and $G_l'(\alpha,\rho)$ are available only in table of low accuracy with large intervals in α and ρ (Bloch, Hull, Broyles, Bouricious, Freeman and Breit, 1951). Instead of the series solutions (7.43), it is often more convenient to use integral formulae as follows (cf. Bloch *et al.* 1950):

$$F_l(\alpha,\rho) = \frac{\mathrm{e}^{-\pi\rho}\rho^{l+1}}{(2l+1)!\,C_l} \int_0^\infty (1 - \tanh^2 u)^{l+1} \cos\,(2\alpha u - \rho\tanh u)\,du,$$

$$\tag{7.46}$$

$$G_l(\alpha,\rho) = \frac{\mathrm{e}^{-\pi\rho}\rho^{l+1}}{(2l+1)!\,C_l} \left[\int_0^\infty (1 - \tanh^2 u)^{l+1} \sin\,(2\alpha u - \rho\tanh u)\,du \right.$$

$$\left. + \mathrm{e}^{\pi\alpha} \int_0^\infty (1 + u^2)^l \exp\left(-\rho u - 2\alpha\arctan\frac{1}{u}\right)du \right]. \quad (7.47)$$

These expressions are much more straightforward to evaluate, for the integrands converge to zero very rapidly as u increases.

For a comparison of regions of validity of series, integral and asymptotic expansion methods, see Fröberg (1955).

7.7. Collisions below 3 MeV.

The earliest accurate experiments were carried out (Tuve *et al.* 1936; Hafstad *et al.* 1938) at 220–900 keV., over an angular range of 15–45°(Lab), i.e. $\theta = 30$–90° (CM). Because of the identity of the two protons the scattering is symmetrical about $\theta = 90°$. The ratio R of actual scattering to Mott scattering in fig. 59 shows that at the lowest energies, the scattering is almost pure Coulomb, as the protons do not get close enough to undergo nuclear interaction. However, as the energy increases to about 400 keV., a strong interference effect between Coulomb and nuclear scattering reduces

R to 0·08 at 0 = 90°. Above this energy R increases rapidly, reaching a value of about 7 at 90° and 900 keV. At 400 keV. the Coulomb penetration factor $C_0^2 \rightarrow 1$ so that nuclear scattering becomes predominant, the incident protons easily penetrating the comparatively weak Coulomb barrier. The critical energy for interference has been measured (1954) at 90° as 383·9 ± 1·5 keV. (cf. § 3.5.2).

It is found that $R \rightarrow 1$ for small scattering angles at all energies. This is due to the incident proton passing outside the nuclear force range, a small Coulomb deflexion only occurring. The scattering very nearly obeys Mott's formula in this region.

Best fits to experimental scattering distributions are shown in fig. 60 (Worthington, McGruer and Findley, 1953), assuming pure s-wave scattering. The scattering is predominantly nuclear except at small and large angles; for $E = 2·4$ MeV. and $\theta = 90°$ the ratio of actual to Coulomb scattering is about 44. At intermediate angles small interference minima occur.

Experiments prior to 1939 were extensively analysed by Breit and co-workers (Breit, Condon and Present, 1936; Breit, Thaxton and Eisenbud, 1939), who calculated the phase shifts K_0 and K_1 for a number of different assumptions about the range, magnitude and radial dependence of the nuclear potential. As K_1 turns out to be completely negligible, equation (7.28) gives an equation in K_0 which leads to the experimental s-phase shift. Actually two solutions for K_0 are possible, one positive and one negative, the first corresponding to an attractive and the second to a repulsive potential. However, the latter is ruled out by the results at 400 keV., as strong destructive interference can take place only if the nuclear and Coulomb potentials are opposite in sign and evenly balanced.

Results show that any one of the usual well shapes (cf. equations (1.1)–(1.5)) gives satisfactory agreement with the data, provided its range and depth is suitably chosen. The p-p data has the same shape-independent character as low energy n-p scattering, so the detailed approach usual in earlier work should not be necessary if a formula similar to equation (6.33) can be found for the phase K_0.

7.8. The shape-independent formula

A formula for the s-phase shifts δ_0 in terms of two interaction constants, the zero energy scattering length and the effective range, was first stated by Landau and Smorodinsky (1944). A variational proof was given by Schwinger (1950), but simpler non-variational approaches are available (Chew and Goldberger, 1949; Hanson

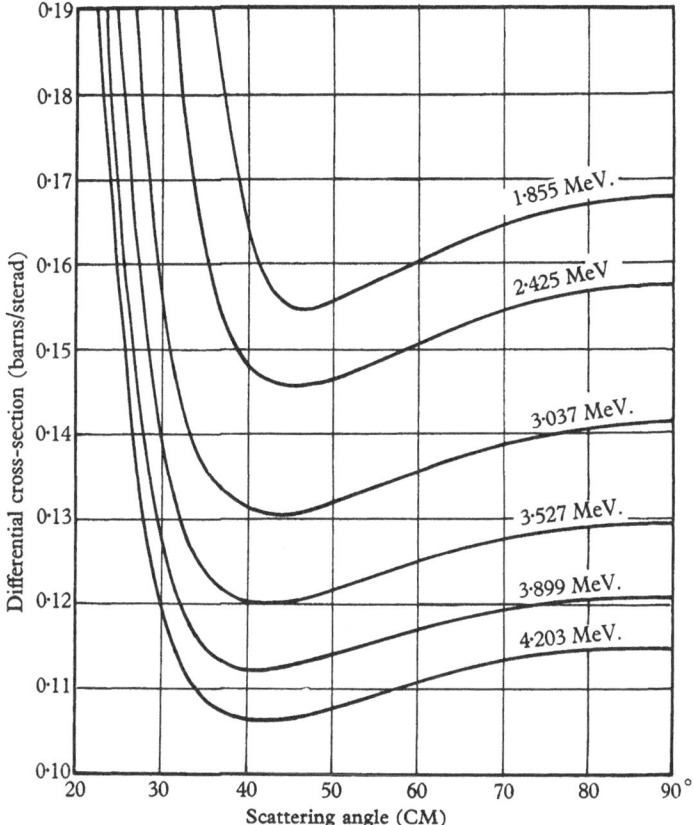

Fig. 60. Best fits to experimental differential cross-sections for *p*-*p* scattering in the CM system. Results are at laboratory energies of 1·855 MeV. (Herb *et al.* 1939; Worthington *et al.* 1953), 2·425 MeV. (Herb *et al.* 1939; Blair, Freier, Lampi, Sleator and Williams, 1948; Worthington *et al.* 1953), 3·037 and 3·527 MeV. (Herb *et al.* 1939; Worthington *et al.* 1953) and 3·899 and 4·203 MeV. (Worthington *et al.* 1953). The curves are calculated assuming a pure *s*-wave anomaly (Worthington *et al.* 1953).

et al. 1949; Bethe, 1949). Bethe's proof (Bethe, 1949), which is based on the properties of the Schrödinger equation (cf. §6.7), is followed here.

Let $u(r)$ be the radial wave function of the p-p system in the S-state, satisfying equation (7.48):

$$u''(r) + \left[k^2 - \frac{M}{\hbar^2} V(r) - \frac{e^2}{r} \right] u(r) = 0. \qquad (7.48)$$

At zero energy, the wave function $u_0(r)$ satisfies

$$u_0''(r) - \left[\frac{m}{\hbar^2} V(r) + \frac{e^2}{r} \right] u_0(r) = 0. \qquad (7.49)$$

Multiplying equation (7.48) by u_0, equation (7.49) by u, subtracting and integrating between the lower and upper limits r and R, then

$$[uu_0' - u_0 u']_r^R = k^2 \int_r^R uu_0 \, dr. \qquad (7.50)$$

The asymptotic form $v(r)$ of $u(r)$ satisfies a similar relation, viz.

$$[vv_0' - v_0 v']_r^R = k^2 \int_r^R vv_0 \, dr. \qquad (7.51)$$

Subtracting equation (7.50) from equation (7.51) gives

$$[vv_0' - v_0 v' - uu_0' + u_0 u']_r^R = k^2 \int_r^R (vv_0 - uu_0) \, dr. \qquad (7.52)$$

If R is chosen large compared to the range of nuclear force, the upper limit of the left-hand side of equation (7.52) gives no contribution, as $u(R)$ is equal to its asymptotic form $v(R)$. Hence we may put $R = \infty$. Taking $r = 0$ and using $u(0) = u_0(0) = 0$ leads to

$$v_0(0) v'(0) - v(0) v_0'(0) = k^2 \int_0^\infty (vv_0 - uu_0) \, dr. \qquad (7.53)$$

Following equations (7.34) and (7.43), we define $v(r)$ but with the normalization chosen anew:

$$v(r) = C_0 G(r) + C_0 \cot \delta . F(r), \qquad (7.54)$$

so that from equation (7.45) for r small:

$$v'(r) = C_0^2 k \cot \delta + \frac{1}{R} \left[\ln \frac{r}{R} + 2\gamma + h(\alpha) \right], \qquad (7.55)$$

where R is the Bohr orbit of a proton, $R = \hbar^2/me^2 = 2 \cdot 88(15) \times 10^{-12}$ cm. Equation (7.54) becomes

$$C_0^2 k \cot \delta + \frac{1}{R} h(\alpha) + \frac{1}{a} = k^2 \int_0^\infty (vv_0 - uu_0) \, dr, \qquad (7.56)$$

where $a = -\lim_{r \to \infty} (k \cot \delta)$ is the zero energy scattering length defined in § 6.6. Equation (7.56) is an exact relation, but replacing $v(r)$ by $v_0(r)$ and $u(r)$ by $u_0(r)$, we obtain the shape independent approximation formula:

$$C_0^2 k \cot \delta + \frac{1}{R} h(\alpha) = -\frac{1}{a} + \tfrac{1}{2} r_0 k^2 + O(k^4),$$

or $\quad K = \dfrac{\pi \cot \delta}{e^{2\pi\alpha} - 1} + h(\alpha) = R \left(-\dfrac{1}{a} + \tfrac{1}{2} r_0 k^2 - P r_0^3 k^4 + \ldots \right), \quad$ (7.57)

where P is a small correcting term as in equation (6.36), $2kR\alpha = 1$ and the effective range r_0 is defined as usual:

$$r_0 = 2 \int_0^\infty [v_0(r)^2 - u_0(r)^2] \, dr. \tag{7.58}$$

For high energies, $\alpha \to 0$, $h(\alpha) \to 0$, and equation (7.57) reduces to the form of equation (6.32) familiar in n-p scattering:

$$k \cot \delta = -1/a + \tfrac{1}{2} r_0 k^2 + O(k^4). \tag{7.59}$$

Assuming the n-p and p-p potentials are the same, apart from the Coulomb repulsion, a relation may be found between the p-p and n-p zero energy scattering lengths, a_p and a_n respectively. If the Coulomb potential is switched on, in addition to the nuclear potential with effective range r_0, then

$$W(r) \to W(r) + \epsilon W'(r), \tag{7.60}$$

where $W(r) = -(m/\hbar^2) V(r)$, $W'(r) = -1/r$. We require the first order change in a_n due to the Coulomb potential, where for n-p scattering $\quad 1/a_n = -\lim_{k \to 0} (k \cot \delta) = v'(0)/v(0).$

However, the logarithmic derivative of the wave function $v(r)$ for $r < r_0$ is nearly independent of energy, at least for low energies, since the strength of the nuclear potential is much greater than the kinetic energy outside the range of nuclear forces. Hence to a good approximation, $\quad 1/a_n = v'(r_0)/v(r_0) \approx u'(r_0)/u(r_0),$ (7.61)

where it is assumed that the nuclear potential is small outside r_0, so that the logarithmic derivative of the outside and inside functions $v(r)$ and $u(r)$ may be fitted at $r = r_0$. To find an expression for a_p, we use equation (7.55) for $v'(r)$ when $r \ll R$ and also $v(r_0) \approx 1$, obtaining

$$v'(r_0)/v(r_0) = kC_0^2 \cot \delta + \frac{1}{R} \left[\ln \frac{r_0}{R} + 2\gamma + h(\alpha) \right]. \qquad (7.62)$$

Hence $1/a_p = -\lim_{k \to 0} (k \cot \delta)$

$$= -[v'(r_0)/v(r_0)]_{k=0} + \frac{1}{R} \left[\ln \frac{r_0}{R} + 2\gamma \right]. \qquad (7.63)$$

However as the R^{-1} terms in equation (7.63) represent first order effects due to the Coulomb field, the first order change in the logarithmic derivative must be included also.

To find $\partial/\partial\epsilon \, [u'(r_0)/u(r_0)]$, consider the wave equation (7.48), rewritten as

$$u''(r) + [k^2 + W(r) + \epsilon W'(r)] \, u(r) = 0. \qquad (7.64)$$

Following the procedure prior to equation (7.50), two functions u_1 and u_2 corresponding to different ϵ but the same k satisfy

$$u_1 u_2 [u_1'/u_1 - u_2'/u_2]_0^r + (\epsilon_1 - \epsilon_2) \int_0^r W'(r) u_1 u_2 \, dr = 0. \qquad (7.65)$$

In the limit as $\epsilon_1 \to \epsilon_2 = \epsilon$, equation (7.65) becomes

$$\frac{\partial}{\partial\epsilon} (u'/u) = -\frac{1}{u^2} \int_0^r W'(r) u^2(r) \, dr. \qquad (7.66)$$

Hence from equations (7.60) and (7.61),

$$\frac{\partial}{\partial\epsilon} \left(\frac{1}{a_n} \right) = \frac{1}{u^2(r_0)} \int_0^{r_0} \frac{u^2(r)}{r} \, dr. \qquad (7.67)$$

Thus approximately

$$[v'(r_0)/v(r_0)]_{k=0} = -1/a_n + \frac{1}{R} \frac{1}{u^2(r_0)} \int_0^{r_0} \frac{u^2(r)}{r} \, dr, \qquad (7.68)$$

where the quantities on the left and right of equation (7.68) are p-p and n-p wave functions respectively. We obtain a good approximation by taking the n-p wave function as $u(r) \approx \sin(\pi r/2r_0)$, which is exact for a square well potential such that $1/a_n = 0$, i.e. a resonance at zero energy. Then

$$\int_0^{r_0} \frac{u^2(r)}{r} \, dr \approx \tfrac{1}{2} [\ln \pi + \gamma - C_i(\pi)] = 0 \cdot 824,$$

where $C_i(r)$ is the cosine integral. Equation (7.64) becomes

$$1/a_p \approx 1/a_n + \frac{1}{R}\left[\ln\frac{r_0}{R} + 2\gamma - 0.824\right]$$

$$\approx 1/a_n + \frac{1}{R}\left[\ln\frac{r_0}{R} + 0.330\right]. \tag{7.69}$$

Estimates show that equation (7.69) is valid to within about 2·5 % for short range potentials.

7.9. The analysis of p-p scattering data

A detailed analysis of the earlier work in fitting empirical short-range potentials is given by Breit, Thaxton and Eisenbud (1939),† but a simpler approach based on the shape-independent formula is due to Jackson and Blatt (1950).‡ Adopting the latter procedure, four steps are involved.

(a) Experimental angular distributions must be reduced to CM co-ordinates (Appendix A) and corrected for geometrical effects, multiple scattering, etc., as discussed in chapter 3.

(b) The angular distribution as a function of scattering angle at a given energy is used to determine the phase shifts $\delta_0, \delta_1, \delta_2, \ldots$, where it is assumed that $\delta_1, \delta_2, \ldots$, are small corrections on s-wave scattering. If s-wave scattering only is assumed, equation (7.28) for the angular distribution in the CM system may be written as

$$\sigma(\theta) = \sigma_M(\theta)\left[1 + \tfrac{1}{9}\{\sin\omega - \sin(2\delta_0 + \omega)\}\right], \tag{7.70}$$

where $\sigma_M(\theta)$ is the Mott cross-section (7.17), and

$$\tan\omega = (2/\alpha + \mathscr{Y})\,\mathscr{X}, \quad q = (\alpha\mathscr{M}\cos\omega)/\mathscr{X}, \tag{7.71}$$

where, in the notation of B.T.E.,

$$\begin{aligned}
\mathscr{X} &= \frac{\cos(\alpha\ln\sin^2\tfrac{1}{2}\theta)}{\sin^2\tfrac{1}{2}\theta} + \frac{\cos(\alpha\ln\cos^2\tfrac{1}{2}\theta)}{\cos^2\tfrac{1}{2}\theta}, \\
\mathscr{Y} &= \frac{\sin(\alpha\ln\sin^2\tfrac{1}{2}\theta)}{\sin^2\tfrac{1}{2}\theta} + \frac{\sin(\alpha\ln\cos^2\tfrac{1}{2}\theta)}{\sin^2\tfrac{1}{2}\theta}, \\
\mathscr{M} &= (mv^2/e^2)^2\,\sigma_M.
\end{aligned} \tag{7.72}$$

Solving equation (7.70) for δ_0 leads to

$$2\delta_0 = \arcsin\left[\sin\omega - q\left\{\frac{\sigma(\theta)}{\sigma_M(\theta)} - 1\right\}\right] - \omega. \tag{7.73}$$

† Hereafter referred to as B.T.E.   ‡ Hereafter referred to as J.B.

J.B. give rapidly convergent series for $\sigma_M(\theta)$, q and $\sin \omega$ in inverse powers of the energy, by which means δ_0 may be rapidly computed from the experimental cross-section. The apparent s-wave phase shift δ_a obtained is equal to the true s-wave phase shift if all the higher phases are zero, and is therefore independent of θ at a given energy. If the higher phase shifts are non-zero, δ_a is a slowly varying function of θ at a given energy.

Assuming $\delta_l \ll \delta_0$ ($l \geqslant 1$), we write

$$\sigma = \sigma(E, \theta, \delta_0, \delta_1, \delta_2, \ldots)$$
$$= \sigma(E, \theta, \delta_a, 0, 0, \ldots). \qquad (7.74)$$

Using Taylor's series, equation (7.74) becomes to first order

$$\sigma = \sigma(E, \theta, \delta_0, 0, 0, \ldots) + \delta_1(\partial\sigma/\partial\delta_1) + \delta_2(\partial q/\partial\delta_2) + \ldots$$
$$= \sigma(E, \theta, \delta_0, 0, 0, \ldots) + (\delta_a - \delta_0)(\partial\sigma/\partial\delta_0), \qquad (7.75)$$

where the partial derivatives are taken at the correct values of δ_0 but with $\delta_1 = \delta_2 = \ldots = 0$. Equating the two expansions in equation (7.75) gives to first order

$$\delta_a = \delta_0 + p_1\delta_1 + p_2\delta_2 + \ldots, \qquad (7.76)$$

where $p_n(E, \theta, \delta_0) = (\partial\sigma/\partial\delta_n)/(\partial\sigma/\partial\delta_0)$. The p_n do not depend very critically on δ_0, so that an approximate value of δ_0 still leads to good approximations to higher phases. Values for p_n are found by partial differentiation of equation (7.31). Assuming $\delta_2, \delta_3, \ldots$ are negligible, the procedure is to plot δ_a against p_1, giving a straight line of intercept δ_0 and slope δ_1. However, if the line has curvature due to δ_2, one plots $\delta_a - p_1\delta_1$ against p_2, choosing trial values of δ_1 until a best straight line is obtained, with intercepts δ_0 and slope δ_2. This treatment assumes that the ^3P state can be represented by a central force, whereas actually tensor forces would split it into the $^3P_0$, $^3P_1$ and $^3P_2$ states (cf. § 10.8). However, the experimental data are not accurate enough to make the distinction important.

(c) The experimental s-phase shifts δ_0 for a number of energies are used to compute the function K defined in equation (7.57) as a function of energy. The plot of K against k^2 is very nearly a straight line up to an energy of 10 MeV., with a slope of $\frac{1}{2}r_0 R$ and an intercept with the K axis of $-R/a$. Any small curvature is a measure of the coefficient P of the k^4 term, which in turn is related to the detailed shape of the S-state potential as illustrated in Table X.

Results obtained with Van de Graaff generators cover the range $0 \cdot 2$–$4 \cdot 2$ MeV.[†] Cyclotron data covers the rest of the low energy region from $4 \cdot 2$ to $18 \cdot 2$ MeV.[‡] For discussion of experiments prior to 1950, see J.B. (1950).

The most accurate Van de Graaff data are at $1 \cdot 855$–$4 \cdot 203$ MeV. (Worthington *et al.* 1953). The scattering is almost pure s-wave, with a very small p-wave component, phases at $1 \cdot 855$ and $3 \cdot 899$ MeV. being $\delta_0 = 44 \cdot 21°$, $\delta_1 = -0 \cdot 05 \pm 0 \cdot 02°$ and $\delta_0 = 53 \cdot 26°$, $\delta_1 = -0 \cdot 11 \pm 0 \cdot 002°$ respectively (Hall and Powell, 1953). Cyclotron results up to $9 \cdot 7$ MeV. confirm the p-wave component, with $\delta_0 = 56 \cdot 5 \pm 0 \cdot 5°$, $\delta_1 = -0 \cdot 55°$ at $E = 9 \cdot 7$ MeV., corresponding to a small repulsive P-potential of uncertain magnitude (due to experimental error). Alternatively a small ^1d-phase could be used, agreeing with evidence at much higher energies (32 MeV.), where d-wave together with little or no p-wave scattering is necessary to explain the angular distribution obtained (cf. § 10.11). However, distributions at $18 \cdot 2$ MeV. have been fitted with $\delta_0 = 54 \cdot 1°$, $\delta_1 = +1 \cdot 0°$, $\delta_2 = +0 \cdot 4°$, which appears not to fit into either scheme (Hartzler and Siegel, 1954). About half of the small ^3P phase found at $1 \cdot 8$–$4 \cdot 2$ MeV. (Hall and Powell, 1953) is due to vacuum polarization effects. The corrected ratio of ^3P to ^1S interaction is $0 \cdot 13 \pm 0 \cdot 07$ (Eriksen, Foldy and Rarita, 1956).

A plot of K against k^2 for the Van de Graaff data is shown in fig. 61, with a best linear approximation to the more accurate experiments at seven energies of Worthington *et al.* (1953). This was obtained from the phase analysis of Hall and Powell (1953) by a least square analysis weighted by the reciprocal of

[†] Tuve *et al.* (1936); Ragan, Kanne and Taschek (1941); Hafstad *et al.* (1938); Herb *et al.* (1939); Heydenburg, Hafstad and Tuve (1939); Cooper *et al.* (1954); Blair *et al.* (1948); Ralph, Worthington and Herb (1950); Worthington *et al.* (1953).

[‡] May and Powell (1947); Meagher (1950); Dearnley *et al.* (1948); Wilson and Kreutz (1947); Wilson (1947); Wilson, Lofgren, Wright and Shankland (1947); Mather (1951); Zimmerman, Kerman, Singer, Kruger and Jentschke (1954); Allred *et al.* (1952*b*); Cork and Hartsough (1954*a*); Hall (1954); Yntema and White (1954).

the standard error. A very good straight line fit is obtained, the parameters being

$$a = -7\cdot692 \pm 0\cdot001 \times 10^{-13}\,\text{cm.}, \quad r_0 = 2\cdot652 \pm 0\cdot009 \times 10^{-13}\,\text{cm.}$$

These agree well with less accurate values obtained by J.B. by analysis of Van de Graaff data prior to 1950, viz.

$$a = -7\cdot65 \pm 0\cdot05 \times 10^{-13}\,\text{cm.}, \quad r_0 = 2\cdot65 \pm 0\cdot07 \times 10^{-13}\,\text{cm.}$$

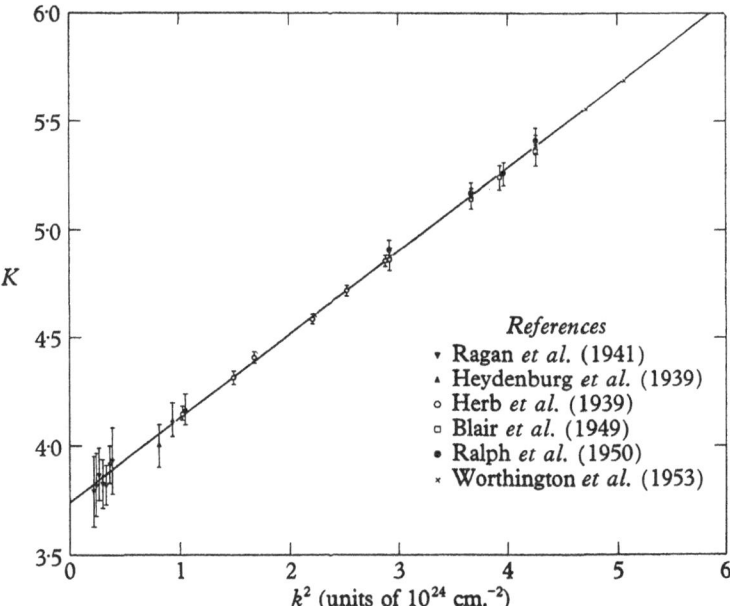

Fig. 61. Experimental values of K plotted against k^2 for the Van de Graaff p-p data. The straight line is the best linear approximation to the points of Worthington et al. (1953), which are of higher accuracy than earlier experiments. It is almost identical with the line fitting the latter experiments (Jastrow, 1950).

However, both sets of values are increased somewhat on allowing for a vacuum polarization potential (interaction between the protons and virtual electron-positron pairs) (Foldy and Eriksen, 1954, 1955).

To obtain a parabolic fit to the data, points at higher energies are needed than are supplied by electrostatic generators. A weighted least square analysis of all Van de Graaff and cyclotron data up to 18·2 MeV. (see footnote, p. 231) leads to the values (fig. 62)

$$a = -7\cdot77 \pm 0\cdot02 \times 10^{-13}\,\text{cm.}, \quad r_0 = 2\cdot80 \pm 0\cdot03 \times 10^{-13}\,\text{cm.},$$

$$P = 0\cdot034 \pm 0\cdot008.$$

However, the general accuracy of experiments prior to 1950 (as dealt with in the analysis of J.B.) is lower by a factor of 10 than that

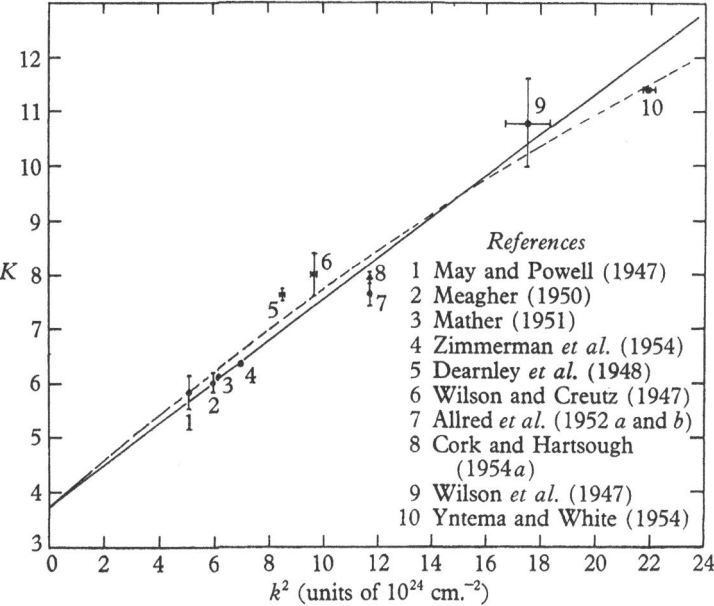

Fig. 62. Experimental values of K for the cyclotron data plotted against k^2. The straight line is the best linear approximation to the Van de Graaff points of Worthington et al. (1953) at 1·8–4·2 MeV., extrapolated to higher energies. The broken line is the best parabolic fit to the points of Worthington et al. (1953) and the listed cyclotron experiments, determined by a weighted least square fit.

of later experiments. Moreover, some of the earlier cyclotron points (Dearnley, Oxley and Perry, 1948; Wilson and Creutz, 1947; Wilson, 1947; Wilson, Lofgren, Wright and Shankland, 1947) at energies 7–14·5 MeV. give K values which are 3–6 % too large in view of later more accurate experiments. This tends to displace the shape parameter P towards negative values. Hence a more reliable procedure is to use the best Van de Graaff values (Worthington *et al.* 1953) together with the best cyclotron points, the latter extending from 5·07 to 18·2 MeV. (Mather, 1951; Zimmerman *et al.* 1954; Allred *et al.* 1952*a,b*; Cork and Hartsough, 1954*a*; Hartzler and Siegel, 1954; Yntema and White, 1954). The K values employed vary in accuracy from 0·05 to 0·3% and 1 to 2% for Van de Graaff and cyclotron data respectively, the accuracy generally decreasing with energy. The weighted least square analysis gives

$$a = -7\cdot75 \pm 0\cdot02 \times 10^{-13}\,\text{cm.}, \quad r_0 = 2\cdot78 \pm 0\cdot03 \times 10^{-13}\,\text{cm.},$$

$$P = 0\cdot041 \pm 0\cdot007.$$

Vacuum polarization corrections are unimportant at the higher energies considered.

TABLE X. *Calculated values of the variational parameters and their derivatives for various potential well shapes*

(All lengths are in units of 10^{-13} cm.)

Well shape	s	b	a	r_0	P	Q
Square	0·890	2·626	−7·7930	−2·6388	−0·03313	0·00179
Gaussian	0·900	2·540	−7·7797	−2·6055	−0·01936	−0·00073
Exponential	0·900	2·500	−7·4235	−2·6776	0·00907	0·00089
Yukawa	0·924	2·400	−7·6512	−2·6756	0·05540	0·019

Well shape	$\partial a/\partial s$	$\partial r_0/\partial s$	$\partial a/\partial b$	$\partial r_0/\partial b$
Square	−33·521	−1·5300	−2·0824	0·94975
Gaussian	−34·654	−1·8553	−2·0966	0·98193
Exponential	−33·813	−2·5690	−2·0510	1·0439
Yukawa	−39·399	−3·6387	−2·0002	1·1157

(*d*) The values of the parameters a and r_0 enable fits of alternative well shapes to the data. J.B. have obtained expressions for the four usual well shapes in terms of the effective range (in units of 10^{-13} cm.) and the well depth parameter s. The potential $V(r)$ is measured in MeV.

Square well: $V(r) = -sb^{-2}(102\cdot35), \quad r < b,$

$\qquad\qquad\quad = 0 \qquad\qquad\qquad\quad r > b.$

Gaussian well: $V(r) = -sb^{-2}(229\cdot37)\exp\left[-2\cdot06049(r/b)^2\right].$

Exponential well: $V(r) = -sb^{-2}(752\cdot06)\exp\left(-3\cdot5412r/b\right).$

Yukawa well: $V(r) = -sb^{-2}(147\cdot69)(b/r)\exp\left(-2\cdot1196r/b\right).$

Results of calculations by J.B. are shown in Table X. The first column specifies the potential shape and the second and third columns the values of s and b assumed. The last four columns give the derivatives needed to compute the effect of small changes in the well parameters.

The shape parameter $P = 0\cdot041 \pm 0\cdot007$ found above appears to agree best with the Yukawa well. The correct values for a and r_0 may be obtained using $s = 0\cdot921$, $b = 2\cdot500$, but this probably increases P somewhat. It therefore appears that the correct singlet potential has a shape intermediate between that of the exponential and Yukawa wells, probably nearer the latter.

If we assume an n-p singlet potential with the same intrinsic range and shape as the p-p potential, the charge independence of nuclear forces can be tested. Bethe (1949) has shown that the value $a = -2\cdot375 \pm 10 \times 10^{-12}$ cm. obtained from the coherent and incoherent n-p cross-sections leads to a well depth parameter s, which is greater than that found for the p-p interaction by $3\cdot3\%$ for the square well, $1\cdot6\%$ for the Yukawa well and intermediate amounts for other well shapes. However, the inclusion of electromagnetic interactions between dipoles produces a change of order 1% in the strength of the nuclear force and improves its charge independence. The maximum effect is obtained for a Yukawa well and the minimum for a square well, exact charge independence being attained for the former within the experimental uncertainty of $\pm 0\cdot5\%$ (Schwinger, 1950). An additional correction for vacuum polarization produces a 12% discrepancy between the corrected p-p and n-p scattering lengths. However this may be due to the assumption of *point* magnetic dipoles (Foldy and Eriksen, 1955).

CHAPTER 8

SCATTERING OF NEUTRONS AND PROTONS BY VERY LIGHT NUCLEI

8.1. Wheeler's resonating group structure method

The collision of nuclear particles with light nuclei involves at least several nucleon pairs for which all interactions are comparable, so that a perturbation treatment is inapplicable. However approximate methods exist for writing down a wave function for the system, leading in general to a system of integro-differential equations from which phase shifts for scattering may be deduced.

The three-body system, embracing the negative ion H^-, the ^3H and ^3He nuclei and the H_2^+ ion, is the simplest to deal with. For both the nuclear systems and H^-, there is no single dominating centre of force, so a time-average field cannot be used to represent the action of any two of the particles on the third because their speeds are comparable. Thus the Hartree central field approximation is not applicable. However the molecular ion H_2^+ combines two heavy, slow-moving protons with a rapidly moving light particle (electron), so that the system can be approximately separated into electronic motion with fixed force centres and slow nuclear motion in the time-average field of the electron.

In a system containing two heavy particles and a light particle, the potential may be considered as acting between a pair comprising a heavy and a light particle, and the other heavy particle. The relative velocity of the two parts of the system has an increasing effect on this interaction as the mass of the light particle increases. This is equivalent to dividing the potential into a static part $U(r)$ and a velocity-dependent potential represented by an integral operator. A suitable wave equation is then

$$\left(\frac{\hbar^2}{2\mu}\nabla^2 + E_i\right) F(\mathbf{r}) = U(r) F(\mathbf{r}) + \int K(\mathbf{r},\mathbf{r}') F(\mathbf{r}') \, d\mathbf{r}', \quad (8.1)$$

where the co-ordinate $\mathbf{r}$ joins the CM of the particle pair with the other heavy particle, and $\mathbf{r}'$ is the co-ordinate obtained from $\mathbf{r}$ by exchanging the two heavy particles. E_i is the energy of the heavy particle (CM system), μ its reduced mass and $F(\mathbf{r})$ its wave function. $K(\mathbf{r}, \mathbf{r}')$ is a kernel function symmetric in $\mathbf{r}$ and $\mathbf{r}'$.

In the limiting case where the mass of the light particle approaches zero, $\mathbf{r}$ approaches $-\mathbf{r}'$, and $K(\mathbf{r}, \mathbf{r}')$ becomes representable as a delta function:

$$K(\mathbf{r}, \mathbf{r}') \to W(r) \, \delta(\mathbf{r} + \mathbf{r}'). \tag{8.2}$$

The potential is then $U(r) + W(r)$ and is no longer velocity-dependent, the H_2^+ ion being an example of such a system. Thus a smooth transition may be made between the group idea as applied to a nucleus and the static potentials involved in molecules.

The concept of resonance is introduced to nuclei by analogy with the H_2^+ ion, which is pictured as resonating between a state of ionic and atomic binding. The three nucleons in ^3H are regarded as spending part of their time in the grouping $(12, 3)$ and part in the grouping $(32, 1)$, the interchange being governed by Pauli's principle, so that the total wave function must be built up out of a combination of the two groupings anti-symmetric in the neutrons 1 and 3:

$$\Psi(12, 3) = [\psi(12, 3) - \psi(32, 1)], \tag{8.3}$$

where ψ involves both spatial and spin co-ordinates.

So far, the three-body systems considered have involved one configuration only, but systems of four or more nucleons may involve a number of configurations. For a problem like n-^3He scattering, only the n-^3He configuration involves free waves, the ^2H-^2H and p-^3H configurations being considered as bound, so that they are not detected by measuring instruments (considered to be relatively at infinity).

In some cases, e.g. α-α scattering, one can often ignore all except one of the groups. The α-group is so stable compared to all others that it is safe to assume that this system spends nearly all its time in the state of lowest energy, viz. two α-groups.

8.2. Elastic scattering of nucleons by deuterons

Although n-p scattering provides one of the simplest ways of obtaining information about nuclear forces, it has the disadvantage of being shape-independent for energies less than about 15 MeV. and of not showing up the presence of exchange forces until relatively high energies. This does not apply to low energy n-d scattering, as the large size of the deuteron leads to important p-wave effects at energies of only several MeV. The p-wave actually refers to angular motion relative to the CM of the deuteron and not to its component nucleons, so that most of the p-scattering arises from the finite size of the deuteron and not from p-wave effects between the nucleons. Nevertheless, interference effects make the p-phases for n-d and p-d collisions quite sensitive to the existence of exchange forces. Above 3·3 MeV., a nucleon can dissociate a deuteron into a neutron and proton, although this is more important at high than low energies (Bransden and Burhop, 1950).

The theory of elastic n-d scattering was given by Wheeler (1937), but early calculations included s-wave effects only (Schiff, 1937; Ochiai, 1937; Flugge, 1938). The most extensive calculations are due to Buckingham, Hubbard and Massey (1952),† who extended the earlier work of Buckingham and Massey (1941), and to Christian and Gammel (1953).‡

Approximations used in the theory are as follows:

(*a*) The potential energy of two nucleons is given by an exchange operator involving a central force:

$$^{3,1}\mathscr{V}_i(r) = (mM + hH + bB + w)\, V(r), \qquad (8.4)$$

where M, H and B are the Majorana, Heisenberg and Bartlett exchange operators which interchange the spatial, spatial and spin, and spin co-ordinates respectively of the interacting nucleons, and $V(r)$ is the triplet n-p potential. m, h, b, w are constants satisfying

$$m + h + b + w = 1 \qquad (8.5)$$

and giving the correct ratio $x \approx 0\cdot 6$ of singlet to triplet n-p interaction, which requires
$$m - h - b + w = x. \qquad (8.6)$$

The constants m, h, b, w may be assigned sets of values leading to the exchange forces customary in meson field theory. These are

(i) W.B. case, *ordinary force* or *neutral interaction*:

$$m = h = 0, \quad w = \tfrac{1}{2}(1 + x), \quad b = \tfrac{1}{2}(1 - x).$$
$$^{3}\mathscr{V}(r) = V(r), \quad ^{1}\mathscr{V}(r) = xV(r).$$

† Hereafter referred to as B.H.M.   ‡ Hereafter referred to as C.G.

(ii) M.H. case, *Majorana–Heisenberg exchange force* or *charge exchange interaction*:

$$w = b = 0, \quad m = \tfrac{1}{2}(1+x), \quad h = \tfrac{1}{2}(1-x).$$

$$^3\mathscr{V}_l(r) = (-1)^l V(r), \quad ^1\mathscr{V}_l(r) = (-1)^l x V(r).$$

(iii) M.H.W.B. case, *symmetrical exchange forces*:

$$m = 2b = \tfrac{1}{3}(1+3x), \quad h = 2w = \tfrac{1}{3}(1-3x).$$

$$^3\mathscr{V}_l(r) = \tfrac{1}{3}[1+2(-1)^l]\,V(r), \quad ^1\mathscr{V}_l(r) = [-1+2(-1)^l]\,xV(r).$$

(iv) *Serber force*:

$$m = w = \tfrac{1}{4}(1+3x), \quad h = 2w = \tfrac{1}{3}(1-3x).$$

$$^3\mathscr{V}_l(r) = \tfrac{1}{2}[1+(-1)^l]\,V(r), \quad ^1\mathscr{V}_l(r) = \tfrac{1}{2}[1+(-1)^l]\,xV(r).$$

The Serber force is empirical, giving no interaction between nucleons in states of odd angular momentum.

It has been assumed that any non-central interaction present can be replaced by an equivalent central force.

(*b*) The radial interaction $V(r)$ has been given various simple shapes by different authors. B.H.M. (1952) use the exponential well

$$V(r) = -A\,\mathrm{e}^{-2r/a}, \tag{8.7a}$$

where $A = 242 m_e C^2$, $a = 1.73 \times 10^{-13}$ cm. Although these values lead to good agreement between the ^{2}H, ^{3}H and ^{3}He binding energies, results of Christian (1952) for nucleon-nucleon scattering lengths require a smaller range $a = 1.36 \times 10^{-13}$ cm.

Verde (1949) and Troesch and Verde (1951) use the gaussian interaction

$$V(r) = -V_0\,\mathrm{e}^{-r^2/\lambda^2}, \tag{8.7b}$$

where $V_0 = 45$ MeV., $\lambda = 1.9 \times 10^{-13}$ cm. This again does not lead to the correct zero energy scattering lengths.

C.G. employ the gaussian interaction $(8.7b)$ with

$$V_0 = 86.4\,\text{MeV.}, \quad \lambda = 1.332 \times 10^{-13}, \tag{8.7c}$$

and also the Yukawa interaction

$$V(r) = -V_0 \exp(-r/\lambda)/(r/\lambda), \tag{8.7d}$$

with $V_0 = 68.0$ MeV., $\lambda = 1.18 \times 10^{-13}$ cm. Both $(8.7c)$ and $(8.7d)$ lead to the correct *n-p* zero energy scattering lengths.

(c) The three-body wave function is approximated by the product of a known deuteron wave function and a scattering wave
function to be found by calculation. As only one configuration
exists, the resonating group structure method merely requires the
total wave function to be anti-symmetric with regard to exchange
of two neutrons. For the deuteron wave function, B.H.M. use the
accurate solution of the deuteron ground-state wave equation with
the potential (8.7 a). Verde (1949) and Troesch and Verde (1951)
use a gaussian wave function of form $\phi(r) = N e^{-\mu^2 r^2}$, with constants
determined by the variational method to give a best fit to the
binding energy of the deuteron. However, this is a very poor
approximation as the variational binding energy is only 0·59 MeV.

C.G. represent the deuteron wave function for the gaussian interaction (8.7 c) by the sum of three gaussian terms:

$$\phi(r) = 0\cdot02133 \exp(-0\cdot003r^2) + 0\cdot08582 \exp(-0\cdot16r^2)$$
$$+ 0\cdot18115 \exp(-0\cdot76r^2).$$

This approximates to within 3 % the actual deuteron function out
to three times the deuteron radius. For the Yukawa function
(8.7 d) they use the approximate function

$$\phi(r) = [\alpha\beta/2\pi(\alpha+\beta)]^{\frac{1}{2}} (e^{-\alpha r} - e^{-\beta r})/r,$$

with $\alpha = 0\cdot2316$, $\beta = 1\cdot268$.

(d) As an unperturbed deuteron wave function is used, polarization is neglected although it may be partly allowed for by the
resonating group structure method (cf. §8.1).

8.2.1. *Equations of motion for elastic neutron-deuteron scattering.*
It can easily be shown (Schiff, 1949) that a three-body system is
described by a quartet spin state with total spin $S = \frac{3}{2}$ and a doublet
state with $S = \frac{1}{2}$. The corresponding spin wave functions possessing
the correct eigenvalues of spin S and z-component of spin S_z may
be found using the spin operator technique. Denoting the deuteron
by (23), particles 1 and 2 being neutrons, appropriate spin functions
are:

Doublet:

$$\left. \begin{aligned} {}^2\sigma_{\frac{1}{2}} &= 3^{-\frac{1}{2}}[\alpha(1)\,\alpha(2)\,\beta(3) + \alpha(1)\,\beta(2)\,\alpha(3) - 2\beta(1)\,\alpha(2)\,\alpha(3)], \\ {}^2\sigma_{-\frac{1}{2}} &= 3^{-\frac{1}{2}}[\beta(1)\,\beta(2)\,\alpha(3) + \beta(1)\,\alpha(2)\,\beta(3) - 2\alpha(1)\,\beta(2)\,\beta(3)]. \end{aligned} \right\} \quad (8.8)$$

Quartet:

$$
\left.
\begin{aligned}
{}^4\sigma_{\frac{3}{2}} &= \alpha(1)\,\alpha(2)\,\alpha(3), \\
{}^4\sigma_{\frac{1}{2}} &= 2^{-\frac{1}{2}}[\alpha(1)\,\alpha(2)\,\beta(3)+\alpha(1)\,\beta(2)\,\alpha(3)+\beta(1)\,\alpha(2)\,\alpha(3)], \\
{}^4\sigma_{-\frac{1}{2}} &= 2^{-\frac{1}{2}}[\beta(1)\,\beta(2)\,\alpha(3)+\beta(1)\,\alpha(2)\,\beta(3)+\alpha(1)\,\beta(2)\,\beta(3)], \\
{}^4\sigma_{-\frac{3}{2}} &= \beta(1)\,\beta(2)\,\beta(3),
\end{aligned}
\right\}
\qquad (8.9)
$$

where α and β are the Pauli spin matrices denoting spins of $\frac{1}{2}$ and $-\frac{1}{2}$ respectively.

For central forces, the result of using any one of the spin functions describing a fixed spin S is independent of S_z. Hence we may simplify proceedings by employing one representative case only for each of the doublet and quartet states.

To satisfy Pauli's principle, the total wave function Ψ, given as a product of spin and spatial wave functions, must be anti-symmetric in the neutrons 1 and 2, so denoting the spatial wave function by $\Phi(23, 1)$, the resonating group functions are

Doublet:

$$
\begin{aligned}
\Psi = 12^{-\frac{1}{2}}[\{\alpha(1)\,\alpha(2)\,\beta(3)+\alpha(1)\,\beta(2)\,\alpha(3)-2\beta(1)\,\alpha(2)\,\alpha(3)\}\,\Phi(23, 1) \\
-\{\alpha(2)\,\alpha(1)\,\beta(3)+\alpha(2)\,\beta(1)\,\alpha(3)-2\beta(2)\,\alpha(1)\,\alpha(3)\}\,\Phi(13, 2)].
\end{aligned}
$$

$$(8.10)$$

Quartet:

$$
\Psi = 2^{-\frac{1}{2}}\alpha(1)\,\alpha(2)\,\alpha(3)\,[\Phi(23, 1)-\Phi(13, 2)].
\qquad (8.11)
$$

The approximation is made that the incident particle 3 does not polarize the deuteron appreciably, i.e. the relative displacement of the deuteron particles is unaltered. Hence

$$
\Phi(23, 1) = \phi(23)\,F(1), \qquad (8.12)
$$

where $\phi(23)$ is the deuteron ground state wave function and $F(1)$ is the scattering wave function of neutron 1 relative to the CM of the deuteron. The wave equation is then

$$
[T + \mathscr{V}(12) + \mathscr{V}(23) + \mathscr{V}(31)]\Psi = (E_n + E_0)\Psi, \qquad (8.13)
$$

where E_0 is the binding energy of the deuteron and E_n is the incident energy of the neutron in the CM system,

$$
E_n(\mathrm{CM}) = \tfrac{2}{3}E_n(\mathrm{Lab})
$$

by equation (2.19). T is the kinetic energy operator of 1,

$$T = -\frac{\hbar^2}{2m}(2\nabla_{23}^2 + \tfrac{3}{2}\nabla_{23-1}^2), \qquad (8.14)$$

where ∇_{23}^2 refers to the relative co-ordinates of 2 and 3, and ∇_{23-1}^2 to those of 1 relative to the CM of 2 and 3.

We substitute expressions (8.10), (8.11) and (8.14) for Ψ and T respectively in the wave equation (8.13), perform the exchange operations implicit in the definition (8.4) of $\mathscr{V}(r)$, and eliminate the spin wave functions by multiplying both sides of equation (8.13) by the Hermitian conjugate spin function $\sigma^*(23, 1)$, thus obtaining doublet and quartet wave equations for $F(3)$. Denoting the vectors from a fixed origin o to 1, 2, 3 by $\mathbf{r}_1, \mathbf{r}_2, \mathbf{r}_3$, respectively, we define

$$\mathbf{R} = \mathbf{r}_2 - \mathbf{r}_3, \quad \mathbf{r} = \tfrac{1}{2}(\mathbf{r}_2 + \mathbf{r}_3) - \mathbf{r}_1.$$

The wave equation of the deuteron becomes:

$$\left\{ -\frac{\hbar^2}{m}\nabla_R^2 + U(\mathbf{R}) \right\} \phi(\mathbf{R}) = E_D\,\phi(\mathbf{R}). \qquad (8.15)$$

On substituting equation (8.15) in the wave equations for $F(3)$, multiplying both sides by $\phi(R)$ and integrating over $\mathbf{R}$ space, one finally obtains two partial integro-differential equations for the doublet and quartet states.

Doublet:

$$(\nabla^2 + k^2)\,F(\mathbf{r}) = (2w - b - \tfrac{1}{2}m - h)\,U(r)\,F(\mathbf{r})$$

$$+ \int \Bigg[(2m - h - \tfrac{1}{2}w - b)\,Q(\mathbf{r}, \mathbf{r}') - \tfrac{1}{2}P(\mathbf{r}, \mathbf{r}')$$

$$+ \left(\frac{E_n}{E_D} - \frac{5}{3}\right) N(\mathbf{r}, \mathbf{r}') \Bigg] F(\mathbf{r}')\,d\mathbf{r}'. \qquad (8.16)$$

Quartet:

$$(\nabla^2 + k^2)\,F(\mathbf{r}) = (2w + 2b - m - h)\,U(r)\,F(\mathbf{r})$$

$$+ \int \Bigg[(2m + 2h - w - b)\,Q(\mathbf{r}, \mathbf{r}') + P(\mathbf{r}, \mathbf{r}')$$

$$+ \left(\frac{E_n}{E_D} - \frac{5}{3}\right) N(\mathbf{r}, \mathbf{r}') \Bigg] F(\mathbf{r}')\,d\mathbf{r}', \qquad (8.17)$$

where

$$P = \frac{64}{81} \frac{m}{\hbar^2} \left[\phi(23) \{V(23) + V(13)\} \phi(13) \right.$$

$$\left. + 4 \frac{\hbar^2}{m} \phi'(23) \phi'(13) (\mathbf{r}_{23} \cdot \mathbf{r}_{13}/r_{23}r_{13}) \right],$$

$$N = \frac{64}{27} \frac{m}{\hbar^2} E_D \phi(23) \phi(13), \quad U = \frac{4}{3} \frac{m}{\hbar^2} \int \phi^2(R) V(12) \, d\mathbf{R},$$

$$Q = \frac{64}{27} \frac{m}{\hbar^2} \phi(23) V(12) \phi(13), \quad \mathbf{r}' = -\tfrac{1}{2}\mathbf{r} - \tfrac{3}{4}\mathbf{R}, \quad k^2 = \frac{4}{3} \frac{m}{\hbar^2} E_n.$$

$$\text{(8.18)}$$

Both the quartet and doublet integro-differential equations can be written in the form

$$(\nabla^2 + k^2) F(\mathbf{r}) = \alpha U(r) F(\mathbf{r}) + \int \left[\beta Q(\mathbf{r}, \mathbf{r}') + \gamma \left\{ P(\mathbf{r}, \mathbf{r}') \right. \right.$$

$$\left. \left. + \left(\frac{E_n}{E_D} - \frac{5}{3} \right) N(\mathbf{r}, \mathbf{r}') \right\} \right] F(\mathbf{r}') \, d\mathbf{r}'$$

$$= \alpha U(r) F(\mathbf{r}) + \int K(\mathbf{r}, \mathbf{r}') F(\mathbf{r}') \, d\mathbf{r}'. \quad \text{(8.19)}$$

To reduce equation (8.19) to radial form, we use the expansions

$$F(\mathbf{r}) = \frac{1}{r} \sum_{l=0}^{\infty} f_l(r) P_l(\cos \theta),$$

$$\left. \begin{matrix} Q(\mathbf{r}, \mathbf{r}') \\ P(\mathbf{r}, \mathbf{r}') \\ N(\mathbf{r}, \mathbf{r}') \end{matrix} \right\} = \sum_s \frac{(2s+1)}{4\pi r r'} P_s(\cos \mathscr{V}) \left\{ \begin{matrix} q_s(r, r') \\ p_s(r, r') \\ n_s(r, r'), \end{matrix} \right.$$

where

$$\left. \begin{matrix} q_s(r, r') \\ p_s(r, r') \\ n_s(r, r') \end{matrix} \right\} = 2\pi r r' \int_0^\pi P_s(\cos \mathscr{V}) \sin \mathscr{V} \, d\mathscr{V} \left\{ \begin{matrix} Q(\mathbf{r}, \mathbf{r}') \\ P(\mathbf{r}, \mathbf{r}') \\ N(\mathbf{r}, \mathbf{r}') \end{matrix} \right. \quad \text{(8.20)}$$

and $\cos \mathscr{V} = \mathbf{r} \cdot \mathbf{r}'/rr'$. The orthogonal property of the harmonics reduces equation (8.19) to

$$\frac{d^2 f_l}{dr^2} + \left\{ k^2 - \frac{l(l+1)}{r^2} \right\} f_l = \alpha U f_l$$

$$+ \int_0^\infty \left[\beta q_l + \gamma \left\{ p_l + \left(\frac{E_n}{E_D} - \frac{5}{3} \right) \right\} n_l \right] f_l(r') \, dr'. \quad \text{(8.21)}$$

The right-hand side of equation (8.21) approaches o as $r \to \infty$ so that the asymptotic solution is $f_l(r) \sim \sin{(kr - \frac{1}{2}l\pi + \delta_l)}$. The corresponding solution to equation (8.19) has the asymptotic form

$$F(r) \sim e^{ikz} + \frac{1}{r}g(\theta)\, e^{ikr},$$

where $g(\theta)$ is the scattering amplitude,

$$g(\theta) = \frac{1}{2ik} \sum_{l=0}^{\infty} (2l+1)(e^{2i\delta_l} - 1)P_l(\cos\theta).$$

The intensity of scattering is found by assigning the correct statistical weights to the doublet and quartet states:

$$I(\theta) = \tfrac{2}{3}|g_Q(\theta)|^2 + \tfrac{1}{3}|g_D(\theta)|^2.$$

The total elastic cross-section becomes

$$\sigma = \int_0^{\pi} I(\theta)\sin\theta\, d\theta$$

$$= \frac{4\pi}{k^2} \sum_{l=0}^{\infty} (2l+1)[\tfrac{2}{3}\sin^2\delta_l^Q + \tfrac{1}{3}\sin^2\delta_l^D]. \tag{8.22}$$

8.2.2. *Equations of motion for elastic proton-deuteron scattering.* The equations of motion for p-d scattering can be found quite easily from those describing n-d collisions. Denoting the protons by 1, 2 respectively and the neutron by 3, equation (8.13) is modified to

$$\left[T + \mathcal{V}(12) + \frac{e^2}{r_{12}} + \mathcal{V}(23) + \mathcal{V}(13) \right]\Psi = (E_p + E_0)\Psi. \tag{8.23}$$

The wave equation (8.19) thus becomes

$$(\nabla^2 + k^2)F(r) = [\alpha U(r) + C(r)]F(r)$$

$$+ \int [K(\mathbf{r}, \mathbf{r}') + H(\mathbf{r}, \mathbf{r}')]F(\mathbf{r}')\, d\mathbf{r}'. \tag{8.24}$$

The integro-differential equation corresponding to equation (8.21) is

$$\frac{d^2 f_l}{dr^2} + \left\{ k^2 - \frac{l(l+1)}{r^2} \right\} f_l = [\alpha U(r) + C(r)]f_l(r)$$

$$+ \int_0^{\infty} (k_l(r, r') + h_l(r, r'))f_l(r')\, dr', \tag{8.25}$$

where

$$C(r) = \frac{4}{3}\frac{m}{\hbar^2} \int \psi^2(R) \frac{e^2}{r_{12}} d\mathbf{R}, \quad H(r,r') = \left(\frac{4}{3}\right)^3 \frac{m}{\hbar^2} \psi(23)\psi(13) \frac{e^2}{r_{12}},$$

$$\left.\begin{matrix} h_l(r,r') \\ k_l(r,r') \end{matrix}\right\} = 2\pi r r' \int_0^\pi P_l(\cos\mathscr{V}) \sin\mathscr{V}\, d\mathscr{V} \begin{cases} H(\mathbf{r},\mathbf{r}') \\ K(\mathbf{r},\mathbf{r}'). \end{cases}$$

(8.26)

$C(r)$ has a Coulomb form for large r, which makes equation (8.25) more difficult to solve than equation (8.21).

To find the differential cross-section for p-d scattering, we must use the theory for charged particle scattering given in chapter 7. The asymptotic form of the solution of equation (8.24) is

$$f_l(r) \sim \sin\left(kr - \alpha \ln 2kr - \tfrac{1}{2}l\pi + \sigma_l + \delta_l\right), \tag{8.27}$$

where $\alpha = e^2/\hbar v$ (CM system) and $\sigma_1 = \arg\Gamma(l+1+i\alpha)$. The differential cross-section is given by equation (7.31), provided E_p is taken as the proton energy in the CM system of the three nucleons. Allowing for spin states, the angular distribution becomes

$$I = \tfrac{2}{3}I_Q + \tfrac{1}{3}I_D.$$

8.2.3. *Comparison with experiment.* From equation (8.22), the observed n-d cross-section follows in terms of doublet and quartet spin states:

$$\sigma = \tfrac{1}{3}\sigma_D + \tfrac{2}{3}\sigma_Q. \tag{8.28}$$

At thermal energies, σ_D and σ_Q may be written in terms of the doublet and quartet scattering lengths a_D and a_Q:

$$\sigma_{Q,D} = 4\pi a_{Q,D}^2. \tag{8.29}$$

Thus the measured total cross-section of deuterons for thermal neutrons gives the quantity $2a_Q^2 + a_D^2$. A second relation between a_D and a_Q is found by measuring the cross-section for scattering by ortho- and para-deuterium. For temperatures below $20°$ K., interference effects occur between the neutron waves scattered from the two nuclei (Hamermesh and Schwinger, 1946), so that the scattering is strongly dependent on the total nuclear spin of the molecule, and hence on the ratio a_Q/a_D. Interference effects are also found in the angular distribution of slow neutrons scattered by deuterium or by crystals such as $N_a D$ (Hurst and Alcock, 1950; Wollan, Shull and Koehler, 1951). Use of these results leads to the alternative set of values for the zero energy scattering lengths:

$$a_D = 8 \cdot 3 \times 10^{-12} \, \text{cm.}, \quad a_Q = 2 \cdot 4 \times 10^{-12} \, \text{cm.}, \quad (8.30)$$

$$\text{or} \quad a_D = 0 \cdot 8 \times 10^{-13} \, \text{cm.}, \quad a_Q = 6 \cdot 2 \times 10^{-13} \, \text{cm.} \quad (8.31)$$

The variational method of solving the scattering equation (8.25) has been employed by Troesch and Verde (1951) and by Clementel (1951) to obtain the zero energy scattering lengths a_D and a_Q. They assumed the gaussian interaction (8.7 b), except that Clementel used a slightly greater range, $2 \cdot 0 \times 10^{-13}$ cm. Troesch and Verde obtained $a_D = 6 \cdot 8$, $7 \cdot 0 \times 10^{-12}$ cm.; $a_Q = 2 \cdot 8$, $3 \cdot 2 \times 10^{-13}$ cm. for ordinary and symmetric theories respectively, whilst Clementel found the values $a_D = 3 \cdot 70$, $3 \cdot 65$, $3 \cdot 88 \times 10^{-13}$ cm.; $a_Q = 3 \cdot 30$, $3 \cdot 47$, $3 \cdot 51 \times 10^{-13}$ cm. for ordinary, symmetric and Serber potentials respectively. The latter values disagree with the experimental figures, probably because the deuteron wave function used is a bad approximation. Also the potential parameters do not lead to the correct zero energy n-p scattering lengths. The same arguments apply to results found by Troesch and Verde, which are closer to the figures (8.30). However, the latter turns out not to be the correct set of values.

C.G. have used the Yukawa potential (8.7 d) in a variational treatment to obtain

$$a_D = 1 \cdot 5 \times 10^{-13} \, \text{cm.}, \quad a_Q = 5 \cdot 9 \times 10^{-13} \, \text{cm.}, \quad (8.32)$$

agreeing fairly well with equation (8.31). As their potential has dimensions giving correct n-p zero energy scattering lengths, we conclude that equation (8.31) describes the experimental n-d scattering lengths.

B.H.M. used iterative and difference equation procedures to find s- and p-phases for scattering. Values extrapolated to zero energy gave $a_D = 4 \cdot 08 \times 10^{-13}$ cm., $a_Q = 4 \cdot 76 \times 10^{-13}$ cm., in very poor agreement with the experimental values (8.31). The approximate equality of quartet and doublet phases found at higher energies is therefore doubtful, probably due to the potential parameters used. Phases for $l \geqslant 2$, being small, were found by Born's approximation (§ 10.7). All four exchange forces of § 8.2 were used.

On the other hand, C.G. solved an integral equation variant of equation (8.25) by the difference equation method to give values of the quartet and doublet effective ranges. Use of the shape independent formula (7.55) led to the s-phase shifts, these agreeing with results of a phase analysis of 14 MeV. n-d experiments (Allred, Armstrong and Rosen, 1953). The quartet phases were much larger than the doublet phases for energies up to about 9 MeV. Higher phases were found by Born's approximation. This was valid for p-phases only because Serber exchange forces were used, ensuring a small $l = 1$ interaction.

A comparison with experiment of the theoretical results for n-d scattering of B.H.M. and C.G. is shown in fig. 63. At 3·27 and 5·5 MeV., both the symmetric curve of B.H.M. and the Serber curve of C.G. agree reasonably well with experiment. At the higher energy of 14 MeV., the Serber force curve of C.G. agrees very well with the data whereas the symmetric curve of B.H.M. gives too little forward scattering and the scattering minimum is too small. However, De Borde and Massey (1955) have used the B.H.M. potential together with Serber forces, obtaining good agreement with experiment, although the back scattering is a little too large.

Ordinary forces are characterized by a very large peak of forward scattering which is at least several times the experimental value at all energies. Thus the evidence is strongly in favour of an exchange force of symmetric or Serber type. It is difficult to distinguish between these two as calculations with symmetric exchange forces have not been performed using a potential with parameters leading to the correct n-p and n-d scattering lengths.

Fig. 64 shows a comparison with experiment of the total n-d cross-section vs. incident neutron energy for ordinary, symmetric and Serber forces. The theoretical curves are due to B.H.M. and C.G. except for two points at 20 MeV. due to Verde (1949). The ordinary force cross-sections of B.H.M. are too large by a factor of two, whereas their symmetric and Serber curves agree roughly with experiment down to an energy of 2 MeV. However, an extrapolation to zero energy using the shape-independent formula (6.33) shows the cross-section for slow neutrons to be much too small for all the exchange theories. On the other hand the Serber cross-section computed from the angular distributions and scattering lengths due to C.G. gives good agreement over the entire experimental range.

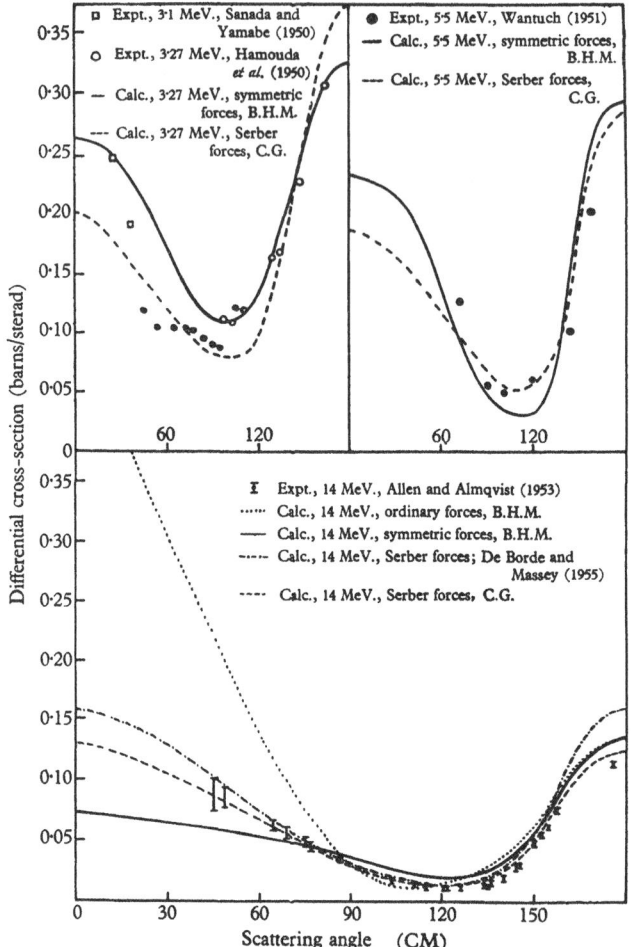

Fig. 63. Comparison of calculated and observed angular distributions for *n-d* scattering at several energies. Theoretical curves are due to B.H.M. (Buckingham *et al.* 1952), C.G. (Christian and Gammel, 1953) and de Borde and Massey (1955). The curve of B.H.M. at 3·27 MeV. has beeninterpolated from values at neighbouring energies.

The experimental data for *p-d* scattering are known to considerably greater accuracy than for *n-d*. Both B.H.M. and C.G. have obtained angular distributions at a number of energies, but whereas the former use theory alone, the latter base their work on *s*-phase shifts adjusted to give best fits with the experimental data.

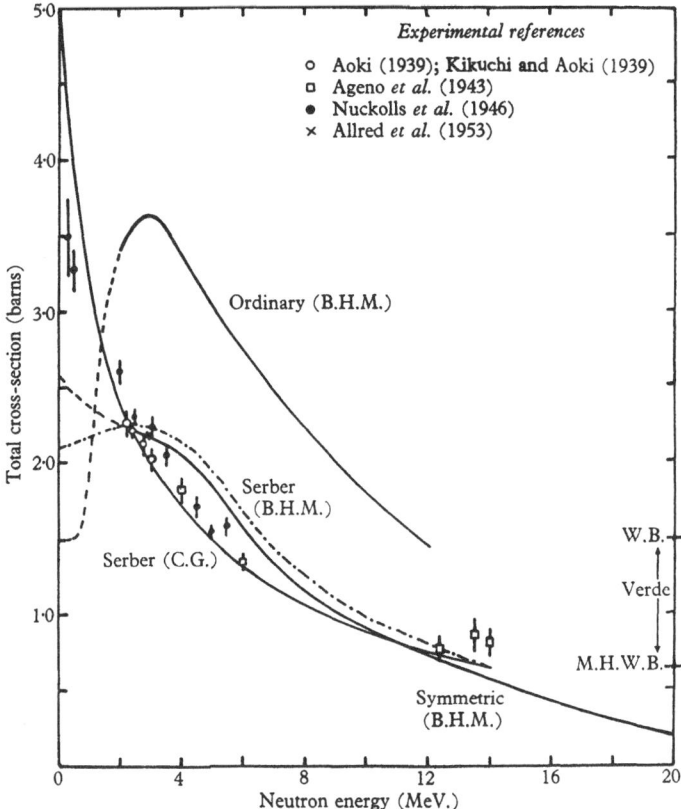

Fig. 64. Comparison of calculated and observed total cross-sections for elastic *n-d* collisions. Calculated curves are for ordinary, Serber and symmetric forces, due to B.H.M. (Buckingham *et al.* 1952) and to C.G. (Christian and Gammel, 1953) (the latter for Serber forces only). The B.H.M. curves have been extended below 2 MeV., using the shape independent formula for *s*-phases. Calculated values at 20 MeV. are due to Verde (1949). *Note added in proof.* For more accurate experimental data in the range 0·2–22 MeV., see Seagrave and Henkel (1955).

Fig. 65 shows the angular distributions for *p-d* scattering at 2·53, 5·0 and 9·7 MeV., each interpolated from results of B.H.M. using symmetric forces. The agreement with experiment is good up to 5 MeV., but at 9·7 MeV. the symmetric force curve noticeably underestimates both forward and backward scattering and differs somewhat in shape. B.H.M. state this to be due to the neglect of phases for $l \geqslant 2$. C.G. have performed the calculations

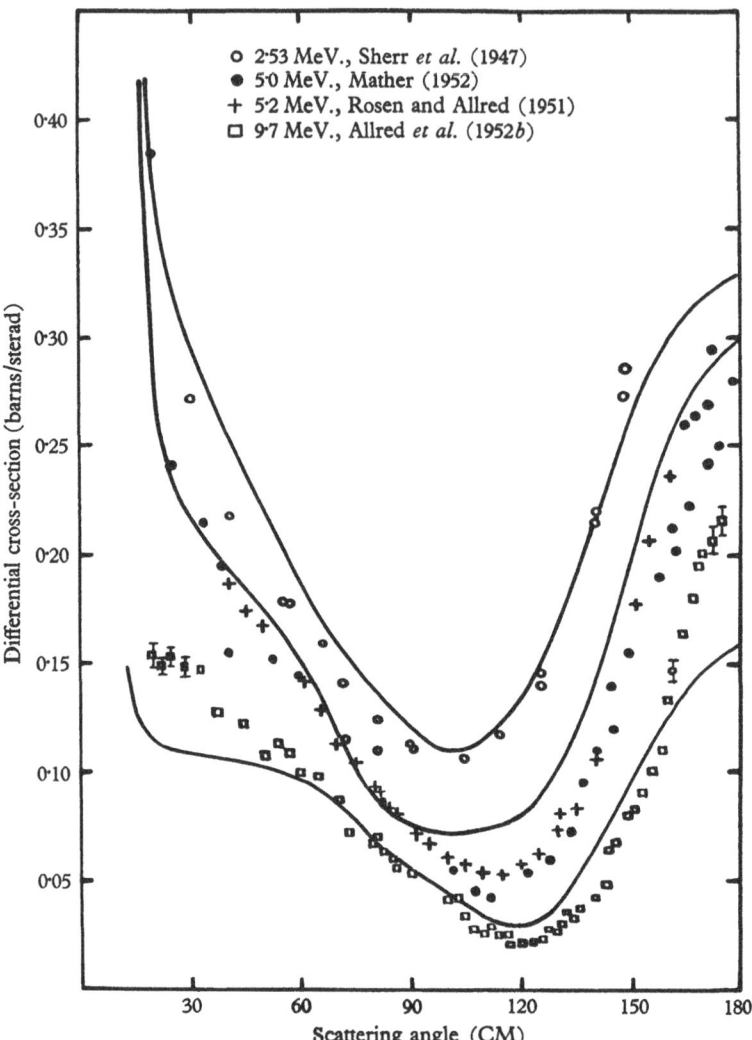

Fig. 65. Comparison of calculated and observed angular distributions for p-d scattering at neutron energies of 2·53, 5·0 and 9·7 MeV. The theoretical curves are interpolated from results of B.H.M. using symmetric forces.

for higher phases and obtain good agreement with experiment to within 4 % over the whole angular range. The differences between the two calculations may be due to higher phases and the different well shapes used (Massey, 1953).

8.2.4. *Polarization of the deuteron.* In all calculations, polarization of the deuteron by the incident nucleon has been neglected. The effect is due to the presence of excited states of the deuteron, i.e. distortion of the charge distribution of the deuteron from spherical symmetry by the incident nucleon. This is unlikely to be important except at low energies, say $E < 2\,$MeV., as faster nucleons are not in the vicinity of the deuteron long enough to distort it appreciably. Also, polarization can only be important for s-wave scattering, as interactions at low energies for $l \geqslant 1$ take place at large distances, so that only small corrections to already small phases are involved. It is almost certainly larger for doublet than quartet scattering, as in the latter case the Pauli principle tends to keep the particles apart (the doublet state corresponds to the bound state of ^{3}H). Werner (1956) has shown that the ratio of the stripping theory cross-sections (§11.2) for the (d, p) and (t, d) processes, when equated to its measured value, fixes an upper limit of 0·11 for the probability of ^{3}H existing in the $d + n$ configuration. This indicates a large polarization effect for n-d and p-d scattering in the doublet state.

C.G. have made rough estimates of the effect of polarization on their zero energy scattering lengths by taking account of excited states of the deuteron group in the three particle system. They find a possible change of 20% for the quartet scattering length, but quote the corresponding doublet change as unreliable. Probably the absolute correction is about $1 \times 10^{-13}\,$cm. for both the states, which would still leave the theoretical figures (8.32) in rough agreement with the experimental values (8.31). Actually at zero energy only the quartet scattering length is important, for from equations (8.28) and (8.31) this contributes about 99% of the cross-section. A phase analysis of p-d experiments by C.G. indicates the cross-section is 90% quartet even at 2·53 MeV., the doublet s-phase being considerably smaller than the quartet for energies up to about 9 MeV. Thus the low energy cross-section is largely quartet with probably only a small effective polarization.

Another method for dealing with polarization would be to use a variational trial function dependent on the inter nucleonic distances to describe the scattering. If the deuteron particles are denoted by 1 and 2, and the co-ordinate from their CM to the incident particle 3 by $\mathbf{r}$, then a suitable form could be

$$\Psi = \chi(r_{12})\,[\sin kr + \{M + (b + cr_{13} + dr_{23})\,\mathrm{e}^{-\nu r}\}\,(1 - \mathrm{e}^{-\nu r})\cos kr], \quad (8.33)$$

where $\chi(r)$ is the wave function for the deuteron ground state. A computation to determine the coefficients in equation (8.33) by using the variational methods of Hulthén and Kohn would be very laborious and has not been performed. However, calculations have been made for polarization involved in the scattering of electrons by hydrogen, using a wave function dependent on the inter-electronic distance (Massey and Moiseiwitsch, 1951). The modification to the scattering (symmetric case) turns out to be important, so that one might expect the corresponding state in n-d scattering (the doublet) to lead to a similar result at energies less than the binding energy of the deuteron. This is in accord with Werner's result above.

8.2.5. *Summary of elastic n-d and p-d scattering.* The scattering can be quantitatively explained up to energies of $14\,\mathrm{MeV}$. by a central potential with parameters giving the correct zero energy n-p scattering lengths and effective ranges, provided exchange forces of Serber or possibly symmetric type are used. The data is in definite disagreement with an ordinary type exchange force. For a summary of experimental n-d and p-d data, see Seagrave and Cranberg (1957) and Caldwell and Richardson (1955).

Polarization of the deuteron may be important below $2\,\mathrm{MeV}$. for s-wave scattering only, and then mainly for the less important doublet case. At zero energy the correction to the quartet scattering length may be of the order of $20\,\%$ and the doublet correction even larger, but the latter contributes only about $1\,\%$ of the cross-section. At energies up to several MeV., the doublet contribution to the cross-section is less than about $10\,\%$, so that the polarization effect may be $\lesssim 40\,\%$. See however, footnote, p. 276.

8.3. Inelastic collision of neutrons with deuterons

Disintegration of the deuteron by neutron impact is possible at energies above $3\cdot3\,\mathrm{MeV}$. but the cross-section remains small below about $10\,\mathrm{MeV}$. Thus, inelastic cross-sections are smaller than the corresponding elastic cross-sections at $5\,\mathrm{MeV}$. and $14\,\mathrm{MeV}$. by factors of ~ 50 and 3 respectively. Nevertheless, some attempt has been made to explain the small amount of experimental information available by an extension of the resonating group structure method (Bransden and Burhop, 1950).

The three particle spin functions appropriate to the problem consist of the quartet (8.9) with total spin $\frac{3}{2}$ and the doublet (8.8) with total spin $\frac{1}{2}$, both symmetric in the deuteron particles (23), and the doublet (8.34) with spin $\frac{1}{2}$ which is antisymmetric in the deuteron particles.

$$\left.\begin{array}{l} \chi_7 = 2^{-\frac{1}{2}}(\alpha_1\alpha_2\beta_3 - \alpha_1\beta_2\alpha_3), \\ \chi_8 = 2^{-\frac{1}{2}}(\beta_1\beta_2\alpha_3 - \beta_1\alpha_2\beta_3). \end{array}\right\} \qquad (8.34)$$

The quartet and the symmetric doublet can describe both elastic and inelastic scattering, but the antisymmetric doublet (8.34) can only describe the virtual ^1S state of the deuteron and hence corresponds to inelastic collisions only.

Development of the theory in terms of a total wave function written as the sum of elastic and inelastic parts leads to a pair of coupled integro-differential equations:

$$(\nabla^2 + k_l^2)F_l(\mathbf{r}) + G_{\kappa\kappa}(r)F_l(\mathbf{r}) = G_{0\kappa}(r)F_k(\mathbf{r}), \qquad (8.35)$$

$$(\nabla^2 + k^2)F_k(\mathbf{r}) + G_{00}(r)F_k'(\mathbf{r}) = G_{\kappa 0}(r)F_l(\mathbf{r}), \qquad (8.36)$$

where $F_k(\mathbf{r})$ and $F_l(\mathbf{r})$ are the elastically and inelastically scattered waves respectively, and k and k_l their wave numbers. The kernel functions $G_{00}(r)$ and $G_{\kappa\kappa}(r)$ refer to elastic scattering by the ground and continuum states of ^2H respectively, and $G_{\kappa 0}(\mathbf{r})$ and $G_{0\kappa}(\mathbf{r})$ to the corresponding inelastic scattering.

As there is only a small probability of the deuteron particles remaining together after disintegration, we might expect the scattering to be roughly plane wave so that $G_{\kappa\kappa}(r)$ may be neglected to a first approximation. The asymptotic solution of (8.35) then gives the scattering amplitude for inelastic collisions (Mott and Massey, 1949):

$$f_l(\theta, \phi) = -\frac{1}{4\pi}\int e^{-i\mathbf{k}_l \cdot \mathbf{r}} G_{0\kappa}(r)F_k(\mathbf{r})\,d\mathbf{r}. \qquad (8.37)$$

However it is inconsistent to neglect the effect of the mean field of the excited deuteron, acting through $G_{\kappa\kappa}(r)$, on the final wave function F_l. A better result is obtained replacing (8.4b) by the distorted wave approximation

$$f_l(\theta, \phi) = -\frac{1}{4\pi} \int \mathscr{F}_l^*(\mathbf{r}) \, G_{0\kappa}(r) F_k(\mathbf{r}) \, d\mathbf{r}, \qquad (8.38)$$

where $F_l(\mathbf{r})$ is the solution of

$$(\nabla^2 + k_l^2) \mathscr{F}_l(\mathbf{r}) + G_{\kappa\kappa}(r) \mathscr{F}_l(\mathbf{r}) = 0. \qquad (8.39)$$

The differential cross-section for scattering into an element of solid angle $d\omega$ of a neutron of initial momentum $\mathbf{k}$, and final momentum between $\mathbf{k}_l$ and $\mathbf{k}_l + d\mathbf{k}_l$, is given in terms of the scattering amplitudes by

$$I(\theta, \phi, k_l) \, d\omega \, d\kappa = \frac{k_l}{k} [\tfrac{2}{3} | f_l^Q(\theta, \phi) |^2 + \tfrac{1}{3} | f_l^{D, \text{sym}}(\theta, \phi) |^2$$
$$+ \tfrac{1}{3} | f_l^{D, \text{asym}}(\theta, \phi) |^2], \qquad (8.40)$$

where $f_l^Q, f_l^{D, \text{sym}}$ and $f_l^{D, \text{asym}}$ refer to the quartet, symmetric doublet and antisymmetric doublet final states respectively of the target nucleus.

The total cross-section of the disintegration process follows by integrating over angles and values from o to κ_{max} of the energy of the inelastically emitted neutron:

$$\sigma(k) = \int_0^{2\pi} d\phi \int_0^{\pi} d\theta \sin\theta \int_{\kappa=0}^{\kappa_{\text{max}}} I(\theta, \phi, k_l) \, d\kappa. \qquad (8.41)$$

For numerical calculations, Bransden and Burhop (1950) used the exponential potential (8.7a) and deuteron wave function employed in elastic n-d calculations. The scattering amplitude was found from equation (8.37), with $F_k(\mathbf{r})$ obtained by solution of equation (8.36). In the method of distorted waves, the term $G_{\kappa 0}(r)$ on the right-hand side of equation (8.36) was neglected by analogy with similar electron problems, since products of the type $G_{\kappa 0}(r) F_l(\mathbf{r})$ are in the latter cases generally much less important than products of the type $G_{0\kappa}(r) F_k(\mathbf{r})$. In this approximation, $F_k(\mathbf{r})$ is the wave function for elastic n-d scattering. Calculated results (Bransden and Burhop, 1950) using symmetric exchange forces are compared in Table XI with experimental values. Theoretical cross-sections are much too large at all energies.

Two explanations suggest themselves. The first is the inadequacy of the distorted wave approximation, the nucleonic interaction being so strong that probably $G_{\kappa 0}(r)$ cannot be neglected as small, unlike analogous electron scattering cases. Secondly, the assumption $G_{\kappa\kappa}(r) = 0$ involved taking $F_l(\mathbf{r})$ as $\exp(i\mathbf{k}_l.\mathbf{r})$, in spite of the fact that for slow collisions, the three particles are liable to strongly perturb each other's motion. Clearly equation (8.38) should be used instead of equation (8.37) to allow for this distortion of the scattered wave. However a really reliable solution of equations (8.35) and (8.38) could probably only be obtained by use of the difference equation method in conjunction with an electronic computer.

TABLE XI. *Experimental and theoretical data on inelastic n-d and p-d scattering, ^{2}H(n, 2n) ^{1}H and ^{2}H(p, 2p) n*

Incident energy (MeV.)	Process	(mb.)	Details
4·1	n-d	22	Theor. (Bransden and Burhop, 1950).
5·1	p-d	14	Exp. (Barkas and White, 1939).
	n-d	(20–30)	Calculated from p-d scattering (Barkas and White, 1939), allowing for Coulomb effects (Hocker, 1942).
	n-d	(50)	Theor. Interpolated from Bransden and Burhop (1950).
7·4	n-d	230	Theor. (Bransden and Burhop, 1950).
9·7	p-d	114	Exp. (Frank and Gammel, 1954). Emission of protons.
		126	Theor. (Frank and Gammel, 1954). $E_p > 0\cdot3$ MeV., 20–130° (Lab).
		233	Theor. (Frank and Gammel, 1954). All proton energies and angles.
11·5	n-d	640	Theor. (Bransden and Burhop, 1950).
14·1	n-d	40–90	Exp. (Ageno et al. 1947).
		50	Exp. (Coon and Taschek, 1949), 0–80° (Lab).
		53 ± 15	Exp. (Allred et al. 1953). Emission of protons, $E_p > 2$ MeV., 0–100° (Lab).
		80	Theor. (Frank and Gammel, 1954). Emission of protons, $E_p > 2$ MeV., all angles.
		112 ± 15	Exp. (Seagrave, 1955). Emission of protons, 0–80° (Lab).
		190 ± 30	Exp. Elastic n-d (610 ± 30) subtracted from transmission cross-section (803 ± 14) (Seagrave, 1955; Poss et al. 1952).
		(1030)	Theor. Interpolated from Bransden and Burhop (1950).
		144	Theor. Frank and Gammel, (1954).
16·6	n-d	1600	Theor. Brandsen and Burhop (1950).

Frank and Gammel (1954) have used a cruder semi-empirical approach instead of the distorted wave approximation (8.38). Essentially they replaced $F_k(\mathbf{r})$ and $F_l(\mathbf{r})$ in equation (8.38) by the plane waves $e^{-i\mathbf{k}\cdot\mathbf{r}}$ and $e^{-i\mathbf{k}_l\cdot\mathbf{r}}$ respectively, amounting to use of the Born approximation for low incident energies. It proved possible

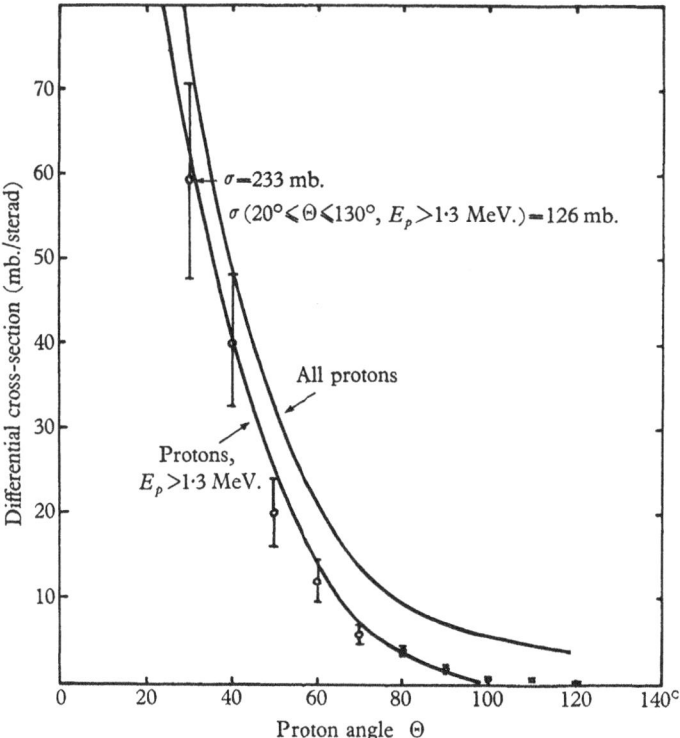

Fig. 66. Angular distribution of protons emitted in 9·66 MeV. inelastic p-d scattering. Experimental points for $E_p > 1·3$ MeV. are due to J. H. Gammel and theoretical curves to J. L. Gammel (Frank and Gammel, 1955).

to relate this to the n-p parameters a and r_0 in a very rough way. The semi-empirical cross-section formula resulting for inelastic n-d scattering is

$$\frac{d\sigma_{1n}}{d\omega_l} = \frac{d\sigma_{el}}{d\omega} \cdot \frac{1}{\pi\alpha} \frac{k_t}{k} \kappa^2 \left[\frac{1}{(\kappa\cot{}^2\delta)^2 + \kappa^2} + \tfrac{1}{3}(^1V_{np})^2 \frac{1}{(\kappa\cot{}^1\delta)^2 + \kappa^2} \right] d\kappa,$$

$$(8.42)$$

where $\kappa \cot \delta = -1/a + \frac{1}{2}r_0 \kappa^2$ employs experimental values of a and r_0 and $^3\delta$ and $^1\delta$ are triplet and singlet phase shifts respectively.

Table XI shows that inelastic cross-sections calculated from equation (8.42) agree surprisingly well with experiment, as do likewise angular distribution calculations at $9 \cdot 66$ MeV. (fig. 66). However, the agreement may be fortuitious in view of the many drastic approximations made in the theory.

8.4. Neutron capture by deuterons

The capture cross-section for the $n + {}^2\mathrm{H} \to {}^3\mathrm{H} + h\nu$ process may be obtained from that for the photodisintegration of the triton, using the principle of detailed balance as employed for the $n(p, \gamma)\,d$ reaction (§ 6.13):

$$\sigma_{\mathrm{cap}} = \frac{64\pi^4\nu^3}{3hc^3v} \,|\,H\,|^2. \qquad (8.43)$$

As in § 6·13, H is the matrix element describing the photoelectric transition,

$$H = \int \Psi_f^* \mathbf{m} \Psi_i \, d\tau, \qquad (8.44)$$

where Ψ_i and Ψ_f are the initial and final wave functions for the three-body system, given by the doublet and quartet functions (8.10) and (8.11) respectively. The radial wave functions needed for Ψ_f may be found numerically as solutions of the equation (8.19) describing elastic n-d scattering, but the radial function in Ψ_i describes the ground state of $^3\mathrm{H}$, also satisfying the same equation for the doublet state. In the electric dipole case, $\mathbf{m} = \frac{1}{2}\mathbf{r}$, but in the magnetic dipole case, $\mathbf{m} = \mu_n \boldsymbol{\sigma}_1 + \mu_n \boldsymbol{\sigma}_2 + \mu_p \boldsymbol{\sigma}_3$.

Burhop and Massey (1947) find that at thermal energies the calculated capture cross-section has the very small value of $0 \cdot 73$ mb. for the symmetric exchange theory. This compares with the experiments of Borst and Harkins (1940), who find an upper limit for the thermal cross-section of $0 \cdot 2$–$0 \cdot 3$ mb. Fast neutrons have much smaller capture cross-sections, calculated values at $0 \cdot 26$ and $11 \cdot 5$ MeV. being $0 \cdot 013$ and $0 \cdot 0307$ mb. respectively, but no experimental results are available for comparison.

8.5. Scattering of nucleons by ^{3}He and ^{3}H

Information similar to that deduced from n-d and p-d collisions may be obtained from the elastic scattering of nucleons by $^3\mathrm{H}$

and ^{3}He. However, as the latter are mirror nuclei, only two problems need be considered. Thus, the neutron scattering equations may be found by omitting terms of Coulomb origin in the wave equations describing the corresponding cases of proton scattering. This is subject, in n-^{3}He scattering, to the addition of a Coulomb term internal to the ^{3}He nucleus, as compared with p-^{3}H scattering.

The ^{3}H and ^{3}He nuclei with their high binding energies of 8·49 and 7·73 MeV. respectively are more compact than the deuteron (2·226 MeV.). Thus, any polarization effect should be smaller than in n-d or p-d scattering, where the resultant effect is believed to be small, at least for energies of more than several MeV. (§ 8.2.4). Other approximations made involve neglect of non-central forces and use of Wheeler's resonating group structure method.

8.5.1. *Elastic p-^{3}He and n-^{3}H scattering.* In p-^{3}He scattering, we consider the ^{3}He nucleus as comprised of protons 1, 2 and neutron 3, with a proton 4 as projectile. The total wave function $\Psi(123, 4)$ may be written in the form

$$\Psi(123, 4) = 6^{-\frac{1}{2}}[\sigma_4\phi(123, 4) + \sigma_1\phi(243, 1) + \sigma_2\phi(413, 2)], \quad (8.45)$$

where $\phi(123, 4) = \phi_0(123) F(4)$ is assumed symmetric in any pair of the ^{3}He particles (123). Values of σ_4, σ_1 and σ_2 for triplet and singlet states are

$$\text{Triplet:} \quad \begin{aligned} \sigma_4 &= 2^{-\frac{1}{2}}(\alpha_1\beta_2 - \beta_1\alpha_2)\,\alpha_3\alpha_4, \\ \sigma_1 &= 2^{-\frac{1}{2}}(\alpha_2\beta_4 - \beta_2\alpha_4)\,\alpha_1\alpha_3, \\ \sigma_2 &= 2^{-\frac{1}{2}}(\alpha_4\beta_1 - \beta_4\alpha_1)\,\alpha_2\alpha_3. \end{aligned} \quad (8.46)$$

$$\text{Singlet:} \quad \begin{aligned} \sigma_4 &= \tfrac{1}{2}(\alpha_1\beta_2 - \beta_1\alpha_2)(\alpha_3\beta_4 - \beta_3\beta_4), \\ \sigma_1 &= \tfrac{1}{2}(\alpha_2\beta_4 - \beta_2\alpha_4)(\alpha_3\beta_1 - \beta_3\alpha_1), \\ \sigma_2 &= \tfrac{1}{2}(\alpha_4\beta_1 - \beta_4\alpha_1)(\alpha_3\beta_2 - \beta_3\alpha_2). \end{aligned} \quad (8.47)$$

The wave equation describing the p-^{3}He interaction is then

$$\left[T + \mathscr{V}(12) + \frac{e^2}{r_{12}} + \mathscr{V}(23) + \mathscr{V}(31) + \mathscr{V}(14) + \frac{e^2}{r_{14}} + \mathscr{V}(24) \right.$$
$$\left. + \frac{e^2}{r_{24}} + \mathscr{V}(34) \right] \Psi(123, 4) = (E_p + E_0)\,\Psi(123, 4), \quad (8.48)$$

where E_0 is the ^{3}He binding energy and E_p is the incident proton energy in the CM system ($E_{CM} = \frac{3}{4}E_{Lab}$). T is the kinetic energy operator,

$$T = -(\hbar^2/2m)(2\nabla^2_{23} + \tfrac{3}{2}\nabla^2_{23-1} + \tfrac{4}{3}\nabla^2_{123-4}). \qquad (8.49)$$

The exchange operator $\mathscr{V}(r)$ is again taken with the form (8.4).

Two procedures are now possible. First, we may simplify equation (8.48) by making use of the equation for the ground state of ^{3}He:

$$\left[-\frac{\hbar^2}{2m}(2\nabla^2_{23} + \tfrac{3}{2}\nabla^2_{23-1}) + \mathscr{V}(12) + \frac{e^2}{r_{12}} + \mathscr{V}(23) + \mathscr{V}(31) \right]$$

$$\times \phi_0(123) = E_0\, \phi_0(123), \qquad (8.50)$$

afterwards multiplying both sides by $\sigma_4^*\,\phi_0(123)$ and integrating over $\mathbf{r}_{23}$ and $\mathbf{r}_{23-1}$ space. This type of procedure was used for n-d collisions (§8.2.1) and is valid if the ground state wave function $\phi_0(123)$ is an accurate solution of (8.50). The resulting integro-differential equation can always be put in a form such that its kernels are symmetric, so that this version is called the *symmetrized theory*. Alternatively, we could multiply equation (8.48) by $\sigma_4^*\,\phi_0(123)$ and integrate over $\mathbf{r}_{23}$ and $\mathbf{r}_{23-1}$ space first, afterwards simplifying by using not equation (8.50) but the approximate binding energy expression

$$E_0 = \int \phi_0(123) \left[-\frac{\hbar^2}{2m}(2\nabla^2_{23} + \tfrac{3}{2}\nabla^2_{23-1}) + \mathscr{V}(12) + \frac{e^2}{r_{12}} \right.$$

$$\left. + \mathscr{V}(23) + \mathscr{V}(31) \right] \phi_0(123)\, d\tau. \qquad (8.51)$$

This procedure is valid even if the ground state wave functions $\phi_0(123)$ are only crude approximations, leading to a wave equation, the kernels of which are asymmetric. The resultant theory is called the *unsymmetrized theory*. The symmetrized and unsymmetrized theories do not in general lead to the same wave equations unless exact ground state wave functions are used. If an approximate form is used for $\phi_0(123)$, e.g. as determined by a variational calculation of the binding energy, then a good fit to the binding energy usually does not mean even a rough fit to the wave function. Equation (8.50) is therefore invalid and the symmetrized theory incorrect. On the other hand, equation (8.51) still holds so that the unsymmetrized theory is consistent and valid in so far as the properties of the ^{3}He nucleus are reproduced by $\phi_0(123)$.

The mathematical complexity of four-body problems necessitates use of the special properties of gaussian function for the two-body potential and wave function:

$$V(r) = V_0 \exp[-\mu r^2], \quad \phi_0(123) = 3^{\frac{3}{4}}(\lambda/\pi)^{\frac{3}{2}} \exp[-\tfrac{1}{2}\lambda(r_{12}^2 + r_{23}^2 + r_{31}^2)].$$
$$(8.52)$$

Detailed application of methods similar to those used for n-d scattering (§8.2.1) leads to integro-differential equations similar to those encountered in p-d scattering (§8.2.2), but involving more complicated kernel sums. The latter are expressible as products of gaussian and hyperbolic functions.

8.5.2. *Elastic* p-^{3}H *and* n-^{3}He *scattering.* The methods of §8.5.1 may be applied to p-^{3}H scattering, provided the alternative d-d configuration is assumed unimportant compared to the p-^{3}H configuration. This is reasonable, as the energy of the latter is considerably greater than that of the former; moreover, variational calculations of the binding energy of the α-particle have shown that it spends almost all of its time in the p-^{3}H and n-^{3}He configurations.

Taking 1, 2 as neutrons, (123) as the triton, and 4 as the incident proton, the total wave function is

$$\Psi(123,4) = \tfrac{1}{2}[\sigma_4\phi(123,4) + \sigma_3\phi(124,3)]. \qquad (8.53)$$

The wave equation (8.48) is replaced by

$$\left[T + \mathscr{V}(12) + \mathscr{V}(23) + \mathscr{V}(31) + \mathscr{V}(14) + \mathscr{V}(24) + \mathscr{V}(34) + \frac{e^2}{r_{34}}\right]$$
$$\times \Psi(123,4) = (E_p + E_0)\Psi(123,4). \quad (8.54)$$

The binding energy equation (8.54) and the expectation energy (8.51) are modified by deletion of the coulomb term e^2/r_{12}. Otherwise the theory of §8.5.1 holds unchanged, resulting in wave equations of either the symmetrized or unsymmetrized kind.

8.5.3. *Comparison of theory with experiment.* Calculations on n-^{3}H and p-^{3}He scattering employing the symmetrized theory (Swan, 1953), assume the gaussian interaction (8.52) with $V_0 = -45$ MeV., $\mu = 2 \cdot 669 \times 10^{25}$ cm.$^{-2}$ and range

$$b = 1/\sqrt{\mu} = 1 \cdot 936 \times 10^{-13} \text{ cm.},$$

together with the variation parameters

$$\lambda = 1\cdot436 \times 10^{25}\,\text{cm.}^2, \quad E_0 = -5\cdot49\,\text{MeV.}$$

(actually these values refer to ^{3}H, the ^{3}He values being assumed approximately the same). Results obtained using phases calculated by the variational methods of Hulthén and Kohn gave agreement with n-^{3}H experiments at 14 MeV. (Coon, Bockelman and Barschall, 1951), but later p-^{3}He experiments showed the p-phases were of the wrong sign and magnitude, leading to incorrect angular distributions at lower energies. Subsequent calculations (Bransden, Robertson and Swan, 1956) on n-^{3}H and n-^{3}He scattering employed the unsymmetrized equations, which were solved on the A.C.E. digital computer using the difference equation method.

Fig. 67 shows n-^{3}H angular distributions calculated for the symmetric theory, agreeing well with experimental values at 14 MeV. Fig. 68 gives the same comparison for Serber forces. At 14 MeV., Serber forces result in twice the peak of forward scattering predicted by symmetric forces, but the scattering minimum in the region of 120° (CM) is smaller by a factor of 2·5. The latter fact favours the symmetric theory, but unfortunately measurements do not extend below 67° where a proper comparison could be made (Coon et al. 1951). The back-scattering is very similar in both cases. An unusual feature is that decreasing the energy shifts the scattering minimum in the forward direction, the minimum actually reaching 0° for $E \sim 1$ MeV. Total cross-sections are shown as a function of energy in fig. 69.

Apart from n-^{3}H scattering at 14 MeV. (Coon et al. 1951), no other experimental results are available for either n-^{3}H or n-^{3}He elastic collisions. However, a number of experiments on proton scattering have been carried out. Angular distributions for p-^{3}He scattering cover energies of 1·01, 1·60, 2·25 and 3·52 MeV. (Famularo, Brown, Holmgren and Stratton, 1954), 5·78 MeV. (Kreger, Jentschke and Kruger, 1954) and 5·0, 8·6 and 9·3 MeV. (Sweetman, 1955). Phase shift analyses have been carried out in the energy range 1·01–3·52 MeV. (Lowen, 1954; Frank and Gammel, 1955).

p-^{3}H experiments include energies of 1·59 and 2·01 MeV. (Taschek et al. 1949), 0·7–2·5 MeV. (Hemmendinger et al. 1949)

262   NUCLEAR SCATTERING

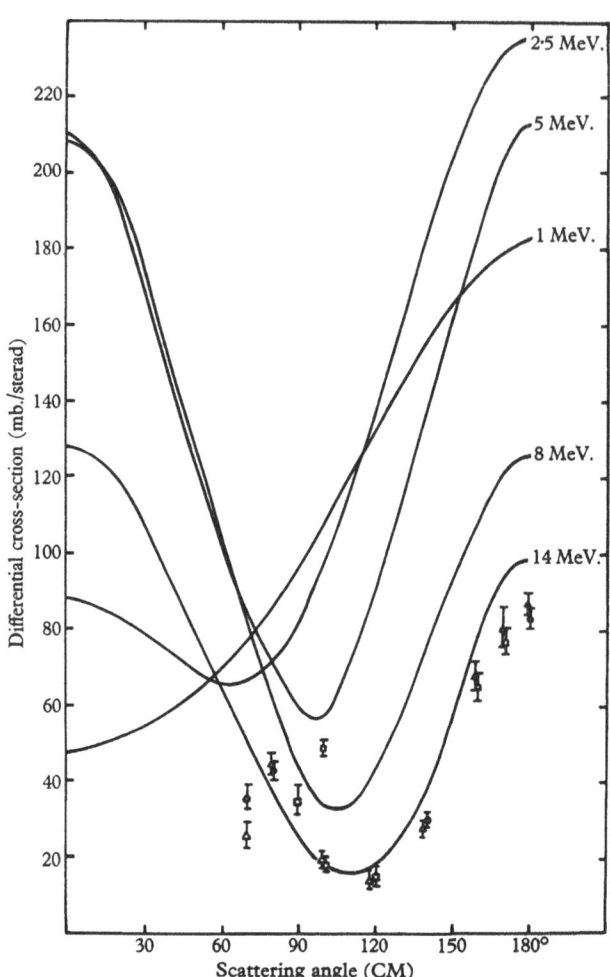

Fig. 67. Angular distributions for n-^{3}H scattering (Bransden, Robertson and Swan, 1956) using symmetric exchange forces with the unsymmetrized theory. The experimental points at 14 MeV. are due to Coon, Bockelman and Barschall, (1951).

2·54–3·50 MeV. (Claassen *et al.* 1951) and 1·0–2·55 MeV. (Ennis and Hemmendinger, 1954). A phase shift analysis has been made (McIntosh, Gluckstern and Sack, 1952) at 0·708–2·548 MeV However, theoretical results for p-^{3}H and p-^{3}H scattering are not available.

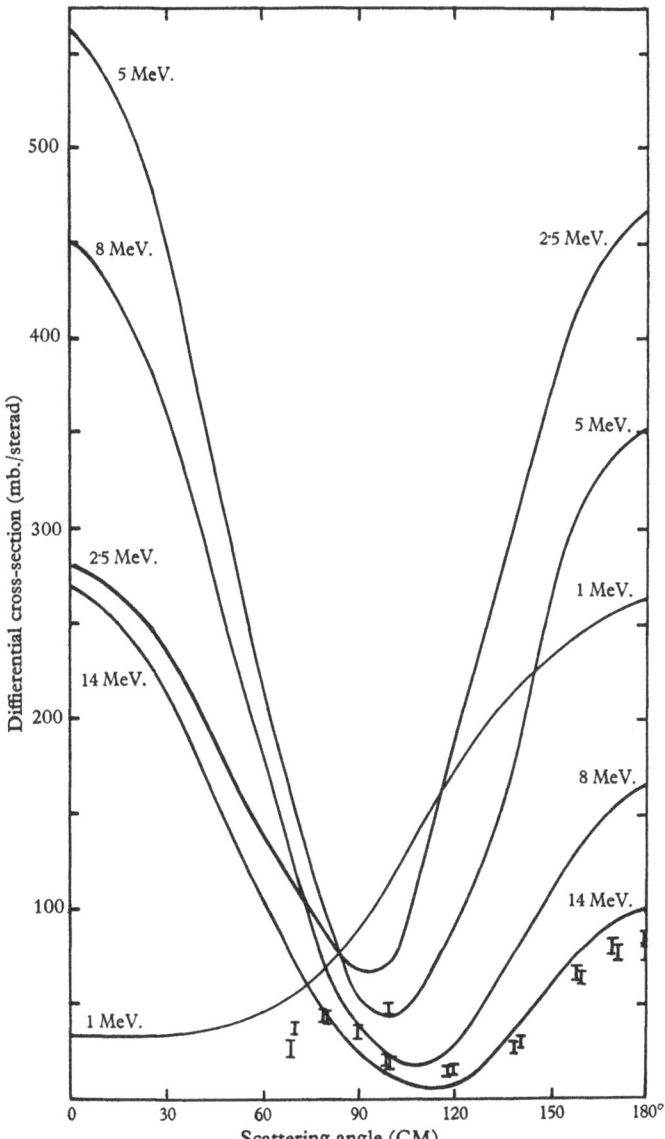

Fig. 68. Angular distributions for n-^{3}H scattering using Serber exchange forces and the unsymmetrized theory (Bransden *et al.* 1956). Experimental points at 14 MeV. are due to Coon *et al.* (1951).

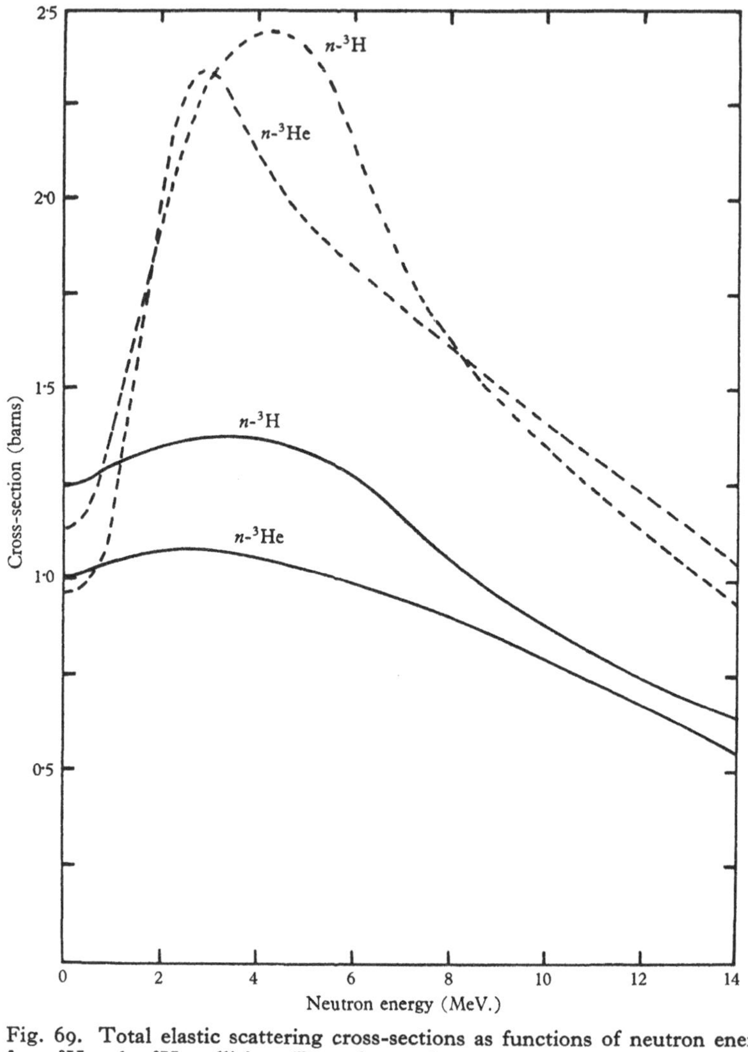

264   NUCLEAR SCATTERING

Fig. 69. Total elastic scattering cross-sections as functions of neutron energy for n-^{3}H and n-^{3}He collisions (Bransden *et al.* 1956). The full curves are for symmetric exchange forces and the dotted curves for Serber exchange forces. The unsymmetrized theory is used throughout.

8.6. Elastic scattering of nucleons by α-particles

The scattering of nucleons by ^{4}He has especial interest for a number of reasons.

(1) The high α-particle binding energy of 28·2 MeV. means that

there are no inelastic processes to compete with elastic scattering below an energy of 19·7 MeV.

(2) It also follows that configurations other than n-α or p-α may be safely neglected, reducing theoretical work to the solution of one integro-differential equation.

(3) The high average binding energy per nucleon in the α-particle implies that exchange forces and Pauli exchange are not very important; thus the scattering can be approximated by two-body scattering between non-identical particles (nucleon plus α-particle).

(4) ^{4}He is the lightest nucleus with zero spin. Hence only one doublet spin state is involved in n-α or p-α scattering and angular distributions may be analysed unambiguously to give one set of phase shifts.

(5) n-α and p-α interactions are subject to a strong spin-orbit coupling (cf. chapter 9) which produces a splitting of the p-phase, resulting in a resonance peak of scattering, thus offering a means of determining the strength of the spin-orbit component.

p-α experiments cover the energy range 1–10 MeV. (Heydenberg and Ramsey, 1941; Freier *et al.* 1949; Braden, 1951; Kreger, Kerman and Jentschke, 1952; Putnam, 1952; Cork and Hartsough, 1954*b*; Freemantle, Grotdal, Gibson, McKeague, Prowse and Roblat, 1954; Kruse *et al.* 1954; Williams and Rasmussen, 1955; Putnam, Brolley and Rosen, 1956), 17·5 MeV. (Brockman, 1956), 28 MeV. (Wickersham, 1954) and 40 MeV. (Brussel and Williams, 1957).

n-α experiments cover the range 0–15·7 Me.V (Staub and Stephens, 1939; Gaerttner, Pardue and Streib, 1939; Staub and Tatel, 1940; Barschall and Kanner, 1940; Carroll, 1941; Hall and Koontz, 1947; Freier *et al.* 1949; Harris, 1950; Bashkin, Mooring and Petrie, 1951; Huber and Baldinger, 1952; Adair, 1952; Seagrave, 1953; Allred *et al.* 1951; Shaw, 1955), 32 MeV. (Cork, 1952), 50 MeV. (Swartz, 1952) and 90 MeV. (Tannenwald, 1953).

Phase shift analyses of both n-α and p-α data exist (Critchfield and Dodder, 1949*b*; Dodder and Gammel, 1952) and also of n-α scattering alone at 2·5 and 3·1 MeV. (Wheeler and Barschall, 1940) and 0·7–4 MeV. (Huber and Baldinger, 1952)

Differential cross-sections for p-α scattering provide the most extensive and reliable measurements. Analysis of results at 1–3 MeV. proton energy (Critchfield and Dodder, 1949b) gives two sets of phase shifts (s, $p_{\frac{1}{2}}$, $p_{\frac{3}{2}}$) allowed by the data (the subscripts refer to $j = l \pm \frac{1}{2}$), corresponding to normal and inverted doublets of ^{5}Li according to whether the level is below or above the $^3P_{\frac{3}{2}}$ level. Measurements of the polarization of scattered protons (Heusink-veld and Freier, 1952) have shown that the $^2P_{\frac{1}{2}}$ and $^2P_{\frac{3}{2}}$ levels in fact constitute an inverted doublet. Analysis of other measurements at 5·81 and 9·48 MeV. (Dodder and Gammel, 1952) yields s, $p_{\frac{1}{2}}$, $p_{\frac{3}{2}}$, $d_{\frac{3}{2}}$, $d_{\frac{5}{2}}$ phase shifts, the p and d doublets being inverted. The s-phases are apparently negative, increasing slowly in magnitude with energy, although from Appendix B it may be seen that a correct definition of the phase shift makes the s-phase equal to π at zero energy, decreasing with increasing energy. The s-phase may be fitted qualitatively with a hard sphere interaction of range $2·6 \times 10^{-13}$ cm.

Fig. 70 shows the s, $p_{\frac{1}{2}}$, $p_{\frac{3}{2}}$, $d_{\frac{3}{2}}$ and $d_{\frac{5}{2}}$ phase shifts plotted as functions of energy (Juveland and Jentschke, 1956a) for p-α scattering at 0–10 MeV. The $p_{\frac{1}{2}}$ and $p_{\frac{3}{2}}$ phase curves show a markedly different energy dependence. Analysis of this data by the single level dispersion theory (§ 12.7) gives the resonance energies and half-widths as $E_{\frac{3}{2}} = 2·1$ MeV., $\Gamma_{\frac{3}{2}} = 1·0$ MeV.; $E_{\frac{1}{2}} \sim 8·5$ MeV., $\Gamma_{\frac{1}{2}} \sim 7·5$ MeV. The large half-widths account for the anomaly in the scattering observed at energies as low as 1 MeV.

To explain the doublet splitting, different size potentials are required in the two cases, interpreted as an effective spin-orbit coupling between the proton and the α-particle of magnitude $\sim$ several MeV. Tensor forces alone cannot account for more than about 30 % of the observed doublet splitting (Dancoff, 1940; Sugie, Hodgson and Robertson, 1957). Sack, Biedenharn and Breit (1954) have investigated the effect of a p-α interaction of the Thomas spin-orbit coupling type,

$$U(r) = -\left[1 - \beta \boldsymbol{\sigma} . \mathbf{L} \frac{1}{r}\frac{d}{dr} \right] V(r), \qquad (8.55)$$

for $V(r)$ of square, gaussian and exponential shapes. The gaussian well $V(r) = -A \, e^{-r^2/a^2}$, with $a = 2·30 \times 10^{-13}$ cm., $A = 47·32$ MeV. and $\beta = 7·40(\hbar/mc)^2$, gave the best fit to the phase curves (see also

Van der Spuy, 1956). However, a more fundamental approach involves the use of a two-body interaction including a spin-orbit coupling term, employing resonating group structure and leading to an integro-differential equation for the scattering wave functions.

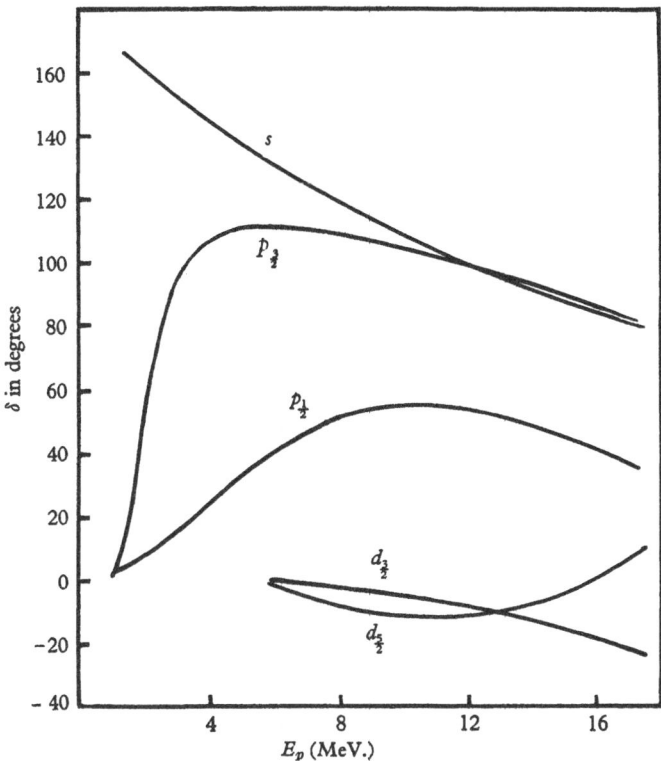

Fig. 70. Phase shifts for p-^{4}He scattering as a function of proton bombardment energy, found by analysis of experiments (Juveland and Jentschke, 1956a).

Theoretical calculations fall into two groups. Computations of the s-phase shifts have been made using both the symmetrized theory (Bransden and McKee, 1954) and the unsymmetrized theory (Hochberg, Massey and Underhill, 1954), assuming an interaction and ^{4}He wave function similar to (8.52):

$$V(r) = V_0 \exp[-\mu r^2],$$
$$\phi_0(1234) = \exp[-\tfrac{1}{2}\lambda(r_{12}^2 + r_{23}^2 + r_{31}^2 + r_{14}^2 + r_{24}^2 + r_{34}^2)], \quad (8.56)$$

with $V_0 = -45$ MeV., $r_0 = 1/\sqrt{\mu} = 1.936 \times 10^{-13}$ cm. λ is determined by the variation method as $\lambda = 1 \cdot 578 \times 10^{-25}$ cm.^{2}, giving a binding energy for the α-particle of $-26 \cdot 68$ MeV., compared to the correct value $-28 \cdot 2$ MeV.

TABLE XII. *The n-α elastic cross-section at zero energy in* mb.

Observed (Hibdon and Muehlhause, 1951)	—	780
Calculated using unsymmetrized theory (Hochberg *et al.* 1954)	Symmetric exchange forces	94
	Serber exchange forces	509
Calculated using symmetrized theory (Bransden and McKee, 1954) by variational method	Symmetric exchange forces	500
	Serber exchange forces	190

The calculations for the unsymmetrized theory with symmetric forces give good agreement with the curve of s-phase vs. energy in the region 0–4 MeV., both for n-α and p-α collisions. Agreement is less satisfactory for Serber forces. On the other hand, the symmetrized theory leads to much smaller phases, the s-phase curve for n-α scattering being in definite disagreement with experimental values. Total cross-sections at zero energy for n-α scattering are compared in Table XII for alternative theories. The unsymmetrized theory with symmetric exchange forces is in excellent agreement with experiment, but both Serber-forces and the symmetrized theory seem to be excluded.

The second group of calculations concerns the computation of the $p_{\frac{1}{2}}$, $p_{\frac{3}{2}}$ and the $d_{\frac{3}{2}}$, $d_{\frac{5}{2}}$ phases. The former were evaluated using the two-body interaction

$$U(r) = (1 + \gamma \boldsymbol{\sigma} . \mathbf{L}) \, V(r), \qquad (8.57)$$

where $V(r)$ is the gaussian potential (8.52). With the unsymmetrized theory and symmetric exchange forces it was not possible to fit the p-phase for any value of γ. Use of Serber exchange forces in the p-state gave a better fit, but final agreement was obtained with $\gamma = 0 \cdot 10$ and an arbitrary mixture of 10 % symmetric and 90 % Serber exchange forces. Although empirical, this exchange mixture is not in disagreement with evidence for a small p-interaction indicated by the observed asymmetry in n-p scattering at high energies (§ 10.10). When recalculated on this basis, the s-phases were only very slightly decreased from the symmetric force results. However, it is possible that including tensor forces could alter the result considerably.

8.7. Polarization of nucleons in collision processes

The spin-orbit coupling force involved in n-α and p-α collisions can cause an incident nucleon to reverse its spin direction on being scattered, resulting in a partially polarized scattered beam with some net spin direction corresponding to each scattered angle. The polarization effect can be detected in a double scattering experiment as discussed in § 2.4.

8.7.1. *Polarization by scattering from nuclei of spin zero.* Lepore (1950), and Sjölander and Köhler (1954), have given derivations of the polarization formula, applying to neutral and charged particles respectively scattered by a nucleus of zero spin and charge Z, e.g. ^{4}He.† Denoting the Pauli spin vector of the incident particle by σ and its direction of incidence and scattering by $\mathbf{k_0}$ and $\mathbf{k}$ respectively, the incident beam may be taken as

$$\psi_{\text{inc}} = \exp i(\mathbf{k_0}.\mathbf{r} - \alpha \ln 2kr) \chi_{\text{inc}}, \qquad (8.58)$$

where χ_{inc} is the spin function, $\alpha = Ze^2 . \mu/k\hbar^2$ and μ is the reduced mass.

The polarization of the incident beam is defined as the expectation value of its spin:

$$\mathbf{P}_{\text{inc}} = \langle \sigma \rangle = \Sigma \chi_{\text{inc}}^* \sigma \chi_{\text{inc}} = \langle \chi_{\text{inc}}, \sigma \chi_{\text{inc}} \rangle. \qquad (8.59)$$

The scattering solution is composed of the incident wave ψ_{inc} plus a diverging spherical wave having the asymptotic form

$$\psi \sim \psi_{\text{inc}} + \frac{1}{r} \exp i(kr - \alpha \ln 2kr) f(\theta) \chi_{\text{inc}}. \qquad (8.60)$$

Because of the spin-orbit coupling component in the interaction between the two particles, the orbital angular momentum is no longer a constant of motion but must be described in terms of the total angular momentum, $J = L + \frac{1}{2}$. The required set of commuting constants of motion is then J^2, J_z^2, L^2 and σ^2.

The incident beam (8.58) can now be decomposed asymptotically into its orbital angular momenta components:

$$\psi_{\text{inc}} \sim \sum_{l=0}^{\infty} (2l+1) i^l \frac{1}{kr} \sin(kr - \tfrac{1}{2}l\pi - \alpha \ln 2kr) P_l(\cos \theta) \chi_{\text{inc}}. \qquad (8.61)$$

† Scattering by nuclei with non-zero spin is dealt with in § 10.14.1.

The scattered wave function is written in the asymptotic form

$$\psi \sim \sum_{l=0}^{\infty} (2l+1) i^l \frac{1}{kr} [A_l^+ \pi_l^+ u_l^+(kr) + A_l^- \pi_l^- u_l^-(kr)] P_l(\cos\theta) \chi_{\text{inc}},$$
$$(8.62)$$

where A_l^+ and A_l^- are constants to be determined by the condition that ψ have the asymptotic form (8.60). π_l^+ and π_l^- are projection operators constructed so that when applied to functions of the type $P_l(\cos\theta)\chi_{\text{inc}}$, they destroy those states within the function for which $J = l - \frac{1}{2}$ and $J = l + \frac{1}{2}$ respectively. Employing the properties of $\mathbf{\sigma}.\mathbf{L}$, appropriate expressions are

$$\pi_l^+ = (l+1+\mathbf{\sigma}.\mathbf{L})/(2l+1), \quad \pi_l^- = (1-\mathbf{\sigma}.\mathbf{L})/(2l+1), \quad (8.63)$$

satisfying the condition

$$\pi_l^+ + \pi_l^- = 1. \tag{8.64}$$

The radial wave functions are

$$u_l^{\pm}(kr) = \sin(kr - \tfrac{1}{2}l\pi - \alpha \ln 2kr + \delta_l^{\pm} + \sigma_l), \tag{8.65}$$

where

$$\delta_l^{\pm} = \beta_l^{\pm} + \gamma_l^{\pm}. \tag{8.66}$$

σ_l is the Coulomb phase shift,

$$\sigma_l = \arg \Gamma(1+l+i\alpha), \tag{8.67}$$

$\beta_l^{\pm}$ is the phase shift due to ordinary potential scattering and $\gamma_l^{\pm}$ is the phase due to resonance scattering, written as

$$\tan\gamma_l^{\pm} = \Gamma_l^{\pm}/(E_l^{\pm} - E). \tag{8.68}$$

Here E is the incident energy, $E_l^{\pm}$ the resonance energy and $\Gamma_l^{\pm}$ the resonance width.

The scattered wave $\psi_{\text{scatt}} = \psi - \psi_{\text{inc}}$ is found by subtracting equation (8.61) from equation (8.62). Substitution of the values of $A_l^{\pm}$ in the result gives

$$\psi_{\text{scatt}} \sim \frac{1}{kr} \exp i(kr - \alpha \ln 2kr) \left[-\frac{\alpha}{2k} \text{cosec}^2 \frac{\theta}{2} \right.$$

$$\times \exp\left\{ -i\alpha \ln\left(\sin^2\frac{\theta}{2}\right) + 2i\sigma_0 \right\}$$

$$+ \sum_{l=0}^{\infty} \{(l+1)\exp(i\delta_l^+)\sin\delta_l^+ + l\exp(i\delta_l^-)\sin\delta_l^-\} P_l(\cos\theta)$$

$$\left. + (\exp(i\delta_l^+)\sin\delta_l^+ - \exp(i\delta_l^-)\sin\delta_l^-)\mathbf{\sigma}.\mathbf{L}P_l(\cos\theta)\} e^{2i\sigma_l} \right] \chi_{\text{inc}}.$$
$$(8.69)$$

If $\mathbf{n}$ is the normal to the scattering plane, defined by

$$\mathbf{k} \times \mathbf{k}_0 = \mathbf{n}k^2 \sin\theta, \tag{8.70}$$

then $\quad \boldsymbol{\sigma}.\mathbf{L}P_l(\cos\theta) = -i\sin\theta \dfrac{\partial}{\partial(\cos\theta)} P_l(\cos\theta)\,\boldsymbol{\sigma}.\mathbf{n}. \qquad$ (8.71)

The scattered amplitude $f(\theta)$ follows from equation (8.68):

$$f(\theta) = A(\theta) + \boldsymbol{\sigma}.\mathbf{n}B(\theta), \tag{8.72}$$

where $\quad A(\theta) = -\dfrac{\alpha}{2k}\operatorname{cosec}^2\dfrac{\theta}{2}.\exp\left\{-i\alpha\ln\left(\sin^2\dfrac{\theta}{2}\right)+2i\sigma_0\right\}$

$$+\frac{1}{k}\sum_{l=0}^{\infty}[(l+1)\exp(i\delta_l^+)\sin\delta_l^+$$

$$+l\exp(i\delta_l^-)\sin\delta_l^-]\,\mathrm{e}^{2i\sigma_l}P_l(\cos\theta), \tag{8.73}$$

$$B(\theta) = -\frac{i\sin\theta}{k}\sum_{l=0}^{\infty}[\exp(i\delta_l^+)\sin\delta_l^+$$

$$-\exp(i\delta_l^-)\sin\delta_l^-]\,\mathrm{e}^{2i\sigma_l}\frac{\partial}{\partial(\cos\theta)}P_l(\cos\theta). \tag{8.74}$$

From equation (8.72), the differential cross-section becomes

$$I(\theta) = (A^*A + B^*B)\left[1 + \frac{A^*B + B^*A}{A^*A + B^*B}\mathbf{P}_{\mathrm{inc}}.\mathbf{n}\right], \tag{8.75}$$

where the term in square brackets gives the effect of a polarized incident beam.

The polarization of the scattered beam is defined by

$$\mathbf{P} = \langle\psi_{\mathrm{scatt}}, \boldsymbol{\sigma}\psi_{\mathrm{scatt}}\rangle = \langle f(\theta)\chi_{\mathrm{inc}}, \boldsymbol{\sigma}f(\theta)\chi_{\mathrm{inc}}\rangle/\langle f(\theta)\chi_{\mathrm{inc}}, f(\theta)\chi_{\mathrm{inc}}\rangle. \tag{8.76}$$

On employing the relations

$$\boldsymbol{\sigma}(\boldsymbol{\sigma}.\mathbf{n}) = i(\mathbf{n}\times\boldsymbol{\sigma}) + \boldsymbol{\sigma}, \quad (\boldsymbol{\sigma}.\mathbf{n})\boldsymbol{\sigma} = -i(\mathbf{n}\times\boldsymbol{\sigma}) + \mathbf{n},$$

$$(\boldsymbol{\sigma}.\mathbf{n})\boldsymbol{\sigma}(\boldsymbol{\sigma}.\mathbf{n}) = 2(\boldsymbol{\sigma}.\mathbf{n}) - \boldsymbol{\sigma},$$

we find for the polarization

$$\mathbf{P} = [A^*A\mathbf{P}_{\mathrm{inc}} + (A^*B + B^*A)\,\mathbf{n} + B^*B(2\mathbf{P}_{\mathrm{inc}}.\mathbf{n}\mathbf{n} - \mathbf{P}_{\mathrm{inc}})$$

$$+ i(B^*A - A^*B)\,\mathbf{P}_{\mathrm{inc}}\times\mathbf{n}]/[A^*A + B^*B + (A^*B + B^*A)\,\mathbf{P}_{\mathrm{inc}}.\mathbf{n}] \tag{8.77}$$

An unpolarized incident beam has $\mathbf{P}_{\mathrm{inc}} = 0$, so the scattered beam has a polarization

$$\mathbf{P} = \left(\frac{A^*B + B^*A}{A^*A + B^*B}\right)\mathbf{n} = P(\theta)\,\mathbf{n}. \tag{8.78}$$

Taking the differential cross-section for single scattering as

$$I_1(\theta) = A_1^* A_1 + B_1^* B_1, \tag{8.79}$$

the polarization produced is accordingly

$$P_1(\theta) = 2(\operatorname{Re} A_1^* B_1)/I_1(\theta), \qquad (8.80)$$

where Re denotes 'real part'. The polarization is directed along the normal to the scattering plane and depends on the interference between the two parts of the scattered wave, so that large polarizations require this interference term to be comparable with the cross-section itself.

Double scattering may be treated by comparing equations (8.78) and (8.75), leading to the expression for the differential cross-section

$$I_2'(\theta) = I_1(\theta)\,[1 + P_1(\theta_1)P_2(\theta_2)\,\mathbf{n}_1 . \mathbf{n}_2], \qquad (8.81)$$

where the unit normal vectors $\mathbf{n}_1$ and $\mathbf{n}_2$ define the two planes of scattering and θ_1 and θ_2 are the two scattering angles. A quantity often used is the asymmetry 2ϵ, defined by

$$2\epsilon = \frac{I_2'(\phi_0) - I_2'(\phi_0 + \pi)}{I_2'(\phi_0) + I_2'(\phi_0 + \pi)} = 2\mathbf{P}_1 . \mathbf{P}_2, \qquad (8.82)$$

where ϕ_0 is the angle between $\mathbf{P}_1$ and $\mathbf{P}_2$, $\cos \phi_0 = \mathbf{n}_1 . \mathbf{n}_2$. In double scattering experiments (cf. § 2.4), the maximum asymmetry is measured with $\phi_0 = 0$, giving $\epsilon = P_1(\theta_1)P_2(\theta_2)$. In general, however, the second scatterer may be regarded as an analyser, with the asymmetry measuring the projection of the incident polarization on the normal $\mathbf{n}_2$ to the analysing plane, multiplied by the analysing power P_2 of the second scattering.

There are two important cases of double scattering:

(1) *First and second scattering in the same plane.* Here either $\mathbf{n}_1 = \mathbf{n}_2$ or $\mathbf{n}_1 = -\mathbf{n}_2$, the resultant polarization P_2' obtained from equations (8.77) and (8.78) being

$$P_2'(\theta) = \frac{P_1(\theta) \pm P_2(\theta)}{1 \pm P_1(\theta)P_2(\theta)}, \qquad (8.83)$$

where $P_1(\theta)$ and $P_2(\theta)$ refer to single scattering at each plane. The ratio of the scattered intensities for the two cases is

$$R = \frac{1 + P_1(\theta)P_2(\theta)}{1 - P_1(\theta)P_2(\theta)} = \frac{1 + \epsilon}{1 - \epsilon}, \qquad (8.84)$$

which becomes large for large polarizations ($P \approx 1$).

(2) *First and second scattering planes orthogonal.* Here $\mathbf{n}_1 . \mathbf{n}_2 = 0$, so from equation (8.77) the polarization resulting after second scattering is

$$\mathbf{P}_2' = P_2(\theta_2)\,\mathbf{n}_2 + \left(\frac{|A_2|^2 - |B_2|^2}{I_2(\theta)}\right) P_1(\theta_1)\,\mathbf{n}_1 + \frac{2\mathrm{Im}A^*B}{I_2(\theta)} P_1(\theta)\,(\mathbf{n}_1 \times \mathbf{n}_2).$$

(8.85)

The polarization has three components, one being in the direction of the normal to the second scattering plane, with a magnitude equal to that which would be acquired by an unpolarized beam. The other two components lie in the second scattering plane with their resultant rotated with respect to $\mathbf{n}_1$ by an angle β, defined as a positive rotation about $\mathbf{n}_2$. From equation (8.85), this resultant has a magnitude equal to $P_1(\theta)\,[1 - P_2^2(\theta)]^{\frac{1}{2}}$ and β is defined by

$$\sin\beta = -\frac{2\mathrm{Im}A^*B}{I_2(1 - P_2^2)^{\frac{1}{2}}},$$

(8.86)

where Im stands for 'imaginary part'.

The differential cross-section for second scattering from orthogonal planes is $I_2'(\theta) = I_1(\theta)$, equal to the scattered intensity at the first plane. Clearly $\epsilon = 0$ and $R = 1$.

Triple scattering, for which the first two scattering planes are orthogonal and the third orthogonal to the second, permits a measurement of the amount of rotation of the polarization produced in the second scattering (fig. 71). From equation (8.82), the asymmetry obtained is

$$\epsilon_3 = P_1(1 - P_2^2)^{\frac{1}{2}} \cos(\theta - \beta) P_3.$$

(8.87)

If a double scattering calibration experiment using targets 1 and 3 is performed, giving the calibration asymmetry $\epsilon_3' = P_1 P_3$, then the ratio

$$R' = \epsilon_3/\epsilon_3' = (1 - P_2^2)^{\frac{1}{2}} \cos(\theta - \beta)$$

(8.88)

is a symmetrical function of θ having maximum and minimum values of $\pm(1 - P_2^2)^{\frac{1}{2}}$ for $\theta = \beta$ and $\pi + \beta$ respectively. The rotation angle β may thus be measured indirectly (Wolfenstein, 1954).

8.7.2. *Application to the scattering of nucleons by α-particles.* The formulae of §8.6 are not immediately applicable to n-α and p-α scattering. Analysis of single scattering experiments was found

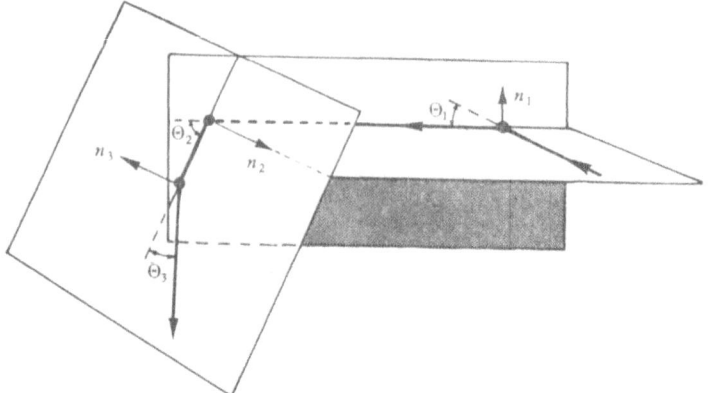

Fig. 71. Triple scattering vectors for second scattering plane perpendicular to the first, and third scattering plane perpendicular to the second. The polarization vector $\mathbf{P}(\Theta)$ is directed along the unit normal vector $\mathbf{n}$, $\mathbf{P}(\Theta) = P(\Theta)\,\mathbf{n}$.

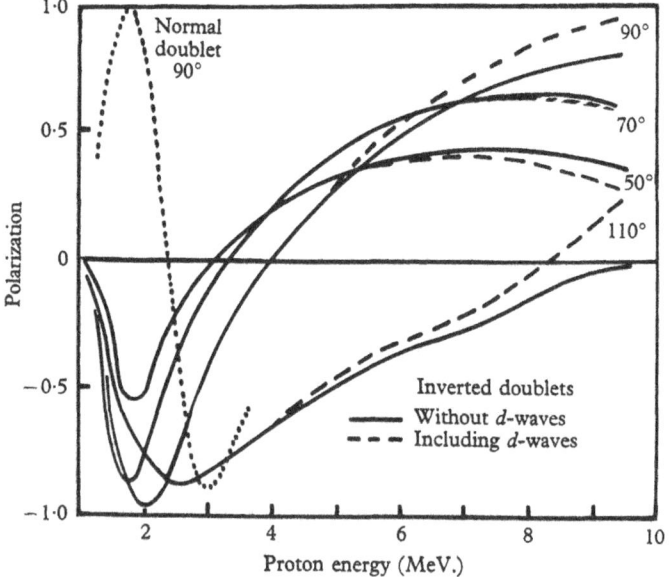

Fig. 72. The polarization of protons scattered from ⁴He at various CM angles as a function of incident proton energy. Curves are given at all angles assuming an inverted $P_{\frac{1}{2}}$-$P_{\frac{3}{2}}$ doublet (Juveland and Jentschke, 1956a) and at $\theta = 90°$ for a normal doublet (Heusinkveld and Freier, 1952).

to lead to two alternative sets of $p_{\frac{1}{2}}$, $p_{\frac{3}{2}}$ phase shifts, one corresponding to normal levels and the other to inverted energy levels of ^{5}Li*. The former correspond to resonance energies both in the

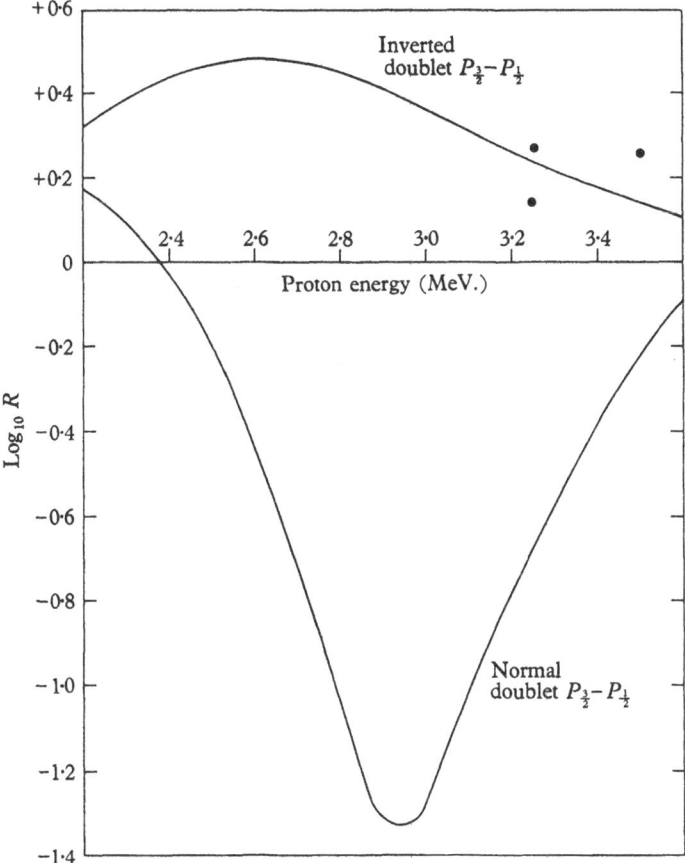

Fig. 73. Logarithm of ratio R of intensities on backward plane to forward plane, for 90° CM double scattering of protons by ^{4}He, as a function of incident proton energy before the first scattering. The experimental points are for 3·25 and 3·50 MeV. (Heusinkveld and Freier, 1952).

neighbourhood of 1–2 MeV., while the latter gives $E_{\frac{3}{2}} = 2\cdot 1$ MeV. and $E_{\frac{1}{2}} \sim 8\cdot 5$ MeV. (§ 8.6).

Double scattering experiments enable the two sets of levels to be distinguished as they lead to quite different polarizations. This is illustrated in fig. 72, showing the expected polarization for the two cases at 90° CM plotted as functions of the proton energy. The curves are based on phase shift analysis at 0·95–3·58 MeV. of p-α scattering (Critchfield and Dodder, 1949b) and 1–10 MeV. (Juveland and Jentschke, 1956a). Fig. 73 shows the logarithm of the ratio R of equation (8.84) plotted against energy, also for $\theta_1 = \theta_2 = 90°$. Measurements at 3·25 and 3·50 MeV. (Heusinkveld and Freier, 1952) and 5·3 MeV. (Juveland and Jentschke, 1956a) are in agreement with the inverted $P_{\frac{3}{2}}$-$P_{\frac{1}{2}}$ doublet and exclude the normal $P_{\frac{3}{2}}$-$P_{\frac{1}{2}}$ case. Results for n-α scattering lead to the same conclusion (Huber and Baldinger, 1952; Seagrave, 1953; Levintov et al. 1957b).

Note added in proof. $n-d$ elastic scattering calculations have been extended down to 0·6 MeV. (Burke and Robertson, 1957), using approximate wave functions, Kohn's variational method, and electronic computers. A trial function similar to (8.33) with $c = d = 0$ was employed, giving good agreement with other methods at higher energies. However, success depended critically on the value taken for the parameter ν, usually unimportant in variational calculations. Moreover the normally more accurate Hulthén method failed, even giving imaginary phases.

Subsequent machine calculations, using a full polarization trial function of type (8·33) for energies below 2 MeV., gave doublet and quartet phases practically unaltered, indicating almost zero polarization in both spin states (private communication, 1958). However, the reliability of the variational calculation is questionable in the absence of an independent check using a different method (cf. §§ 8.2.4, 8.2.5).

277

CHAPTER 9

NON-CENTRAL FORCES AND
SPIN-ORBIT COUPLING

9.1. Introduction

In previous chapters the discussion of collision phenomena has been based on the assumption of central forces only, and this has proved adequate for handling interactions at low energies. However, there is considerable evidence that an all-embracing law of force must include effects due to coupling between the spins and orbital angular momenta of nucleons, e.g. that there must be non-central forces and spin-orbit coupling. These spin-orbit forces can be shown to influence scattering at high energies; this aspect of the problem is treated in subsequent chapters.

Meson field theories provide one line of approach to spin-orbit forces. A non-central force is predicted, resulting in an electric quadrupole moment of the deuteron. Unfortunately, this law of force contains a $1/r^3$ term, leading to an infinite binding energy of this nucleus, so that it must be subjected to an arbitrary cut-off within some critical radius r_c, an example being the potential of Levy (1952 a, b). Levy employed an infinite barrier or hard core of radius r_c (§ 1.5.5) with an attractive, non-central meson potential outside this hard core. Kligman (1940) and Rarita and Schwinger (1941 a, b) have given a semi-empirical theory which avoids the difficulties encountered in field theory by using an empirical shape (rectangular well) for the non-central potential. These calculations have been extended (Hu and Massey, 1949) to Yukawa, exponential and gaussian well shapes.

The essential feature of the meson potentials of § 1.5 is the occurrence of a non-central or tensor term S_{12} in the potential:

$$S_{12} = \frac{3\sigma_1 \cdot \mathbf{r}\, \sigma_2 \cdot \mathbf{r}}{r^2} - \sigma_1 \cdot \sigma_2. \tag{9.1}$$

This term also appears in the interaction energy of two electric dipoles. The n-p force is dependent on both the radial distance r between neutron and proton and on the angles between $\mathbf{r}$ and the

Pauli spin vectors σ_1, σ_2. It distorts the charge distribution of the deuteron from spherical symmetry to an ellipsoidal shape which turns out to be prolate along the spin axis.

9.2. General form of the non-central interaction

If we restrict attention to a conservative, static potential, severe limitations are placed upon its form as it must be a scalar, i.e. $V(r)$ is invariant under rotations and reflexions of the co-ordinate system. Possible combinations of the Pauli spin vectors σ_1, σ_2 and the relative co-ordinate $\mathbf{r}$ are limited for the following reasons:

(a) Polynomials in σ can be simplified to expressions linear in σ by using spin identities such as $(\sigma.\mathbf{r})^2 = r^2$.

(b) The σ's are axial vectors so that $\mathbf{r}$ is the only polar vector. Hence the invariance condition requires r to occur in even powers only. Possible functions are discussed below:

(i) $V(r)$ is a simple scalar.

(ii) $\sigma_1.\sigma_2 V(r)$ is a scalar representing a spin-dependent but central force.

(iii) $\sigma_1.\mathbf{r}$, $\sigma_2.\mathbf{r}$ and $\sigma_1 \times \sigma_2.\mathbf{r}$ are pseudoscalars as they change sign under reflexion ($\mathbf{r} \to -\mathbf{r}$, $\sigma \to +\sigma$). Also they are not bilinear in r.

(iv) $(\sigma_1.\mathbf{r})(\sigma_2.\mathbf{r})$ and $(\sigma_1 \times \mathbf{r}).(\sigma_2 \times \mathbf{r})$ are suitable bilinear expressions, but $(\sigma_1 \times \mathbf{r}).(\sigma_2 \times \mathbf{r}) = -(\sigma_1.\mathbf{r})(\sigma_2.\mathbf{r}) + r^2\sigma_1.\sigma_2$. Hence $(\sigma_1.\mathbf{r})(\sigma_2.\mathbf{r})/r^2$ is the only non-redundant, bilinear, scalar expression. In practice, equation (9.1) for S_{12} is employed because its average over all directions of $\mathbf{r}$ is zero. Thus the most general static potential has the form

$$V = V_1(r) + \sigma_1.\sigma_2 V_2(r) + S_{12} V_3(r). \tag{9.2}$$

9.3. Multipole moments of a charged-particle system

The interaction energy $\mathscr{V}_{em}$ of an arbitrary charge distribution in a slowly varying electromagnetic field, e.g. a nucleus in the field of the atom, can be expanded in terms of the potentials of a series of multipoles in the order total charge, electric dipole, magnetic dipole and electric quadrupole, and higher negligible terms like octupoles. The vector expansion with respect to the point o is (Rosenfeld, 1948, p. 340)

$$\mathscr{V}_{em} = eB_0 - \mathbf{P}.\mathbf{E}_0 - \mathbf{M}.\mathbf{H}_0 - (Q\nabla).\mathbf{E}_0..., \tag{9.3}$$

where the total charge $e = \int \rho \, dV$, electric dipole moment $\mathbf{P} = \int \rho \mathbf{r} \, dV$,

magnetic dipole moment $\mathbf{M} = \frac{1}{2} \int \mathbf{r} \times \mathbf{I} \, dV$,

and electric quadrupole moment

$$Q_{ik} = \frac{1}{2} \int \rho(x_i x_k - \tfrac{1}{3} \delta_{ik} r^2) \, dV. \tag{9.4}$$

ρ and $\mathbf{I}$ are the charge and the vector current density respectively. The quadrupole moment is a tensor, and $(Q\mathbf{V})$ is defined as a vector with components $\sum_k Q_{ik}(\partial/\partial x_k)$. For nuclei, the electric dipole moment is identically zero as the expectation value (or average value) of $\mathbf{P}$ is

$$\langle \mathbf{P} \rangle = \int \sum_i e_i \mathbf{r}_i \, | \, \psi(\mathbf{r}_k) \, |^2 \, dV_k, \tag{9.5}$$

where $\psi(\mathbf{r}_k)$ denotes a nucleonic wave function and k is summed over all values. On making the variable change $\mathbf{r}_i = -\mathbf{r}_i$ (inversion), then $\psi(-\mathbf{r}_k) = \pm \psi(\mathbf{r}_k)$ because of parity, and hence $\mathbf{P} = -\mathbf{P}$, or $\mathbf{P} = 0$.

9.4. Spectroscopic magnetic moment μ

We require the expectation value of $\mathbf{M}$ for the ground state of the nucleus, and use the property that only the projection of $\mathbf{M}$ in the direction z of the total angular momentum $\hbar\mathbf{J}$ contributes to $\langle \mathbf{M} \rangle$ in any state for which $\mathbf{J}$ is fixed. Thus

$$\langle \mathbf{M} \rangle = \mu \mu_0 \langle \mathbf{J} \rangle / J, \tag{9.6}$$

where the nuclear magneton is $\mu_0 = e\hbar/2m_p$ and the magnetic moment μ is defined as the expectation value in units μ_0, of the projection M_z of $\mathbf{M}$ on some fixed direction in the substate of magnetic quantum number $M_J = J$.

9.5. Spectroscopic electric quadrupole moment Q

The expectation value of Q may be split into two parts, the one independent of the magnetic quantum number M_J, the other being the tensor $\langle J_i J_k - \tfrac{1}{3} \delta_{ik} J^2 \rangle$. To determine the former part, any component of Q may be used. Casimir (1936) defines the spectroscopic quadrupole moment as the expectation value of $\dfrac{6}{e} Q_{33}$ in the substate $M_J = J$, i.e.

$$Q = \left\langle \frac{1}{e} \int \rho (3z^2 - r^2)\, dV \right\rangle_{M_J = J}. \qquad (9.7)$$

For the deuteron, this reduces to

$$Q(^2\mathrm{H}) = \frac{1}{4} \int (3z^2 - r^2)\, \psi^2(r)\, dr, \qquad (9.8)$$

having the measured value $Q = (2\cdot74 \pm 0\cdot02) \times 10^{-27}\,\mathrm{cm.}^2$. The definitions (9.7) and (9.8) depend only on the spatial distribution of the particles and not on the charge density, so that the quadrupole moment gives a measure of the departure from spherical symmetry of the nucleus. As $\langle J_i J_k - \frac{1}{3}\delta_{ik} J^2 \rangle$ vanishes for $J = 0$ and $J = \frac{1}{2}$, only nuclei with $J \geqslant 1$ have electric quadrupole moments. Octupole and higher moments have proved too small to be detected, which is not surprising as in equation (9.3) successive terms are smaller by the ratio of nuclear to atomic dimensions (10^{-5}).

9.6. The conservation of quantum numbers

The conservation of quantum numbers is affected by the introduction of the non-central term S_{12} into the interaction. Actually S_{12} and hence the Hamiltonian H of the system are invariant only under the coupled rotation of spin and spatial co-ordinates (Rarita and Schwinger, 1941). As $\mathbf{L}$ and $\mathbf{S}$ may both be expressed as infinitesimal rotation operators, it follows that $\mathbf{J} = \mathbf{L} + \mathbf{S}$ commutes with H, i.e. H is a good quantum number. However the orbital quantum number L is not a good quantum number. Similarly the magnetic quantum number M is a constant of motion, but the total spin S is a good quantum number only for the two-body system (due to the symmetry of H in this case). Thus the deuteron is described by the four eigenvalues J, M, S and parity $(-)^L$. The singlet state of the n-p system involves central forces only, as $S = 0$ implies $\sigma_1 = -\sigma_2$, so that

$$S_{12} = -3(\sigma_1 . \mathbf{r})^2 / r^2 + \sigma_1 . \sigma_2 = -3 + 3 = 0.$$

9.7. The ground state of the deuteron

A triplet state of total angular momentum J is regarded as a mixture of all possible states of orbital angular momentum L consistent with the rules for compounding angular momenta, $L = J$ or $L = J \pm 1$. The quantum number parity $(-)^L$ allows a complete classification of triplet states by separating even and odd states of L (Table XIII).

TABLE XIII. *Classification of two-body triplet states*

J	Even parity	Odd parity
0	—	$^3P_0$
1	$^3S_1 + {}^3D_1$	$^3P_1$
2	$^3D_2$	$^3P_2 + {}^3F_2$
3	$^3D_3 + {}^3G_3$	$^3F_3$

In the case of central forces, the $^3S_1$ state is the ground state of the deuteron and hence when non-central forces are introduced it becomes the $^3S_1 + {}^3D_1$ state. As purely central forces give a good approximation in scattering and binding energy calculations, the percentage of $^3D_1$ state in the ground state must be small. However, the fact that the $^3D_1$ wave function is not spherically symmetric should explain the electric quadrupole moment of the deuteron.

The part of the $^3D_1$ wave function involving angular and spin co-ordinates is $(4\pi)^{-\frac{1}{2}}\chi_1^M$, where χ_1^M is the spin wave function of a triplet state having a z-component of J equal to M and $(4\pi)^{-\frac{1}{2}}$ is a normalizing factor. The corresponding part of the $^3D_1$ wave function is proportional to $S_{12}\chi_1^M$, for the rotational invariance and spin symmetry of S_{12} indicate that it represents a triplet state with $J = 1$, while the fact that $\nabla^2(r^2 S_{12}\chi_1^M) = 0$ reveals it to be a D-state. The normalized angular spin part of the $^3D_1$ wave function is then $\frac{1}{4}(2\pi)^{-\frac{1}{2}} S_{12}\chi_1^M$, so that denoting the radial wave functions of the ^3S and ^3D states by $u(r)/r$ and $w(r)/r$ respectively, the wave function of the coupled $^3S_1 + {}^3D_1$ state becomes

$$\psi = (4\pi)^{-\frac{1}{2}}\chi_1^M \left(\frac{u}{r} + 2^{-\frac{3}{2}} S_{12} \frac{w}{r} \right). \tag{9.9}$$

A suitable form for the nuclear potential is

$$\mathscr{V}(r) = -(1 - \tfrac{1}{2}g + \tfrac{1}{2}g\boldsymbol{\sigma}_1 . \boldsymbol{\sigma}_2 + \gamma S_{12}) V(r), \tag{9.10}$$

which reduces to $-(1-2g)\,V(r)$ and $-(1+\gamma S_{12})\,V(r)$ for singlet and triplet spin states respectively. g and γ are arbitrary parameters fixing the proportion of central and non-central interaction and $V(r)$ is a central potential.

Substitution of equations (9.9) and (9.10) in the Schrödinger wave equation

$$\nabla^2\psi + \frac{m}{\hbar^2}[E - \mathscr{V}(r)]\,\psi = 0 \tag{9.11}$$

leads to two coupled, second order differential equations for the ^3S and ^3D radial wave functions:

$$\left.\begin{array}{l} \dfrac{d^2u}{dr^2} + [\epsilon + U(r)]\,u = -2^{\frac{3}{2}}\gamma U(r)\,w, \\[2mm] \dfrac{d^2w}{dr^2} - \dfrac{6w}{r^2} + [\epsilon + (1-2\gamma)\,U(r)]\,w = -2^{\frac{3}{2}}\gamma U(r)\,u, \end{array}\right\} \tag{9.12}$$

where $\epsilon = (m/\hbar^2)\,E$ and $U(r) = (m/\hbar^2)\,V(r)$. The expression (9.8) for the quadrupole moment becomes, on substituting equation (9.9) for ψ,

$$Q = \frac{2^{\frac{1}{2}}}{10}\int_0^\infty r^2(uw - 2^{-\frac{3}{2}}w^2)\,dr. \tag{9.13}$$

The fraction of time spent by the deuteron in the D-state is

$$w_D = \int_0^\infty w^2\,dr \Big/ \int_0^\infty (u^2 + w^2)\,dr. \tag{9.14}$$

Using a rectangular well of range $2\cdot 80 \times 10^{-13}$ cm., Rarita and Schwinger solved equations (9.12) for the deuteron ground state by use of infinite series, writing the binding energy ϵ_0 as $\epsilon = -\epsilon_0$. They found $U_0 = 6\cdot 40$, $\gamma = 0\cdot 775$, $Q = (2\cdot 73 \pm 0\cdot 05) \times 10^{-27}$ cm.^{2}, $w_D = 0\cdot 039$, and taking the singlet well depth as $11\cdot 9$ MeV., obtained $g = 0\cdot 0715$. An experimental check on the value of w_D is found using the expression for the magnetic moment μ of the deuteron:

$$\mu = 0\cdot 87971(1 - w_D) + 0\cdot 31015 w_D, \tag{9.15}$$

where $0\cdot 87971$ and $0\cdot 31015$ are the deuteron moments in the S- and D-states respectively, using the sum of the measured neutron and proton moments for the S-state and the Landé formula to calculate the D-state moment. Taking Rabi's experimental value

$$\mu = (0\cdot 85741 \pm 0\cdot 00001)$$

nuclear magnetons leads to $w_D = (3\cdot915 \pm 0\cdot006)\%$ in good agreement with the theoretical value.

Calculations have been extended (Hu and Massey, 1949) to rectangular, Yukawa, exponential and gaussian wells with a number of equal triplet and singlet ranges. In each case the range values needed to obtain the correct ϵ_D, Q and w_D were somewhat larger than the unequal ranges deduced from the experimental effective ranges for n-p and p-p singlet scattering. However Padfield (1949) has shown for the rectangular well that a central force range of $1\cdot60 \times 10^{-13}$ cm. and a non-central force range of $3\cdot07 \times 10^{-13}$ cm. satisfies all low energy data, including effective ranges for triplet n-p and singlet p-p scattering and the binding energy, quadrupole moment and required percentage of D-state for the deuteron.

9.8. Low energy n-p scattering and exchange forces

Rarita and Schwinger calculated n-p scattering using S- and P-waves at 0–15 MeV. and found the cross-section was decreased by only 2% from the central force result at zero energy and was increased slightly at 15 MeV. The theory of n-p scattering with non-central forces is given in chapter 10, for arbitrary angular momenta, only S-and P-waves being needed at low energies. Thus, the two-body potential can be represented to a good approximation by central forces in the energy region in which S-wave scattering predominates, 0–15 MeV.

The phase shift δ^1 for the $^3S_1$ scattering of neutrons by protons may be determined from the shape-independent approximations (6.32) and (6.34), rewritten as

$$k \cot \delta^1 = -\alpha + \tfrac{1}{2}(k^2 + \alpha^2)\, r_1 + O\{(k^2 + \alpha^2)\, r_1^3\}, \qquad (9.16)$$

where $\alpha = (m/\hbar^2)\, W$, W is the deuteron binding energy and r_1 is the triplet effective range, given in terms of the deuteron wave function by

$$r_1 = 2 \int_0^\infty [e^{-2\alpha r} - (u^2 + w^2)]\, dr. \qquad (9.17)$$

Here, $(u^2 + w^2)$ has been normalized to approach $e^{-2\alpha r}$ as $r \to \infty$.

Feshbach and Schwinger (1951) have used equations (9.16) and (9.17) to compute values of the triplet effective range and scattering length a_1. These may be compared with the experimental values (6.58):

$$r_1 = (1 \cdot 70 \pm 0 \cdot 03) \times 10^{-13} \text{cm.}, \quad a_1 = -(5 \cdot 39 \pm 0 \cdot 03) \times 10^{-13} \text{cm.}$$

$$(9.18)$$

Using a Yukawa potential, they found that the tensor range must be significantly greater than the central range to fit all the low energy data.

9.9. Spin-orbit coupling

It is well known that one of the consequences of the Dirac relativistic theory of the electron is an interaction between the spin and orbital motion of each electron, having the form

$$\sum_k \mathbf{L}_k \cdot \mathbf{S}_k \frac{1}{2m^2c^2} \frac{1}{r} \frac{dV}{dr},$$

where $V(r)$ is the potential energy of the central field. In nuclei, spin-orbit coupling results from two effects:

(a) The magnetic dipole due to the spin of a nucleon moves in the electric field $\mathbf{E}$ of the nucleus. On carrying out a Lorentz transformation to the system in which the nucleon is at rest, the dipole is acted upon by a magnetic field $\mathbf{E} \times \mathbf{v}$ and so performs a Larmor precession.

(b) On transforming to the original reference frame, the acceleration of the nucleus leads to a Thomas precession due partly to the nuclear and partly to the electric potential (Dancoff and Inglis, 1936).

LS coupling is zero for S-wave and all singlet ($S = 0$) interactions but for other L-states leads to a velocity-dependent or non-static force. We have the operator relation for the total angular momentum of the nucleus,

$$\mathbf{J}^2 = (\mathbf{L} + \mathbf{S})^2 = \mathbf{L}^2 + \mathbf{S}^2 + 2\mathbf{L}.\mathbf{S}, \qquad (9.19)$$

so that

$$\mathbf{L}.\mathbf{S} = \tfrac{1}{2}\hbar^2[J(J+1) - L(L+1) - S(S+1)]. \qquad (9.20)$$

If $L > 0$, J can equal $L+S$, $L+S-1$, $L+S-2$, ..., $L+S-n$, ..., $L-S$, leading to a splitting of the J states of multiplicity $(2S+1)$, $\mathbf{L}.\mathbf{S}$ having the value $[SL - n(L+S) + \tfrac{1}{2}n(n-1)]\hbar^2$, $n = 0, 1, 2, ...,$ $2S$. Thus $\mathbf{L}.\mathbf{S}$ ranges from $SL.\hbar^2$ if $J = L+S$, to $-S(L+1)\hbar^2$ if $J = L-S$.

Inglis (1936) showed the main effect of LS coupling is to make $\mathbf{L}$ and $\mathbf{S}$ line up parallel for mass numbers $A < 10$ and anti-parallel for $A > 10$. The only (odd N, odd Z) nucleus tested is ^{14}N, which gives $J = 1 = |\mathbf{L} - \mathbf{S}|$ as predicted, while for the odd nuclei ^{7}Li and ^{9}Be we find $J = \frac{3}{2} = L + S$ and for ^{13}C and ^{15}N, $J = \frac{1}{2} = |\mathbf{L} - \mathbf{S}|$. The one exception is ^{11}B with $J = \frac{3}{2} = L + S$.

An important application of spin-orbit coupling has been in helping to explain the occurrence of 'magic numbers' among the N and Z values of stable nuclei (see shell structure, § 1.2.5). The strength of spin-orbit coupling has been estimated from n-α and p-α scattering (§ 8.7) and there are indications that it may prove important in the calculation of binding energies of the light nuclei ^{2}H, ^{3}H, ^{3}He and ^{4}He (Abraham, Cohen and Roberts, 1955; Abraham, 1956).

CHAPTER 10

HIGH ENERGY NEUTRON-PROTON AND PROTON-PROTON SCATTERING

10.1. Introduction

Low energy n-p scattering was defined in chapter 6 as the energy range 0–10 MeV. of isotropic scattering for which the s-wave interaction predominates. However, at energies above 20 MeV., n-p scattering becomes increasingly anisotropic, indicating the growing importance of higher orbital angular momentum components, so that this range is called 'high energy'. The range 10–20 MeV. in which any scattering anisotropy is small is often referred to as 'intermediate'.

The classical argument of §6.2 states that interaction takes place only if the distance of closest approach is less than the range b, so that $1 < b/\lambda$, or $E(\text{Lab.}) > l^2$ (22 MeV.) and l^2(14 MeV.) for triplet and singlet ranges $b = 2 \cdot 05$ and $2 \cdot 58 \times 10^{-13}$ cm. respectively (cf. §6.2). This gives $E(\text{Lab.}) > l^2$ (20 MeV.) for the average phases (neglecting spin). However, wells such as the Yukawa potential $e^{-\kappa r}/r$ have a long tail outside the range $1/\kappa$, which is detectable to about three times the range. This results in higher phases affecting the angular distribution appreciably at intermediate energies. Calculations of the n-p differential cross-section at an energy of $13 \cdot 25$ MeV. show that the anisotropic ratio $d\sigma(\pi)/d\sigma(\tfrac{1}{2}\pi)$ is increased by 30 % if phase shifts higher than s and p are included (fig. 74). The main difference is near $\theta = 0°$ and $180°$, but $d\sigma$ has a weighting factor $\sin\theta$ unfavourable to forward and backward scattering contributions. Also compensatory effects occur at intermediate angles so that the total cross-section increases by only a few per cent.

These effects are illustrated by elastic n-p scattering experiments at $14 \cdot 1$ MeV. (Barschall and Taschek, 1949; Allred et al. 1953; Seagrave, 1955), $17 \cdot 9$ MeV. (Galonsky and Judish, 1955) and $27 \cdot 2$ MeV. (Brolley et al. 1951), the measured anisotropic ratios being $1 \cdot 05 \pm 0 \cdot 02$, $1 \cdot 08 \pm 0 \cdot 03$ and $1 \cdot 28 \pm 0 \cdot 10$ respectively. A rough analysis using average phase shifts fits these distributions with the values $\delta_0 \approx 1 \cdot 31$, $\delta_1 \approx 0$, $\delta_2 \approx 0 \cdot 012$ (14·1 MeV.), $\delta_0 \approx 1 \cdot 31$, $\delta_1 \approx 0$,

$\delta_2 \approx 0.018$ (17·9 MeV.) and $\delta_0 \approx 1.30$, $\delta_1 \approx -0.06$, $\delta_2 \approx 0.03$ (27·2 MeV.). In spite of the observed peaks of forward and backward scattering, the contributions of partial cross-sections for $l \geqslant 1$ are only 0·1, 0·2 and 2 % of the total cross-section respectively.

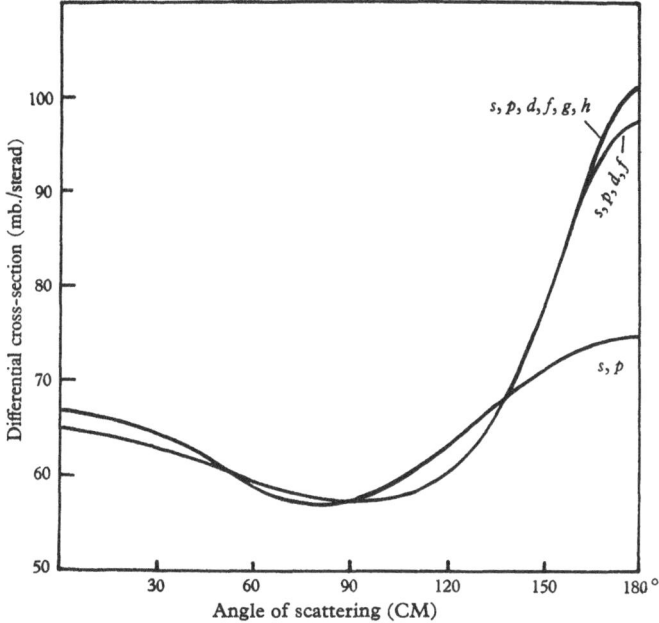

Fig. 74. Contributions of different order waves to n-p scattering at 13·25 MeV., using a symmetric scalar Yukawa potential with a meson mass of $200m_e$ (Frohlich, Ramsey and Sneddon, 1946).

p-p scattering at high energies is nearly isotropic after allowing for Coulomb effects, but at several hundred MeV., the cross-section is greater than the maximum allowed by the conservation theorem for s-state scattering, so that states of higher angular momentum must also be present (cf. § 12.1).

10.2 Spin and isotopic spin

A formal simplification of ideas is obtained if neutron and proton are regarded as eigenstates of the one particle, or nucleon. For this purpose, we introduce an isotopic spin quantum number defined in Table XIV by analogy with spin. As neutrons and protons are no longer separate particles, Pauli's exclusion principle for nucleons states that the total wave function for two or more particles, written as the product of spatial, spin and isotopic spin functions,

$$\Psi = \psi(\mathbf{r})\,\psi(\boldsymbol{\sigma})\,\psi(\boldsymbol{\tau}), \qquad (10.1)$$

must be antisymmetric with respect to the exchange of all coordinates of any two nucleons. Table XV shows the possible two-particle isotopic spin functions and their description. The total

TABLE XIV. *Analogous quantities for spin and isotopic spin*

Spin	Isotopic spin
Spin	Isotopic spin
Total spin $S=\frac{1}{2}$ for neutron and proton	Total isotopic spin $T=\frac{1}{2}$ for neutron and proton
z-component of total spin for a nucleon,	Component of $\boldsymbol{\tau}$ in the direction of positive charge,
$\uparrow$, $M=+\frac{1}{2}$, spin function α	$M=+\frac{1}{2}$ for the proton, isotopic spin function γ
$\downarrow$, $M=-\frac{1}{2}$, spin function β	$M=-\frac{1}{2}$ for the neutron, isotopic spin function δ
Spin wave function $\psi(\boldsymbol{\sigma})$	Isotopic spin wave function $\psi(\boldsymbol{\tau})$

TABLE XV. *Two-particle spin functions*

(s. = symmetric; a.s. = antisymmetric.)

State	Function	Symmetry	Charge	Nucleus	T	M
I	$\gamma(1)\,\gamma(2)$	s.	$2e$	^{2}He	1	1
II	$\delta(1)\,\delta(2)$	s.	0	^2n	1	-1
III	$2^{-\frac{1}{2}}[\gamma(1)\,\delta(2)+\gamma(2)\,\delta(1)]$	s.	e	^{2}H	1	0
IV	$2^{-\frac{1}{2}}[\gamma(1)\,\delta(2)-\gamma(2)\,\delta(1)]$	a.s.	e	^{2}H	0	0

isotopic spin T is 1 for symmetric and 0 for antisymmetric functions and M is the sum of the charge components M for the two nucleons. Also from § 6.10, the Pauli spin vector is $\boldsymbol{\sigma}=2\mathbf{S}$ in units of $\hbar$, where

$$\boldsymbol{\sigma}^2 = 4S(S+1),$$
$$\boldsymbol{\sigma}_1\cdot\boldsymbol{\sigma}_2 = +1 \text{ or } -3 \quad \text{for} \quad S=1 \text{ or } 0 \text{ respectively.} \quad (10.2)$$

Similarly we define the isotopic spin vector $\boldsymbol{\tau}$ with the properties

$$\boldsymbol{\tau}^2 = 4T(T+1),$$
$$\boldsymbol{\tau}_1\cdot\boldsymbol{\tau}_2 = +1 \text{ or } -3 \quad \text{for} \quad T=1 \text{ or } 0 \text{ respectively.} \quad (10.3)$$

10.3. Saturation and exchange forces

The approximate proportionality of nuclear binding energies to the mass number A necessitates the use of a short-range potential in order to prevent a nucleon interacting with all except its nearest neighbours. However, a force attractive in all angular momentum states leads to increased packing as A increases, resulting in non-saturation.

One possible solution involves the use of an infinite barrier inside the potential well to prevent the particles from getting too close together. Unfortunately, a barrier of width sufficient to give saturation contradicts the requirements of low energy scattering and deuteron bound state data.

Another possibility involves the use of many-body forces, whereby the force between two nucleons is altered by the presence of a third. If a substantial proportion of the two-body force is due to meson pair exchange, a three-body force could be due to the absorption of one member of an emitted pair by a third nucleon. It has been shown that if the even body forces are attractive, then the odd body forces are repulsive (Clementel and Villi, 1955). Qualitative estimates indicate that this might be sufficient to explain saturation with an ordinary type force.

The more usual postulate to explain saturation is that of exchange forces, for which the states of odd angular momentum are repulsive. Nuclei of mass up to ^{4}He can satisfy the Pauli principle by having two neutrons and two protons respectively paired with opposite spins in the S-state. A fifth nucleon must be placed in the P-state where it is repelled by the other particles, so that ^{5}He and ^{5}Li should be unstable, a first sign of saturation. Still more nucleons added restore stability (^{6}He, ^{6}Li, ^{6}Be) as the attractions within a shell are more important than the repulsion between particles in different shells.

For a total wave function of two particles in the CM system,

$$\Psi(\mathbf{r}_{12}, \boldsymbol{\sigma}_{12}) = \psi(\mathbf{r}_{12})\chi(\boldsymbol{\sigma}_{12}), \qquad (10.4)$$

we have the expansion in spherical polar co-ordinates

$$\psi(\mathbf{r}) = P_l^m(\cos\theta)\,e^{im\phi}f_l(r). \qquad (10.5)$$

Exchange of spatial co-ordinates of 1 and 2, $\mathbf{r}_{12} \rightarrow \mathbf{r}_{21}$ or $\mathbf{r} \rightarrow -\mathbf{r}$, is equivalent to replacing θ, ϕ by the opposite direction, $\theta \rightarrow \pi - \theta$, $\phi \rightarrow \pi + \phi$. Thus

$$\psi(-\mathbf{r}) = (-1)^l \psi(\mathbf{r}). \tag{10.6}$$

Similarly, if the spin co-ordinates of 1 and 2 are exchanged, $\sigma_{12} \rightarrow \sigma_{21}$, the spin function is either symmetric or antisymmetric according as the spin S is 1 or 0:

$$\chi(\sigma_{21}) = (-1)^{S+1} \chi(\sigma_{12}). \tag{10.7}$$

We shall define the operator $O = [(\hbar^2/m)\nabla^2 + E]$, so that the Schrödinger equation is $O\Psi = V(r)\Psi$. There are then four simple types of exchange force possible:

(a) The Wigner, ordinary or non-exchange force,

$$O\Psi(\mathbf{r}_{12}, \sigma_{12}) = V(r)\Psi(\mathbf{r}_{12}, \sigma_{12}). \tag{10.8}$$

(b) The Majorana or spatial exchange force,

$$O\Psi(\mathbf{r}_{12}, \sigma_{12}) = V(r)\Psi(\mathbf{r}_{21}, \sigma_{12})$$
$$= (-1)^l V(r)\Psi(\mathbf{r}_{12}, \sigma_{12}). \tag{10.9}$$

The Majorana force $(-1)^l V(r)$ can be alternatively written using equations (10.2), (10.3) and Table XV:

$$(-1)^l V(r) = \tfrac{1}{4}(1 + \sigma_1 . \sigma_2)(1 + \tau_1 . \tau_2) V(r). \tag{10.10}$$

The Majorana force leads to saturation.

(c) The Bartlett or spin exchange force,

$$O\Psi(\mathbf{r}_{12}, \sigma_{12}) = V(r)\Psi(\mathbf{r}_{12}, \sigma_{21})$$
$$= (-1)^{S+1} V(r)\Psi(\mathbf{r}_{12}, \sigma_{12}), \tag{10.11}$$

where $\qquad (-1)^{S+1} V(r) = \tfrac{1}{2}(1 + \sigma_1 . \sigma_2) V(r).$ $\qquad$ (10.12)

The Bartlett force gives different signs for ^3S and ^1S potentials so that the nuclear interaction cannot be wholly of this type.

(d) The Heisenberg or spin-spatial exchange force,

$$O\Psi(\mathbf{r}_{12}, \sigma_{12}) = V(r)\Psi(\mathbf{r}_{21}, \sigma_{21})$$
$$= (-1)^{l+S+1} V(r)\Psi(\mathbf{r}_{12}, \sigma_{12}), \tag{10.13}$$

where $\qquad (-1)^{l+S+1} V(r) = \tfrac{1}{2}(1 + \tau_1 . \tau_2) V(r).$ $\qquad$ (10.14)

The Heisenberg force is unacceptable as it leads to opposite signs for triplet and singlet potentials which alternate in sign with l.

However, it is possible to mix the exchange types (a)–(d) (as in §8.2) to produce cases such as W.B., M.H. and M.H.W.B. mixtures, which gives the right sign and magnitude of the interaction for even parity states, denoted by V_{even}. The magnitude and sign of V_{odd} determines the saturation character of nuclear forces and the angular distributions for high energy n-p and p-p scattering.

10.4. Meson field exchange theories

Calculation to second order of the potential energy of a pseudoscalar field with pseudoscalar coupling (cf. §1.5.3) gives the expression

$$V_{\text{even}}^{(2)}(\mathbf{I}) = -g_1 \left[\boldsymbol{\sigma}_1 \cdot \boldsymbol{\sigma}_2 + \left(\mathbf{I} + \frac{3\beta}{r} + \frac{3\beta^2}{r^2} \right) S_{12} \right] e^{-r/\beta}/(r/\beta), \quad (10.15)$$

where g_1 is the effective nucleon 'charge' or coupling constant for nucleon 1. The potential acting between two nucleons 1 and 2 is

$$V_{\text{even}}^{(2)}(12) = -g_1 g_2 \left[\boldsymbol{\sigma}_1 \cdot \boldsymbol{\sigma}_2 + \left(\mathbf{I} + \frac{3\beta}{r} + \frac{3\beta^2}{r^2} \right) S_{12} \right] e^{-r/\beta}/(r/\beta). \quad (10.16)$$

Taking g_n and g_p as coupling constants for neutron and proton respectively with their meson fields, two relations follow:

(1) Arguments based on the $N \approx Z$ rule for light nuclei, the binding energies of ^{3}H and ^{3}He, and the semi-empirical binding energy formula of Weizsacker indicate that p-p and n-n forces are at least approximately equal. Hence

$$g_n^2 = g_p^2, \quad \text{i.e.} \quad g_n = \pm g_p. \quad (10.17)$$

Equation (10.17) only holds if the meson theory employed leads to p-p and n-n forces in second order.

(2) Low energy p-p scattering experiments show that the ^1S n-p and p-p forces are at least approximately equal. Therefore $g_n g_p = g_p^2$ for the singlet state, giving

$$g_n = g_p = g. \quad (10.18)$$

Equation (10.18) holds only if the meson theory used leads to ^1S n-p and p-p forces in second order.

10.4.1. 'Neutral' theory.

We assume nucleons interact by exchanging neutral mesons only. Then the n-p, p-p and n-n interactions all exist so that equation (10.18) holds. The neutron and proton have identical meson fields, leading to the potential for odd l,

$$V_{\text{odd}} = V_{\text{even}}. \qquad (10.19)$$

The potential determined by equations (10.16) and (10.19) is a mixture of Bartlett and Wigner forces having the exchange character of the W.B. case of §8.2, and cannot lead to saturation.

10.4.2. *'Charged' theory.* When a nucleon emits a charged virtual meson of zero spin, the nucleonic charge changes so that the second order nucleon-nucleon interaction may be pictured for particles 1 and 2 as

$$n_1 \to p_1 + \pi^-, \quad p_2 + \pi^- \to n_2,$$

or

$$p_2 \to n_2 + \pi^+, \quad n_1 + \pi^+ \to p_1.$$

Thus there can be no second order n-n or p-p interaction, although one can occur in fourth order theory using meson pair exchange. Hence equations (10.17) and (10.18) are invalid for charge exchange theory. Adopting the notation of §10.3, the wave equation is

$$O\Psi(\mathbf{r}_{12}, \boldsymbol{\sigma}_{12}, \boldsymbol{\tau}_{12}) = V(r)\Psi(\mathbf{r}_{12}, \boldsymbol{\sigma}_{12}, \boldsymbol{\tau}_{21}). \qquad (10.20)$$

Pauli's principle requires Ψ to be antisymmetric with respect to exchange of all co-ordinates. The resultant charge symmetry of the wave function is shown in Table XVI. Equation (10.20) may then be rewritten

$$O\Psi(\mathbf{r}_{12}, \boldsymbol{\sigma}_{12}, \boldsymbol{\tau}_{12}) = (-1)^{l+S} V(r)\Psi(\mathbf{r}_{12}, \boldsymbol{\sigma}_{12}, \boldsymbol{\tau}_{12}). \qquad (10.21)$$

The charge exchange force $(-1)^{l+S} V(r)$ is the same, apart from an arbitrary minus sign, as the Heisenberg or spin-spatial exchange force (10.14), so that the charged theory potential is

$$(-1)^{l+S} V(r) = -\tfrac{1}{2}(1 + \boldsymbol{\tau}_1 \cdot \boldsymbol{\tau}_2) V(r). \qquad (10.22)$$

This theory fails because of its change of sign between the ^3S and ^1S states and also the absence of n-n and p-p forces.

There is a confusion in the literature in as much as the term 'charged theory' is often used for the Majorana-Heisenberg force of §8.2. In any numerical computation, charged theory always denotes this phenomenological exchange mixture.

TABLE XVI. *Charge symmetry of n-p wave functions*

(s. = symmetric; a.s. = antisymmetric.)

Spin state	$\psi(r)$	$\psi(\sigma)$	$\psi(\tau)$	$\psi(\tau_{21})/\psi(\tau_{12})$
^3S	s.	s.	a.s.	-1
^1S	s.	a.s.	s.	$+1$
^3P	a.s.	s.	s.	$+1$
^1P	a.s.	a.s.	a.s.	-1

10.4.3. Symmetric theory. To obtain n-n and p-p forces as well as n-p forces in a second order exchange type theory requires both neutral and charged mesons. Equations (10.17), viz. $g_n = \pm g_p$, then holds, but the solution (10.18), $g_n = +g_p$, is invalid. This is because the part of the interaction due to neutral mesons is equal for like and unlike particles in the ^1S state, whereas the part due to charged mesons is unequal for the two cases (cf. § 10.4.2). Thus on the basis of equation (10.18), the total neutral-charged interaction is unequal for like and unlike particles, in contradiction to experiment. The alternative is to choose $g_n = -g_p$, so that the total ^1S (or ^3P) interaction for neutral mesons becomes $+ W_n$ for like and $- W_n$ for unlike particles. There is an equal contribution W_c from both positively and negatively charged mesons, leading to a total ^1S (or ^3P) interaction of W_n for like particles and $- W_n + 2W_c$ for unlike particles. Hence $W_n = - W_n + 2W_c$ from experiment, so that $W_n = W_c$. The absolute value of g must therefore be identical for positive, negative and neutral mesons, making the theory symmetrical in all three kinds of mesons, $|g^+| = |g^-| = |g^0| = |g|$ (Bethe and Placzek, 1940). The interaction energy for both like and unlike particles is $W = + W_n$ for states with a symmetric $\psi(\tau)$, i.e. ^1S and ^3P states. Similarly for states with an anti-symmetric $\psi(\tau)$, i.e. ^3S and ^1P states, $W = -3W_n$ (cf. equation (10.22)). Hence $W = \tau_1 . \tau_2 W_n$, the potential (10.16) becoming

$$V(12) = \tfrac{1}{3}\tau_1 . \tau_2 g^2 \left[\sigma_1 . \sigma_2 + \left(1 + \frac{3\beta}{r} + \frac{3\beta^2}{r^2}\right) S_{12} \right] e^{-r/\beta}/(r/\beta)$$

$$= \tfrac{1}{3}\tau_1 . \tau_2 \sigma_1 . \sigma_2 g^2 \left[1 + \left(1 + \frac{3\beta}{r} + \frac{3\beta^2}{r^2}\right) S_{12} \right] e^{-r/\beta}/(r/\beta), \quad (10.23)$$

as $S_{12} = 0$ for singlet states and $\sigma_1 . \sigma_2 = +1$ for triplet states.

10.5. Phenomenological exchange potentials

Rarita and Schwinger ($1941a$) have shown that the ground state properties of the deuteron follow by use of the phenomenological potential form (9.10):

$$\mathscr{V}_{\text{even}}(r) = -(1 - \tfrac{1}{2}g + \tfrac{1}{2}g\sigma_1.\sigma_2 + \gamma S_{12})\, V(r). \qquad (10.24)$$

The potential for states of odd parity is suggested by the three meson exchange types above:

(i) Ordinary forces or neutral meson theory,

$$\mathscr{V}_{\text{odd}} = \mathscr{V}_{\text{even}}. \qquad (10.25)$$

(ii) Exchange forces or charged meson theory,

$$\mathscr{V}_{\text{odd}} = (-1)^l\, \mathscr{V}_{\text{even}} = \tfrac{1}{4}(1 + \tau_1.\tau_2)(1 + \sigma_1.\sigma_2)\, \mathscr{V}_{\text{even}}. \qquad (10.26)$$

(iii) Symmetric forces,

$$\mathscr{V}_{\text{odd}} = -\tfrac{1}{3}\tau_1.\tau_2\,\sigma_1.\sigma_2 \mathscr{V}_{\text{even}},$$

equivalent to $^3\mathscr{V}_{\text{odd}} = -\tfrac{1}{3}\,{}^3\mathscr{V}_{\text{even}}, \quad {}^1\mathscr{V}_{\text{odd}} = 3\,{}^1\mathscr{V}_{\text{even}}.$ (10.27)

Both exchange forces II and symmetric forces III give saturation.

(iv) Serber forces, $\mathscr{V}_{\text{odd}} = 0.$ (10.28)

This force is empirical, corresponding to a field mixture of two neutral mesons to each pair comprising a positive and negative meson. The absence of odd harmonics in the cross-section formula leads to an angular distribution symmetric about $\theta = 90°$ (CM). However, the absence of attractive forces in states of odd l is not sufficient to give saturation. For this, repulsive forces of sufficient magnitude are needed as in (ii) and (iii).

10.6. n-p scattering with non-central forces and spin-orbit coupling

At high energies, we might expect the non-central force of §10.5 to lead to results appreciably different from those of the equivalent central force which reproduces n-p scattering in the low energy range 0–10 MeV. A spin-orbit coupling term in the interaction should also lead to important effects at high energies, as it results in a velocity dependent, central force $\mathbf{L}.\mathbf{S}$ having for a given l the alternative values

$$l\hbar^2,\ -\hbar^2 \text{ and } -(l+1)\hbar^2 \quad \text{for} \quad j = l+1,\ l \text{ and } l-1 \qquad (10.29)$$

respectively. We take the interaction as

$$\mathscr{V}_{\text{even}} = -\left[1 - \tfrac{1}{2}g + \tfrac{1}{2}g\boldsymbol{\sigma}_1 \cdot \boldsymbol{\sigma}_2 + \gamma S_{12} + \frac{2\lambda}{\hbar^2}\, \mathbf{S}.\mathbf{L} \right] V(r), \quad (10.30)$$

where $\mathbf{S} = \tfrac{1}{2}(\boldsymbol{\sigma}_1 + \boldsymbol{\sigma}_2)$, λ is an arbitrary constant and $\mathscr{V}_{\text{odd}}$ is determined by one of the exchange operators P of §10.5 Thus we have from equation (9.20):

$$^3\mathscr{V}(r) = -P[U(r) + \gamma S_{12} V(r)], \quad (10.31)$$

where $\quad U(r) = [1 + \lambda\{j(j+1) - l(l+1) - 2\}]\, V(r),$

and $\quad\quad\quad ^1\mathscr{V}(r) = (1 - 2g)\, V(r). \quad (10.32)$

For a particular triplet state there are three good quantum numbers, the total angular momentum j, its z-component M and the parity $(-1)^l$. There are three possible l values, $l = j, j \pm 1$ and three possible values of the magnetic quantum number m (the z-component of L), $m = M$, $M \pm 1$. Taking the relative polar co-ordinates of the two nucleons as (r, θ, ϕ) and s_1, s_2 as their spin co-ordinates, the wave function describing the scattering may be expanded as (Hughes and Burgy, 1949)

$$\Psi = \frac{(4\pi)^{\frac{1}{2}}}{kr} \sum_{j=0}^{\infty} \sum_{l=j-1}^{l=j+1} f_{jlM}(r) F_{jlM}(\theta, \phi, s_1, s_2) = \frac{(4\pi)^{\frac{1}{2}}}{kr} \sum_{j=0}^{\infty} \psi_{jM}, \quad (10.33)$$

where $\quad F_{jlM} = \sum_{m=M-1}^{m=M+1} c_{jlMm} Y_{lm}(\theta, \phi) \chi_{M-m}(s_1, s_2) \quad (10.34)$

and $\quad Y_{lm}(\theta, \phi) = \left\{ \frac{2l+1}{4\pi} \cdot \frac{(l-m)!}{(l+m)!} \right\}^{\frac{1}{2}} P_l^m(\cos\theta)\, e^{im\phi} \quad (10.35)$

is a normalized tesseral harmonic. χ_{M-m} comprises the triplet spin functions for $M - m = \pm 1, 0$:

$$\chi_1 = \alpha(1)\,\alpha(2), \quad \chi_0 = 2^{-\frac{1}{2}}\{\alpha(1)\,\beta(2) + \alpha(2)\,\beta(1)\}, \quad \chi_1 = \beta(1)\,\beta(2).$$

The constants c_{jlMm} are determined by the condition that ψ_{jM} is a proper function for the operator $\mathbf{J}^2$ and the normalization condition $\sum_{m=M-1}^{m=M+1} c_{jlMm}^2 = 1$. The values of these Clebsch–Gordan coefficients are shown in Table XVII.

The radial functions $f_{jlM}(r)$ have the asymptotic form for large r

$$f_{jlM}(r) \sim \kappa_{jlM} \sin(kr - \tfrac{1}{2}l\pi + \eta_{jlM}),$$

where η_{jlM} is a phase shift and χ_{jlM} an unknown constant. The incident wave is the product of a plane wave and a spin wave function and may be expanded in the form

$$e^{ikz} \chi_M = \frac{(4\pi)^{\frac{1}{2}}}{kr} \sum_{j=0}^{\infty} \sum_{l=j-1}^{l=j+1} i^l c_{jlM0} (2l+1)^{\frac{1}{2}} g_l(r) F_{jlM}, \quad (10.36)$$

where $g_l(r) = (\frac{1}{2}\pi kr)^{\frac{1}{2}} J_{l+\frac{1}{2}}(kr) \sim \sin(kr - \frac{1}{2}l\pi)$, $r \to \infty$.

An alternative expression for the wave function (10.33) follows from equation (10.35):

$$\Psi = e^{ikz} \chi_M + \frac{(4\pi)^{\frac{1}{2}}}{kr} \sum_{j=0}^{\infty} \sum_{l=j-1}^{l=j+1} [f_{jlM} - i^l(2l+1)^{\frac{1}{2}} c_{jlM0} g_l] F_{jlM}. \quad (10.37)$$

As the first term represents the incident wave, the second must represent asymptotically an outgoing or scattered spherical wave of the form $B e^{ikr}/r$. Substitution of the asymptotic forms of f_{jlM} and g_l leads to the necessary condition

$$\kappa_{jlM} = i^l(2l+1)^{\frac{1}{2}} \exp(i\eta_{jlM}) c_{jlM0}. \quad (10.38)$$

The triplet differential cross-section $^3I_M(\theta)$ is then given by

$$^3I_M(\theta) = \frac{\pi}{k^2} \sum_s \left| \sum_{l=0}^{\infty} \sum_{j=l-1}^{j=l+1} (2l+1)^{\frac{1}{2}} c_{jlM0} \{\exp(2i\eta_{jlM}) - 1\} F_{jlM} \right|^2.$$

$$(10.39)$$

Substitution of the c_{jlM0} and f_{jlM} into (10.39) leads to

$$^3I_0(\theta) = {}^3I_{00}(\theta) + {}^3I_{01}(\theta), \quad {}^3I_{\pm 1}(\theta) = {}^3I_{12}(\theta) + {}^3I_{11}(\theta) + {}^3I_{10}(\theta),$$

where

$$^3I_{00}(\theta) = (1/4k^2) \left| \sum_{l=0}^{\infty} \{l\epsilon_{l-1,l,0} + (l+1)\epsilon_{l+1,l,0}\} P_l(\cos\theta) \right|^2,$$

$$^3I_{01}(\theta) = (1/4k^2) \left| \sum_{l=0}^{\infty} \{\epsilon_{l-1,l,0} - \epsilon_{l+1,l,0}\} P_l^1(\cos\theta) \right|^2,$$

$$^3I_{11}(\theta) = (1/8k^2) \left| \sum_{l=1}^{\infty} \left\{ \frac{l-1}{l} \epsilon_{l-1,l,1} + \frac{2l+1}{l(l+1)} \epsilon_{l,l,1} - \frac{l+2}{l+1} \epsilon_{l+1,l,1} \right\} \right.$$
$$\left. \times P_l^1(\cos\theta) \right|^2,$$

$$^3I_{12}(\theta) = (1/16k^2) \left| \sum_{l=2}^{\infty} \left\{ \frac{1}{l} \epsilon_{l-1,l,1} - \frac{2l+1}{l(l+1)} \epsilon_{l,l,1} + \frac{1}{l+1} \epsilon_{l+1,l,1} \right\} \right.$$
$$\left. \times P_l^2(\cos\theta) \right|^2,$$

$$^3I_{10}(\theta) = (1/16k^2) \left| \sum_{l=0}^{\infty} \{(l-1)\epsilon_{l-1,l,1} + (2l+1)\epsilon_{l,l,1} + (l+2)\epsilon_{l+1,l,1} \} \right.$$
$$\left. \times P_l(\cos\theta) \right|^2,$$

$$\epsilon_{jlM} = \exp(2i\eta_{jlM}) - 1.$$

TABLE XVII. *Values of the coefficients* c_{jlMm}

	$m = M - 1$	$m = M$	$m = M + 1$
$l = j - 1$	$\left\{\dfrac{(j+M)(j+M-1)}{2j(2j-1)}\right\}^{\frac{1}{2}}$	$\left\{\dfrac{j^2 - M^2}{j(2j-1)}\right\}^{\frac{1}{2}}$	$\left\{\dfrac{(j-M)(j-M-1)}{2j(2j-1)}\right\}^{\frac{1}{2}}$
$l = j$	$-\left\{\dfrac{(j+M)(j-M+1)}{2j(j+1)}\right\}^{\frac{1}{2}}$	$\left\{\dfrac{M^2}{j(j+1)}\right\}^{\frac{1}{2}}$	$\left\{\dfrac{(j-M)(j+M+1)}{2j(j+1)}\right\}^{\frac{1}{2}}$
$l = j + 1$	$\left\{\dfrac{(j-M+2)(j-M+1)}{2(j+1)(2j+3)}\right\}^{\frac{1}{2}}$	$-\left\{\dfrac{(j+1)^2 - M^2}{(j+1)(2j+3)}\right\}^{\frac{1}{2}}$	$\left\{\dfrac{(j+M+2)(j+M+1)}{2(j+1)(2j+3)}\right\}^{\frac{1}{2}}$

The total differential cross-section is then

$$I(\theta) = \tfrac{1}{4}[{}^1I(\theta) + {}^3I_{M=0}(\theta) + {}^3I_{M=+1}(\theta) + {}^3I_{M=-1}(\theta)], \quad (10\cdot41)$$

where ${}^1I(\theta)$ is the singlet differential cross-section. The total cross-section σ turns out to be

$$\sigma = \tfrac{1}{4}[{}^1\sigma + {}^3\sigma_{M=0} + {}^3\sigma_{M=+1} + {}^3\sigma_{M=-1}], \quad (10.42)$$

where ${}^1\sigma$ is the singlet cross-section:

$$\,^1\sigma = (4\pi/k^2) \sum_{l=0}^{\infty} (2l+1)\sin^2 \eta_l$$

and

$$\left.\begin{aligned}
{}^3\sigma_{M=0} &= (\pi/k^2) \sum_{l=0}^{\infty} \{l\,|\,\epsilon_{l-1,l,0}\,|^2 + (l+1)\,|\,\epsilon_{l+1,l,0}\,|^2\}, \\
{}^3\sigma_{M=\pm1} &= (\pi/2k^2) \sum_{l=0}^{\infty} \{(l-1)\,|\,\epsilon_{l-1,l,1}\,|^2 + (2l+1)\,|\,\epsilon_{l,l,1}\,|^2 \\
&\qquad\qquad + (l+2)\,|\,\epsilon_{l+1,l,1}\,|^2\}.
\end{aligned}\right\}$$

$$(10.43)$$

In general the phases η_{jlM} are found by solution of the wave equation for the radial wave functions $f_{jlM}(r)$. As there is no coupling between states of different parity, the state $j = l$ is not coupled to the states $j = l \pm 1$. Thus for $j = l$, $S_{12} = +2$ and $f_{jlM}(r)$ satisfies the wave equation

$$\left[\frac{d^2}{dr^2} - \frac{l(l+1)}{r^2} - k^2\right] f_{jlM} + P\frac{m}{\hbar^2}[U(r) + 2\gamma V(r)] f_{jlM} = 0. \quad (10.44)$$

If $j = l \pm 1$, the states are coupled and two coupled second-order equations must be solved:

$$\left[\frac{d^2}{dr^2} - \frac{l(l+1)}{r^2} - k^2\right] f_{jlM} + P\frac{m}{\hbar^2}[\{U(r) + \gamma V(r) t_{jl}\} f_{jlM} + t_{jl'} f_{jl'M}] = 0,$$

$$(10.45)$$

where the coefficients t are given by Bethe ($1940b$):

$$t_{j=l+1,l} = -\frac{2(j-1)}{2j+1}, \quad t_{j=l-1,l} = -\frac{2(j+2)}{2j+1}, \quad t_{jll'} = \frac{6j^{\frac{1}{2}}(j+1)^{\frac{1}{2}}}{2j+1}. \quad (10.46)$$

For the special case $j=0$, equation (10.45) reduces to a single equation describing the uncoupled $^3P_0$ state, with $S_{12} = -4$.

From 10 to 20 MeV. only the phases for $l=0, 1$ are important, the problem reducing to solution of the coupled equations (10.45) for the $^3S_1 + {}^3D_1$ state and evaluation of the P-wave scattering. The $^3P_2$ state has a fairly definite value of $S_{12} \approx -\frac{2}{5}$ since at medium energies it is only weakly coupled to the $^3F_2$ state. Thus P-wave scattering for $^3P_0$, $^3P_1$ and $^3P_2$ states is found by solving Schrödinger equations for potentials $U(r)-4V(r)$, $U(r)+2U(r)$, $U(r)-\frac{2}{5}V(r)$ respectively. Scattering calculations are available at 15·3 MeV. for the spherical well (Rarita and Schwinger, 1941) and at 2·1–20·8 MeV. for the spherical, exponential and Yukawa wells using a number of alternative ranges and exchange forces (Hu and Massey, 1949). The cross-sections and scattering distributions obtained are not, except for ordinary forces, markedly different from those resulting from central potentials of the same range with depths adjusted to give the deuteron binding energy. Thus the effect of a non-central exchange potential should only become obvious at energies above 20 MeV.

10.7. The Born approximation for n-p scattering with non-central forces

At energies large compared to the nuclear interaction, the neutron wave may be approximated by a plane wave and the Born approximation can be used for the angular distribution for n-p scattering. We consider the potential (10.24) such that

$$^3\mathscr{V}(r) = -{}^3P(1+\gamma S_{12}) V(r), \quad {}^1\mathscr{V}(r) = -{}^1P(1-2g) V(r), \quad (10.47)$$

where ^3P and ^1P are exchange operators for triplet and singlet states respectively. These are now rewritten in terms of the Majorana spatial exchange operator $P_M = (-1)^l$, where

$$P_M f(r, \theta, \phi) = f(-r, \theta, \phi) = f(r, \pi-\theta, \pi+\phi). \quad (10.48)$$

(i) Ordinary forces, $\quad ^3P = {}^1P = 1.$

(ii) Exchange forces, $^3P = {}^1P = P_M$.

(iii) Symmetric forces,

$$^3P = \tfrac{1}{3}(1 + 2P_M), \quad {}^1P = -(1 - 2P_M).$$

(iv) Serber forces, $^3P = {}^1P = \tfrac{1}{2}(1 + P_M)$. (10.49)

The differential cross-section for scattering by a central force $-V(r)$ is given in terms of the scattering amplitude $F(\theta)$ by

$$d\sigma/d\omega = |F(\theta)|^2, \tag{10.50}$$

where

$$F(\theta) = -(m/4\pi\hbar^2) \int_0^\infty e^{-ik\mathbf{n}\cdot\mathbf{r}} V(r) e^{ik\mathbf{n}_0\cdot\mathbf{r}} d\mathbf{r}$$

$$= -(m/\hbar^2) \int_0^\infty \frac{\sin Kr}{Kr} V(r) r^2 dr. \tag{10.51}$$

Here $K = 2k\sin\tfrac{1}{2}\theta$ and $\mathbf{n}$ and $\mathbf{n}_0$ are unit vectors along the incident and scattered beams respectively.

The corresponding scattering amplitude for a non-central force $-S_{12}V(r)$ is (Ashkin and Wu, 1948)

$$S(\theta) = -(m/4\pi\hbar^2) \int_0^\infty e^{-ik\mathbf{n}\cdot\mathbf{r}} S_{12} V(r) e^{ik\mathbf{n}_0\cdot\mathbf{r}} d\mathbf{r}$$

$$= C(\theta)\tau(\theta, \phi), \tag{10.52}$$

where

$$C(\theta) = -(m/\hbar^2) \int_0^\infty r^2 V(r) \left[\frac{\sin Kr}{Kr} - 3 \left\{ \frac{\sin Kr}{Kr} - \frac{Kr\cos Kr}{(Kr)^3} \right\} \right] dr \tag{10.53}$$

and

$$\tau(\theta, \phi) = \boldsymbol{\sigma}_1 \cdot \boldsymbol{\sigma}_2 - 3\sigma_{1,\mathbf{n}_0-\mathbf{n}} \, \sigma_{2,\mathbf{n}_0-\mathbf{n}}. \tag{10.54}$$

$\tau(\theta, \phi)$ is the angular average of S_{12} weighted by $e^{ik(\mathbf{n}_0-\mathbf{n})\cdot\mathbf{r}}$ and $\sigma_{1,\mathbf{n}_0-\mathbf{n}}$ and $\sigma_{2,\mathbf{n}_0-\mathbf{n}}$ are the components of $\boldsymbol{\sigma}_1$ and $\boldsymbol{\sigma}_2$ in the direction $(\mathbf{n}_0 - \mathbf{n})$ of the momentum transfer. If m_s denotes the z-component of spin of a particular triplet wave function, then the result of applying the operator τ to the latter is

$$\tau \chi_{m_s} = \sum_{m_{s'}=-1}^{+1} \tau_{m_{s'}m_s}(\theta, \phi) \chi_{m_{s'}}. \tag{10.55}$$

The matrix elements of $\tau_{m_{s'}m_s}(\theta, \phi) \chi_{m_{s'}}$ are given in Table XVIII.

The scattering amplitudes $S_{m_{s'}m_s}(\theta, \phi)$ for the combined central and non-central triplet force (10.47) become

$$S_{m_{s'}m_s}(\theta, \phi) = {}^3P[F(\theta) \delta_{m_{s'}m_s} + \gamma C(\theta) \tau_{m_{s'}m_s}(\theta, \phi)], \tag{10.56}$$

where ^3P and ^1P are the values (10.49) for the forces (i)–(iv) and $S_{m_s m_{s'}} = 1, 0$ for $m_s = m_{s'}$, $m_s \neq m_{s'}$ respectively. The singlet scattering amplitude is

$$S(\theta) = (1 - 2g)\,{}^1PF(\theta). \tag{10.57}$$

On employing equation (10.48) and the expressions for $\tau_{m_s m_{s'}}(\theta, \phi)$ and squaring equation (10.56), we obtain the differential cross-section by assigning statistical weights of 3 and 1 to triplet and singlet states respectively:

$$(d\sigma/d\omega) = 0\cdot75(d\sigma_T/d\omega) + 0\cdot25(d\sigma_S/d\omega). \tag{10.58}$$

TABLE XVIII. *Matrix elements of* $\tau_{m_{s'} m_s}$

	$m_s = 1$	$m_s = 0$	$m_s = -1$
$m_{s'} = 1$	$-\tfrac{1}{2}(1 - 3\cos\theta)$	$(3/2^{\frac12})\sin\theta\,.\,e^{-i\phi}$	$-3\cos^2\theta/2\,.\,e^{-2i\phi}$
$m_{s'} = 0$	$(3/2^{\frac12})\sin\theta\,.\,e^{i\phi}$	$(1 - 3\cos\theta)$	$-(3/2^{\frac12})\sin\theta\,.\,e^{-i\phi}$
$m_{s'} = -1$	$-3\cos^2\theta/2\,.\,e^{2i\phi}$	$-(3/2^{\frac12})\sin\theta\,.\,e^{i\phi}$	$-\tfrac{1}{2}(1 - 3\cos\theta)$

The differential cross-sections for the four exchange types are

(i) Ordinary forces,

$$d\sigma/d\omega = \tfrac{1}{4}[3 + (1 - 2g)^2]\,F^2(\theta) + 6\gamma^2 C^2(\theta). \tag{10.59}$$

(ii) Exchange forces,

$$d\sigma/d\omega = \tfrac{1}{4}[3 + (1 - 2g)^2]\,F^2(\pi - \theta) + 6\gamma^2 C^2(\pi - \theta). \tag{10.60}$$

(iii) Symmetric forces,

$$d\sigma/d\omega = [\tfrac{1}{3} + (1 - 2g)^2]\tfrac{1}{4}F^2(\theta) + F^2(\pi - \theta)$$
$$+ [\tfrac{1}{3} - (1 - 2g)^2]\,F(\theta)\,F(\pi - \theta) + \tfrac{8}{3}\gamma^2[\tfrac{1}{4}C^2(\theta) + C^2(\pi - \theta)]$$
$$- \tfrac{4}{3}\gamma^2 C(\theta)\,C(\pi - \theta). \tag{10.61}$$

(iv) Serber forces,

$$d\sigma/d\omega = \tfrac{1}{16}[3 + (1 - 2g)^2]\,[F(\theta) + F(\pi - \theta)]^2$$
$$+ \tfrac{3}{2}\gamma^2[C^2(\theta) + C^2(\pi - \theta) - C(\theta)\,C(\pi - \theta)]. \tag{10.62}$$

In the Born approximation, the angular distributions for ordinary and exchange forces are reflexions of each other about 90° (CM). Thus ordinary forces give a strong forward peak of scattering, and exchange forces a strong backward scattering peak due to the neutron exchanging its energy with the proton. Symmetric forces give strong backward and weaker forward scattering peaks as the mixture of exchange forces is in the ratio $2:1$. Serber forces lead to a symmetric distribution about $90°$ (CM) due to the absence of odd harmonics in the scattering distribution, actually a characteristic at all energies.

Experience has shown that although the Born approximation with central forces gives reliable results for n-p scattering for neutron energies $\gtrsim 200$ MeV., it is a poor approximation when tensor forces are included. In the case of a singular tensor potential, e.g. of Yukawa type, a very fast neutron may penetrate the inner region where the potential is not small compared to the incident energy. The scattered wave is not then approximated by a plane wave and the Born result may be wrong by a factor of two for back scattering. Higher Born approximations have been given by Wu (1948), Dalitz (1951) and Barker (1956).

At energies higher than 200 MeV., relativistic corrections become important so that the Born approximation is of limited reliability.

10.8. p-p scattering with non-central forces

Analysis of p-p scattering is simplified by the fact that only $^1S, ^3P, ^1D, ...$, states occur due to the exclusion principle. ^1S and ^1D scattering is necessarily central so that the non-central force affects scattering mainly through the $^3P_0$, $^3P_1$ and $^3P_2$ phases. The effect for $l \geqslant 3$ of higher orbital angular triplet phases (assumed small) may be taken into account using a Born approximation expression for the phase similar to equation (6.11) for n-p scattering phases, but with the plane wave replaced by the logarithmically varying asymptotic solution (7.32) for a pure Coulomb field:

$$\delta_l \approx -(m/\hbar^2 k) \int_0^\infty V(r) F^2(kr) \, dr, \qquad (10.63)$$

where

$$F(kr) = \sin(kr - \tfrac{1}{2}l\pi - \alpha \ln 2kr + \sigma_l) \qquad (10.64)$$

and $V(r)$ is an equivalent central triplet force (cf. equation (10.72)).

The cross-section is given by Breit, Kittel and Thaxton (1940):

$$P = P_M + (\Delta P)_0 + (\Delta P)_1 + (\Delta P)_2 + (\Delta P)_3 + ...$$

as in §7.4, with P_M, $(\Delta P)_0$, $(\Delta P)_2$, ..., given by equation (7.31) but with $(\Delta P)_1$ replaced by

$$(\Delta P)_1 = \sum_i g_i \frac{18}{\alpha} P_1(\cos\theta) \left(\frac{\cos\alpha_1}{s^2} - \frac{\cos\beta}{c^2}\right) \sin K_i \cos K_i$$
$$+ \left\{\frac{108}{\alpha^2} P_1^2(\cos\theta) + \frac{18}{\alpha} P_1(\cos\theta) \left(\frac{\sin\alpha_1}{s^2} - \frac{\sin\beta_1}{c^2}\right)\right\} \sin^2 K_i$$
$$+ \frac{12}{\alpha^2} \left[-\tfrac{3}{4}(\sin K_1 - \sin K_2)^2 - \tfrac{1}{3}(\sin K_0 - \sin K_2)^2 \right.$$
$$- 3 \sin K_1 \sin K_2 \sin^2 \tfrac{1}{2}(K_1 - K_2)$$
$$\left. - \tfrac{4}{3} \sin K_0 \sin K_2 \sin^2 \tfrac{1}{2}(K_0 - K_2)\right](3\cos^2\theta - 1), \qquad (10.65)$$

where K_0, K_1, K_2 are the $^3P_0$, $^3P_1$ and $^3P_2$ phases respectively and

$$\alpha_1 = \alpha \ln s^2 + 2(\eta_1 - \eta_0), \quad \beta_1 = \alpha \ln c^2 + 2(\eta_1 - \eta_0),$$
$$s = \sin\theta, \quad c = \cos\theta, \quad g_0 = \tfrac{1}{9}, \quad g_1 = \tfrac{3}{9}, \quad g_2 = \tfrac{5}{9}.$$

10.9. The Born approximation for p-p scattering with non-central forces

p-p scattering by the non-central potential (10.47) may be evaluated from the corresponding n-p expressions by introducing the Pauli principle requirement that the triplet spatial wave function be antisymmetric (odd l) and the singlet spatial function be symmetric (even l). We thus replace the triplet scattering amplitudes $S_{m_s'm_s}(\theta, \phi)$ of equation (10.56) by $S_{m_s'm_s}(\theta, \phi) - S_{m_s'm_s}(\pi - \theta, \pi + \phi)$ and the singlet n-p amplitude $S(\theta)$ of equation (12.57) by $S(\theta) + S(\pi - \theta)$. After squaring and applying appropriate statistical weights, the angular distributions for the various exchange types are

(i) Ordinary forces,

$$d\sigma/d\omega = \tfrac{1}{4}[3 + (1 - 2g)^2] [F^2(\theta) + F^2(\pi - \theta)]$$
$$+ 6\gamma^2[C^2(\theta) + C^2(\pi - \theta)] - \tfrac{1}{2}[3 - (1 - 2g)^2] F(\theta) F(\pi - \theta).$$
$$+ 6\gamma^2 C(\theta) C(\pi - \theta). \qquad (10.66)$$

(ii) Exchange forces, the same as for ordinary forces (i).

(iii) Symmetric forces,

$$d\sigma/d\omega = [\tfrac{1}{12} + \tfrac{1}{4}(1 - 2g)^2] [F^2(\theta) + F^2(\pi - \theta)]$$
$$+ [\tfrac{1}{2}(1 - 2g)^2 - \tfrac{1}{6}] F(\theta) F(\pi - \theta)$$
$$+ \tfrac{2}{3}\gamma^2[C^2(\theta) + C^2(\pi - \theta) + C(\theta) C(\pi - \theta)]. \qquad (10.67)$$

(iv) Serber forces,

$$d\sigma/d\omega = (1 - 2g)^2 [F(\theta) + F(\pi - \theta)]^2. \qquad (10.68)$$

Thus the Serber force result involves pure singlet scattering only. The Born approximation is valid under the same conditions as for n-p scattering, viz. $E > 200$ MeV., but is likewise not reliable where singular tensor forces are concerned.

10.10. Analysis of high energy n-p scattering

High energy scattering experiments with unpolarized beams are difficult to interpret, for there is no means of finding a unique set of phase parameters. At 90 MeV., the s-, p- and d-phases should account for most of the scattering, but the division into singlet and triplet states, with possible splitting of the latter by non-central forces into $m = 0$, ± 1 magnetic substates, means that ten phase parameters may be involved. Even if the angular distribution were known exactly, the determination of a unique set of phases would not be possible. However, the scattering may be empirically fitted by a linear sum of Legendre polynomials involving $(2n + 1)$ parameters:

$$d\sigma/d\omega = \sum_{n=0}^{l} a_{2n} P_{2n}(\cos\theta). \qquad (10.69)$$

The phenomenological approach involves fitting the observed scattering over a wide range of energies with an exchange potential of the form (10.30), containing central, non-central and possible spin-orbit coupling terms and an arbitrary radial dependence. The magnitude and range of the potential is subject to low energy conditions, viz. the binding energy E_D and quadrupole moment Q of the deuteron, the zero energy scattering lengths a_1, a_0 and the effective ranges r_1, r_0 for low energy scattering as given by the shape independent formula (6.32). This involves six parameters whereas the usual non-central force without spin-orbit coupling contains only four. It does not appear possible to satisfy numerically all these conditions with such a potential. Thus calculations for high energy scattering generally use either central forces satisfying all requirements except that of the quadrupole moment Q, or a non-central force which gives Q, E_D, r_1 and r_0 correctly, but may lead to an incorrect zero energy scattering length.

Christian (1952) gives the following low energy parameters:

(a) $n\text{-}p$ data,

$$a_0 = -23 \cdot 68 \;\pm\; 0 \cdot 06 \times 10^{-13}\,\text{cm.}, \quad r_0 = 2 \cdot 7 \;\pm\; 0 \cdot 5 \times 10^{-13}\,\text{cm.},$$

$$a_1 = \quad 5 \cdot 388 \pm 0 \cdot 04 \times 10^{-13}\,\text{cm.}, \quad r_1 = 1 \cdot 71 \pm 0 \cdot 04 \times 10^{-13}\,\text{cm.},$$

$$E_D = \quad 2 \cdot 223 \pm 0 \cdot 004\,\text{MeV.}, \quad\quad Q = 2 \cdot 73 \pm 0 \cdot 05 \times 10^{-27}\,\text{cm.}^2,$$

percentage of D-state admixture $w_D = 3 \cdot 9 \%$.

(b) $p\text{-}p$ data,

$$a_0 = -7 \cdot 67 \pm 0 \cdot 05 \times 10^{-13}\,\text{cm.}, \quad r_0 = 2 \cdot 65 \times 0 \cdot 07 \times 10^{-13}\,\text{cm.}$$

These figures determine the parametric ranges of the ^3S $n\text{-}p$ and ^1S $p\text{-}p$ states as shown in Table XIX for various well shapes, and offer strong support for the charge independence hypothesis, in particular the equality of the ^1S $n\text{-}p$ and ^1S $p\text{-}p$ interactions. The triplet and singlet potentials have different ranges, although the difference is smaller for long-tailed potentials.

TABLE XIX. *Parametric ranges*

Model	$p\text{-}p$ singlet (10^{-13} cm.)	$n\text{-}p$ triplet (10^{-13} cm.)
Spherical	$2 \cdot 58 \pm 0 \cdot 06$	$2 \cdot 05 \pm 0 \cdot 20$
Gauss	$1 \cdot 24 \pm 0 \cdot 03$	$1 \cdot 06 \pm 0 \cdot 12$
Exponential	$0 \cdot 71 \pm 0 \cdot 02$	$0 \cdot 67 \pm 0 \cdot 08$
Yukawa	$1 \cdot 17 \pm 0 \cdot 03$	$1 \cdot 35 \pm 0 \cdot 17$

Experiments at 90 MeV. (fig. 75) have shown near symmetry of the angular distribution about 90° for $n\text{-}p$ scattering. Results at 40 MeV. are not inconsistent with this but do not extend to sufficiently small angles to give clear cut support. The expansion of the 90 MeV. distribution according to equation (10.69) is

$$d\sigma/d\omega = C[\mathrm{I} - 0 \cdot 14 P_1(\cos\theta) + 0 \cdot 68 P_2(\cos\theta) + 0 \cdot 02 P_3(\cos\theta)$$
$$+ 0 \cdot 11 P_4(\cos\theta) + \ldots]. \quad (10.70)$$

The odd harmonics are small, indicating that the forces in odd states are considerably smaller than in even states. The same conclusion is reached by neglecting spin in analysing the distribution for the average phase shifts. These are $\delta_s = 55°$, $\delta_p = -1°$, $\delta_d = 5°$, with higher phases less than $1°$. The small size of the p-phase indicates the odd state interaction is about one-tenth that for even states.

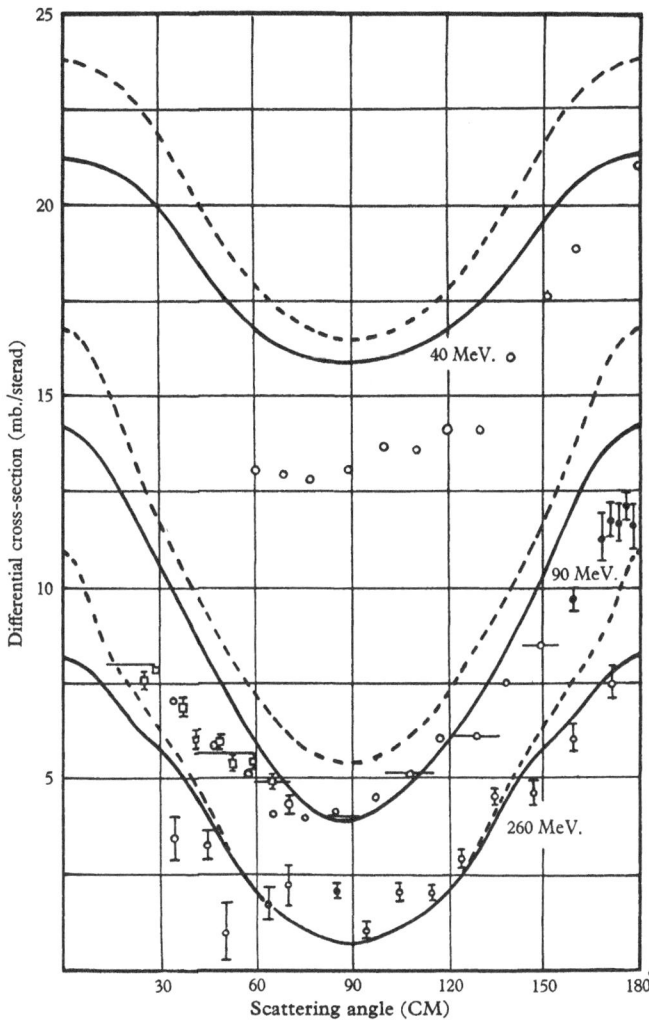

Fig. 75. Experimental angular distributions for *n-p* scattering normalized to give the measured total cross-sections.

Full lines: *n-p* scattering predicted by an exponential non-central potential with Serber exchange dependence,

$$V = (-69 \text{ MeV.}) (1 + 1.845 S_{12}) \exp(-r/R), \quad \text{with} \quad R = 0.75 \times 10^{-13} \text{ cm.}$$

Broken lines: *n-p* scattering predicted by a Yukawa non-central potential with Serber exchange dependence,

$$V = (-25.3 \text{ MeV.}) (1 + 1.91 S_{12}) \exp(-r/R)/(r/R), \quad \text{with} \quad R = 1.35 \times 10^{-13} \text{ cm.}$$

The total cross-sections at 40, 90 and 260 MeV. are 203 ± 7, 79 ± 7 and 35 ± 3 mb. as compared to the values predicted by the exponential and Yukawa models, 217, 87, 36 mb. and 231, 98, 37 mb. respectively.

Attempts to reproduce the distribution (10.70) generally neglect relativistic effects, the correction for which is proportional to $(v/c)^2$, equal to 0·05 at 90 MeV. Probably the coefficient of this term does not make the correction at 90 MeV. much larger than 5 % (Breit, 1955; Ebel and Hull, 1955; Breit, 1957).

Using central forces for simplicity, we may determine the nature of the radial dependence of the potential by plotting the s-wave cross-section against energy for several well shapes, together with the experimental average cross-sections found from the average phase shifts at 90 MeV. and lower energies. The spherical well results are too low at 90 MeV. and the Yukawa well results slightly too large, but the exponential well results agree closely. This is due to the Yukawa potential having the deepest hole near the origin, the exponential well a shallower hole and the spherical well the shallowest hole of the trio. The wavelength of the neutron is so short at 90 MeV. ($\sim 1 \times 10^{-13}$ cm.) that it can penetrate into the region of strong attraction, so that the potential with the deepest hole leads to the largest s-phase at high energies.

If the average d-phase is plotted against energy, it is found to vary nearly linearly with energy above 10 MeV. On the other hand a spherical well gives a characteristic $E^{\frac{5}{2}}$ energy dependence up to 40 MeV., as the penetration of the d-wave into the well is governed largely by the centrifugal barrier $l(l+1)/r^2$. However, with the Yukawa potential, the d-wave is affected by the tail at low energies, the gradual penetration into the region of higher potential leading to a linear energy variation of the d-phase from 10 MeV. up. The d-phase of course largely determines the shape of the angular distribution.

The total cross-section data indicates that the potential should have a long tail but should not have a deep hole at the origin, e.g. the exponential potential. However, the angular distribution evidence largely favours a long-tailed potential with a deep hole, e.g. the Yukawa potential. Using an exchange force with no odd states (Serber potential), the spherical, exponential and Yukawa potentials lead to anisotropic ratios $d\sigma(180°)/d\sigma(90°)$ at 40 MeV. and 90 MeV. of 1·1, 1·3, 1·4 and 4.2, 4·0 and 3·2 respectively. These compare with experimental values of 1·55 ± 0·2 and 3·0 ± 0·1, favouring the Yukawa potential (Christian, 1952).

The inclusion of non-central forces in the interaction leads to an effective range theory (Christian, 1949) for the $^3S + {}^3D$ ground state (cf. §9.8), the effective range in terms of s- and d-waves being

$$r_1 = 2\int_0^\infty [(1 - r/a)^2 - u^2(r) - w^2(r)]\, dr, \qquad (10.71)$$

where $\qquad u(r) \to 1 - r/a, \quad w(r) \to 0 \quad$ as $\quad r \to \infty$.

The parametric ranges may then be computed subject to the requirement of the deuteron quadrupole moment. For a spherical well, this gives a range of $2\cdot3 \times 10^{-13}$ cm. for the non-central force, the central force vanishing entirely. The exponential well has a range of $0\cdot75 \times 10^{-13}$ cm., with a triplet central depth of 69 MeV. and a non-central depth of 128 MeV. The corresponding figures for the Yukawa potential are $1\cdot35 \times 10^{-13}$ cm., $25\cdot3$ and $48\cdot2$ MeV.

These ranges are not very different from the central force values of Table XIX, so that the use of non-central forces does not change the low energy scattering properties very much. In fact it is found that the non-central potential $V_c(r) + S_{12}V_t(r)$ may be replaced by an equivalent central potential for the s-wave:

$$V(r) = V_c(r) + 2^{\frac{3}{2}}\gamma V_t(r)\, R(r), \qquad (10.72)$$

where $R(r)$ is the ratio of the d-wave to the s-wave. The ratio of the equivalent potential (10.72) to the central potential which gives the correct binding energy, increases to about unity in the region of the range and decreases asymptotically outside it.

Another form for the equivalent potential uses the W.K.B. approximation (Christian and Hart, 1950). Here $R(r)$ is independent of energy and decreases asymptotically to zero:

$$V(r) = V_c(r) - \gamma V_t(r) - \frac{3}{r^2} + \left[\left\{\gamma V_t(r) + \frac{3}{r^2}\right\}^2 + 8\{\gamma V_t(r)\}^2\right]^{\frac{1}{2}}, \qquad (10.73)$$

where the centrifugal potential is usually large compared to $V_t(r)$. Thus approximately,

$$V(r) = V_c(r) + \tfrac{4}{3}[r\gamma V_t(r)]^2. \qquad (10.74)$$

Calculations of scattering which use the W.K.B. approximation to find the equivalent central potentials appropriate to each L and J value, neglect the coupling between states and do not differentiate between states of different magnetic quantum number m_s.

308   NUCLEAR SCATTERING

However, results for the Yukawa potential differ from more accurate calculations by only 10–20 %, so that the W.K.B. approximation provides an easy method of evaluating scattering by non-central forces.

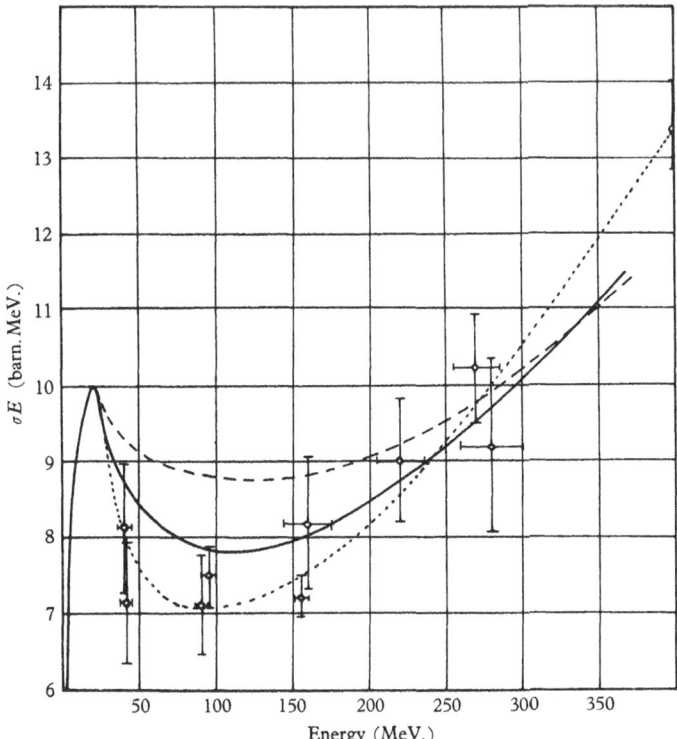

Fig. 76. (Energy) × (total cross-section) plotted as a function of energy for *n-p* scattering.
Dotted line: smoothed out experimental curve. The data below 24 MeV. is from Sleator (1947); for the data above 40 MeV. see § 10.10.
Full line: curve for an exponential potential,

$$V = (-69 \text{ MeV.}) (1 + 1 \cdot 84 S_{12}) \exp (-r/R), \quad \text{with} \quad R = 0 \cdot 75 \times 10^{-13} \text{ cm.}$$

Broken line: curve for a Yukawa non-central potential,

$$V = (-25 \cdot 3 \text{ MeV.}) (1 + 1 \cdot 91 S_{12}) \exp (-r/R)/(r/R), \quad \text{with} \quad R = 1 \cdot 35 \times 10^{-13} \text{ cm.}$$

Fig. 75 shows the *n-p* scattering resulting from non-central Serber type potentials of exponential and Yukawa shape. Just as for the central force calculations, the exponential potential gives smaller cross-sections than the Yukawa, although the scattering distributions of the latter have a better general shape. The agreement is best at the higher energies as both potentials predict too much scattering at 40 MeV. On the whole, the more detailed agreement with experiment is obtained with the exponential potential. This is further illustrated in fig. 76, where σE is plotted against E in the range 5–400 MeV. The exponential non-central potential

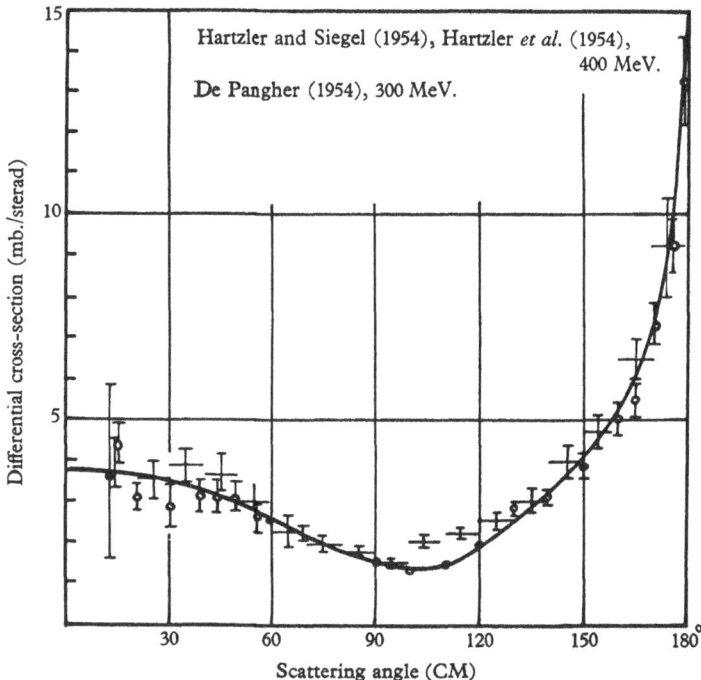

Fig. 77. Experimental angular distributions for n-p scattering, normalized to give a total cross-section of 35 mb. at 300 MeV. and 33·6 mb. at 400 MeV.

gives a much better overall fit to the mean experimental curve than the Yukawa non-central model. It is worth noting that for any given well shape, the minimum cross-sections occur with Serber exchange forces, owing to the absence of odd state potentials.

Experiments at 300 and 400 MeV., illustrated in fig. 77, involve more accurate measurements extending to angles as small as 13° and show a marked departure from the scattering symmetry around 90° characteristic of Serber type forces. In particular, differential cross-sections at 400 MeV. (Hartzler and Siegel, 1954; Hartzler, Siegel and Opitz, 1954) have a value of about 13·3 mb./sterad. at 180° and a probable extrapolated value of 3·8 mb./sterad. at 0°, leading to a ratio of forward to back scattering of about 3·5. The minimum of scattering is in the neighbourhood of 100°. The curve for 300 MeV. shows similar characteristics but with a scattering minimum near 90° (De Pangher, 1954).

10.11. High energy p-p scattering

The p-p scattering distribution at 32 MeV. is characterized by a dominant Coulomb term for $\theta < 20°$ and Coulomb-nuclear inter-ference for $20° < \theta < 50°$. The distribution near 90° has a Coulomb scattering amplitude so small that the result is almost pure nuclear scattering (fig. 78). The 32 MeV. data are in good agreement with pure s-wave scattering, a mean fit to the experiments (Panofsky and Fillmore, 1950; Cork et al. 1950) being given by $\delta_0 = 50·37°$. This cannot be reconciled with any exchange potential, for the large potential range required by the low energy data gives a d-phase leading to a cross-section with a dip between 50 and 90°, instead of the experimental rise. Inclusion of either a repulsive or attractive central triplet force for odd l only aggravates the disagreement, for near 90° where Coulomb forces are unim-portant, the p-wave partial cross-section has the form $\cos^2 \theta$ and is thus always additive.

Results at higher energies up to about 400 MeV. display an almost isotropic distribution (see fig. 78). The 340 MeV. result gives a differential cross-section of almost 4 mb./sterad., which is con-siderably greater than the conservation theorem maximum for elastic s-wave scattering, viz. $1/k^2 = \lambda^2 = 2·6$ mb./sterad. The explanation of pure s-wave scattering is thus ruled out. The results are similar to hard sphere scattering, but at 340 MeV. the necessary condition that the wavelength of the proton be very much smaller than the size of the scattering region is not satisfied, for the hard sphere radius required ($0·8 \times 10^{-13}$ cm.) is barely twice the proton wavelength. Hence 340 MeV. is too small an energy for hard sphere scattering; moreover the isotropic angular distribution extends down to energies as low as 75 MeV., completely invalidating this argument.

Both charge independence and central triplet forces appear to be ruled out by the 32 MeV. data. However, a non-central force leads to cross-sections which can be peaked at 90° for any potential model by choosing the range correctly. Thus addition of its con-

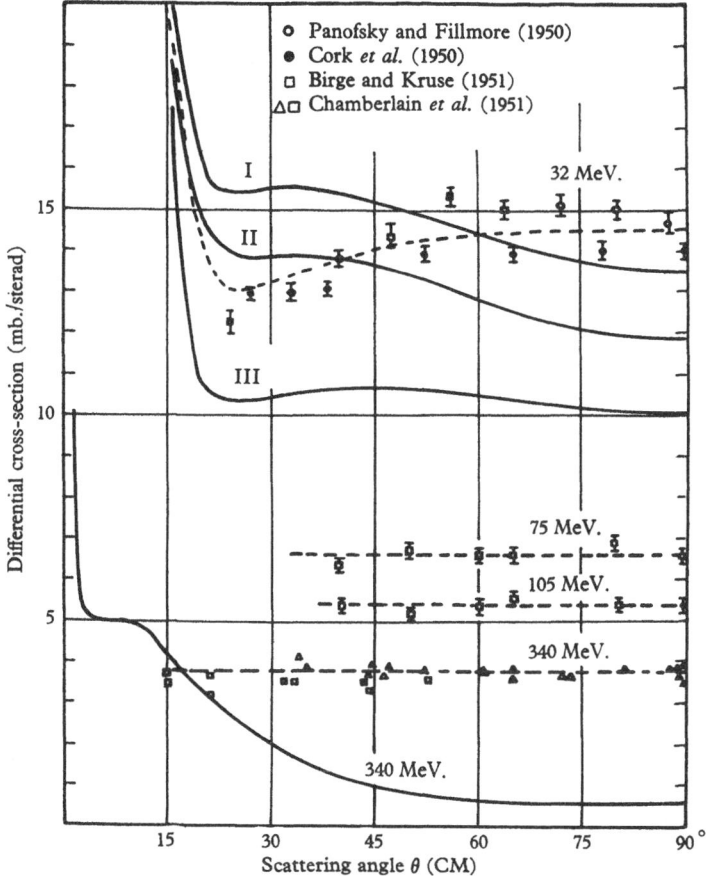

Fig. 78. Experimental p-p angular distributions. The broken curve at 32 MeV. represents pure s-wave scattering with a phase shift of $51 \cdot 15°$ as predicted by a Yukawa potential of range $1 \cdot 1417 \times 10^{-13}$ cm. and depth $49 \cdot 350$ MeV.
I. $s+d$ scattering for the above Yukawa potential, $\delta_0 = 51 \cdot 15°$, $\delta_2 = 1 \cdot 40°$.
II. $s+d$ scattering for an exponential potential, range $0 \cdot 7088 \times 10^{-13}$ cm., depth $108 \cdot 27$ MeV., $\delta_0 = 47 \cdot 54°$, $\delta_2 = 1 \cdot 20°$.
III. $s+d$ scattering for a spherical well potential, range $2 \cdot 615 \times 10^{-13}$ cm., depth $13 \cdot 273$ MeV., $\delta_0 = 41 \cdot 99°$, $\delta_2 = 0 \cdot 77°$.
The singlet scattering illustrated at 340 MeV. is similar for all potentials and is much smaller than experiment for $\theta > 30°$.

tribution to the singlet cross-section, which dips at 90°, can yield an almost flat distribution.

Fig. 79 shows the singlet p-p scattering at 32 and 340 MeV. for Yukawa, exponential and spherical well shapes adjusted to fit the low energy scattering. The disagreement with experiment is particularly large at 340 MeV., as the predicted cross-section is peaked in the forward direction and falls away to a small fraction of the experimental value at 90°. The spherical well gives the smallest cross-section, so any attempt to mask the singlet behaviour by using a non-central triplet force should employ a potential of this cut-off type.

For the ^3P state, either a Yukawa or singular tensor force, the latter of the form $\exp\left(-r/R\right)/(r/R)^2$ as used by Christian and Noyes (1950), will reproduce the scattering (fig. 79), although results below 30° are larger than experimental values. The tensor force computations are made in the Born approximation only and may be in error by about 20 %. Exact calculations usually predict more nearly isotropic results than the Born approximation. The sign of the ^3P singular tensor force is undetermined in the latter.

10.12. The charge-independence hypothesis

The models so far used to explain high energy scattering have been different for n-p and p-p interactions, the former requiring a non-central, non-singular Serber potential with a long tail and the latter a cut-off singlet potential together with a singular tensor triplet potential. However, the presence of some odd state scattering in the 400 MeV. results makes it possible that an alternative, charge-independent model may yet be found. One possibility is to apply to n-p scattering the even singlet cut-off potential and odd triplet singular tensor forces used for p-p scattering, using them together with the zero odd singlet and Yukawa tensor even triplet state of fig. 75. As the singular tensor contribution at 340 MeV. is an almost constant cross-section of 4 mb./sterad. and the differential cross-section for a particular state is four times larger for the p-p system, the contribution to the n-p total cross-section is $4 \times 4\pi \times \frac{1}{4}$ mb. Thus the n-p cross-section, already 102 mb. at 90 MeV. for the Yukawa tensor model with Serber forces, is

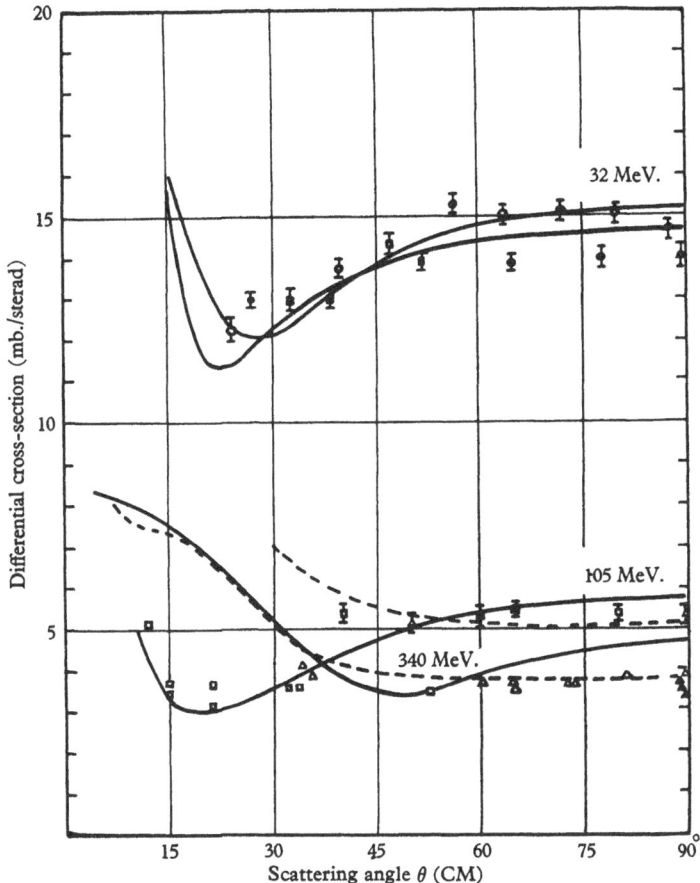

Fig. 79. p-p scattering at various energies.

Broken lines: scattering for the combination of a spherical well, range $2\cdot615 \times 10^{-13}$ cm. and depth $13\cdot273$ MeV. in the singlet state, with a singular tensor triplet,

$$V = (-18 \text{ MeV.}) \, S_{12} \exp (-r/R)/(r/R)^2, \quad R = 1\cdot6 \times 10^{-13} \text{ cm.}$$

Full lines: scattering for a repulsive core potential in the singlet state,

$$V = \infty, \quad r < r_0, \quad V = (-375 \text{ MeV.}) \exp \{-(r+r_0)/r_s\}, \quad r > r_0,$$

where $\qquad r_0 = 0\cdot60 \times 10^{-13}$ cm., $\quad r_s = 0\cdot40 \times 10^{-13}$ cm.

In odd triplet states the potential is

$$V = (27\cdot6 \text{ MeV.}) \, S_{12} \exp (-r/r_t), \quad r_t = 0\cdot75 \times 10^{-13} \text{ cm.}$$

The repulsive core of radius $0\cdot2 \times 10^{-13}$ cm. is left out (Christian, 1952).

increased by a further 12 mb., in contrast with an experimental value of 79 mb. Thus although the angular distribution largely retains its shape, the Serber result of fig. 75 is a minimum cross-section already and the introduction of odd triplet forces only worsens the agreement with total cross-section measurements.

Jastrow (1951) has suggested that at short distances the central potential becomes strongly repulsive, represented by a hard sphere of radius 0.6×10^{-13} cm. in singlet states and 0.2×10^{-13} cm. in triplet states. By using a singlet Serber potential outside the hard sphere and a non-central triplet potential of arbitrary exchange type, both with exponential radial dependence, Jastrow obtains agreement with the p-p results at 32, 105 and 340 MeV. (fig. 79). At 340 MeV., this is due to cancellation of a sharp minimum of the singlet scattering at about 30–60° by a maximum of triplet tensor scattering. The n-p scattering at 260 MeV. is fitted as far as the measurements extend ($\theta > 60°$) but an unconvincing scattering peak is obtained at about 25°. However, introduction of relativistic corrections indicates that the cancellation of the singlet scattering minimum by a triplet tensor scattering maximum may no longer occur, so that the scattering isotrophy is destroyed.

Case and Pais (1950) have suggested another model involving spin-orbit coupling in the triplet odd parity states, together with a strongly singular potential. Born approximation calculations show that agreement with p-p experiments is at best no better than with the singular tensor model; moreover satisfactory calculations on n-p scattering have not been carried out.

Fig. 80 illustrates experimental differential cross-sections for p-p scattering at very high energies (Smith et al. 1955; Cork, 1955). Even at 440 MeV. a small rise in cross-section towards forward scattering is noticeable, but at higher energies a very strong peak of forward scattering develops. Measurements by Shapiro et al. (1954) show the total p-p cross-section, including inelastic contributions from meson production, is roughly twice the elastic cross-section. This, together with the strong forward peaking of the elastic p-scattering, suggests that the latter is primarily shadow scattering associated with the strong absorption of the incident proton wave via meson production. The latter greatly reduces the large scale scattering near 90° where large momentum transfers are possible. Such an effect could be roughly accounted for by the optical model (§ 12.4).

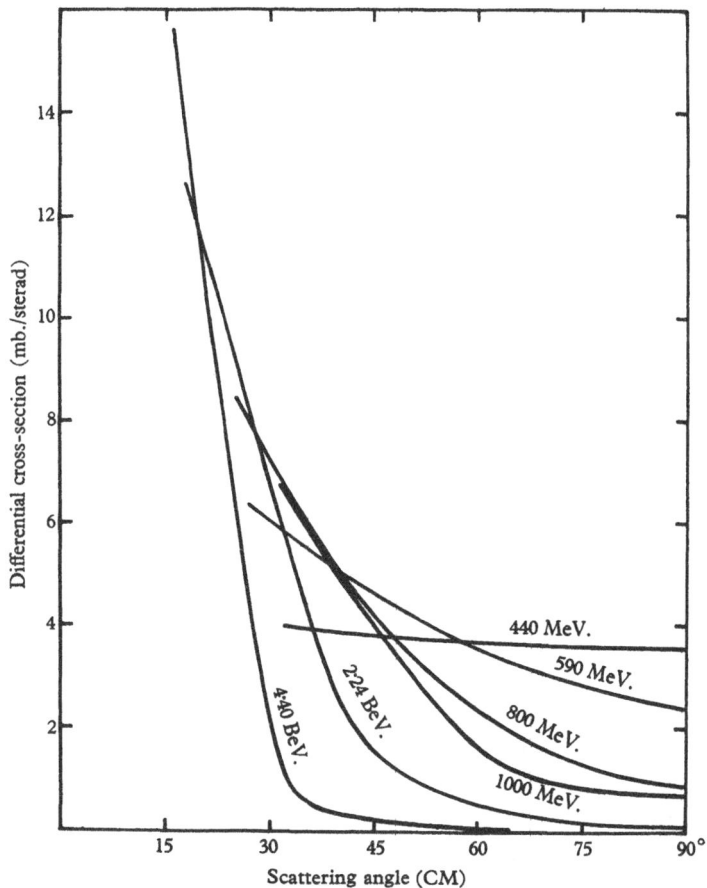

Fig. 80. Angular distributions for elastic p-p scattering for proton energies of 0·44, 0·59, 0·8 and 1·0 BeV. (Smith *et al.* 1955) and 2·24, 4·40 BeV. (Cork and Wenzel, 1955). The curves are normalized to the value 3·49±0·17 mb./sterad. at 90° (Smith *et al.* 1955) as measured by Sutton *et al.* (1955) at 437 MeV.

An interesting consequence of the charge independence hypothesis follows from § 10.8, where it is seen that if the n-p scattering amplitudes in triplet and singlet states are $f(\theta, \phi)$ and $g(\theta)$ respectively, then the corresponding p-p amplitudes are

$$f(\theta, \phi) - f(\pi - \theta, \pi + \phi) \quad \text{and} \quad g(\theta) + g(\pi - \theta).$$

The maximum possible amplitudes at $\theta = 90°$ are thus $2f(\tfrac{1}{2}\pi, \phi)$ and $2g(\tfrac{1}{2}\pi)$ respectively so that the following inequality holds:

$$d\sigma_{p-p}(90°)/d\sigma_{n-p}(90°) \leqslant 4. \qquad (10.75)$$

At 400 MeV. the ratio is 4 mb./1·5 mb. $\approx 2\cdot 7$. All other measurements have likewise been found to obey (10.75).

10.13. Summary of high energy results and conclusions

For summaries of high energy n-p data at $\sim$ 100 MeV., see Stahl and Ramsey (1954); Randle, Skyrme, Snowden, Taylor, Uridge and Wood (1956); Chih and Powell (1957); 300–600 MeV., De Pangher (1955); Hartzler and Siegel (1954); Hartzler et al. (1954); Dzhelepov, Kazarinov, Golovin, Fljagin and Satarov (1956); 0·41–2·6 BeV., Chen, Leavitt and Shapiro (1956).

High energy p-p data exists at 30 MeV., Panofsky and Fillmore (1950); Cork et al. (1950); 40–100 MeV., Kruse, Teem and Ramsey (1954); 100–400 MeV., Chamberlain and Wiegand (1950); Chamberlain et al. (1951); Towler (1952); Marshall et al. (1953); Mott et al. (1953); Chamberlain and Garrison (1954); Chamberlain, Pettengill, Segrè and Wiegand (1954); Fischer and Goldhaber (1954); Marion, Bonner and Cook (1955); Sutton et al. (1955); Taylor and Wood (1956); Batson and Riddiford (1956); 0·4–1·0 BeV., Smith et al. (1955); Mescerjakov, Bogacev and Neganov (1956); Morris, Fowler and Garrison (1956); Chamberlain and Clark (1956); Duke, Lock, March, Gibson, McEwen, Hughes and Muirhead (1957) and 1–3 BeV., Smith et al. (1955); Chamberlain and Clark, (1956); Cester, Hoang and Kernan (1956); Fowler, Shutt, Thorndike and Whittemore (1956); Block, Harth, Coccini, Hart, Fowler, Shutt, Thorndike and Whittemore (1956).

Attempts at phase shift analysis have been singularly uninformative as they have invariably led to a number of sets of alternative phase values (cf. Garren (1953); Beretta, Clementel and Villi (1955); Ohnuma and Feldman (1956); Stapp, Ypsilantis and Metropolis (1957)). For details of attempts at theoretical interpretation of high energy scattering, see Christian (1952); Fowler, Shutt, Thorndike, Whittemore, Cocconi, Hart, Blochs, Harth, Fowler, Garrison and Morris (1956); Gammel, Christian and Thaler (1956);

Geissler (1956); Randle *et al.* (1956); Gelernter (1957); Signell and Marshak (1957).

Rough agreement with experimental n-p results up to 260 MeV. has been obtained using static potentials with a Serber exchange force. However differential cross-sections at the lower end of the high energy range (40 MeV.) are 20 % too high in absolute magnitude if theory is adjusted to fit experiment at 260 MeV.; moreover, experimental results at the latter energy do not extend to sufficiently small angles. In fact results at 300 and 400 MeV. show a ratio of backward to forward scattering of about 3, so that the Serber exchange model is not correct at these energies and odd harmonic components must be present. Fourier analysis indicates that at least an eighth harmonic term, probably even a tenth, is necessary to fit the 400 MeV. data. Hence phase shifts up to $l = 4$ are present. However, a complete analysis is not available.

Similar agreement to within 20 % with p-p scattering experiments is found using static potentials, the best results being obtained using a singular tensor triplet potential. However, it has not proved possible to find a charge-independent model giving as good an agreement as with charge-dependent models, but this may be due to insufficient theoretical work.

The overall picture is that high energy experiments have not enabled a comprehensive understanding of nuclear forces. Neither the radial dependence nor exchange nature of the forces has been settled satisfactorily; also saturation is not explained in a manner consistent with high energy scattering. No model gives complete agreement with experiment from low energies to relativistic energies.

10.14. Polarization of nucleons by high energy scattering

If the n-p or p-p interactions include appreciable spin-orbit terms, examples being spin-orbit coupling $(\mathbf{S}.\mathbf{L})$ and tensor (S_{12}) forces, then the spin-orbit force can cause a neutron or proton incident with a given spin direction to reverse this direction on being scattered. The scattered wave consists of two parts representing particles with unchanged and reversed spin directions respectively. Their amplitudes have different angular dependence so that for a given scattering angle there is some net spin direction. Polarization is limited for tensor forces to triplet states and for spin-orbit coupling to triplet states with $l \geqslant 1$. Its effect on scattering has been treated by many authors (Wolfenstein, 1949 a, b; Lepore, 1950; Wolfenstein, 1951; Dalitz, 1952; Goldfarb and Feldman, 1952; Wolfenstein and Askin, 1952; Swanson, 1953; Hull and Saperstein, 1954; Breit, Ehrman and Hull, 1955; Tamor, 1955; Eriksson, 1956; Barker, 1956; Takano and Hull, 1957).

10.14.1. *Polarization by spin-orbit forces in n-p scattering.* n-p scattering can be treated by a theory similar to that of §8.7.1 for particles of zero spin. The incident beam is described by equation (8.58) and the scattering solution by equation (8.60), taking $\alpha = 0$. The polarization of the incident beam is given by equation (8.59). From equation (10.33), the incident beam (8.58) may be expanded in terms of its constituent orbital angular momenta:

$$\psi_{\rm inc} = \frac{(4\pi)^{\frac{1}{2}}}{kr} \sum_{l=0}^{\infty} \sum_{j=l-1}^{l+1} i^l c_{jlM0} (2l+1)^{\frac{1}{2}} g_l(r) F_{jlM}, \qquad (10.76)$$

where $g_l(r) \sim \sin(kr - \tfrac{1}{2}l\pi)$, $r \to \infty$.

There are three possible j values for a given l, viz. $j = l+1$, l, $l-1$, so we express the incident and scattered beams in terms of the following projection operators:

$$\left.\begin{array}{l}
\pi_l^+ = \dfrac{1}{(2l+1)} \left[1 + \dfrac{1}{2}\left(\dfrac{l+2}{l+1}\right)(\boldsymbol{\sigma}.\mathbf{L}) + \dfrac{1}{4(l+1)}(\boldsymbol{\sigma}.\mathbf{L})^2 \right], \\[3mm]
\pi_l^0 = \left[1 - \dfrac{1}{2l(l+1)}(\boldsymbol{\sigma}.\mathbf{L}) + \dfrac{1}{4l(l+1)}(\boldsymbol{\sigma}.\mathbf{L})^2 \right], \\[3mm]
\pi_l^- = \dfrac{1}{(2l+1)} \left[-1 - \left(\dfrac{l-1}{2l}\right)(\boldsymbol{\sigma}.\mathbf{L}) + \dfrac{1}{4l}(\boldsymbol{\sigma}.\mathbf{L})^2 \right],
\end{array}\right\} \qquad (10.77)$$

where $\boldsymbol{\sigma}.\mathbf{L} = l$, -1, and $-(l+1)$ in units of $\hbar^2$ for $j = l+1$, l and $l-1$ respectively. π_l^+ is so constructed that when applied to functions of the type $c_{jlM0} F_{jlM}$, it destroys all states for which $j = l$, $l-1$ and selects out states for which $j = l$ and $l-1$ respectively.

We write the wave function (10.76) in the asymptotic form

$$\psi = \frac{(4\pi)^{\frac{1}{2}}}{kr} \sum_{l=0}^{\infty} i^l (2l+1)^{\frac{1}{2}} [A_l^+ \, \pi_l^+ f_l^+ \, c_{l+1,l,M,0} \, F_{l+1,l,M}$$
$$+ A_l^0 \pi_l^0 f_l^0 c_{l,l,M,0} F_{l,l,M} + A_l^- \pi_l^- f_l^- c_{l-1,l,M,0} F_{l-1,l,M}], \qquad (10.78)$$

where A_l^+, A_l^0 and A_l^- are constants determined by the conditions that equation (10.78) have the same asymptotic form as equation (8.60) with $\alpha = 0$, and the scattering wave function satisfy

$$f_l(r) \sim \sin(kr - \tfrac{1}{2}l\pi + \delta_l), \quad r \to \infty.$$

The scattered wave $\psi_{\text{scatt}} = \psi - \psi_{\text{inc}}$ becomes

$$\psi_{\text{scatt}} \sim \frac{(4\pi)^{\frac{1}{2}}}{kr} \exp(ikr) \sum_{l=0}^{\infty} \frac{1}{(2l+1)^{\frac{1}{2}}} \left[\{H^+ + (2l+1)H^0 - H^-\} \right.$$

$$+ (\boldsymbol{\sigma}.\mathbf{L}) \left\{ \frac{1}{2}\left(\frac{l+2}{l+1}\right)H^+ - \frac{(2l+1)}{2l(l+1)}H^0 - \left(\frac{l-1}{2l}\right)H^- \right\}$$

$$\left. + (\boldsymbol{\sigma}.\mathbf{L})^2 \left\{ \frac{1}{4(l+1)}H^+ - \frac{(2l+1)}{4l(l+1)}H^0 + \frac{1}{4l}H^- \right\} \right], \quad (10.79)$$

where $H = c_{jlM0} Q_{jlM} F_{jlM}$,

$$Q_{jlM} = [\exp(2i\delta_{jlM}) - 1]/2i = \exp(i\delta_{jlM}) \sin\delta_{jlM}. \quad (10.80)$$

If $\mathbf{k}$ is the direction in which scattering is observed, then writing $\mathbf{L}$ in operator form in units of $\hbar$,

$$\mathbf{L} = \mathbf{r} \times \mathbf{p} = -i\mathbf{r} \times \frac{\partial}{\partial \mathbf{r}},$$

we have the relation

$$\boldsymbol{\sigma}.\mathbf{L} = -i\sin\theta \frac{\partial}{\partial(\cos\theta)} \boldsymbol{\sigma}.\mathbf{n}, \quad (10.81)$$

where $\mathbf{n}$ is the normal to the scattering plane defined by

$$\mathbf{k} \times \mathbf{k}_0 = \mathbf{n}k^2 \sin\theta. \quad (10.82)$$

The scattering amplitude defined by equation (8.60) becomes

$$f(\theta) = A_1(\theta) + (\boldsymbol{\sigma}.\mathbf{n}) B(\theta) + (\boldsymbol{\sigma}.\mathbf{n})^2 A_2(\theta)$$

$$= A(\theta) + (\boldsymbol{\sigma}.\mathbf{n}) B(\theta), \quad (10.83)$$

where $(\boldsymbol{\sigma}.\mathbf{n})^2 = 1$, $A(\theta) = A_1(\theta) + A_2(\theta)$ and

$$A_1(\theta) = \frac{1}{k}(4\pi)^{\frac{1}{2}} \sum_{l=0}^{\infty} \frac{1}{(2l+1)^{\frac{1}{2}}} [H^+ + (2l+1)H^0 - H^-],$$

$$A_2(\theta) = -\frac{1}{k}(4\pi)^{\frac{1}{2}} \sin^2\theta . \frac{\partial^2}{\partial z^2} \sum_{l=0}^{\infty} \frac{1}{(2l+1)^{\frac{1}{2}}}$$

$$\times \left[\frac{1}{4(l+1)}H^+ - \frac{(2l+1)}{4l(l+1)}H^0 + \frac{1}{4l}H^- \right],$$

$$A_3(\theta) = -\frac{i}{k}(4\pi)^{\frac{1}{2}} \sin\theta . \frac{\partial}{\partial z} \sum_{l=0}^{\infty} \frac{1}{(2l+1)^{\frac{1}{2}}}$$

$$\times \left[\frac{1}{2}\left(\frac{l+2}{l+1}\right)H^+ - \frac{1}{2l}\left(\frac{2l+1}{l+1}\right)H^0 - \left(\frac{l-1}{2l}\right)H^- \right],$$

$$(10.84)$$

where $z = \cos \theta$. The differential cross-section is given by equation (8.75) and the polarization of the scattered beam by equations (8.77) and (8.78). Application of polarization theory to double and triple scattering follows on identical lines to §8.7.1.

A more explicit expression for the polarization (8.77) follows using equations (10.84), (10.80), (10.33) and (10.34). Thus

$$H = c_{jlM0} Q_{jlM} \sum_{m=M-1}^{m=M+1} c_{jlMm}$$

$$\times \left\{ \frac{(2l+1)}{4\pi} \frac{(l-m)!}{(l+m)!} P_l^m(\cos \theta) e^{im\phi} \chi_{M-m}(s_1, s_2) \right\}, \quad (10.85)$$

where the Clebsch–Gordan coefficients are given in Table XVII. Breit *et al.* (1955) obtain the polarization by single scattering of an initially unpolarized beam as

$$k^2 P(\theta) = \tfrac{1}{2} \sin \theta . \operatorname{Im} (\alpha_1 \alpha_2^* - \sin^2 \theta . \alpha_3^* \alpha_1 + \alpha_4^* \alpha_5), \quad (10.86)$$

where the α's are summations defined below in §12.14.2, extending over all triplet states for both odd and even l. Im denotes the imaginary part of the expression in brackets.

10.14.2. *Polarization by spin-orbit forces in p-p scattering.* The polarization of protons scattered by hydrogen may be treated in a similar manner. Using the notation of §10.14.1, the incident wave becomes

$$\psi_{\mathrm{inc}} = \{ \exp i(\mathbf{k}_0 . \mathbf{r} - \ln 2kr) - \exp -i(\mathbf{k}_0 . \mathbf{r} - \ln 2kr) \} \chi_{\mathrm{inc}}, \quad (10.87)$$

where χ_{inc} is a triplet wave function. The full wave function is

$$\psi \sim \psi_{\mathrm{inc}} + \frac{1}{r} \exp i(\mathbf{k}_0 . \mathbf{r} - \ln 2kr) [f(\theta) - f(\pi - \theta)] \chi_{\mathrm{inc}}. \quad (10.88)$$

The formulae of §10.14.1 may be employed, subject to adding the Coulomb amplitude $f_c(\theta)$ to eq. (10.82), by making the transition

$$g_l(r) \to g_l'(r) \sim \sin (kr - \tfrac{1}{2}l\pi - \alpha \ln 2kr + \sigma_l),$$

$$f_l(r) \to f_l'(r) \sim \sin (kr - \tfrac{1}{2}l\pi - \alpha \ln 2kr + \sigma_l),$$

$$F_{jlM} \to F_{jlM}(\theta, \phi) - F_{jlM}(\pi - \theta, \pi + \phi),$$

$$\exp (ikr) \to \exp i(kr - \alpha \ln 2kr).$$

Breit *et al.* (1955) give as the polarization by single scattering the expression

$$k^2 P(\theta) = 2 \sin \theta . \operatorname{Im} \{ \alpha_1(\alpha_2^* + \alpha_c^*) - \sin^2 \theta . \alpha_3^* \alpha_1 + \alpha_4^*(\alpha_5 + \alpha_c) \}, \quad (10.89)$$

where the six quantities α_1, α_2, ..., α_5, α_c are as follows:

$$\alpha_1 = -\Sigma e_{l0}[l(l+2) Q_{l+1,l} - (2l+1) Q_{l,l} - (l^2-1) Q_{l-1,l}] P'_l/l(l+1)$$
$$- 2B[(l'+1) P'_{l'+2} - (l+2) P'_l],$$

$$\alpha_2 = \Sigma \tfrac{1}{2} e_{l0}[(l+2) Q_{l+1,l} + (2l+1) Q_{l,l} + (l-1) Q_{l-1,l}] P_l$$
$$- (l'+1)(l'+2) B(P_{l'+2} + P_{l'}),$$

$$\alpha_3 = \Sigma \tfrac{1}{2} e_{l0}[l Q_{l+1,l} - (2l+1) Q_{l,l} + (l+1) Q_{l-1,l}] P''_l/l(l+1)$$
$$- B(P''_{l'+1} + P''_l),$$

$$\alpha_4 = \Sigma e_{l0}[Q_{l+1,l} - Q_{l-1,l}] P'_l - 2B[(l'+1) P'_{l'+2} - (l'+2) P'_{l'}],$$

$$\alpha_5 = \Sigma e_{l0}[(l+1) Q_{l+1,l} + l Q_{l-1,l}] P_l + 2B(l'+1)(l'+2)(P_{l'+2} + P_{l'}),$$

$$\alpha_c = \tfrac{1}{4}\alpha[-s^2 \exp(-i\alpha \ln s^2) + c^{-2} \exp(-i\alpha \ln c^2)],$$

$$(10.90)$$

where $Q_{j,l}$ is given by equation (10.80) and

$$e_{l0} = \exp(2i\eta_{l0}) = \exp 2i[\arctan \alpha/l + \arctan \alpha/(l-1) + ... + \arctan \alpha],$$

$$s = \sin \tfrac{1}{2}\theta, \quad c = \cos \tfrac{1}{2}\theta, \quad P'_l = dP_l/dz, \quad P''_l = d^2P_l/dz^2.$$

If there is no coupling between states of different l for the same j then the constant $B = 0$. If states with $j = l'+1$ having values $l = l'$ and $l = l'+2$ are coupled to each other, then

$$B = \tfrac{1}{2}C_{l',l'+2}(l'+1)^{-\frac{1}{2}}(l'+2)^{-\frac{1}{2}} \exp[i\eta_{l',0} + i\eta_{l'+2,0}], \quad (10.91)$$

where

$$C_{l,l+2} = \frac{(\tan \delta_l \tan \delta_{l+2})^{\frac{1}{2}}}{(\tan \delta_l + \tan \delta_{l+2})} \exp i(\delta_l + \delta_{l+2})$$

and $Q_{l+1,l}$, $Q_{l+1,l+2}$ are modified by replacing them by $Q_{l+1,l} + C_{l,l}$, $Q_{l+1,l+2} + C_{l+2,l+2}$ respectively. The sums over l are taken over odd values only.

The formula for polarization may be applied to n-p scattering by multiplying equation (10.86) by a factor of $\tfrac{1}{4}$, extending the sum over both even and odd triplet values and putting $\alpha_c = 0$, $e_{l0} = 1$. A more specific expression for polarization in single scattering, taking into account p- and f-waves, is also available (Hull and Saperstein, 1954).

10.14.3. *High energy scattering of polarized beams.* Polarization experiments (Oxley *et al.* 1954; Simmons, 1956; Chamberlain, Segrè, Tripp, Wiegand and Ypsilantis, 1957; Baskir, Hafner,

Roberts and Tinlot, 1957) provide a test for nuclear force models supplementary to the angular distribution curves for high energy n-p and p-p scattering. Oxley et al. (1954) have found for p-p scattering at 200 MeV. the values

$$P_H = 22 \pm 4\%, \quad 2\delta = 9\cdot6 \pm 3\cdot5\%.$$

Theoretical calculations have also been carried out for specific models of the p-p interaction (Goldfarb and Feldman, 1952; Swanson, 1953). At 240 MeV., the asymmetry 2ε, allowing for the angular acceptance of the apparatus, is $0\cdot5\%$ for the Jastrow hard core model, 13% for the Christian–Noyes tensor model and 30% for the Case–Pais $\mathbf{L}.\mathbf{S}$ coupling model.

Thus the Jastrow hard core model is inconsistent with polarization results, which appear to need the singular tensor forces characteristic of the Christian–Noyes and Case–Pais models. Unfortunately, accurate calculations (Goldfarb and Feldman, 1952) show that only the hard core model is capable of giving a flat cross-section down to small angles. The other models give too large a peak of forward scattering for $\delta < 40°$.

Fig. 81 shows the asymmetry 2ε plotted as a function of θ at 240 MeV. for the several models, assuming $\theta_1 = \theta_2 = \theta$ and $k_1 = k_2 = k$. Thus $2\varepsilon = P(k_1, \theta_1) P(k_2, \theta_2) = P^2(k, \theta)$. The asymmetry is only roughly symmetrical about $45°$ but is negligible for the hard core model. The singular tensor model has a $1/r^2$ singularity at the origin, necessitating a cut-off procedure if a solution to the wave equation is to be obtained. This leads to an asymmetry increasing slightly with cut-off radius. The $\mathbf{L}.\mathbf{S}$ model gives nearly twice the asymmetry for an attractive potential as for a repulsive potential of the same magnitude, which is not surprising as the former always gives much larger scattering phases than the latter.

Experiments with high energy neutrons are more difficult owing to the small polarizations obtainable from (p, n) reactions at a few hundred MeV. However Siegel et al. (1956) have used a 16% polarized neutron beam to measure the asymmetries in n-p scattering at 350 MeV. Fig. 82 shows the percentage polarization obtained as a function of CM neutron scattering angle.

Chamberlain et al. (1954, 1957) have used the alternative approach of bombarding deuterons with a highly polarized proton beam,

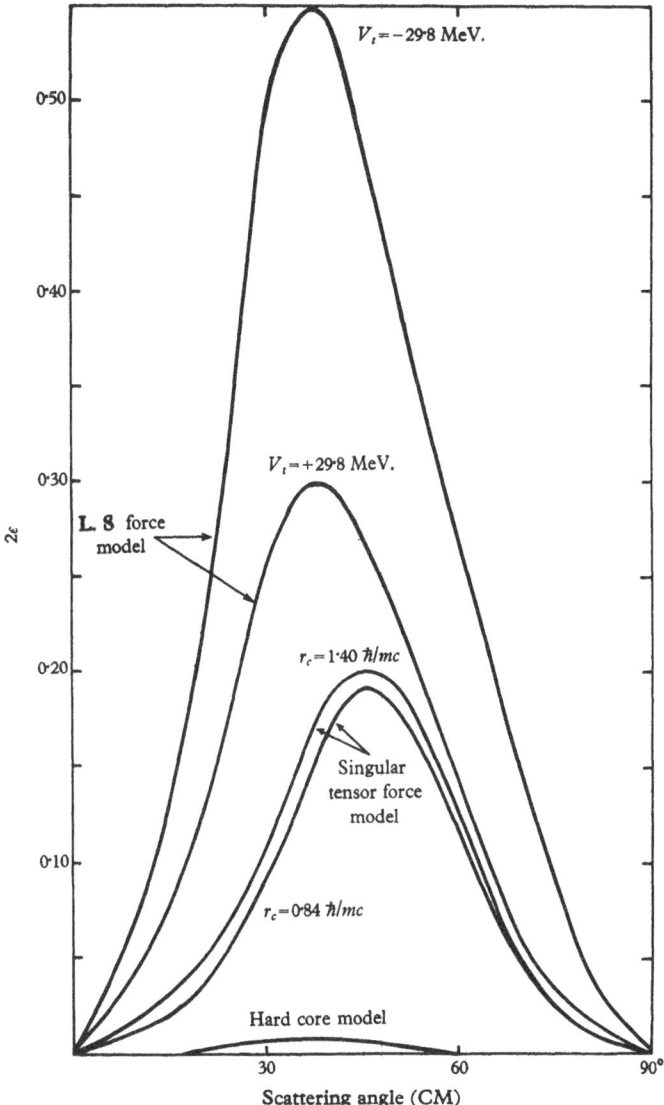

Fig. 81. Asymmetries 2ϵ in double p-p scattering at 240 MeV. for the hard core model, the singular tensor model with cut-offs $r_c = 0.84\hbar/mc$ and $1.40\hbar/mc$, and the L.S model for attractive and repulsive potentials $V = \pm 29.8$ MeV. (Goldfarb and Feldman, 1952; Swanson, 1953).

NUCLEAR SCATTERING

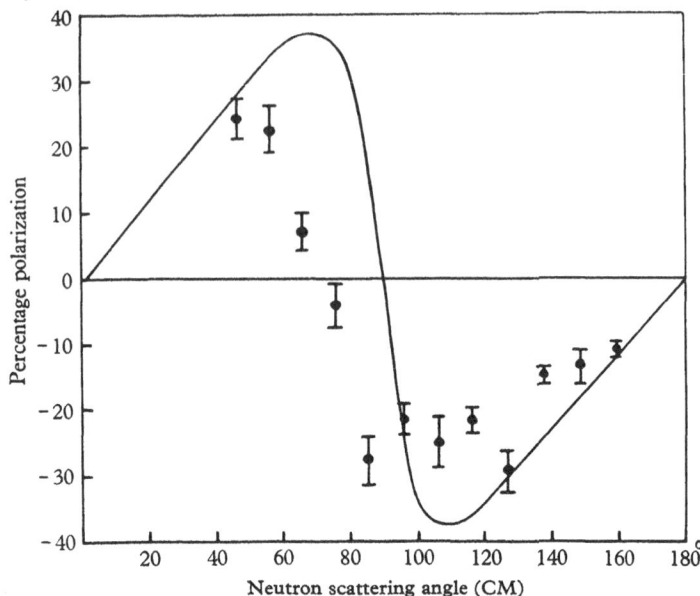

Fig. 82. Percentage polarization vs. CM scattering angle for n-p scattering at 350 MeV. (Siegel, Hartzler and Love, 1956). The theoretical curve (Swanson, 1953) uses the non-central Serber interaction of Christian and Hart (1950).

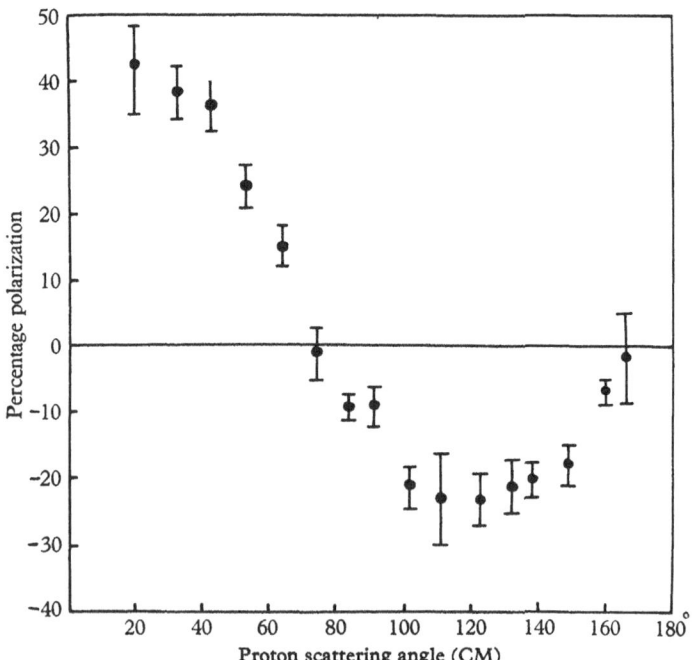

Fig. 83. Percentage polarization vs. scattering angle for p-n scattering at 310 MeV. (Chamberlain et al. 1954).

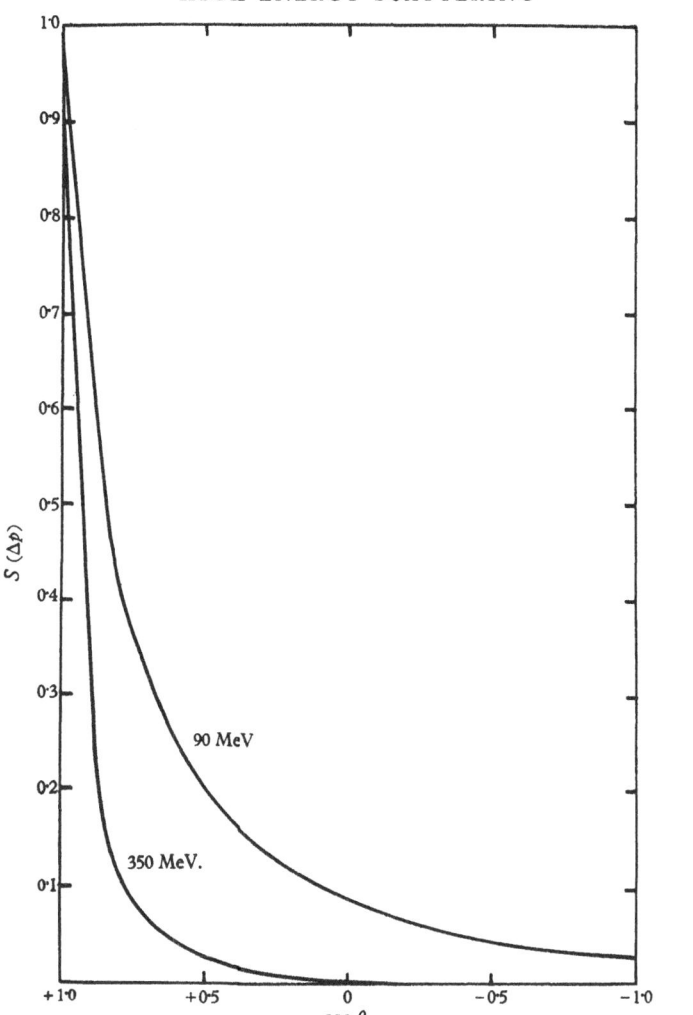

Fig. 84. The 'sticking factor' $S(\Delta p)$ as a function of the cosine of the CM scattering angle. A Yukawa potential of range $1 \cdot 18 \times 10^{-13}$ cm. is assumed in the calculation (Chew, 1948).

neglecting the deuteron binding energy. They assume the p-p and n-p scattering involved is such that a p-p or n-p pair escapes at approximately 90° (Lab), apart from elastic scattering of neutrons by the deuteron. The percentage of polarization obtained for p-n scattering at 310 MeV. is shown in fig. 83 as a function of CM scattering angle.

The polarization curves for the two experiments show a striking similarity, interpreted as evidence for charge symmetry in nucleon-nucleon scattering. Some theoretical results are available for comparison, notably calculations of Swanson (1953) using the Serber interaction of Christian and Hart (1950). A curve for 250 MeV. interpolated from these calculations is shown in fig. 84, agreeing in magnitude if not in detailed shape with experiment.

CHAPTER 11

HIGH ENERGY SCATTERING OF NEUTRONS, PROTONS AND DEUTERONS BY LIGHT NUCLEI

11.1. High energy n-d and p-d scattering

At energies of the order of 100 MeV., n-d and p-d scattering is mostly inelastic owing to the large momentum transfers possible. Nevertheless, the ratio of elastic to inelastic collisions does not decrease indefinitely, but approaches a finite limit due to the most probable encounters being those involving an energy transfer of the order of $\mu c^2 \sim 140$ MeV., μ being the rest mass of the π-meson responsible for nuclear forces.

The passage of a beam of high energy neutrons through deuterium results in the production of fast and slow groups of projected deuterons and the inelastic production of two groups comprising fast and slow protons. These are explained as follows:

(a) The fast deuterons result from neutron exchange or pick-up reactions in which the incident neutron captures the proton, the resulting deuterons being observed at angles less than $41°$ in the Lab. system.

(b) The slow deuteron group is due to elastic scattering.

(c) The fast protons arise from charge exchange between the incident neutrons and the component protons within the deuterons. They are therefore produced at small angles in the forward direction.

(d) The slow protons result from the head-on collision of neutrons with the diffuse structure of the deuteron, stripping off the neutron and leaving the proton with the small momentum appropriate to its existence in the bound state.

High energy p-d collisions involve similar effects, although in general two groups of charged particles are produced. Moreover some of the projected particles are of the same type as the incident particles so that the distinction between fast and slow groups of protons is not so clear-cut as for n-d collisions. The classification thus becomes:

(a) Fast deuterons resulting from proton exchange accompanied by a slow proton group.

(b) A slow deuteron group together with fast elastically scattered protons.

(c) Two groups of slow protons resulting from charge exchange between the incident proton and a nuclear neutron, a beam of fast neutrons being produced. The proton which has gained its charge by exchange moves in general at a large angle to the incident beam, but the nuclear proton is left with the momentum it had at the moment of impact and therefore has a nearly isotropic distribution.

(d) A close encounter between an incident proton and a nuclear proton results in a high energy and a low energy proton, the former being the incident particle after small angle scattering and the latter the struck particle moving at a large angle to the incident beam with an energy just sufficient to break the deuteron bond.

Most of the work at high energies has been denoted to expressing the three-body cross-sections in terms of the two-body cross-sections. It is found that the total cross-section σ_{tot} for n-d collisions may be written as
$$\sigma_{\text{tot}} = (\mathrm{I} - \epsilon)\, \sigma_{n-p} + \sigma_{n-n} + I, \qquad (\mathrm{II.I})$$

where σ_{n-p} and σ_{n-n} are the cross-sections for scattering of neutrons by free protons and neutrons respectively at the same energy. I is an interference term arising mainly from elastic collisions, the scattering from the two nucleons being coherent in this case. ϵ is a correction due to the Pauli principle, occurring because of the exchange production of fast protons and slow neutrons by collision between the incident neutron and the nuclear proton. The low energy scattered neutron is restricted in phase space by the presence of the low energy nuclear neutron. At 90 MeV., ϵ is about 0·15.

One of the main purposes of high energy n-d and p-d scattering experiments is to see whether charge symmetry (the equality of p-p and n-n forces) holds in this region. At the comparatively low energies ($\sim$ 20 MeV.) characteristic of nucleons bound in nuclei, the study of binding energies and of mirror nuclei is well known to indicate charge symmetry (§§ 1.2.4, 1.4). However, at high energies so many surprises have occurred with p-p and n-p scattering that charge symmetry needs confirming in this region. If true, the n-d and p-d cross-sections should be equal, apart from small modifications introduced by Coulomb scattering in the latter case.

At energies above 200 MeV., the Born approximation is valid, although it is often used of necessity at energies as low as 90 MeV., where it is not very reliable. A number of theoretical calculations have been made (De Hoffman, 1950; Chew, 1948; Gluckstern and Bethe, 1951; Bransden, 1950, 1952), but only those of Bransden (1950, 1952) deal with non-central forces. Another method, known as the impulse approximation, has been developed by Chew (Chew and Steinberger, 1950; Chew, 1951 a, b) which is claimed as valid at energies as low as 20 MeV. However, Brueckner (1953) has argued that the neglect of multiple scattering involved in the impulse approximation may limit its usefulness to the Born approximation region. Nevertheless, in § 11.1.8 we shall show that above 30 MeV. the multiple scattering correction is probably not important.

11.1.1. *Simplified model for elastic scattering.* In the n-d scattering problem it is desirable to separate features involving the nature of nuclear forces from those reflecting the presence of two of the three particles in the bound state. If this separation is possible, it will be found by making the simplest possible assumptions about the interaction, combining these with the use of the Born approximation. One might then hope that the results would hold at least qualitatively in a more complete theory, which proves to be the case.

The elastic scattering probability may be considered as the product of two separate factors:

(a) The chance that the incident neutron may collide with either of the two deuteron particles, transferring a momentum $\mathbf{q}$ to that particle. This should be related to the n-p and n-n cross-sections.

(b) The probability of finding the two deuteron particles still in the ground state after the collision. This is called the 'sticking factor', $S(q)$.

The simplest model is one in which symmetry, spin and exchange forces are ignored. We take the incident and final neutron momenta as $\mathbf{k_0}$ and $\mathbf{k}$ respectively, with $\mathbf{q} = \mathbf{k_0} - \mathbf{k}$, where $q = 2k \sin \frac{1}{2}\theta$ and $|\mathbf{k_0}| = |\mathbf{k}| = k$. The initial state can be written in the CM system as the product of a plane wave and the ground state wave function $\Phi_0(R)$,

$$\Phi_0(R, r) = L^{-\frac{3}{2}} \phi_0(R) \exp(i\mathbf{k}_0 . \mathbf{r}), \qquad (11.2)$$

where $\mathbf{R} = \mathbf{r}_n - \mathbf{r}_p$, $\mathbf{r} = \mathbf{r}_i - \frac{1}{2}(\mathbf{r}_n + \mathbf{r}_p)$, $L = 2\pi$, $\mathbf{r}_n$, $\mathbf{r}_p$ are the coordinates of the component particles of the deuteron and $\mathbf{r}_i$ that of the incident particle.

The final state is similarly described by

$$\Phi_f(R, r) = L^{-\frac{3}{2}} \phi_0(R) \exp(-i\mathbf{k} . \mathbf{r}). \qquad (11.3)$$

The success of the plane wave approximation depends on the incident energy being sufficiently large compared to the scattering potential, so that the wave function of the neutron is not appreciably distorted by the latter. We shall use the first order perturbation expression for the differential cross-section, viz.

$$\frac{d\sigma}{d\omega} = \frac{\mu^2 L^3}{4\pi^2 \hbar^4} |H'_{f0}|^2, \qquad (11.4)$$

where $\mu = \frac{2}{3}m$ is the reduced mass of the system, and H'_{f0} is the matrix element describing the transition from the initial to the final state. H'_{f0} is directly proportional to the scattering amplitude f of Born approximation theory if plane waves are employed, viz. $f = (L^{\frac{3}{2}}/4\pi)(2\mu/\hbar^2) H'_{f0}$, and is given by

$$H'_{f0} = \int \Psi_f^* H_1 \Psi_0 \, d\tau. \qquad (11.5)$$

Here H_1 is the small perturbation of the incident energy producing the scattering, $H_1 = V_{nn}(|\mathbf{r} - \frac{1}{2}\mathbf{R}|) + V_{np}(|\mathbf{r} + \frac{1}{2}\mathbf{R}|). \qquad (11.6)$

Thus from equations (11.2) to (11·6), we obtain

$$H'_{f0} = L^{-3} \int d\mathbf{R} \int d\mathbf{r} \exp(i\mathbf{q} . \mathbf{r}) (V_{nn} + V_{np}) \phi_0^2(R). \qquad (11.7)$$

Writing $\mathbf{q} = \mathbf{r} - \frac{1}{2}\mathbf{R}$ and $\mathbf{r} + \frac{1}{2}\mathbf{R}$ for V_{nn} and V_{np} respectively,

$$H'_{f0} = L^{-\frac{3}{2}} [V_{nn}(\mathbf{q}) + V_{np}(\mathbf{q})] S^{\frac{1}{2}}(\mathbf{q}), \qquad (11.8)$$

where

$$S^{\frac{1}{2}}(\mathbf{q}) = \int d\mathbf{R} \exp(i\tfrac{1}{2}\mathbf{q} . \mathbf{R}) \phi_0^2(R), \qquad (11.9)$$

$$\left. \begin{aligned} V_{nn}(q) &= \int d\mathbf{y} \exp(i\mathbf{q} . \mathbf{y}) V_{nn}(y), \\ V_{np}(q) &= \int d\mathbf{y} \exp(i\mathbf{q} . \mathbf{y}) V_{np}(y). \end{aligned} \right\} \qquad (11.10)$$

The differential cross-section follows from equations (11.4) and (11.8):

$$\frac{d\sigma}{d\omega} = \frac{(\frac{2}{3}m)^2}{4\pi^2\hbar^4} \mid V_{nn}(q) + V_{np}(q) \mid^2 S(q). \tag{11.11}$$

The result contains the essential features needed. The sticking factor $S(\mathbf{q})$ is a function only of the deuteron configuration, approaching zero for very weak binding and unity for very strong binding. It is always unity for zero momentum transfer. The form of $S(\mathbf{q})$ is illustrated in fig. 84 as a function of the cosine of the CM scattering angle at Lab. energies of 90 and 350 MeV., employing a Yukawa potential of range $1 \cdot 18 \times 10^{-13}$ cm. and depth $67 \cdot 8$ MeV. for the deuteron. Forward scattering is clearly favoured.

The term multiplying $S(\mathbf{q})$ in equation (11.11) is the scattering probability and we wish to relate this to the corresponding two-body cross-sections. Using the same simplifications as above, the n-p scattering cross-section is

$$\frac{d\sigma_{n-p}}{d\omega} = \frac{(\frac{1}{2}m)^2}{4\pi^2\hbar^4} \mid V_{np}(\mathbf{q'}) \mid^2, \tag{11.12}$$

where $\mathbf{q'}$ is the momentum transfer for the n-p system. Similarly

$$\frac{d\sigma_{n-n}}{d\omega} = \frac{(\frac{1}{2}m)^2}{4\pi^2\hbar^4} \mid V_{nn}(\mathbf{q'}) \mid^2. \tag{11.13}$$

For a given scattering angle and CM scattering energy, the ratio of the corresponding energies in the Lab system for the n-d and n-p systems is $\frac{4}{3}$. Thus writing the cross-sections as functions of Lab. energy E, we have from equations (11.11) to (11.13),

$$I_{n-d}^{el}(E) = [I_{n-p}(\tfrac{4}{3}E) + I_{n-n}(\tfrac{4}{3}E) + \{I_{n-p}(\tfrac{4}{3}E)\, I_{n-n}(\tfrac{4}{3}E)\}^{\frac{1}{2}}] \tfrac{16}{9} S(\mathbf{q}), \tag{11.14}$$

where $I(E) = d\sigma/d\omega$ at a Lab. energy E.

At 90 MeV., Chew (1951 a) finds on integrating (11.14) that the total elastic scattering cross-section is about half the n-p cross-section. Equation (11.14) actually allows the substitution of experimental scattering values for the two-body cross-sections, only the deuteron ground state entering into the calculation of the sticking factor $S(\mathbf{q})$. The expression (11.14) is, of course, very approximate as all complications due to spin, symmetry, exchange forces and non-central forces are ignored. However, this may be partly compensated for if experimental n-p and n-n differential cross-sections are used in (11.14).

Chew (1951 a) has shown that another simplified model, assuming s-scattering only and no spin dependence except that demanded by the Pauli principle in the n-n system, leads to a formula very similar to (11.14). There is a multiplying factor of $\frac{3}{4}$ for I_{nn} and $\cos \Delta$ for the square root term, Δ being the difference between the n-p and n-n phase shifts.

11.1.2. *The relation between two- and three-body total cross-sections.* A simple model may also be obtained for the Born approximation n-d total cross-section (elastic plus inelastic), again neglecting spin, symmetry, exchange forces and non-central forces. By equation (10.51), the scattering amplitude for n-p scattering is

$$f_{np}(\theta) = -\frac{1}{4\pi} \cdot \frac{2\mu}{\hbar^2} \int V_{np}(r)\, e^{i\mathbf{q} \cdot \mathbf{r}}\, d\mathbf{r}, \qquad (11.15)$$

where

$$q = |\, \mathbf{k_0} - \mathbf{k} \,| = 2k \sin \tfrac{1}{2}\theta, \quad \mu = \tfrac{1}{2}m.$$

The total cross-section is given by

$$\sigma_{n-p} = \int |\, f_{np} \,|^2 \, d\omega, \qquad (11.16)$$

where the neutrons are scattered at angle θ into the element of solid angle $d\omega$ in the CM system. Equation (11.16) may be transformed to

$$\sigma_{n-p} = \frac{2\pi}{k^2} \int_0^{2k} q\, dq \,|\, f_{np} \,|^2$$

$$= \frac{m}{8\pi\hbar^2 E} \int_0^{2k} q\, dq \,|\, V_{np}(\mathbf{q}) \,|^2, \qquad (11.17)$$

where E is the energy in the CM system and $V_{np}(q)$ is defined by equation (11.10). The scattering amplitude and total cross-section for n-n scattering are given by similar expressions with $V_{np}(r)$ and $V_{np}(\mathbf{q})$ replaced by $V_{nn}(r)$ and $V_{nn}(q)$ respectively.

For n-d collisions, if we ignore the internal motions of the nucleons in the deuterons, the scattered amplitude may be found by adding the amplitudes scattered from the two nucleons, allowing for the phase difference. Taking the initial neutron momentum as $\mathbf{k_0}$, the final neutron momentum as $\mathbf{k}$, and the nucleonic separation in the deuteron as $\mathbf{R}$, the phase difference between the waves scattered by the two nucleons becomes $(\mathbf{k}_0 - \mathbf{k}).\mathbf{R} = \mathbf{q}.\mathbf{R}$. The scattered amplitude is then

$$f_{nd} = f_{np} + f_{nn}\, e^{i\mathbf{q} \cdot \mathbf{R}}. \tag{11.18}$$

The total n-d cross-section for a given nucleonic separation is

$$\sigma_{\text{tot}}(R) = \frac{m}{8\pi\hbar^2 E} \int_0^{2k} q\,dq \,|\, V_{np}(\mathbf{q}) + e^{i\mathbf{q} \cdot \mathbf{R}}\, V_{nn}(\mathbf{q})\,|^2. \tag{11.19}$$

Denoting the deuteron ground state wave function by $\phi_0(R)$, the chance that the separation of the nucleons in the deuteron be between R and $R + dR$ is $\phi_0(R)\,dR$, so that the total n-d cross-section becomes

$$\sigma_{\text{tot}} = \frac{m}{8\pi\hbar^2 E} \int d\mathbf{R}\,\phi_0^2(R) \int_0^{2k} q\,dq \,|\, V_{np}(\mathbf{q})\, e^{i\mathbf{q} \cdot \mathbf{R}}\, V_{nn}(\mathbf{q})\,|^2, \tag{11.20}$$

$$= \sigma_{n-p} + \sigma_{n-n} + I, \tag{11.21}$$

where $\quad I = \dfrac{m}{4\pi\hbar^2 E} \displaystyle\int d\mathbf{R}\,\phi_0^2(R) \cos\mathbf{q}.\mathbf{R} \int_0^{2k} q\,dq\,V_{np}(\mathbf{q})\,V_{nn}(\mathbf{q}). \tag{11.22}$

Equation (11.22) is only a very rough approximation which moreover, owing to the neglect of the exclusion principle, does not include the interference term ϵ quoted in equation (11.1). It also assumes that the particle in the deuteron which is not struck remains stationary. However the latter effect may be taken into account by allowing for the momentum distribution of the nucleons within the deuteron. Taking the wave function of the scattered neutron as $(2\pi)^{-\frac{3}{2}} e^{i\mathbf{k} \cdot \mathbf{r}}$, the momentum distribution (Fourier transform) of the final state becomes

$$g_0(\varkappa) = (2\pi)^{-\frac{3}{2}} \int \phi_0(R)\, e^{-i\varkappa \cdot \mathbf{R}}\, d\mathbf{R}. \tag{11.23}$$

The probability that the relative momentum of the neutron and proton comprising the deuteron ground state is between $\varkappa\hbar$ and $(\varkappa + d\varkappa)\hbar$ is then $|g_0(\varkappa)|^2\,d\varkappa$.

The cross-section for n-d collisions may be found in terms of the momentum distribution by means of the following argument. If a neutron of momentum $\mathbf{k}_0$ collides with a deuteron comprising a neutron and proton of momentum $-\varkappa_0$ and $\varkappa_0$ respectively, and emerges with momentum $\mathbf{k} = \mathbf{k}_0 - \mathbf{q}$, then if the nucleonic neutron is struck, it has a momentum $-\varkappa_0 + \mathbf{q}$ after collision and the proton a momentum $\varkappa_0$. The amplitude for scattering in this case is proportional to $V_{nn}(\mathbf{q}) \phi_0(\varkappa)$. However, the same final state can be obtained if the deuteron particles have momenta before collision of $-\varkappa_0 + \mathbf{q}$ and $\varkappa_0 + \mathbf{q}$ for neutron and proton respectively, and if the proton is the struck particle. The amplitude for scattering is then proportional to $V_{np}(\mathbf{q}) \phi_0(|\varkappa_0 - \mathbf{q}|)$. On integrating over all values of q and κ_0 that conserve the overall energy, we find for the total cross-section,

$$\sigma_{\text{tot}} = \frac{m}{8\pi\hbar^2 E} \int_0^{2k} q\,dq \int \varkappa_0 \mid dV_{np}(\mathbf{q}) g_0(|\varkappa_0 - \mathbf{q}|) + V_{nn}(\mathbf{q}) g_0(\varkappa_0) \mid^2.$$

(11.24)

Gluckstern and Bethe (1951) have given a rigorous proof that equations (11.20) and (11.24) actually represent the total n-d cross-section and not the inelastic cross-section. The latter is obtained by subtraction of the elastic cross-section, as found by angular integration of equation (11.14), from the total cross-section.

11.1.3. *Born approximation for elastic scattering.* The expressions of § 11.12 would be more realistic if spin, the Pauli exclusion principle and exchange forces were taken into account.

If we denote an incident neutron by 1 and the deuteron by (23), 2 being a neutron also, then there are two sets of spin functions symmetric in the deuteron particle spin states. These comprise a quartet and doublet, described by equations (8.9) and (8.8), corresponding to a total spin $\frac{3}{2}$ and $\frac{1}{2}$ respectively. Denoting the initial and final spin functions by χ_0 and χ_f respectively, the spatial functions by equations (11.2) and (11.3), and writing P_{mn} as an operator exchanging both spatial and spin co-ordinates of particles m and n, appropriate initial and final wave functions are

$$\Psi_0 = 2^{-\frac{1}{2}}(1 - P_{12})\Phi_0\chi_0, \qquad (11.25)$$

$$\Psi_f = 2^{-\frac{1}{2}}(1 - P_{12})\Phi_f\chi_f. \qquad (11.26)$$

Writing $\int \Psi_f^* H\Psi_0\, d\tau = \langle \Psi_f, H\Psi_0 \rangle$, where $d\tau = d\mathbf{R}\, d\mathbf{r}$, the matrix element H'_{f0} describing the elastic scattering then becomes

$$H'_{f0} = \frac{1}{2}\langle (1 - P_{12})\Phi_f\chi_f,\ (1 - P_{12})\{\mathscr{V}_{nn}(12) + \mathscr{V}_{np}(13)\}\Phi_0\chi_0\rangle,$$

where $\mathscr{V}_{nn}$ and $\mathscr{V}_{np}$ are exchange operators. Since $1 - P_{12}$ is hermitian and $(1 - P_{12})^2 = 2(1 - P_{12})$, the matrix element is

$$H'_{f0} = \langle \Phi_f \chi_f, (1 - P_{12})\{V_{nn}(12) + V_{np}(13)\} \Phi_0 \chi_0 \rangle. \qquad (11.27)$$

For the exchange operators $\mathscr{V}_{np}$ and $\mathscr{V}_{nn}$ we choose the linear combination (8.4) of Wigner, Majorana, Bartlett and Heisenberg forces. However, for the n-p interaction only the Wigner and Majorana forces are necessary owing to the effect of the exclusion principle:

$$\begin{aligned} \mathscr{V}_{np}(r) &= (m_p M + b_p B + h_p H + w_p)\, V(r), \\ \mathscr{V}_{nn}(r) &= (m_n M + w_n)\, V(r). \end{aligned} \right\} \qquad (11.28)$$

On carrying out the symmetry operations, spin products and exchange operation implicit in equation (11.27), one finds

$$H'_{f0} = L^{-\frac{3}{2}} \sum_{n=1}^{6} a_n H_n, \qquad (11.29)$$

where

$$\begin{aligned} H_1 &= H_3 = I_1 = V(\mathbf{q})\, S^{\frac{1}{2}}(\mathbf{q}), \\ H_2 &= H_4 = I_2 = \phi(\tfrac{1}{2}q) \int d\mathbf{y}\, V(y)\, \psi_0(|\,\mathbf{y} - \mathbf{R}\,|) \exp\{-i\mathbf{y}.(\mathbf{q} + \tfrac{3}{2}\mathbf{k})\}, \\ H_5 &= H_6 = I_3 = \phi(|\,\mathbf{q} + \tfrac{3}{2}\mathbf{k}\,|) \int d\mathbf{R}\, \psi_0(R)\, V(R) \exp\{\tfrac{1}{2}i(\mathbf{q} + 3\mathbf{k}).\mathbf{R}\}, \end{aligned} \right\}$$

$$(11.30)$$

where $V(q)$ is defined by equation (11.10), $S^{\frac{1}{2}}(\mathbf{q})$ by (11.9) and $g_0(\mathbf{x})$ by (11.23).

The coefficients a_n depend on the exchange forces used, being functions of m_p, b_p, h_p, w_p, and m_n, w_n (Gluckstern and Bethe, 1951). They depend also on whether the scattering is quartet or doublet; thus

$$\sigma = \tfrac{2}{3}\sigma_Q + \tfrac{1}{3}\sigma_D, \qquad (11.31)$$

where

$$\sigma_{Q,D} = L^3 (2m^2/9\pi\hbar^4 k^2) \int_0^{2k} q\, dq\, |H'_{f0}|^2_{Q,D} \qquad (11.32)$$

$$= (m/8\pi\hbar^2 E) \int_0^{2k} q\, dq\, \left| \sum_{n=1}^{6} a_n H_n \right|^2_{Q,D}, \qquad (11.33)$$

where E is the incident energy in the CM system. By equation (A. 15), $E = \tfrac{2}{3}E_0$, where E_0 is the incident Lab. energy.

Making the simplifying assumption that the triplet and singlet
n-n interactions are equal to the corresponding triplet and singlet
n-p interactions and assuming a Serber exchange force, we find
for the quartet states:

$$H'_{f0} = \tfrac{1}{2}L^{-\frac{3}{2}}\{I_1 + I_2 - 2I_3\}, \qquad (11.34)$$

and for the doublet:

$$H'_{f0} = \tfrac{1}{2}L^{-\frac{3}{2}}\{\tfrac{1}{4}(1 + 9x)(I_1 + I_2) + I_3\}, \qquad (11.35)$$

where x is the ratio of singlet to triplet n-p forces. The term I_1
represents a small momentum transfer to either the neutron or
proton in the deuteron and is therefore the main term. I_2 and I_3
represent the effects of antisymmetrization and exchange forces
and include the possibility of 'pick-up' processes, giving a small
backward peak for the free neutron. Equation (11.14) giving the
n-d scattering intensity in terms of the n-p and n-n scattering intensities must therefore be modified by the addition of symmetry
and exchange terms due to I_2 and I_3.

Gluckstern and Bethe find that assuming a Serber n-p force of
Yukawa shape, the elastic n-d cross-section at 90 MeV. is 80, 60
and 30 mb. for ordinary, Majorana–Heisenberg and Serber n-n
forces respectively. The measured cross-section is 48 mb. (Powell,
1953).

Proton-deuteron scattering involves additional Coulomb potentials between the two identical particles. These may be allowed for
by adding to both doublet and quartet matrix elements the terms

$$L^{-\frac{3}{2}}\{I_1^c - I_2^c\},$$

where I_1^c and I_2^c are obtained from I_1 and I_2 by replacing $V(r)$ by
e^2/r. At the energies concerned, the additional terms have negligible
effect except at very small scattering angles and there only I_1^c need
be considered. At 90 MeV. the difference between n-d and p-d
scattering is noticeable only for scattering angles $\theta < 15°$ (CM),
destructive interference between nuclear and Coulomb scattering
occurring at $\sim 4°$.

Measurements are available at 240 MeV. for p-d differential
cross-sections (Schamberger, 1951) and have been compared with
the Born approximation for a Serber type Yukawa n-p potential
and n-n potentials of the above three exchange types. None of the

n-n exchange types gives agreement with experiment except at small angles, the scattering for angles greater than $30°$ being much too small, especially for exchange type (M.H.) forces. However, this is almost certainly due to the failure to represent the n-n interaction by the singular tensor force found necessary to explain high energy p-p scattering.

On the other hand, elastic d-p measurements at 190 MeV. (Stern and Bloom, 1951) in the angular range 38–$150°$ CM show good agreement with the Serber type, Yukawa potential calculations of Chew (1948) up to angles of $110°$, but at greater angles theoretical values are too large. The agreement is probably fortuitous as Chew's calculations employ approximate expressions, valid only for small angles ($\theta < 25°$), for the measured n-p and p-p cross-sections.

11.1.4. *Born approximation for total scattering.* There are eight linearly independent and orthogonal spin functions which can describe a three-body system. Of these the quartet and doublet (equations (8.9) and (8.8)), corresponding to total spins $\frac{3}{2}$ and $\frac{1}{2}$ respectively, are symmetric in the deuteron particles. They can thus describe either elastic or inelastic scattering. On the other hand, another doublet described by (8.34) exists which is antisymmetric in the deuteron particles (23), thus corresponding to a singlet state of the latter and hence describing only inelastic scattering.

The initial state must be either a symmetric quartet or doublet with statistical weights of 2 and 1 respectively. For the quartet, the spin exchange operators in equation (11.28) leave the state unchanged so that the final spin function must be the same as the initial one to give a non-vanishing matrix element. A symmetric doublet initial state can lead to either symmetric or antisymmetric doublet final states.

To obtain the total cross-section, the matrix element (11.27) may be evaluated as for the elastic scattering, the difference lying in the treatment of the final radial function Ψ_f. In elastic scattering, Ψ is taken as the product of a plane wave and a bound state deuteron wave function. However, the total scattering may be described by three possibilities, each involving the product of a different free particle plane wave by the wave function of the other two particles, this being either free or bound.

(a) $\Psi_f = L^{-\frac{3}{2}} \exp(i\mathbf{k}.\mathbf{r}) \psi_\kappa(\mathbf{R})$, particle 3 free.

(11.36a)

(b) $\Psi_f = L^{-\frac{3}{2}} \exp\{i\mathbf{k}'.(\tfrac{3}{4}\mathbf{R} - \tfrac{1}{2}\mathbf{r})\} \psi_{\kappa'}(-\mathbf{r} - \tfrac{1}{2}\mathbf{R})$, particle 1 free.

(11.36b)

(c) $\Psi_f = L^{-\frac{3}{2}} \exp\{i\mathbf{k}''.(-\tfrac{3}{4}\mathbf{R} - \tfrac{1}{2}\mathbf{r})\} \psi_{\kappa''}(-\mathbf{r} + \tfrac{1}{2}\mathbf{R})$, particle 2 free.

(11.36c)

To evaluate the integrals encountered in the matrix element (11.27), a sum rule is used: If M and N are hermitian operators and

$$_2M_1 = \langle \psi_2, M\psi_1 \rangle = \int \psi_2^*(r) M\psi_1(r) \, dr,$$

then $$\sum_f \langle \psi_f, M\psi_0 \rangle^* \langle \psi_f, N\psi_0 \rangle = \sum_f {}_0 M_f^* \cdot {}_f N_0 = {}_0(M^*N)_0. \quad (11.37)$$

Using this sum rule, Gluckstern and Bethe (1951) succeed in proving that if the $\psi_\kappa(\mathbf{r})$ form a complete set of eigenfunctions, including both bound and free states, then the total cross-section corresponding to possibility (a) is independent of the form of $\psi_\kappa(\mathbf{R})$. The same result is obtained with (b) and (c). Moreover, exactly the same conclusion would have followed from using the set of plane waves $L^{-\frac{3}{2}} \exp(i\boldsymbol{\kappa}.\mathbf{r})$ for the $\psi_\kappa(\mathbf{r})$. We therefore arrive at the result that the total cross-section can be obtained by using plane waves for the final state two particle wave functions everywhere. This is quite startling, since over much of the region of integration the two-particle wave functions are *not* plane waves (elastic scattering). The reason is that the contribution of the elastic scattering is just cancelled by the correction to the plane waves which must be applied for the low energy two-particle bound states. The elastic scattering in a sense 'robs' the neighbouring low energy two-particle states.

It now appears that the total cross-section is a more fundamental quantity than the inelastic cross-section, and calculations should be directed accordingly. To be sure, the total cross-section is made up of an elastic and inelastic part, but these complement one another in such a way that the two-particle wave functions approximate a complete set only if both parts are taken together.

On substituting the expression ($11 \cdot 36a$) for Ψ'_f in the equation (11.27) for H'_{f0}, the cross-section for each spin state follows from equation (11.32). Labelling symmetric and antisymmetric functions by suffix S and A respectively, the total cross-section becomes

$$\sigma_{\text{tot}} = \tfrac{2}{3}\sigma_Q + \tfrac{1}{3}(\sigma_D^S + \sigma_D^A). \qquad (11.38)$$

For details of the very complicated calculations involved, see Gluckstern and Bethe (1951). The final result for the total cross-section is

$$\sigma = \sigma_{n-p}(1 - \epsilon) + \sigma_{n-n} + I, \qquad (11.39)$$

where σ_{np} and σ_{nn} are the Born approximation free n-p and n-n cross-sections, and ϵ is a correction to the n-p cross-section arising from the interference between exchange scattering of the direct and antisymmetrized neutrons by the proton. ϵ is about $0 \cdot 15$ at 90 MeV. I is the interference term between the n-p and n-n scattering. It seems reasonable that the *exact* total cross-section should give the sum of the *exact* two-particle cross-sections, corrected in a similar way. In this case, it would be better to use the measured n-p and n-n cross-sections rather than those calculated by the Born approximation when evaluating the total scattering by equation (11.38).

The measured total cross-section at 90 MeV. (Coon and Taschek, 1949) is 117 mb. and the corresponding n-p cross-section 83 mb., using $\epsilon = 0 \cdot 15$. However, the n-n cross-section is not uniquely determined by these values owing to the presence of the interference term I. Different assumptions on the exchange properties of the n-n force lead to widely varying values of both ϵ and I.

To conclude, the total cross-section (elastic plus inelastic) can be calculated using plane waves for the final state of the three particles involved. No such simple method is available for the inelastic scattering alone, or for the differential cross-section. However, Wu and Ashkin (1948) have given a full treatment on conventional Born approximation lines for the total differential cross-sections for n-d and p-d scattering, including the modifications due to tensor forces.

11.1.5. *Scattering with non-central forces.* The effect of using a non-central potential of form (9.10) in the Born approximation has been investigated (Wu and Ashkin, 1948; Bransden, 1950, 1952), but without any attempt to relate the three-body to the two-body cross-sections. The calculations differ only in mathematical detail from the central force Born approximation.

Bransden (1952) has carried out numerical calculations using the charge independent, hard core model of Jastrow (§ 10.12), which involves a non-central triplet potential together with a Serber exchange potential outside an infinite barrier of width

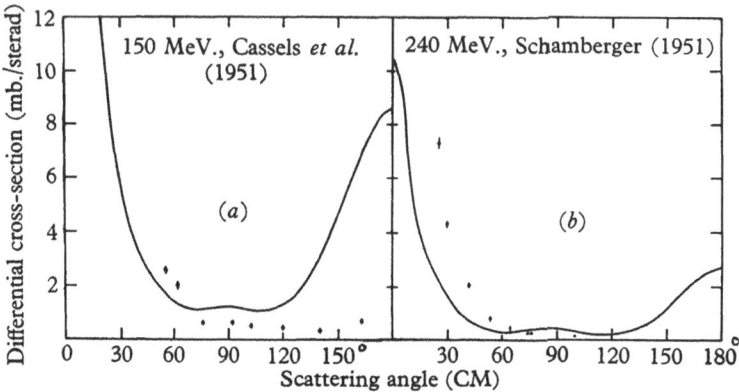

Fig. 85. Comparison of observed and calculated differential cross-sections for elastic p-d scattering, the latter due to Bransden (1952) using the Born approximation for the Jastrow hard core potential.

0.6×10^{-13} cm. in the singlet state. This interaction is not inconsistent with the high energy n-p and p-p scattering data. Calculated results for elastic p-d scattering are compared in fig. 85 with observed angular distributions for scattering at proton energies of 150 MeV. (Cassels, Stafford and Pickavance, 1951) and 240 MeV. (Schamberger, 1951). Only qualitative agreement with experiment is obtained, the forward scattering being too small and the back scattering too large. The discrepancy may be due as much to the method used in allowing for the singlet repulsive barrier as to the inadequacy of the Jastrow interaction.

11.1.6. *The impulse approximation—simplified model.* A method due to Chew (Chew and Steinberger, 1950; Chew, 1951 a, b) enables evaluation of n-d scattering, but differs from the Born approximation in that the scattering potential used need not be small compared with the incident energy of the neutron. The assumptions of the impulse approximation are twofold:

(1) The collision time in high energy n-d scattering is so small compared to the period of the internal motion in the deuteron that the change in the wave function of the latter during the collision can be described by an 'impulse' approximation. This involves the forces between the collision partners being so strong that they overwhelm all the forces binding the struck nucleon in the deuteron. The whole effect of the binding forces is thought of as determining the initial momentum distribution of the struck nucleon in the original deuteron ground state.

If a is the scattering amplitude for two-body collisions and v the relative velocity of neutron and deuteron, the time of collision is of order a/v. The period of internal motion is of order $\hbar/W$, W being the deuteron binding energy; hence the impulse approximation should be valid for $aW/\hbar v = (a/R)(1/kR) \ll 1$, where R, the deuteron 'radius', is $\hbar/(mW)^{\frac{1}{2}} = 4 \cdot 31 \times 10^{-13}$ cm. At 90 MeV., $a \approx 10^{-13}$ cm., $aW/\hbar v \approx 0 \cdot 03$. Calculations (Chew and Wick, 1952) indicate that R should be taken as the mean nucleon separation instead of the deuteron radius, giving $R = 3 \cdot 2 \times 10^{-13}$ cm. This increases the order of magnitude correction from 3 % as above to 6 %. Even so, the condition is well satisfied; in fact experience indicates that it holds down to energies as low as 20 MeV. We shall, however, introduce approximations in the evaluation of the impulse formulae which lead to rather larger errors than those quoted above.

(2) The deuteron has such a diffuse structure compared with the nuclear force range that the incident neutron wave function at either of the two scattering centres within the deuteron is not appreciably perturbed by the presence of the other centre. Outgoing waves from both centres are present and may be added together, but individually they are the same as produced by a single neutron or proton. Thus the three-body problem is reduced to a superposition of two-body problems.

These assumptions are not the same as the Born approximation, which assumes the perturbation of the incident wave to be small. Instead it is only assumed that the perturbing centres act independently and suddenly.

A simplified model of n-d inelastic scattering may be obtained as follows: Consider the deuteron as composed of a proton 3 bound to a neutron 2 by a fixed short range force $U(R)$, the deuteron bound state wave function being $\phi_0(R)$. The incident neutron 1 interacts with the proton via a force $V(r)$, but the interaction with the neutron 2 is ignored for simplicity. If the initial and final momenta of the incident neutron are $\mathbf{k}_0$ and $\mathbf{k}_1$ respectively and likewise the initial and final proton momenta $\varkappa_0$ and $\varkappa_1$, then the wave function describing the scattering may be approximated by

$$\Psi(\mathbf{R}, \mathbf{v}) \approx \Psi_a(\mathbf{R}, \mathbf{v}) = \int d\varkappa_0\, g_0(\varkappa_0)\, \psi^{np}_{\mathbf{k}_0, \varkappa_0}(\mathbf{R}, \mathbf{v}), \qquad (11.40)$$

where

$$g_0(\varkappa_0) = (2\pi)^{-\frac{3}{2}} \int d\mathbf{R}\, \mathrm{e}^{-i\varkappa_0 \cdot \mathbf{R}}\, \phi_0(R),$$

and $\psi^{np}_{\mathbf{k}_0, \varkappa_0}(\mathbf{R}, \mathbf{v})$ is the wave function describing the scattering of a neutron of momentum $\mathbf{k}_0$ by a free proton with the same momentum $\varkappa_0$ as the bound proton. This may be factored as follows:

$$\psi^{np}_{\mathbf{k}_0, \varkappa_0}(\mathbf{R}, \mathbf{v}) = (2\pi)^{-\frac{3}{2}} \exp\left[i\tfrac{1}{2}(\mathbf{k}_{n_0} + \mathbf{k}_{p_0}) \cdot (\mathbf{r}_n + \mathbf{r}_p)\right] \phi^{np}_{\mathbf{k}_0}(\mathbf{v}). \quad (11.41)$$

The approximation (11.40) is such that in the limit of a very weak n-p force,

$$\Psi_a \to (2\pi)^{-\frac{3}{2}} \mathrm{e}^{i\mathbf{k}_0 \cdot \mathbf{v}} \phi_0(R),$$

so that the impulse approximation is at least as good as the Born approximation. Also in the limiting case of small proton binding energy, $g_0(\varkappa)$ approaches a delta function and one obtains free n-p scattering. It thus seems reasonable to regard Ψ_a as an interpolation between these two limits.

The replacement of Ψ by Ψ_a corresponds to assuming that for the duration of the proton's interaction with the neutron, the former does not interact with the force field $U(R)$. The only effect of $U(R)$ is to generate the momentum distribution $g_0(\varkappa)$ much earlier than the time of arrival of the neutron—hence the postulate that the collision time be short compared to the period of the deuteron.

The cross-section for free n-p scattering into the momentum interval $d\mathbf{k}$ is given by the first order perturbation expression (11.4):

$$\sigma_{n-p}\, d\mathbf{k} = \frac{2\pi}{\hbar v} |\langle \mathbf{k} | r_{np} | \mathbf{k}_0 \rangle|^2\, d\mathbf{k}, \qquad (11.42)$$

where $d\mathbf{k} = k^2 d\omega\, dk$ includes the density of states $\rho(\mathbf{k}) = k^2 d\omega$. Equation (11.42) is subject to the energy conservation rule

$$E_n + E_p = E_{n_0} + E_{p_0}.$$

Here the n-p scattering matrix is written as

$$\langle \mathbf{k} \mid r_{np} \mid \mathbf{k}_0 \rangle = \int d\mathbf{v}\, \mathrm{e}^{-i\mathbf{k}\cdot\mathbf{v}}\, V(v)\, \phi_{k_0}^{np}(\mathbf{v}), \qquad (11.43)$$

where $\phi_{k_0}^{np}(v)$ is the scattering wave function defined by equation (11.41) for n-p scattering.

The scattering cross-section for inelastic n-d scattering, describing emission of a neutron into the momentum element $d\mathbf{k}$ while the proton emerges into $d\mathbf{\varkappa}$, is given in terms of the n-d scattering matrix r_{nd} by

$$\sigma_{n-d}\, d\mathbf{k}\, d\mathbf{\varkappa} = \frac{2\pi}{\hbar v_0} \mid r_{nd} \mid^2 d\mathbf{k}\, d\mathbf{\varkappa}. \qquad (11.44)$$

The scattering amplitude is here found by averaging the n-p amplitude (again neglecting the n-n interaction) over the initial and final proton momentum distribution:

$$\langle k, \kappa; K \mid r_{nd} \mid k_0, 0 \rangle = \int d\mathbf{\varkappa}'\, g_\kappa^*(\mathbf{\varkappa}') \left\langle \frac{\mathbf{k} - \mathbf{\varkappa}'}{2} \mid r_{np} \mid \frac{\mathbf{k}_0 - \mathbf{\varkappa}^0}{2} \right\rangle g_0(\mathbf{\varkappa}^0),$$
$$(11.45)$$

where

$$g_\kappa(\mathbf{\varkappa}') = (2\pi)^{-\frac{3}{2}} \int d\mathbf{v}\, \mathrm{e}^{-i\kappa'\cdot\mathbf{v}}\, \phi_\kappa(v), \qquad (11.46)$$

and $\mathbf{\varkappa}^0 = \mathbf{k} + \mathbf{\varkappa}' - \mathbf{k}_0$.

It would be convenient to take the n-p scattering matrix r_{np} outside the integral, which is possible subject to one or other of two conditions:

(1) If the velocity v of the incident neutron relative to the proton varies very little over the important range of κ, we may replace r_{np} by the value it takes when $g_\kappa^*(\kappa^*)$ as defined by equation (11.46) is singular, i.e. $\mathbf{\varkappa} = \mathbf{\varkappa}'$.

(2) If r_{np} depends only on the difference of initial and final momenta rather than on the individual values, then it is independent of κ. Thus the conservation of energy gives

$$\tfrac{1}{2}(\mathbf{k} - \mathbf{\varkappa}') - \tfrac{1}{2}(\mathbf{k}_0 - \mathbf{\varkappa}^0) = \mathbf{k} - \mathbf{k}_0. \qquad (11.47)$$

This may be identified with the formula which Fermi (1936) deduced to describe the scattering of slow neutrons by molecularly bound protons, by writing the second factor in configuration space, employing the properties of the three-dimensional Dirac delta function:

$$\int d\varkappa' g_\kappa^*(\varkappa') g_0(\mathbf{k} + \varkappa' - \mathbf{k}_0) = \int d\mathbf{R} \, \phi_\kappa^*(\mathbf{R}) \, e^{-i(\mathbf{k} - \mathbf{k}_0) \cdot \mathbf{R}} \, \phi_0(R).$$

The scattering length for elastic scattering may be obtained from equations (11.45) or (11.47) by replacing $g^*(\varkappa')$ by $g_0^*(\varkappa)$, so that both elastic and inelastic scattering events may be described in terms of the n-p scattering amplitude without any assumption about the nature of the force involved.

Polarization of high energy particles scattered by complex nuclei has been treated by the impulse approximation (Tamor, 1955).

11.1.7. *Inelastic n-d scattering in the impulse approximation.* To apply the impulse approximation to the actual n-d inelastic scattering it is necessary to remove the simplifying assumptions of §11.1.6. First, the bound neutron is not infinitely heavy, but this fact can be taken into account by using the usual centre of mass transformation. Secondly, the incident neutron interacts with the bound neutron as well as with the proton.

We shall denote the initial and final momenta of the incident neutron 1 by $\mathbf{k}_0$ and $\mathbf{k}$, of the proton 3 by $\varkappa_0$ and $\varkappa$, and of the neutron 2 by $\mathbf{K}_0$ and $\mathbf{K}$ respectively. Take also $\mathbf{r}_{23} = \mathbf{R}$, $\mathbf{r}_{13} = \mathbf{v}$, $\mathbf{r}_{12} = \mathbf{t}$. Then for the neutron 1 close to the proton 3, the wave function is approximated by

$$\Psi_a^{np} = \iint d\varkappa_0 \, d\mathbf{K}_0 \, G_0(\varkappa_0, \mathbf{K}_0) \, e^{i\mathbf{K}_0 \cdot \mathbf{r}_1} \psi_{k_0, \kappa_0}(\mathbf{R}, \mathbf{v}), \qquad (11.48)$$

where $\quad G_0(\varkappa_0, \mathbf{K}_0) = (2\pi)^{-\frac{3}{2}} \iint dr_3 \, dr_2 \, e^{-i(\kappa_0 \cdot \mathbf{r}_3 + \mathbf{K} \cdot \mathbf{r}_2)} \phi_0(\mathbf{r}_3 - \mathbf{r}_2)$

$$= \delta(\varkappa_0 + \mathbf{K}_0) g_0 \left(\frac{\varkappa_0 - \mathbf{K}_0}{2} \right). \qquad (11.49)$$

$\delta(\mathbf{x})$ is the Dirac delta function.

Thus G_0 is given entirely in terms of the Fourier transform of the deuteron wave function $G_0(\varkappa_0, \mathbf{K}_0) = g_0(\varkappa_0)$. This is equivalent to assuming that the smallness of the high energy two-body scattering amplitude, compared to the dimensions of the deuteron, is such that the perturbation of the wave function produced at either one of the two scattering centres by the other can be ignored.

Similarly for neutron 1 close to neutron 2, the wave function is built up from a superposition of n-n scattering functions only. The formula for scattering amplitude analogous to equation (11.45) may now be found by averaging the sum of the n-p and n-n scattering amplitudes over the initial and final momentum distributions of the deuteron particles:

$$\langle k, \kappa, K \,|\, r_{nd} \,|\, k_0, 0 \rangle \approx \iiint d\varkappa'\, d\mathbf{K}'\, d\varkappa^0\, d\mathbf{K}^0 G^*_{\kappa, K}(\varkappa', \mathbf{K})$$

$$\times \{\langle k, \kappa' \,|\, r_{np} \,|\, k_0, \kappa^0 \rangle \delta(\mathbf{K}' - \mathbf{K}^0)$$

$$+ \langle k, K' \,|\, r_{nn} \,|\, k_0, K^0 \rangle \delta(\varkappa' - \varkappa^0)\} G_0(\varkappa^0, \mathbf{K}^0),$$

$$(11.50)$$

where $G_{\kappa, K}(\varkappa', \mathbf{K}')$ is the Fourier transform of the final continuum deuteron wave function defined similarly to (11.48). The Dirac three-dimensional delta functions express the independence of the scattering from each centre.

However, equation (11.50) is not wholly self-consistent. If we insert the additional factors $\delta(\mathbf{k} - \mathbf{k}')$ and $\delta(\mathbf{k}_0 - \mathbf{k}'_0)$ after $G^*_{\kappa, K}$ and G_0 respectively, the value of the integral involved remains unchanged, but in momentum space the expression now describes the transition from a deuteron plus a free neutron to the final state of the three nucleons. Neither term in the integrand involves more than two of the three two-body interactions possible.

Chew proposes that (11.50) should be modified by computing the integral with either

$$G^*_{\kappa, K}(\varkappa', \mathbf{K}')\, \delta(\mathbf{k} - \mathbf{k}') \quad \text{or} \quad G^*_{k, K}(\mathbf{k}', \mathbf{K}')\, \delta(\varkappa - \varkappa'),$$

according to whether $|\varkappa - \mathbf{K}|$ or $|\mathbf{k} - \mathbf{K}|$ is the smaller, i.e. that interaction should be taken into account which is the stronger. Because of conservation of energy and momentum, $|\varkappa - \mathbf{K}|$ and $|\mathbf{k} - \mathbf{K}|$ cannot both be small, and if they are nearly equal, they must be large enough to make both pairs of interactions equal. The second part of (11.50) based on r_{nn} may be modified by a similar choice governed by the relative magnitudes of $|\varkappa - \mathbf{K}|$ and $|\varkappa - \mathbf{k}|$.

Several further modifications are needed to bring the impulse approximation into accord with reality, viz. the inclusion of spin, the Pauli exclusion principle and tensor forces. For simplicity the latter effect is omitted here, although Chew and Steinberger (1950) have indicated the lines to be followed in such a treatment.

With central forces, there is no coupling between S and L, so that S and S_z are constants of motion. Hence instead of considering all six possible initial spin functions and eight possible final spin functions, it suffices to consider two initial states (one quartet and one doublet), and three final states (one quartet and two doublets). Spin wave functions must be included in the arguments of initial and final wave functions and in the matrices r_{np} and r_{nn}, and summations over spin added to the integrations over momenta.

In allowing for spin, it was previously found (§ 11.1.4) that the initial and final quartet spin states for a given S_z are unique: likewise the initial doublet state, because of the symmetry requirements of the deuteron particles. However, the final doublet state for the same S_z is doubly degenerate, this usually being resolved according to the spin symmetry of the particles originally forming the deuteron. It is more consistent here to resolve the degeneracy in terms of the spin symmetry of the two neutrons. The three possibilities are then initial and final quartet states symmetric in neutron spins, and an initial doublet state coupled with either symmetric or antisymmetric neutron spins. Replacing the vectors $\mathbf{k}$, $\mathbf{K}$ (which refer to neutrons 1 and 2 respectively) by $\mathbf{k}_{12} = \frac{1}{2}(\mathbf{k} - \mathbf{K})$, and denoting symmetry by superfix ^{S} or ^{A}, the matrices describing these inelastic scattering possibilities become respectively

$$\langle \mathbf{k}_{nn}, \varkappa \mid {}^4r^S_{nd} \mid \mathbf{k}, \mathrm{o} \rangle, \quad \langle \mathbf{k}_{nn}, \varkappa \mid {}^2r^S_{nd} \mid \mathbf{k}, \mathrm{o} \rangle \quad \text{and} \quad \langle \mathbf{k}_{nn}, \varkappa \mid {}^2r^A_{nd} \mid \mathbf{k}, \mathrm{o} \rangle,$$

but differ from the definition (11.50) including spin functions. The total inelastic cross-section thus becomes

$$\sigma_{n-d} \, d\varkappa \, d\mathbf{k}_{12} = \tfrac{2}{3} {}^4\sigma_{n-d} \, d\varkappa \, d\mathbf{k}_{12} + \tfrac{1}{3} {}^2\sigma_{n-d} \, d\varkappa \, d\mathbf{k}_{12}, \quad (11.51)$$

where quartet inelastic scattering is given by

$$\begin{aligned} {}^4\sigma_{n-d} \, d\varkappa \, d\mathbf{k}_{12} = \frac{2\pi}{\hbar v_0} \mid \langle \mathbf{k}_{12}, \varkappa \mid {}^4r^S_{nd} \mid \mathbf{k}_0, \mathrm{o} \rangle \\ - \langle \mathbf{k}_{21}, \varkappa \mid {}^4r^S_{nd} \mid \mathbf{k}_0, \mathrm{o} \rangle \mid^2 d\mathbf{k}_{12} \, d\varkappa. \quad (11.52) \end{aligned}$$

Doublet scattering is likewise given by

$$^2\sigma_{n-d}\,d\varkappa\,d\mathbf{k}_{12} = \frac{2\pi}{\hbar v_0}\{|\langle\mathbf{k}_{12},\varkappa|\,^2r_{nd}^S\,|\mathbf{k}_0,\mathrm{o}\rangle - \langle\mathbf{k}_{21},\varkappa|\,r_{nd}^S\,|\mathbf{k}_0,\mathrm{o}\rangle|^2$$

$$+\,|\,\langle\mathbf{k}_{12},\varkappa|\,^2r_{nd}^A\,|\mathbf{k}_0,\mathrm{o}\rangle + \langle\mathbf{k}_{21},\varkappa|\,^2r_{nd}^A\,|\mathbf{k}_0,\mathrm{o}\rangle|^2\}.$$

$$(11.53)$$

In all cases, the two-body matrices are taken outside the momentum integrals and evaluated at the final observed values of the momenta. We define the following two-body matrices:

$$\left\langle\mathbf{k}_{1p}\,\Big|\,r_{np}^t\,\Big|\,\frac{\mathbf{k}_0+\mathbf{K}}{2}\right\rangle = r_{1p}^t,\quad \left\langle\mathbf{k}_{1p}\,\Big|\,r_{np}^s\,\Big|\,\frac{\mathbf{k}_0+\mathbf{K}}{2}\right\rangle = r_{1p}^s,$$

$$\left.\left\langle\mathbf{k}_{12}\,\Big|\,r_{nn}^t\,\Big|\,\frac{\mathbf{k}_0+\varkappa}{2}\right\rangle = r_{12}^t,\quad \left\langle\mathbf{k}_{12}\,\Big|\,r_{nn}^s\,\Big|\,\frac{\mathbf{k}_0+\varkappa}{2}\right\rangle = r_{12}^s,\right\} \quad (11.54)$$

where $\mathbf{k}_{1p} = \frac{1}{2}(\mathbf{k}-\varkappa)$ and the superscripts t and s refer to triplet and singlet two-body interactions.

Chew and Steinberger (1950) have expressed the complicated expressions for the matrices involved in equations (11.52) and (11.53) as functions of r_{1p}^t, r_{1p}^s, r_{12}^t, r_{12}^s and integrals over the initial and final two-body wave functions. However, it is simpler to make use of the closure argument (Gluckstern and Bethe, 1951) referred to in § 11.1.4, whereby it is shown that the replacement of the actual final two-body wave functions by plane waves is equivalent to calculating the sum of elastic plus inelastic scattering. Taking

$$\phi_{k_{mn}}(r) \approx (2\pi)^{-\frac{3}{2}}e^{ik_{mn}\cdot\mathbf{r}},$$

we find the matrices in equations (11.52) and (11.53) simplify to

$$\begin{aligned} (\mathbf{k}_{12},\varkappa|\,^4r_{nd}^S\,|\mathbf{k}_0,\mathrm{o}) &\approx r_{np}^t g_0(\mathbf{K}) + r_{12}^t g_0(\varkappa),\\ (\mathbf{k}_{12},\varkappa|\,^2r_{nd}^S\,|\mathbf{k}_0,\mathrm{o}) &\approx -\tfrac{1}{4}(r_{1p}^t - 3r_{1p}^s)g_0(\mathbf{K}) + \tfrac{1}{2}r_{12}^t g_0(\varkappa),\\ (\mathbf{k}_{12},\varkappa|\,^2r_{nd}^A\,|\mathbf{k}_0,\mathrm{o}) &\approx \frac{3^{\frac{1}{2}}}{4}(r_{1p}^t + r_{1p}^s)g_0(\mathbf{K}) + \frac{3^{\frac{1}{2}}}{2}r_{12}^s g_0(\varkappa). \end{aligned}\right\} \quad (11.55)$$

This approach avoids the two-body continuum wave functions, which are difficult to handle analytically. The elastic scattering may be found separately and the inelastic cross-section obtained by subtraction from the total. On changing the integration variables $\mathbf{k}_{1p}$ and $\mathbf{k}_{12}$ to $\mathbf{k}_0$, we have

$$d\mathbf{k}_{1p} = \frac{mk_0}{4\hbar^2} d\omega_{1p}, \quad d\mathbf{k}_{12} = \frac{mk_0}{4\hbar^2} d\omega_{12},$$

where $d\omega$ is an element of solid angle.

The total cross-section becomes

$$\sigma_{\text{tot}} = \frac{2\pi}{\hbar v_0} \int d\mathbf{K} g_0^2(\mathbf{K}) \left(\frac{mk_{1p}}{2\hbar^2}\right) \int d\omega_{1p}[\tfrac{3}{4} \mid r_{1p}^t \mid^2 + \tfrac{1}{4} \mid r_{1p}^s \mid^2]$$

$$+ \frac{2\pi}{\hbar v_0} \int d\varkappa g_0^2(\varkappa) \left(\frac{mk_{12}}{4\hbar^2}\right) \int d\omega_{12}[\tfrac{3}{4} \mid r_{12}^t - r_{21}^t \mid^2 + \tfrac{1}{4} \mid r_{12}^s + r_{21}^s \mid^2]$$

$$- (\text{exclusion principle term}) + (\text{interference terms}). \quad (11.56)$$

The corresponding n-p and n-n cross-sections at the same energy are

$$\left.\begin{array}{l}
\sigma_{n-p} = \dfrac{2\pi}{\hbar v_0} \left(\dfrac{mk_0}{4\hbar^2}\right) \displaystyle\int d\omega_{1p}[\tfrac{3}{4} \mid r_{1p}^t \mid^2 + \tfrac{1}{4} \mid r_{1p}^s \mid^2], \\[12pt]
\sigma_{n-n} = \dfrac{2\pi}{\hbar v_0} \left(\dfrac{mk_0}{8\hbar^2}\right) \displaystyle\int d\omega_{12}[\tfrac{3}{4} \mid r_{12}^t - r_{21}^t \mid^2 + \tfrac{1}{4} \mid r_{12}^s + r_{21}^s \mid^2].
\end{array}\right\} \quad (11.57)$$

The energy-momentum conditions in the two-body problem actually differ from those in the three-body problem, but for the latter, scattering is strongly peaked in the forward direction where initial and final relative momenta are roughly equal to $\tfrac{1}{2}\mathbf{k}_0$. Replacing $\mathbf{k}_{1p}$ and $\mathbf{k}_{12}$ by $\tfrac{1}{2}\mathbf{k}_0$ leads to the expression

$$\sigma_{\text{tot}} = (1 - \epsilon) \sigma_{n-p} + I + P, \quad (11.58)$$

already found in § 11.1.4. using the Born approximation, but with an additional term allowing for capture of the incident neutron.

A similar treatment of the elastic n-d cross-section, using the deuteron ground state function instead of the continuum wave function for the final state, leads to the result

$$\sigma_{\text{el}} = (\sigma_{n-p} + \tfrac{3}{2}\sigma_{n-n}) \bar{S} + I, \quad (11.59)$$

where $\bar{S}$ is the average of the sticking factor $S(q)$ of § 11.1.1 over the interval $0 \leqslant q^2 \leqslant k_0^2$. We thus have

$$\sigma_{\text{inel}} = (1 - \epsilon - \bar{S}) \sigma_{n-p} + (1 - \tfrac{3}{2}\bar{S}) \sigma_{n-n}. \quad (11.60)$$

This result can be further broken down, as the ejected protons comprise a fast and a slow group. According to equation (11.11) we may therefore write

$$\sigma_{\text{fast}} = (1 - \epsilon - \bar{S})\,\sigma_{n-p}, \quad \sigma_{\text{slow}} = (1 - \tfrac{3}{2}\bar{S})\,\sigma_{n-n}. \quad (11.61)$$

At 90 MeV., $\bar{S}$ is calculated as 0·28 and ϵ as 0·15, giving

$$\sigma_{\text{fast}}(90\,\text{MeV.}) = 0\!\cdot\!57\sigma_{n-p}, \quad \sigma_{\text{slow}}(90\,\text{MeV.}) = 0\!\cdot\!58\sigma_{n-n}. \quad (11.62)$$

The total slow proton cross-section (Powell, 1953) is 20 mb., leading to $\sigma_{n-n} \approx 35$ mb., which is not inconsistent with observed p-p cross-sections at the same energy. The measured fast proton cross-section of 51 mb. also agrees well with the value 47 mb. deduced from equation (11.62) using $\sigma_{n-p} = 83$ mb. Thus, agreement for both slow and fast integrated cross-sections is satisfactory.

Chew has also shown that the differential cross-section of the fast protons is given for $\theta < 60°$ by

$$I_f(\theta_p) = \{1 - KF(k_0 \sin \theta_p)\}\, I_{np}(\theta_p), \quad (11.63)$$

where $I_{np}(\theta)$ is the corresponding differential cross-section for free n-p collisions, and the form factor is $F(q) = S(2q)$, S being the sticking factor defined by (11.9).

$K = I_{\text{ex}}/I_{np}$ is the ratio of two differential cross-sections, I_{ex} being the cross-section for the incident neutron to end up with its spin parallel to that of the other neutron. It is a constant which depends not only on the two-body scattering lengths but also on the unknown phase of r^s_{1p}. If $r^s_{1p} = r^t_{1p}$, then $K = \tfrac{1}{2}$, but if $r^s_{1p} = 0$, then $K = \tfrac{5}{8}$. Thus K should lie between $\tfrac{1}{2}$ and $\tfrac{5}{8}$, but its precise value is best found by experiment. For $\theta < 60°$, the proton moves so slowly that it has a good chance of being captured, so that (11.63) is invalid.

Fig. 86 shows the theoretical fast proton angular distribution for several different values of K, compared with the experimental results of Powell (1953). The value $K \approx 0\!\cdot\!6$, $r^s_{1p} \approx 0\!\cdot\!65r^t_{1p}$ can be seen to give reasonable agreement. The value of the Pauli exclusion principle correction ϵ may be found by integrating (11.63) over all angles and equating it with the first expression of equation (11.62). This gives the figure $\epsilon = 0\!\cdot\!15$ already employed above.

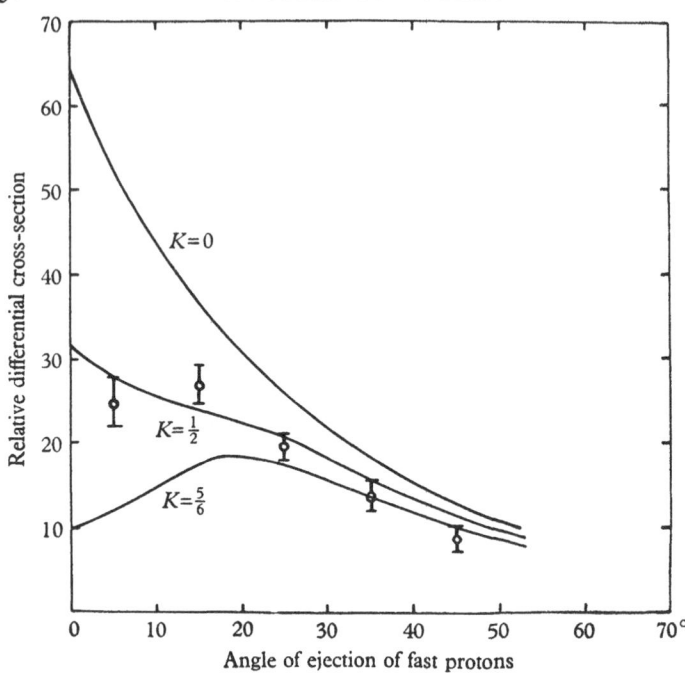

Fig. 86. The theoretical fast proton angular distribution for several different values of K, compared with experimental points due to Powell (1953) for the collision of 90 MeV. neutrons with deuterons (Marshall *et al.* 1953).

11.1.8. *Multiple scattering corrections to the impulse approximation.* Brueckner (1953) has pointed out that the impulse approximation assumes that multiple scattering effects can be neglected if the two-body scattering amplitudes are small compared with the deuteron radius. To test this assumption, he examines the consequences of an *exact* treatment of the impulse approximation as applied to s-state scattering from a two-body system with heavy point scatterers at r_A and r_B. The wave function of the system is

$$\Phi(r) = e^{i\mathbf{k}_0 \cdot \mathbf{r}} + A \frac{e^{ik|\mathbf{r}_A - \mathbf{r}_B|}}{|\mathbf{r}_A - \mathbf{r}_B|} + B \frac{e^{ik|\mathbf{r} - \mathbf{r}_B|}}{|\mathbf{r} - \mathbf{r}_B|}, \qquad (11.64)$$

which is the general solution of the wave equation outside the range of the scatterers. The outgoing amplitude A is related to the total wave amplitude at r_A by the equation

$$A = \eta_A \left\{ e^{i\mathbf{k}_0 \cdot \mathbf{r}_A} + B \frac{e^{ik|\mathbf{r}_A - \mathbf{r}_B|}}{|\mathbf{r}_A - \mathbf{r}_B|} \right\}, \qquad (11.65)$$

where $k\eta_A = e^{i\delta_A}\sin\delta_A$ and similarly for B. Solution of the resulting equations leads to the amplitude of the outgoing wave:

$$f(\theta) = \left[\eta\{\exp i(\mathbf{k}_0 - \mathbf{k})\cdot\mathbf{r}_A + \exp i(\mathbf{k}_0 - \mathbf{k})\cdot\mathbf{r}_B\} \right.$$
$$\left. + \eta^2 \frac{1}{R}\exp ikR\{\exp i(\mathbf{k}_0\cdot\mathbf{r}_A - \mathbf{k}\cdot\mathbf{r}_B) + \exp i(\mathbf{k}_0\cdot\mathbf{r}_B - \mathbf{k}\cdot\mathbf{r}_A)\} \right]$$
$$\times \left(1 - \eta^2 \frac{1}{R^2}\exp ikR \right)^{-1}, \tag{11.66}$$

where $R = |\mathbf{r}_A - \mathbf{r}_B|$ and for simplicity we take $\eta_A = \eta_B = \eta$.

The total cross-section is given in terms of the scattering amplitude by

$$\sigma_{\text{tot}} = \frac{4\pi}{k}\operatorname{Im}f(\mathrm{o}), \tag{11.67}$$

leading to the exact impulse approximation result:

$$\sigma_{\text{tot}}(\delta) = 2\sigma_0 \left[1 + \frac{1}{x^2}\sin x\cdot\sin(x + 2\delta) + \frac{1}{x^2}\cdot\sin^2\delta\cdot\sin(2x + \delta) \right.$$
$$\left. + \frac{1}{x^4}\sin^2\delta\cdot\sin^2 x \right]\cdot\left[\left(1 - \frac{1}{x^2}\sin^2\delta\right)^2 \right.$$
$$\left. + \frac{4}{x^2}\sin^2\delta\cdot\sin^2(x + \delta) \right]^{-1}, \tag{11.68}$$

where we have averaged over angles of R and introduced $x = kR$ and $\sigma_0 = (4\pi/k^2)\sin^2\delta$. In the limit $\delta \to \mathrm{o}$, we find

$$\sigma_{\text{tot}}(\mathrm{o}) = 2\sigma_0\left(1 + \frac{1}{x^2}\sin^2 x \right), \tag{11.69}$$

which is the Born approximation result. This is also obtained by the impulse approximation if multiple scattering is neglected. It is correct only if $\delta \to \mathrm{o}$ and not if $R \to \infty$, i.e. even for widely separated scatterers, neglect of multiple scattering is valid only if the Born approximation is applicable to the free particle case.

Fig. 87 shows the ratio of s-wave scattering to the sum $2\sigma_0$ of free particle cross-sections plotted as a function of $x = kR$ for various values of δ. The exact solution $\sigma_{\text{tot}}(\delta)$ deviates rapidly from the solution $\sigma_{\text{tot}}(\mathrm{o})$, even for rather small values of the phase shift. For large R, the disagreement between the two becomes small, being at a maximum for $\delta = \frac{1}{2}\pi$. The interference is negative for all values of R for phase shifts somewhat larger than $\frac{1}{4}\pi$.

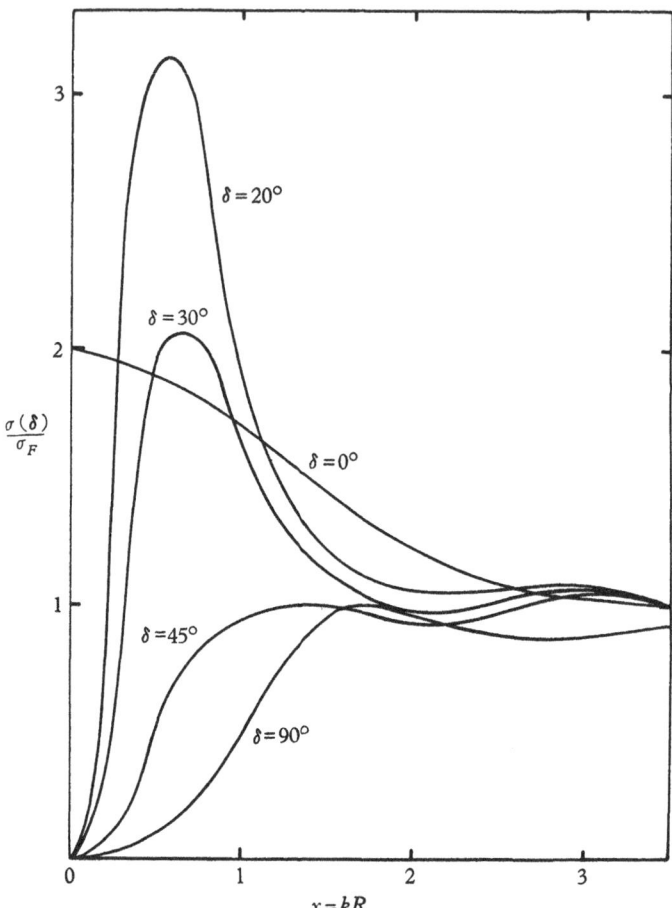

Fig. 87. *s*-wave scattering by two heavy point scatterers plotted as a function of kR, where R is the separation of the scatterers and $k = 1/\lambda$ the wave number (Brueckner, 1953). The ratio of the cross-section $\sigma(\delta)$ to the sum $\sigma_F = 2\sigma_0$ of the free particle cross-sections is given. The phase shifts δ for the two particles have been assumed equal. $\sigma(\delta)$ is the consequence of an exact treatment of the impulse approximation, whereas $\sigma(o)$ is the result obtained neglecting multiple scattering.

If we take R as the average separation of the deuteron particles, $R \approx 3{\cdot}2 \times 10^{-13}$ cm., then $\sigma_{\text{tot}}(o)$ may be regarded as approximating $\sigma_{\text{tot}}(\delta)$ satisfactorily for $x > x_0$, that is $E_{\text{Lab.}} > 4{\cdot}0x_0^2$. Taking $x_0 = 2{\cdot}1$ gives the error in using $\sigma_{\text{tot}}(o)$ as varying from about 15 % for $\delta = 20°$, to 25 % at $\delta = 30°$ and 30 % at $\delta = 45° - 90°$, and leads to

$E_{\text{Lab.}} > 18$ MeV. Similarly $x_0 = 3$ gives the error as $-4\cdot 5\%$ for $\delta = 20°\text{-}30°$, -3% at $45°$ and 16% at $\delta = 90°$; subject to $E_0 > 49$ MeV. Thus the impulse approximation with neglect of multiple scattering is valid for s-wave scattering for energies greater than about 36 MeV., provided the s-phase is less than about $50°$. At energies greater than about 50 MeV. it is satisfactory for all phase angles, so that results previously quoted at $90°$ should be quite valid.

The situation in actual scattering problems where other than s-phases are involved should in practice be more favourable to the impulse approximation (neglecting multiple scattering). If an s-phase should be near $90°$, it usually happens in practice that the p, d and higher phases are comparatively small, yet because of the $(2l+1)$ weighting factor in the partial cross-sections, they make large contributions to the total cross-section. If it is assumed, as seems reasonable, that neglect of multiple scattering is no more important for $l \geqslant 1$ than for $l = 0$, the total cross-section should then be more nearly correct than the s-wave cross-section. As an example, consider calculations (Buckingham *et al.* 1952) at the highest energy available, $16\cdot 6$ MeV., for quartet elastic n-d scattering using symmetric exchange forces. Results for the phase values are $\delta_0 = \pi - 118\cdot 2°$, $\delta_1 = 17\cdot 2°$, $\delta_2 = -11\cdot 1°$. These lead to the partial cross-sections

$$\sigma_0 = 271 \text{ mb.}, \quad \sigma_1 = 91 \text{ mb.}, \quad \sigma_2 = 65 \text{ mb.}$$

The two-body wave number is $k = 0\cdot 6364$, giving $x = 2\cdot 036$ so that the errors of the partial cross-sections are roughly 30, 12 and $7\cdot 5\%$ respectively for $l = 0$, 1 and 2, assuming the s-wave result in fig. 87 holds for all l. The resultant error in the total elastic cross-section is about 23%, or $\frac{3}{4}$ the error in the s-wave cross-section.

Of course $16\cdot 6$ MeV. is too low an energy for the impulse approximation to hold on other grounds, so that the error due to multiple scattering should be smaller at higher energies where s-phases are relatively less important and x is larger. In fact if we assume that the same phases hold at an energy of 25 MeV. $(x \approx 2\cdot 5)$, the errors in s-, p- and d- scattering being $16\cdot 5$, $1\cdot 5$ and $0\cdot 5\%$ respectively, then the final error is about 11% or about $\frac{2}{3}$ the error in the partial cross-section. Similarly at 37 MeV. $(x \approx 3)$, the errors in

s-, p- and d-scattering are $+1\cdot6$, $-0\cdot75$ and $-0\cdot36\%$ respectively, the final error being $-0\cdot5\%$ or less than $\frac{1}{3}$ the error in the s-wave result. If we had taken the variation of the phases into account, the final error would have been even smaller.

Summarizing the situation we may conclude that the error due to the neglect of multiple scattering in the impulse approximation decreases very rapidly with increasing energy for actual n-d scattering. It is less than about 10 % at 25 MeV., 5 % at 30 MeV., and 1 % for energies greater than 34 MeV., so that the impulse approximation should be satisfactory for $E > 30$ MeV.

11.1.9. *The energy spectrum of ejected protons.* Gluckstern and Bethe (1951) have used the Born approximation to find the energy spectrum of fast protons ejected at angles of 0, 10 and 30° in the Lab. system. This calculation is also possible with the impulse approximation.

It is found that at each angle the protons are nearly mono-energetic with an energy spread of the order of 1 MeV., illustrated in fig. 88 for 90 MeV. incident neutrons. The spectra are distinguished by sharp peaks associated with the high value of the singlet scattering length for low energy n-n collisions. However, the usual energy spread of the incident neutrons would make such an effect very difficult to observe. High energy monoenergetic neutron beams could also be produced by the similar process of bombarding deuterium with a high energy monoenergetic proton beam.

11.2. Deuteron stripping processes

If a target nucleus A is bombarded with a light particle a, producing another nucleus B and a very light particle b, in most cases the nuclear reaction concerned can be explained by Bohr's theory of the compound nucleus.

$$A + a \rightarrow C \rightarrow B + b. \qquad (11.70)$$

The nucleus C is conceived as an intermediate state of short life-time, the properties and mode of decay of which are largely independent of the manner of formation. However, if the total number of nucleons involved is very small, or if the particles a

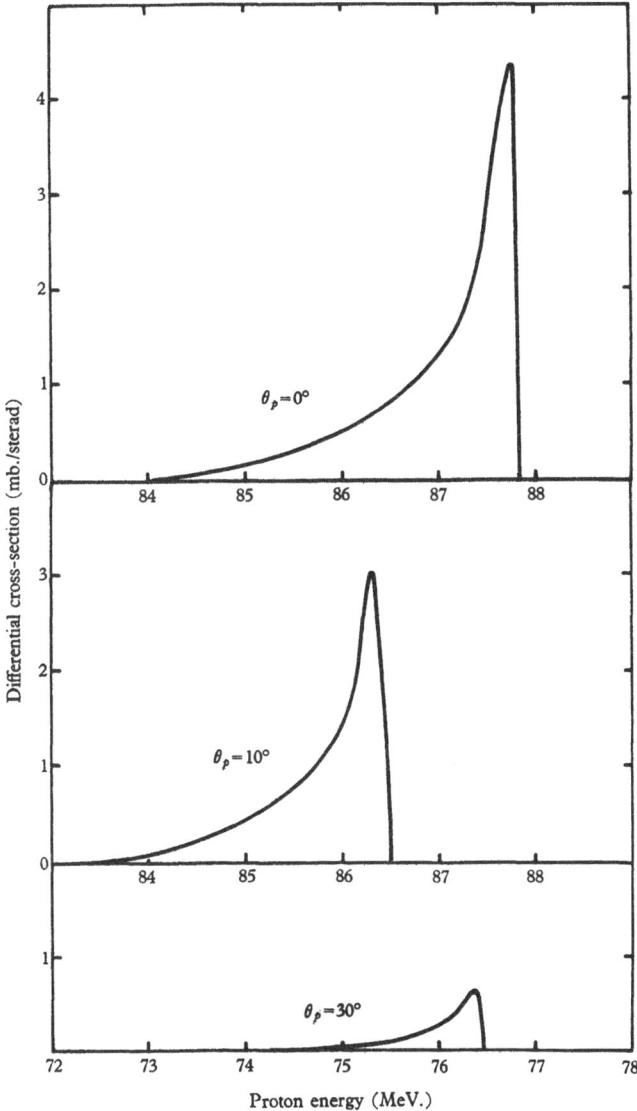

Fig. 88. Born approximation calculations of the energy spectra of the fast proton group resulting from the collision of 90 MeV. neutrons with deuterons (Gluckstern and Bethe, 1951).

have an energy greater than about 50 MeV., nuclear reactions can occur for which Bohr's theory is not valid. Two such reactions are stripping and the Oppenheimer–Phillips reaction.

Stripping can occur if the projectile is a light nucleus, one of whose constituent nucleons is weakly bound and described by a diffusely spread-out wave function, the deuteron being the best example. If the projectile does not make the head-on collision with the target nucleus which would result in formation of an intermediate nucleus C, but merely brushes close to it, a loosely bound nucleon may be stripped off and captured by the target:

$$A + a \to B + b. \qquad (11.71)$$

No intermediate state is involved, and the remainder of the stripped projectile passes on as b. The conditions for stripping are similar to those needed for the impulse approximation.

The inverse process is also possible in which projectile b captures a nucleon from target B, resulting in end products a and A. This is called 'inverse stripping' or the 'pick-up' process.

Almost all existing information relates to the deuteron as the stripped projectile. Its low binding energy of only $2 \cdot 23$ MeV. results in an average distance of separation d given by

$$d = \int | \Phi_0(r) |^2 r \, d\mathbf{r}. \qquad (11.72)$$

If we take $\Phi_0(r) = (\alpha/2\pi)^{\frac{1}{2}} e^{-\alpha r}$, $\alpha = (mW)^{\frac{1}{2}}/\hbar$, we find

$$d = 1/(2\alpha) = 2 \cdot 18 \times 10^{-13} \text{ cm}.$$

Actually the deuteron wave function has this approximate form only outside the range of nuclear forces; in fact if the finite range of nuclear forces is taken into account, d is increased to about $3 \cdot 2 \times 10^{-13}$ cm. However in considering the stripping effect we are interested only in the narrow neutron beam coming off nearly in the forward direction, and these neutrons are produced in collisions in which the neutron and proton are outside the force range at the time of the collision. Collisions occurring with neutron and proton inside the range of forces give rise to a wide angular distribution, similar to that resulting from direct nuclear encounters of the type (11.70), and may be included with the latter effects.

Thus within the limits of uncertainty inherent in the separation of the effects, it is proper to ignore the finite range of nuclear forces and to take $d = 2 \cdot 18 \times 10^{-13}$ cm.

Stripping reactions are described as either $A(d, p) B$ or $A(d, n) B$ according to whether a neutron or proton is stripped off. Other reactions which have also been observed are the pick-up process $B(d, t) A$ and $B(d, \alpha) A$ (Nadi, 1957).

The Oppenheimer–Phillips process concerns an opposite case, viz. the emission of protons by targets bombarded by very slow deuterons, the energy of the latter being well below the Coulomb energy of the target A. This makes the formation of a compound nucleus by barrier penetration very improbable. However Oppenheimer and Phillips (1935) point out that the stripping reaction (d, p) can occur if the neutron penetrates to the target surface, the protons being repelled at a distance of order d. This type of stripping is difficult to compute as the low energies involved necessitate use of the accurate wave functions for a slow deuteron in a Coulomb field.

Serber (1947) introduced the term 'stripping' to describe the observation (Helmholtz *et al.* 1947) that high energy deuterons striking a target produce narrowly collimated beams of neutrons and protons, each having approximately half the deuteron energy. The physical picture of the process is as follows: If a deuteron grazes the edge of a nucleus, one of its two nucleons may strike the nucleus while the other misses and continues on with little change of direction. The condition for this to occur is $E_d \gg E_0$, where E_0 is the deuteron binding energy and E_d its kinetic energy, for the collision time will then be short compared to the period of internal motion of the deuteron.

A small transverse momentum appears in the escaping nucleons due to the internal motion which they possessed in the deuteron. The internal momentum of a nucleon is $p_0 \approx (mE_0)^{\frac{1}{2}}$, where m is the nucleon mass. Its forward momentum is $p_d \approx (mE_d)^{\frac{1}{2}}$ and since p_i is randomly oriented in direction, the width at half maximum intensity of the forward cone of particles is, according to the classical model,

$$\Delta\theta \approx 2p_0/p_d = 2(E_0/E_d)^{\frac{1}{2}}.$$

More accurate theory replaces the factor 2 by $1 \cdot 6$ giving, for instance, a half width of $9 \cdot 9°$ for 190 MeV. deuterons on targets of light elements. This is in satisfactory agreement with theory. Heavier targets cause appreciable Coulomb deflexion of the deuterons, resulting in a somewhat wider spread (§ 4.2.5). Nucleons emerge from the stripping interaction with a range of energies likewise determined by the internal motion:

$$E = \frac{1}{2m}(p_d + p_0)^2 \approx \tfrac{1}{2}E_d \pm (E_d E_0)^{\frac{1}{2}}.$$

Wave mechanical theory obtains $1 \cdot 5(E_d E_0)^{\frac{1}{2}}$ for the width at half maximum.

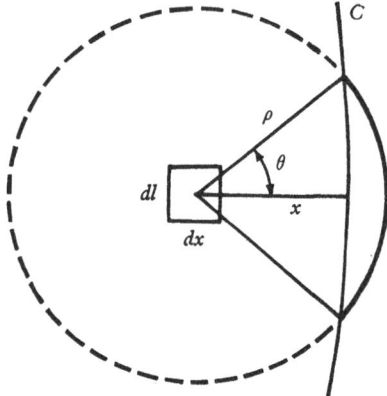

Fig. 89. Schematic diagram in a plane perpendicular to the direction of deuteron motion for the collision of the latter with a nucleus, circumference C.

To calculate the total cross-section for stripping on a semi-classical basis, we need consider only the projected positions of neutron and proton in a plane perpendicular to the deuteron's motion. This involves the probability that at the moment of collision the proton is within a circle C in this plane of radius equal to the nuclear radius R, while the neutron is outside it. Referring to fig. 89, this probability is seen to be θ/π.

Consider a collision in which the neutron-proton separation is ρ and let the nuclear radius R be large compared to ρ so that we can neglect the curvature of the former. The cross-section for the proton hitting at a distance $x = \rho \cos \theta$ inside the nucleus, within an interval dx, and within an interval dl along the circumference of the circle, is $dx\,dl$. The total cross-section for the proton hitting and neutron missing is

$$\sigma(\rho) = \iint \frac{\theta}{\pi} dx\,dl = 2\pi R \cdot \frac{\rho}{\pi} \int_0^{\frac{1}{2}\pi} \theta \sin\theta\, d\theta$$

$$= 2R\rho. \tag{11.73}$$

To find the total cross-section, we must multiply (11.73) by the probability of finding a separation ρ between the deuteron particles and integrate over all ρ. The probability is $|\Phi_0(r)|^2\,d\mathbf{r}$ for finding an n-p separation $\mathbf{r}$ in the volume element $d\mathbf{r}$, so that introducing cylindrical polar co-ordinates, the probability of a separation is

$$2\pi\rho \int_{-\infty}^{\infty} |\Phi_0(r)|^2\,dz,$$

where $r^2 = \rho^2 + z^2$. The total stripping cross-section becomes

$$\sigma = 4\pi R \int_{-\infty}^{\infty} dz \int_0^{\infty} |\Phi_0(r)|^2 \rho^2\,d\rho$$

$$= \frac{\pi}{2} R \int_0^{\infty} R\,|\Phi_0(r)|^2\,dr.$$

Using equation (11.23), we find for the cross-section the energy-independent value

$$\sigma = \tfrac{1}{2}\pi Rd. \tag{11.74}$$

Taking $d = 2\cdot18 \times 10^{-13}$ cm. and putting $R = 1\cdot5A^{\frac{1}{3}} \times 10^{-13}$ cm. gives $\sigma = 5\cdot1A^{\frac{1}{3}} \times 10^{-26}$ cm.^{2} and leads to stripping cross-sections ranging from $0\cdot1$ b. for Be to $0\cdot3$ b. for U (see also §4.2.5). Similar results are found for the (d,p) reaction. The use of a 190 MeV. beam of deuterons actually leaves the target nucleus in a very high state of excitation ($\sim$ 100 MeV.) and many further particles are boiled off subsequently.

An alternative process which is possible is electro-dissociation of the deuteron by the Coulomb charge of the nucleus. However even for the heaviest nuclei and very high deuteron energies, this effect is no more than 25 % of the stripping cross-section (Dancoff, 1947). It increases as the deuteron energy is lowered down to 30 MeV., then decreases rapidly again.

Stripping reactions of the (d,p) type only have thus been found to occur at low deuteron energies, and both the (d,p) and (d,n)

type at high deuteron energies. The conditions for validity of equation (11.74) used above, viz. that deflexion of the ejected deuteron particles by stripping is negligible and also $d \ll R$, may be employed to find the limiting energy for high energy stripping. The former condition, of course, follows from the assumption that

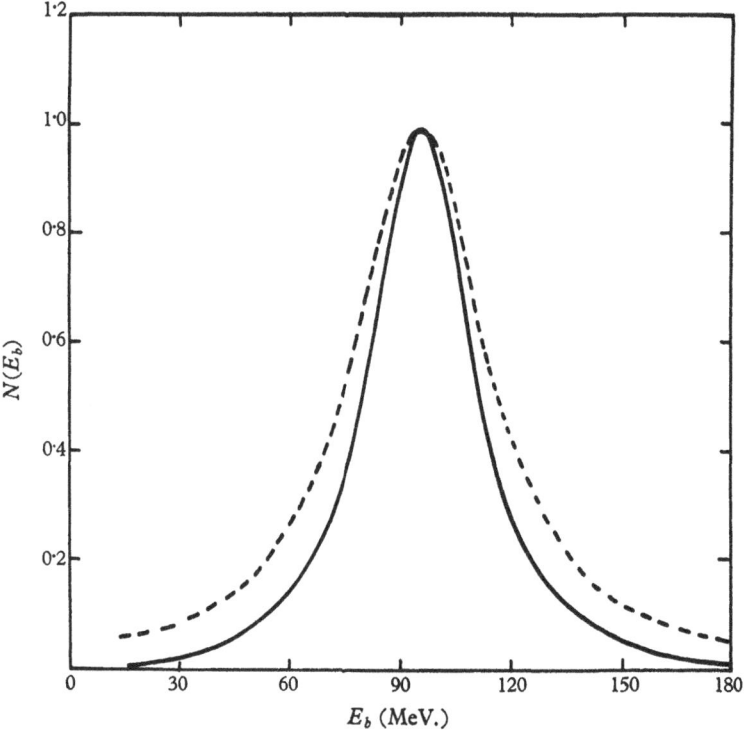

Fig. 90. Energy distribution of neutrons obtained from 190 MeV. deuterons. Solid curve, opaque nucleus; broken curve, transparent nucleus model (Serber, 1947).

the collision time is small compared to the period of the deuteron particles' internal motion.

In a typical impact, the proton will fail to clear the edge of the nucleus by a distance given approximately for $d \ll R$ by $l = (2Rd)^{\frac{1}{2}}$ in front of a plane through the centre of the nucleus. The neutron has a velocity normal to the direction of deuteron motion of the order of $(E_0/m)^{\frac{1}{2}}$, so its displacement in this direction is $(E_0/m)^{\frac{1}{2}} \, l/v$, where $v = (E_d/m)^{\frac{1}{2}}$ is the incident velocity. This displacement is unimportant provided it is small compared to d, $(E_0/m)^{\frac{1}{2}} l/v < d$, giving the condition

$$E_d > 2(R/d) E_0. \qquad (11.75)$$

Numerically this becomes $E_d > 3 \cdot 07 A^{\frac{1}{3}}$, so that the limiting energy varies from $6 \cdot 5$ MeV. for Be to about 19 MeV. for U. Unless this

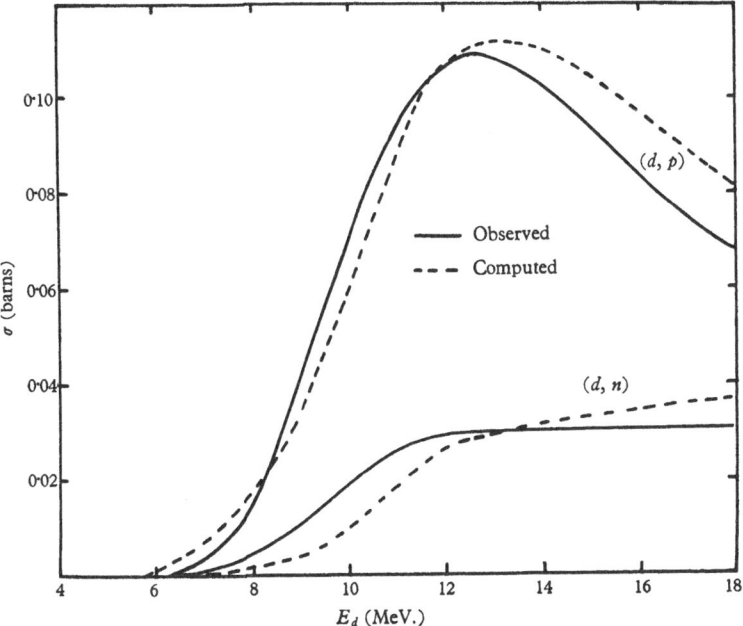

Fig. 91. (d, p) and (d, n) excitation curves for ^{209}Bi (Peaslee, 1948). Calculated values assume $R = 10 \cdot 7 \times 10^{-13}$ cm., with sticking factors of $0 \cdot 23$ and $0 \cdot 25$ for neutron and proton respectively.

condition is satisfied it is not valid to consider one constituent particle of the deuteron as interacting with the nucleus, while the other passes on uninfluenced. However, it is possible to adapt Serber's theory to deal with stripping in the deuteron energy range 2–15 MeV. (Peaslee, 1948). The additional assumption is made that the second particle in the deuteron may, dependent on its position when the first particle is captured, be captured also by the nucleus, to which a sticking factor may be assigned. The effect of the Coulomb field is also allowed for. A survey of the results of this theory is given in the review by Huby (1953). Indications are found for the importance of both (d,p) and (d,n) reactions at all energies.

Fig. 90 shows the energy distribution of neutrons calculated for 190 MeV. deuterons (Serber, 1947) assuming an opaque nucleus and a transparent nucleus respectively. Experimental results (Hanson, 1949) show good general agreement with the opaque nucleus model.

Fig. 91 shows (d,p) and (d,n) excitation curves for ^{209}Bi as calculated by Peaslee (1948) in the 6–18 MeV. region, again in good agreement with experiment. The energy distributions at medium energies are characterized by much broader peaks than at high energies, particularly the (d,n) curves. This would be expected because of the much larger ratio of internal to external momenta of the deuteron particles in the former case.

11.2.1. *Survey of differential stripping cross-sections.* So far the formation of the residual nucleus in a definite energy level and of individual energy groups of emitted particles has been neglected. This phenomenon can have an important effect on the angular distributions of particles emitted in the (d,p) and (d,n) reactions. Observations of this type using intermediate energy deuterons to bombard light nuclei have proved valuable in nuclear spectroscopy, as information about the spin-parity state of the residual nucleus can be deduced from the shape of the angular distribution. Analysis of the angular distribution curves into Legendre polynomials, as in §10.10, should lead to an upper limit on their order l equal to twice the maximum value of orbital angular momentum l_d of the incident deuterons participating in the reaction.

On the basis of the compound nucleus model, l_m is estimated as $p_d R/\hbar$, where p_d is the momentum of the deuteron, but this is found not to yield Legendre polynomials of high enough order to account for the above analysis. However, the stripping model enables the centre of mass of the deuteron particles to be at a distance of up to $\frac{1}{2}d$ or more outside the nucleus, so that $l_d \sim p_d(R+\frac{1}{2}d)\hbar$. Thus the shape of the angular distribution depends on the incident energy E_d. It also depends on the emitted nucleon energy E_b and the orbital angular momentum l_c with which the captured nucleon enters. The latter is limited by two selection rules:

(a) The spin J_B of the product nucleus must be such that it can be formed from the vector sum of the spin J_A of the target, the orbital angular momentum l_c of the absorbed nucleon, and its spin $\frac{1}{2}$. This requires

$$J_A + l_c + \tfrac{1}{2} \geqslant J_B \geqslant \text{Min.} \, | \pm J_A \pm l_c \pm \tfrac{1}{2} |. \qquad (11.76)$$

(b) If the level B has the same parity as A, l_c must be even, but if the parities are different, l_c is odd.

These rules severely limit the permissible values of l_c for given levels of A and B. If more than one value of l_c is permitted, the differential cross-section is composed of additive contributions from each l_c value. The principal maximum usually occurs at a small angle to the deuteron beam direction, the angle increasing with l_c. If $l_c = 0$, the maximum is usually at $0°$, but for $l_c \geqslant 1$ it generally occurs at a small angle. However, exceptions from this rule are known, with curves for $l_c = 1$ or 2 having maxima at $0°$ when E_b is small.

To show this, write $\mathbf{k}_b$ as the momentum of the ejected particle b at the moment of stripping and $\mathbf{k}_c$ as the momentum of the captured particle c at the same time. Then we have

$$\mathbf{k}_d = \frac{m_A}{m_B} \mathbf{k}_b + \mathbf{k}_c, \qquad (11.77a)$$

$$\mathbf{k}_b = \tfrac{1}{2} \mathbf{k}_d + \mathbf{K}_0, \qquad (11.77b)$$

where $\hbar \mathbf{K}_0$ is the internal momentum of the deuteron particle b and $\hbar \mathbf{k}_d$ is the external deuteron momentum in the CM system. From equation (11.77), we find on writing θ as the angle of stripping between $\mathbf{k}_d$ and $\mathbf{k}_b$:

$$K_0 = \left[(k_b - \tfrac{1}{2} k_d)^2 + 2 k_b k_d \sin^2 \frac{\theta}{2} \right]^{\frac{1}{2}}, \qquad (11.78)$$

$$k_c = \left[(k_d - k_b)^2 + 4 \frac{m_A}{m_B} k_b k_d \sin^2 \frac{\theta}{2} \right]^{\frac{1}{2}}. \qquad (11.79)$$

There will be a probability $\pi(K_0)$ for obtaining an internal momentum K_0 in the cross-section, which should fall off rapidly for $\hbar^2 K_0^2 / m > 2 \cdot 23$ MeV., so that $\pi(K_0)$ is a decreasing function of θ, illustrated in fig. 92 (a). Also, in the classical picture, the nucleon c is captured at or near the nuclear surface of radius R, so that

$$k_c(R-\epsilon)=l_c, \tag{11.80}$$

where ϵ is some quantity small compared to R which allows for capture at smaller impact parameters.

From equations (11.79) and (11.80), we see that for a fixed k_d and k_b there is a definite angular range of emission with its maximum at θ_{l_c}, which should be widened further by quantum effects that replace the nuclear boundary R by a probability density. Writing $L_{l_c}(k)$ as the probability density that the nucleon c is described by k_c and l_c at the nuclear boundary R, and P_{l_c} as the capture probability factor, itself dependent on l_c and the levels of A and B, we find for the differential cross-section,

$$d\sigma/d\omega = \pi(K_0) \sum_{l_c} P_{l_c} L_{l_c}(k_c). \tag{11.81}$$

If $l_c = 0$, equation (11.80) has no classical solution unless $k_d = k_b$. However as $\theta \to 0$, it will be more nearly satisfied from equation (11.79), so that the maximum occurs at $\theta = 0$. Similarly even for $l_c = 1$ or 2, it is still possible in the quantum-mechanical problem to get a maximum at 0°, provided k_b is small.

Results obtained using equation (11.81) as developed in § 11.2.3 are shown in fig. 92 for a deuteron energy $E_d = 6.9$ MeV. with emitted nucleon energy $E_b = 10.8$ MeV. and $R = 7.0 \times 10^{-13}$ cm. Fig. 92 (a) gives the momentum probability distribution $\pi(K_0)$ for the captured deuteron particle as a function of θ, while fig. 92 (b) shows the probability L_{l_c} of the nucleon being at the nuclear surface for several values of l_c, also plotted as a function of θ. Fig. 93 shows the resultant differential cross-section with its principal maximum at 0° and small secondary maxima at large angles.

Equation (11.81) gives good agreement with experiment, but in general the small secondary maxima are not observed because of a roughly isotropic background, probably due to contributions from the compound nucleus process. Good agreement is usually obtainable with one value of l_c only (the lowest possible), as cross-sections for larger l_c, although allowed by the selection rule, fall off rapidly with l_c and form only small corrections to the former.

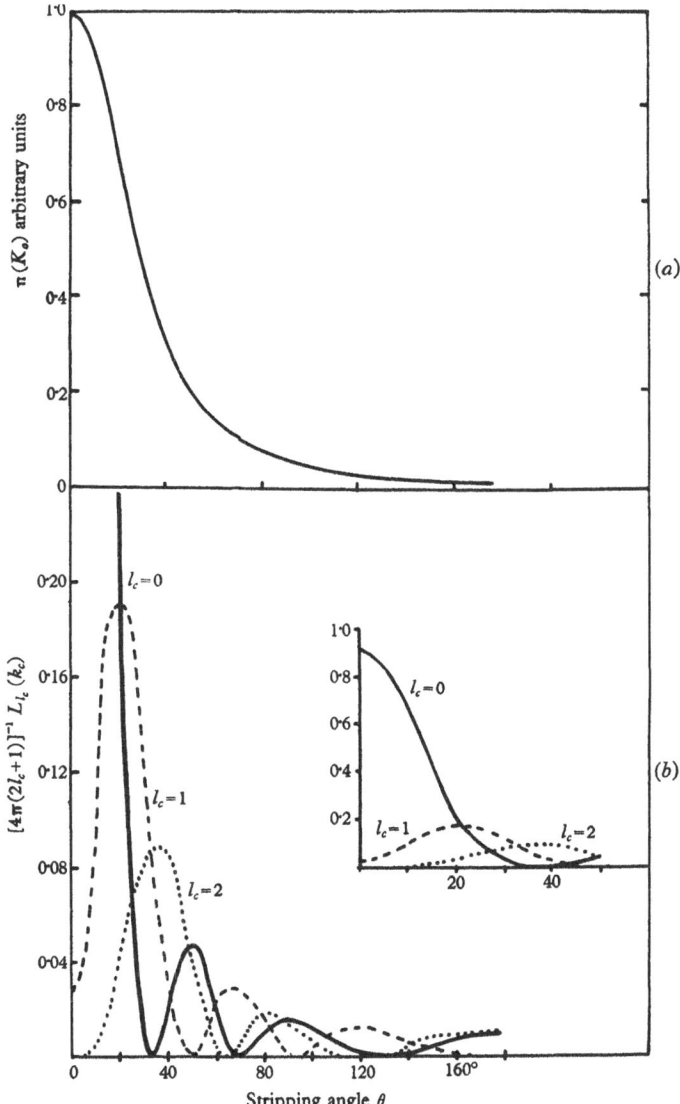

Fig. 92. Probability factors $\pi(K_0)$ and $L_{l_c}(k_c)$ plotted as functions of the stripping angle θ, using $E_d = 6\cdot9$ MeV., $E_0 = 10\cdot8$ MeV., $R = 8\cdot0 \times 10^{-13}$ cm. (Huby, 1953). $\pi(K_0)$ refers to the internal deuteron motion and L_{l_c} to the location of the captured particle b at the nuclear surface R with angular momentum l_c.

Differential stripping cross-sections can be analysed to find l_c so that if the spin-parity of the target A is known, the parity of B follows uniquely, while its spin is limited to certain alternative values. The latter may often be resolved using information from

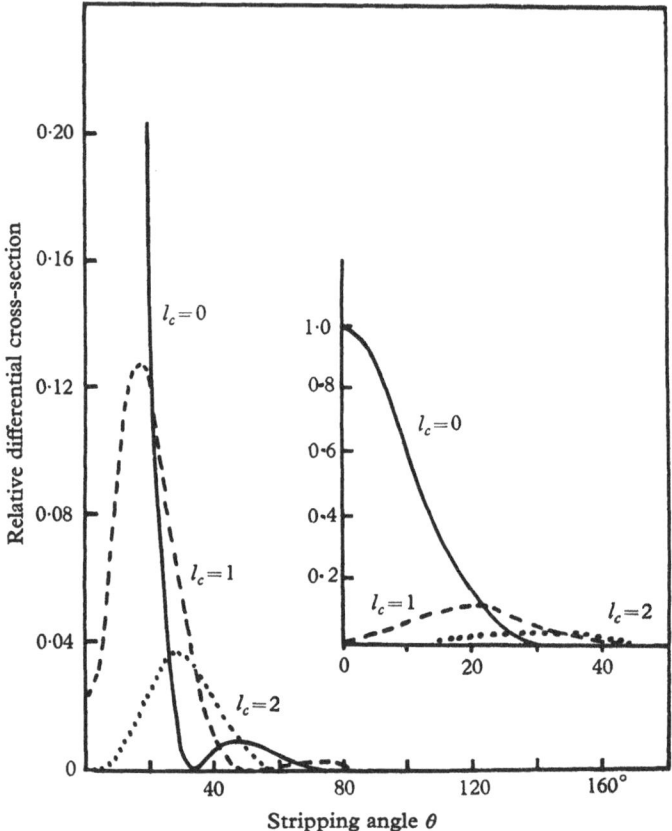

Fig. 93. Theoretical angular distributions for deuteron stripping, $E_d = 6 \cdot 9$ MeV., $E_0 = 10 \cdot 8$ MeV., $R = 7 \cdot 0 \times 10^{-13}$ cm. Target recoil is neglected. The formulae of Bhatia, Huang, Huby and Newns, (1952) are used (Huby, 1953).

other experiments. Moreover, as a weapon of nuclear spectroscopy, stripping has an advantage over the compound nucleus reaction in that it identifies ground or low excited states whereas the latter specifies only highly excited states.

Polarization of nucleons produced in stripping reactions has been investigated assuming the final state nucleon is scattered by a complex potential ('clouded crystal ball') plus an $L . S$ component (Chamberlain and Stern, 1954; Sawacki, 1957).

11.2.2. High energy deuteron stripping. To calculate the angular distribution and energy spectrum of stripped particles, it is simpler to assume the target nucleus as transparent instead of opaque to the incident beam. This means that at the moment of stripping when the captured nucleon c is hitting the target, the other nucleon b can either be inside or outside the target with a probability distribution defined by the deuteron ground state wave function $\phi_0(r)$. Defining the momentum distribution $g_0(K_0)$ of the nucleon b by equation (11.23), the probability $\pi(K_0) dK_0$ that the nucleon b has momentum between $\hbar K_0$ and $\hbar(K_0 + dK_0)$ is given by

$$\pi(K_0) = g_0^2(K_0), \qquad (11.82)$$

where we use the approximate form $\phi_0(r) = (\alpha/2\pi)^{\frac{1}{2}} e^{-\alpha r}/r$, which is valid for small r and hence large K_0. This leads to

$$\pi(K_0) = \frac{\alpha}{\pi^2(\alpha^2 + K_0^2)^2}. \qquad (11.83)$$

The probability that the total momentum $\hbar k_b$ of b on emission lies in the momentum range $dk_b = k_b^2 dk_b d\omega$ is thus

$$\pi(K_0) k_b^2 dk_b d\omega = (m/\hbar^2) \pi(K_0) k_b dE_b d\omega, \qquad (11.84)$$

so that the differential cross-section for particles of definite energy is proportional to $\pi(K_0)$, since $k_b \sim \frac{1}{2}k_d$.

The total differential cross-section for all emitted energies is found by integrating equation (11.84) over k_b, using equations (11.83) and (11.78). Assuming $k_d \gg \alpha$ and θ small, we find

$$d\sigma(\theta) \propto \frac{\epsilon}{2\pi(\epsilon^2 + \theta^2)^{\frac{3}{2}}} d\omega, \qquad (11.85)$$

where $\epsilon = 2\alpha/k_d = (E_0/E_d)^{\frac{1}{2}}$. The beam is peaked in the forward direction with a half-width $\Delta\theta = 1.533\epsilon$. Integration of equation (11.84) over $d\omega$ likewise leads to the energy spectrum

$$d\sigma(E_b) = \frac{1}{\pi} \frac{(E_0 E_d)^{\frac{1}{2}}}{[(E_b - \frac{1}{2}E_d)^2 + E_0 E_d]} dE_b, \qquad (11.86)$$

which has its maximum at $E_b = \frac{1}{2}E_d$, with a half-width

$$\Delta E_b = 2(E_0 E_d)^{\frac{1}{2}}.$$

This treatment has been corrected by Serber to allow for relativistic effects and a broadening of the beam due to multiple scattering in the target. Moreover, the effect of the Coulomb repulsion of the target nucleus on the deuteron has been neglected.

In the case of neutron stripping, the deuteron loses kinetic energy Ze^2/R in approaching the nucleus so that its effective energy on impact is $E_d - Ze^2/R$, and the most probable neutron energy becomes $\frac{1}{2}(E_d - Ze^2/R)$. On the other hand, proton stripping gives a deuteron energy on impact of $E_d - Ze^2/R'$, where $R' = R + d$ allows for the proton being outside the nucleus during dissociation of the deuteron. As the proton recovers this Coulomb energy on the way out, the most probable proton energy becomes

$$\tfrac{1}{2}(E_d - Ze^2/R') + Ze^2/R' = \tfrac{1}{2}(E_d + Ze^2/R'),$$

so that the mean proton energy is increased by the Coulomb effect.

The effect of the Coulomb field on the angular distribution may also be allowed for. The deuteron is found to be deviated from its path by the angle $\quad \theta_c = Ze^2/2RE_d \quad$ (11.87)

before reaching the target nucleus. The momentum the ejected neutron b takes from the deuteron is thus altered from $\frac{1}{2}\hbar k_d$ to $\frac{1}{2}\hbar(k_d + \theta_c k_d \mathbf{v})$, where $\mathbf{v}$ is the outward (radial) unit vector at the impact point. The variation of $\mathbf{v}$ round the circumference of the nucleus results in an angular broadening of the emitted neutron beam.

Proton emission causes a further deflexion $2\theta_c$ after ejection, although the proton approaches closely to the target nucleus. The calculated increase of half-width is roughly proportional to θ_c^2.

Modification of the transparent nucleus calculations to apply to the opaque nucleus case involves reducing the integration range implied in equation (11.82), to a value for which the nucleon b is outside and the nucleon c inside the target nucleus. This results in the half-width of the angular distribution being increased from $1 \cdot 533(E_0/E_d)^{\frac{1}{2}}$ to $1 \cdot 6(E_0/E_d)^{\frac{1}{2}}$. The energy distribution similarly becomes

$$d\sigma(E_b) = \tfrac{1}{4}\pi R d \cdot \frac{E_0 E_d}{[(E_b - \frac{1}{2}E_d)^2 + E_0 E_d]^{\frac{3}{2}}} dE_b \qquad (11.88)$$

so that its half-width is reduced from $2(E_0 E_d)^{\frac{1}{2}}$ to $1 \cdot 533(E_0 E_d)^{\frac{1}{2}}$.

11.2.3. *Angular distributions at intermediate energies, simple stripping theory.* The simplest treatment of the angular distributions in (d, n) and (d, p) stripping involves an adaptation of the Born approximation, due to Bhatia *et al.* (1952) and Daitch and French (1952), assuming a transparent nucleus. It is difficult to justify use of such a perturbation treatment at medium energies, but as the differential cross-section turns out to be mainly determined by general features such as change of angular momentum and parity, it is probable that the results are largely independent of the details of treatment. Moreover the concentration of stripping distributions at small angles even for low energy deuteron beams is in favour of the weak interaction assumed by perturbation theories.

Butler's (1950, 1951, 1952) non-perturbation treatment allows rigorously for strong interaction with the target nucleus by joining up the wave functions for the captured nucleon at a radius r_0 (opaque nucleus). However, approximations are still made, some effects of which are difficult to assess, notably the assumption that inside the radius r_0, one may neglect the interactions of the captured nucleon with the other deuteron particle and with the target nucleus. Butler's formula has the advantage that it relates the capture probability factor P_{l_c} to significant nuclear quantities in a more direct manner. However, it breaks down for a capture probability E_c nearly zero, where

$$E_c = -(E_b - E_d - E_0). \tag{11.89}$$

On the other hand the Born approximation result behaves quite normally in this energy region. Neither formula takes account of Coulomb effects, yet they give good agreement with experiment even for low deuteron energies or heavy nuclei where these effects ought to be large. In both theories the interaction between the target nucleus A and the particle b which emerges is neglected.

The differential cross-section for the $A(d, b)B$ reaction in Born approximation for a transparent nucleus is

$$d\sigma/d\omega = \frac{\mu_b \mu_d}{4\pi^2 \hbar^4} \cdot \frac{k_b}{k_d} |I|^2, \tag{11.90}$$

where $\mu_b = m_b m_B/(m_b + m_B)$ and $\mu_d = m_d m_A/(m_d + m_A)$

are the reduced masses of the emitted particle b and the deuteron respectively, k_b and k_d are their relative wave numbers, and

$$I = \iiint \phi^*_{M_B}(\mathbf{r}_c, \mathbf{r}_A, \sigma_c) \chi^*_b(\sigma_b) \, e^{-i\mathbf{k}_b \cdot \mathbf{r}_b} \, V(r_c, r_A) \, \phi_{M_A}(\mathbf{r}_A)$$

$$\times \phi_0(\mathbf{r}_{bc}) \chi_b(\sigma_b) \chi_c(\sigma_c) \, e^{i\frac{1}{2}\mathbf{k} \cdot (\mathbf{r}_b + \mathbf{r}_c)} \, d\mathbf{r}_b \, d\mathbf{r}_c \, d\mathbf{r}_A. \qquad (11.91)$$

ϕ_0, ϕ_{M_A} and ϕ_{M_B} are the ground state wave functions of the deuteron, the target nucleus A and the target nucleus B respectively, χ is the spin function for a nucleon, and the spins of nuclei A and B are neglected. $V(r_c, r_A)$ is the interaction potential between the nucleons in A (co-ordinates r_A) and the captured nucleon c, and M_A, M_B and m_c are the magnetic quantum numbers describing A, B and c.

We select first from (11.91) the integral over $d\mathbf{r}_A$:

$$I' = \int \phi^*_{M_B}(\mathbf{r}_c, \mathbf{r}_A, \sigma_c) \, V(r_c, r_A) \, \phi_{M_A}(\mathbf{r}_A) \, d\mathbf{r}_A, \qquad (11.92)$$

which is a function of r_c. The capture is assumed to take place at the radius R so that we may expect $I' \propto \delta(r_c - R)/R^2$. Expanding I' in spherical harmonics leads to

$$I' = \frac{\delta(r_c - R)}{R^2} \sum_{l_c, m_c, \mu_c} Y^{m_c^*}_{l_c}(\theta_c, \phi_c) \langle M_B \,|\, V \,|\, M_A, l_c, m_c, s_c \rangle \chi^*(\sigma_c),$$

$$(11.93)$$

where the Dirac brackets $\langle M_B \,|\, V \,|\, M_A, l_c, m_c, \mu_c \rangle$ are numerical coefficients and $s_c = \pm \frac{1}{2}$ is the spin quantum number of c. Substitution of equation (11.93) in (11.91) yields

$$I = \sum_{l_c, m_c, \mu_c} \langle M_B \,|\, V \,|\, M_A, l_c, m_c, s_c \rangle \iint \chi^*_{\mu_c}(\sigma_c) \chi^*_b(\sigma_b) \, e^{-i\mathbf{k}_b \cdot \mathbf{r}_b} \, Y^{m_c^*}_{l_c}(\theta_c, \phi_c)$$

$$\times e^{i\frac{1}{2}\mathbf{k}_d \cdot (\mathbf{r}_b + \mathbf{r}_c)} \, \phi_0(\mathbf{r}_{bc}) \chi_b(\sigma_b) \chi_{\mu_c}(\sigma_c) \, d\mathbf{r}_b \, d\mathbf{r}_c. \qquad (11.94)$$

The volume integral involving $\mathbf{r}_c$ is taken on the surface of the nucleus, $r_c = R$, and may be evaluated easily, assuming the approximate form

$$\phi_0(r) = (\alpha/2\pi)^{\frac{1}{2}} e^{-\alpha r}/r$$

for the ground state wave function and taking $d\mathbf{r}_c = r^2 \, dr_c \, d\omega$. After substitution in equation (11.90), summation over the final magnetic quantum numbers M_B and averaging over the initial M_A, the differential cross-section is of the form (11.81):

$$d\sigma/d\omega = \pi(K_0) \sum_{l_c} P_{l_c}(k_c), \qquad (11.95)$$

where K_0 and k_c are defined by equations (11.78) and (11.79) and $\pi(K_0)$ is the deuteron probability factor (11.82) of Serber's theory. $L_{l_c}(k)$ is the kinetic factor:

$$L_{l_c}(k) = \sum_{m_c} \left| \int Y_{l_c}^{m_c*}(\theta_c, \phi_c) e^{i\mathbf{k}_c \cdot \mathbf{r}_c} d\omega_c \right|^2$$

$$= 4\pi(2l_c + 1)[j_{l_c}(kR)]^2, \qquad (11.96)$$

where $j_l(x)$ is the spherical Bessel function,

$$j_l(x) = (\pi/2x)^{\frac{1}{2}} J_{l+\frac{1}{2}}(x). \qquad (11.97)$$

The capture probability factor P_{l_c} is

$$P_{l_c} = \frac{2\pi\mu_b\mu_d}{\hbar^4} \frac{k_b}{k_d} \frac{(2J_B + 1)}{2(2J_A + 1)(2l_c + 1)} \sum_{M_A, m_c, s_c} |\langle M_B | V | M_A, l_c, m_c, s_c \rangle|^2.$$

$$(11.98)$$

Here the angular momentum factors represent the ratio of the weights of the final level B to that of the initial system $(A + C)$, and the factor of 2 in the denominator gives the statistical weight of the spin of c. P_{l_c} is subject to the selection rules (a) and (b) of §11.22. Writing

$$A_{l_c} = \sum_{M_A, m_c, s_c} |\langle M_B | V | M_A, l_c, m_c, s_c \rangle|^2,$$

we find for the differential cross-section,

$$d\sigma/d\omega = \frac{4\pi^2\mu_b\mu_d}{\hbar^4} \frac{k_b}{k_d} \frac{(2J_B + 1)}{(2J_A + 1)} \pi(K_0) \sum_{l_c} A_{l_c} j_{l_c}^2(kR). \quad (11.99)$$

where A_{l_c} depends on the nuclear model employed.

The alternative treatment developed by Butler (1950, 1951, 1952, 1957; Butler and Hittmair, 1957) and other authors (Friedman and Tobocman, 1952; Huber and Baldinger, 1952; Dalitz, 1953; Hittmair, 1955), although not obstensibly a perturbation one, has since been found to be obtainable from a Born type matrix element. This is discussed more fully by Huby (1953), see also §11.2.4. The formula obtained is of the form

$$d\sigma/d\omega = \frac{4\pi^2\mu_b\mu_d}{\hbar^4} \frac{k_b}{k_d} \frac{(2J_B + 1)}{(2J_A + 1)} \pi(K_0)$$

$$\times \sum_{l_c} B_{l_c}[M_{l_c} j_{l_c}(kr_0) + N_{l_c} \cdot j_{l_c-1}(kr_0)]^2, \quad (11.100)$$

NUCLEAR SCATTERING

where M_{l_c} and N_{l_c} are numerical functions of r_0. For numerical tables of the deuteron factor $\pi(K_0)$ and the kinetic factors following, see Lubitz and Parkinson (1955), Enge and Grave (1956).

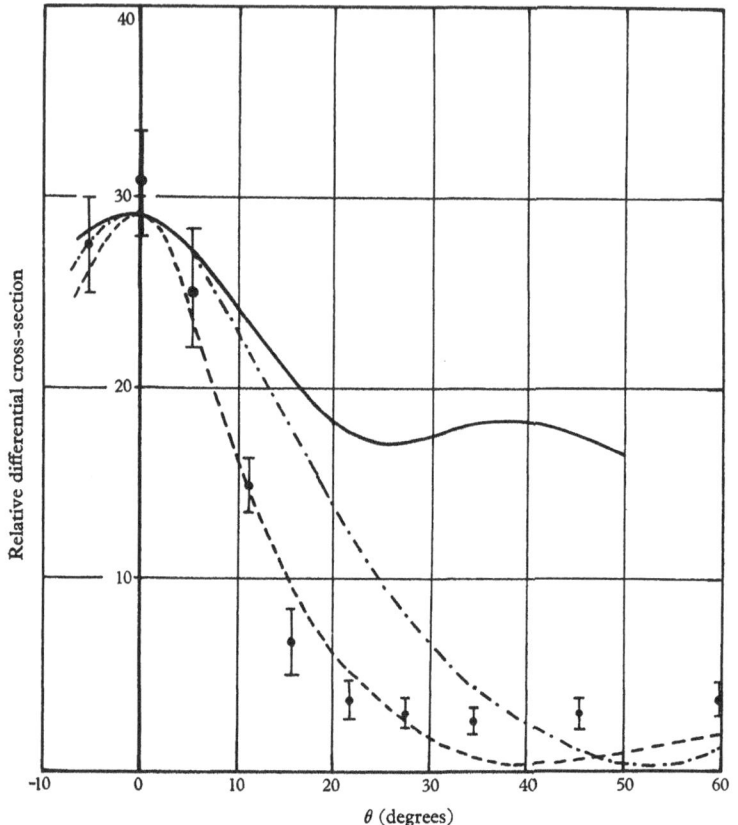

Fig. 94. ^{12}C(d, n) ^{13}N. Measured angular distribution (CM system) of neutrons leaving ^{13}N in excited state at 2·38 MeV. E_d(Lab.)=8·13 MeV., $Q_1 = -2·62$ MeV., $E_c = 0·41$ MeV.
   Full curve: theory of Butler, for $l_c = 0$, $r_0 = 4·5 \times 10^{-13}$ cm.
   Dot-dash curve: theory of Butler, for $l_c = 1$, $r_0 = 4·5 \times 10^{-13}$ cm.
   Broken curve: theory of Bhatia et $al.$, for $l_c = 0$, $R = 6·3 \times 10^{-13}$ cm. (Huby, 1953).

Equation (11.100) differs from (11.99) essentially in the extra term $j_{l_c-1}(kr_0)$ in the square bracket. This generally involves only a small correction and either formula can usually be made to fit an experimental curve equally well if suitable values of r_0 and R are chosen. However, the latter always has to be chosen larger than the former, for whereas the Born approximation formula is best fitted by $R = (1 \cdot 5 A^{\frac{1}{3}}) \times 10^{-13}$ cm, Butler's theory gives the best fit with the empirical formula (Gamow and Critchfield, 1949):

$$r_0 = (1 \cdot 7 + 1 \cdot 22 A^{\frac{1}{3}}) \times 10^{-13} \text{ cm.}$$

However, the latter is a more consistent fit.

If r_0 is taken equal to R, the maxima of the Butler curve are displaced to smaller angles relative to those of the Born approximation, and the secondary maxima at large angles are relatively greater in magnitude. On the whole, Butler's theory tends to overestimate cross-sections. Also when $E_c \sim 0$, Butler's curve becomes non-typical, especially for $l_c = 0$, for the principal maxima may either disappear or lose their dominance over the maxima at large angles. For $E_c = 0$ and $l_c = 0$, the cross-section actually vanishes owing to the vanishing of P_0. Summing up, Butler's theory involves a correction to the Born approximation result, but this correction breaks down for $E_c \sim 0$ and $l_c = 0$.

Fig. 94 shows the measured angular distribution in the CM system for the ^{12}C(d, n) ^{13}N reaction in which $E_d = 8 \cdot 13$ MeV. and ^{13}N is left in an excited state at $2 \cdot 38$ MeV., so that $E_c = 0 \cdot 41$ MeV. is small and therefore unfavourable to Butler's theory. The curve clearly requires $l_c = 0$ rather than $l_c = 1$, and the Born approximation result fits the experiment more closely than the $l_c = 1$ curve of Butler. As $J_A = 0$ for ^{12}C, the rule (11.76) gives $J_B = \frac{1}{2}$. Also as the parity of ^{12}C is even and l_c is even, the excited state of ^{13}N at $2 \cdot 38$ MeV must have even parity.

Fig. 95 gives the measured angular distribution, again for the ^{12}C(d, n) ^{13}N reaction, but using $E_d = 8 \cdot 13$ MeV. and leaving ^{13}N in the ground state. Both the Born approximation and Butler's theory give similar curves in reasonable agreement with experiment, provided $l_c = 1$. This again limits the total angular momentum of the product nucleus to $J_B = \frac{3}{2}$ or $\frac{1}{2}$ and fixes its parity as odd. In both cases the energy levels of the residual nucleus can be found via measurement of the Q values, using the formulae

$$Q_s = E_b - E_d = \mathscr{E}_B - \mathscr{E}_d - E_{B,s}, \qquad (11.101)$$

$$\begin{aligned} E_c &= E_{B,s} - \mathscr{E}_B = -(Q_s + \mathscr{E}_d) \\ &= -(E_b - E_d + \mathscr{E}_d), \qquad (11.102) \end{aligned}$$

where Q_s is the change in kinetic energy, $\mathscr{E}_B$ and $\mathscr{E}_d$ are the binding energies of the ground state levels of the product nucleus B and of the deuteron respectively, and $E_{B,s}$ is the excitation energy of the level in which B is formed. E_c is the energy with which the captured nucleon can escape from the target. Thus if Q_s is found and the ground state energy $\mathscr{E}_B$ is known, the excitation energy $E_{B,s}$ follows immediately.

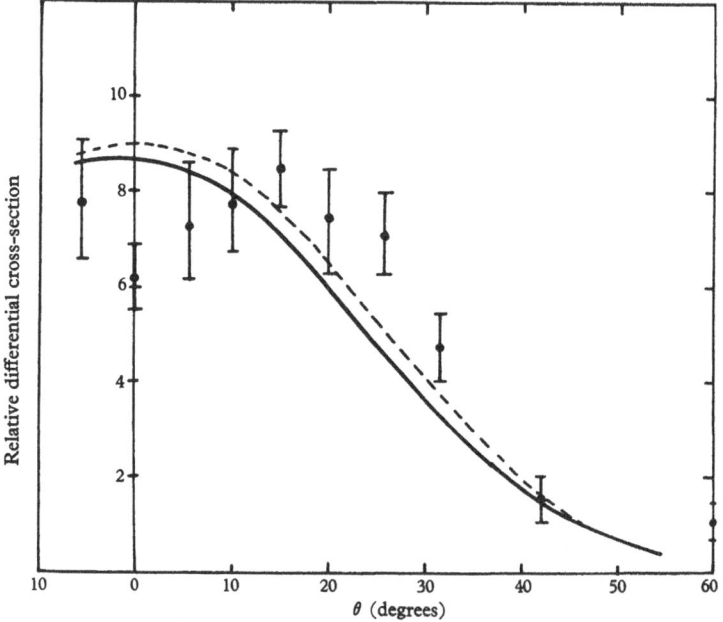

Fig. 95. ^{12}C(d, n) ^{13}N. Measured angular distribution (CM system) of neutrons leaving ^{13}N in ground state. E_d(Lab.) $= 8\cdot13$ MeV., $Q_0 = -0\cdot24$ MeV.
Full curve: theory of Butler, for $l_c = 1$, $r_0 = 4\cdot5 \times 10^{-13}$ cm.
Broken curve: theory of Bhatia *et al.*, for $l_c = 1$, $R = 5\cdot5 \times 10^{-13}$ cm. (Huby, 1953).

11.2.4. *Stripping distributions with nuclear and Coulomb corrections.* Both the transparent nucleus theory ('Born approximation') of Bhatia, Daitch and French and the opaque nucleus theory of Butler neglect the reaction effects of the various outgoing waves in the stripping process. They also ignore all Coulomb effects. Horowitz and Messiah (1953 a, b) have discussed the relation between these theories, showing that Butler's theory may be effectively obtained from a Born approximation matrix element using a linear combination of plane waves for the incident deuteron and the ejected particle b. The difference lies in Butler's theory excluding the range of integration $r \leqslant r_0$, where r_0 is the total range of inter-action between the captured nucleon c and the target nucleus A. It is therefore hardly surprising that the two theories differ so little. However, at high energies (~ 100 MeV.), nuclear trans-parency increases considerably so that it is no longer valid to neglect contributions to the matrix element for $r < r_0$ as in Butler's theory.

As pointed out in § 11.2.3, Butler's opaque nucleus theory breaks down for a capture energy $E_c \sim o$. A like result can be expected for the reverse action, viz. the pick-up process, when the binding energy E_c of the picked-up nucleon in the target nucleus is small. Glashow and Selove (1956) have treated the ^{9}Be(p, d)^{8}Be pick-up reaction, $E_c = 1 \cdot 67$ MeV., showing that the Butler theory fails whereas the 'Born approximation' transparent nucleus theory works well over a wide energy range. The interpretation is that ^{9}Be consists of a tightly bound core with a loosely bound neutron circulating round it, the net picture being inconsistent with a sharp nuclear radius as in Butler's theory but rather with a transparent nucleus. This situation should occur most easily for target nuclei which can be pictured as being built up out of α-particles, i.e. $N = Z = 2n$ (n any positive integer), provided $|E_c| \leqslant 2$ MeV. Table XX shows a number of cases of this type for $A \leqslant 20$, some involving the nth excited state of the product nucleus, capture energy E_c^{n*}. Fig. 94 illustrates the ^{12}C(d, n)^{13}N reaction, $E_c^* = o \cdot 41$ MeV., where Butler's theory breaks down badly. Many other cases exist for nuclei of arbitrary (N, Z), but usually involving excited states of the product nucleus. However, if E_c is large, Butler's opaque nucleus theory may be expected to work better than the transparent nucleus theory; in fact subsequent treatments involving nuclear and Coulomb corrections usually take it as a starting point.

It has proved very difficult to include in Butler's original theory the effects of Coulomb interactions and the neglected interaction between the outgoing stripped nucleon and the product nucleus, owing to the complexity of the integrals involved in matching boundary conditions at a radius r_0. Horowitz and Messiah (1953 a, b) assumed a hard sphere interaction for the latter effect and also

NUCLEAR SCATTERING

allowed for the finite nuclear mass by replacing masses by their corresponding reduced masses, but numerical results proved to be poor. Another outline treatment by Francis and Watson (1954) used the many-body approach to the problem, but does not appear adaptable to numerical calculations.

TABLE XX. *Stripping reactions with* $|E_c^{n*}| \leqslant 2$ MeV., *where* E_c^{n*} *denotes the energy of the captured nucleon in the nth excited state of the product nucleus*

The target nuclei are for $A \leqslant 20$ and all have α-particle type cores, i.e. $N = Z = 2n$ (n any positive integer). Butler's opaque nucleus stripping theory is unsatisfactory in these cases, but the transparent nucleus 'Born approximation' should be valid.

Reaction	E_c	E_c^{1*}	E_c^{2*}	E_c^{3*}	E_c^{4*}	E_c^{5*}	E_c^{6*}
$^4\text{He}(d, p)\,^5\text{He}$	0·8	.	.	.	.	.	.
$^8\text{Be}(d, p)\,^9\text{Be}$	−1·67	−0·08	.	.	.	.	.
$^8\text{Be}(d, n)\,^9\text{B}$	0·15	1·55	.	.	.	.	.
$^{12}\text{C}(d, n)\,^{13}\text{N}$	−1·96	0·41	.	.	.	.	.
$^{12}\text{C}(d, p)\,^{13}\text{C}$	−4·9	−1·81	−1·22	−1·04	−2·0	.	.
$^{16}\text{O}(d, p)\,^{17}\text{O}$	(−4·14)	(−3·27)	−1·08	−0·29	0·42	0·94	1·09
$^{16}\text{O}(d, n)\,^{17}\text{N}$	1·96	.	.	.	.	.	.
$^{20}\text{Ne}(d, p)\,^{21}\text{Ne}$	(−6·62)	−1·91	−1·18	−0·88	0·04	0·68	1·66
$^{20}\text{Ne}(d, n)\,^{21}\text{Na}$	(−3·32)	0·25	0·88	1·0	1·16	1·73	.

On the other hand, a Born approximation treatment of Butler's formula can be modified fairly easily to allow for Coulomb and nuclear corrections. This was done by Tobocman and Kalos (1955), who eliminated the troublesome internal range of integration $r < r_0$ in favour of a set of parameters involving the logarithmic derivative of the wave function at the nuclear surface. This also compensated at least partly for the neglect of distortion of the incident deuteron plane wave used in the Born approximation. The formula obtained reduces to Butler's result on assuming zero corrections and putting the n-p range of force equal to zero. For treatment of heavy target nuclei, see Biedenharn, Boyer and Goldstein (1956).

Numerical examples show that introduction of the Coulomb interaction shifts the angular distribution towards larger angles, broadening the peaks and filling in the valleys. On the other hand, nuclear corrections (e.g. proton absorption) give opposite effects, peaks being narrowed and displaced towards smaller angles, so that partial cancellation with the Coulomb correction may occur.

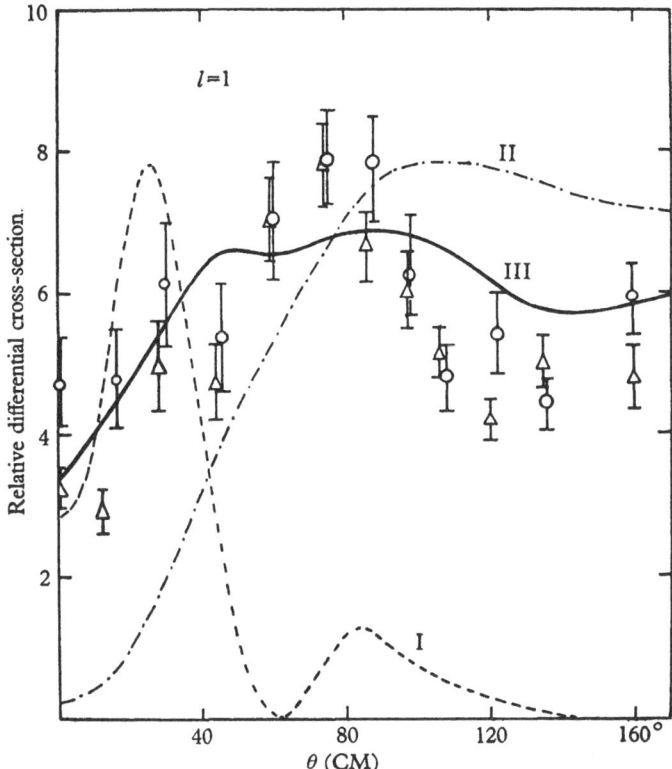

Fig. 96. Measured angular distribution of protons from ^{48}Ti(d, p) ^{49}Ti (Pratt, 1955), leaving ^{49}Ti in the two excited states at 1·40 and 1·74 MeV. Triangles represent the reaction with $Q = 4·46$ MeV. and circles the reaction with $Q = 4·12$ MeV. The incident deuteron energy is $E_d = 2·60$ MeV. (Lab.)

Theoretical curves by Tobocman and Kalos (1955) comprise:

    I. Simple stripping on Butler's opaque nucleus theory.

    II. Butler's theory corrected for Coulomb interactions.

    III. Butler's theory corrected for Coulomb interactions and absorption of protons with $l \leqslant 1$. The nuclear radius used is $R = 6·49 \times 10^{13}$ cm.

The experimental distribution is in arbitrary units, with III normalized to give a best fit.

This is illustrated in fig. 96 for the ^{48}Ti(d, p) ^{49}Ti reaction, where the experimental angular distribution is non-typical in shape, i.e. very far from the sharp peaks and valleys of simple stripping as given by Butler's theory. The fully corrected theory works well in this and other cases which have been evaluated. In some cases, however, e.g. the ^{13}C(d, p) ^{14}C reaction, an additional correction is needed for the competing compound nucleus process and interference effects (Marion and Weber, 1956). Heavy particle stripping has also been invoked in some cases to explain backward components of stripping, e.g. $^{11}B(d, n)$ ^{12}C (Owen and Madansky, 1957).

A measure of the applicability of simple stripping theory is the ratio of maximum to minimum in the scattering distribution, which should be large for the theory to work well. A small ratio indicates large nuclear, Coulomb and compound nucleus corrections. Both nuclear and Coulomb corrections reduce the total cross-section from its simple stripping value. However the corrected total cross-sections are in many cases too small by factors of two or more, constituting an as yet unsolved problem (Butler, 1957). This may be due to the lack of a completely self-consistent nuclear model, necessary to determine the factors A_{l_c} and B_{l_c} (cf. equations (11.99), (11.100)) of simple stripping theory and also the various corrections.

11.3. High energy scattering of neutrons by α-particles

If helium is bombarded with high energy neutrons, the processes listed in the first column of Table XXI may occur. Proton bombardment leads to a similar classification. As cases (4)–(6) require more energy to be given up by the incident particle, they are much less probable and hence only cases (1)–(3) are important. In fact the cross-section for (5) can be estimated as about 1 % of that for (1).

TABLE XXI. *Comparison of experiments at* 90 MeV. (*Tannenwald,* 1953) *with Born approximation theory cross-sections for collision of neutrons with* α-*particles*

	Process	Experiment	Born approximation
(1)	$^4He + n$ (elastic)	96 ± 17 mb.	160 mb.
(2)	$\begin{cases} {}^3He + n + n \\ {}^3H + p + n \end{cases}$	16 (assumed) 42 ± 6	6 60 ± 10
(3)	$^3H + d$	13 ± 2·5	16
(4)	$d + d + n$	7 ± 1·5	0
(5)	$d + n + p + n$	15 ± 2·5	2
(6)	$3n + 2p$	0·8 ± 0·4	0

Heidmann (1950) has applied the Born approximation to calculate cases (1)–(3) for n-α collisions at 90 MeV., assuming a central potential of gaussian form and Serber exchange type, together with gaussian functions for the ground state of the α-particle. Results at 90 MeV. show strong peaks of forward scattering, with elastic scattering predominant, as would be expected from the high binding energy of the α-particle.

Calculations at 90 MeV. are compared in Table XXI with experimental measurements (Tannenwald, 1953), the latter normalized to the estimated value of 190 mb. obtained by interpolation from cross-sections for the neighbouring nuclei, H, D, Li, Be, C, N, O, ... (Cook, McMillan, Peterson and Sewell, 1949). The total Born approximation cross-section is 244 mb., due mainly to over-estimation of the elastic cross-section and the cross-section for $^3H + p$ production. The $^3H + p$ cross-sections agree well, but the other cross-sections tend to be underestimated. However, considering the nature of the approximations made, the agreement is fair.

380

CHAPTER 12

NUCLEAR MODELS AND
RESONANCE SCATTERING

12.1. Maximum cross-section values and shadow scattering

This chapter is concerned with predicting collision cross-sections for nucleons with nuclei in general. With a few exceptions, it is impractical to tackle such problems in terms of two-body interactions. Instead phenomenological models based on simplified assumptions about nuclei as a whole must be employed. The total cross-section may be broken up into two parts, viz. the elastic and reaction cross-sections. The reaction cross-section is defined as the total cross-section (cf. §2.3) minus the elastic cross-section. It includes the inelastic cross-section as a special case.

The partial cross-sections are subject to certain conservation rules which limit their maximum values. A beam of incident particles is represented by a plane wave e^{ikz}, expansion into spherical harmonics leading to the asymptotic form for large r,

$$e^{ikz} \sim \frac{1}{kr} e^{ikr} \sum_{l=0}^{\infty} i^l (2l+1) P_l(\cos\theta) \sin(kr - \tfrac{1}{2}l\pi). \qquad (12.1)$$

The wave function representing the incident and outgoing wave is likewise (Mott and Massey, 1949, p. 24)

$$\Psi(\mathbf{r}) \sim \frac{1}{2ikr} \sum_{l=0}^{\infty} i^l (2l+1) P_l(\cos\theta) . \{\eta_l e^{i(kr-\frac{1}{2}l\pi)} - e^{-i(kr-\frac{1}{2}l\pi)}\}. \qquad (12.2)$$

The scattered wave Ψ_f is the difference between (12.2) and (12.1):

$$\Psi_f = \Psi(\mathbf{r}) - e^{ikz} \sim \frac{1}{2ikr} \sum_{l=0}^{\infty} i^l (2l+1) P_l(\cos\theta)(\eta_l - 1). \qquad (12.3)$$

For pure elastic scattering, η_l is related to the real scattering phase δ_l by $\eta_l = e^{2i\delta_l}$, so that $|\eta_l| = 1$. However, inelastic scattering may be included by assuming δ_l complex, $\delta_l = \lambda_l + i\mu_l$, so that η_l is no longer limited in magnitude. To obtain the elastic scattering cross-section, we divide the elastically scattered flux N_{el} by the incident flux $N = v$, where v is the incident particle velocity. Hence integrating over a sphere of radius $r = r_0$,

$$N_{el} = \frac{\hbar}{2im} \int \left(\frac{\partial \Psi_f'}{\partial r} \Psi_f^* - \frac{\partial \Psi_f'^*}{\partial r} \Psi_f' \right) d\mathbf{r}$$

$$= \pi \frac{v}{k^2} \sum_{l=0}^{\infty} (2l+1) \, |\, 1 - \eta_l \,|^2. \tag{12.4}$$

The partial elastic cross-section σ_{el}^l becomes

$$\sigma_{el}^l = \frac{\pi}{k^2} \cdot (2l+1) \, |\, 1 - \eta_l \,|^2. \tag{12.5}$$

The reaction cross-section σ_r is defined as the number of incident particles taken out of the beam per second, equal to the net flux into the sphere of radius $r = r_0$ computed with the full wave function Ψ' of equation (12.2):

$$N_r = \frac{\hbar}{2im} \int \left(\frac{\partial \Psi'}{\partial r} \Psi'^* - \frac{\partial \Psi'^*}{\partial r} \Psi' \right) d\mathbf{r}$$

$$= \pi \frac{v}{k^2} \sum_{l=0}^{\infty} (2l+1)(1 - |\, \eta_l \,|^2). \tag{12.6}$$

Thus $\qquad\qquad \sigma_r^l = \frac{\pi}{k^2} (2l+1)(1 - |\, \eta_l \,|^2), \tag{12.7}$

subject to $|\, \eta_l \,| \leqslant 1$. Equation (12.5) shows that σ_{el}^l has its maximum value for $\eta_l = -1$:

$$\sigma_{el}^l \leqslant \frac{4\pi}{k^2} (2l+1), \tag{12.8}$$

in which case $\sigma_r^l = 0$. Similarly the maximum value of σ_r^l according to equation (12.7) occurs for $\eta_l = 0$:

$$\sigma_r^l \leqslant \frac{\pi}{k^2} (2l+1), \tag{12.9}$$

in which case σ_{el} has the same value. The total partial cross-section σ_{tot} follows from equation (12.8):

$$\sigma_{tot}^l = \sigma_{el}^l + \sigma_r^l$$

$$\leqslant \frac{4\pi}{k^2} (2l+1). \tag{12.10}$$

However no inequalities are known for the total elastic and inelastic cross-sections (as opposed to the partial cross-sections).

An interesting case occurs if the target nucleus is much larger than the wavelength $\lambda = 1/k$, a condition satisfied for incident energies of order 1 BeV or more. A nucleus of radius R is struck by all neutral particles satisfying the classical condition $l < R/\lambda$. If these are all absorbed (black nucleus condition), then

$$\eta_l = 0 \quad \text{for} \quad l < R/\lambda$$
$$= 1 \quad \text{for} \quad l > R/\lambda. \qquad (12.11)$$

Thus $\qquad \sigma_{el} = \sigma_r = \sum_{l=0}^{\infty} \frac{\pi}{k^2}(2l+1) \approx \pi(R+\lambda)^2. \qquad (12.12)$

Hence $\sigma_{tot} \approx 2\pi R^2$, which is twice the geometrical cross-section of the nucleus. In the limit of very high energies, this relation is a good approximation, but for energies below about 50 MeV. the cross-section is appreciably larger and the expression

$$\sigma_{tot} \approx 2\pi(R+\lambda)^2 = 2\pi\left(R + \frac{1}{kR}\right)^2 \qquad (12.13)$$

should be used. This paradoxical result is interpreted as a diffraction effect due to the nucleus throwing a blurred shadow behind it of length of order R^2/λ. This shadow becomes more clear-cut as λ decreases, but nevertheless involves particles which clear the edge of the nucleus in a deviation of their paths toward the furthermost tip of the shadow. As k increases, the angular deviation decreases so that a pronounced forward peak of 'shadow' scattering develops. The effect is most noticeable for p-p collisions in the energy range $0\cdot4$–$1\cdot0$ BeV. (see § 10.12).

12.2. Resonance and potential scattering

The calculation of collision cross-sections for nuclei involves great difficulties even for the very light nuclei if account is taken of nuclear structure. However, a simpler and more practical approach for all except the very lightest nuclei may be obtained if the internal structure of the target nucleus is ignored, the nucleus being assumed to have a definite radius R (including its range of interaction). If the nuclear projectile has radius a, no interaction should occur unless their distance r is less than $R + a$. Actually this picture is approximate as nuclear surfaces are not really sharp.

If the target nucleus and projectile have charge $Z_R e$ and $Z_a e$ respectively, the Coulomb interaction is

$$V(r) = Z_R Z_a e^2/r. \qquad (12.14)$$

The radial wave function $\psi_l(r)$ satisfies the equation for $r > R$

$$\frac{d^2\psi_l}{dr^2} + \left\{ k^2 - \frac{l(l+1)}{r^2} - \frac{2\mu}{\hbar^2} V(r) \right\} \psi_l(r) = 0, \qquad (12.15)$$

where μ is the reduced mass of the system. There are two linearly independent solutions to (12.15). The regular solution $F_l(r)$ and the irregular solution $G_l(r)$ have asymptotic forms for large r

$$F_l(r) \sim \sin{(kr - \tfrac{1}{2}l\pi - \alpha \ln 2kr + \sigma_l)}, \qquad (12.16)$$

$$G_l(r) \sim \cos{(kr - \tfrac{1}{2}l\pi - \alpha \ln 2kr + \sigma_l)}, \qquad (12.17)$$

where $\alpha = Z_R Z_a \mu e^2/\hbar^2 k$, σ_l is the Coulomb phase shift of equation (7.16) and $F_l(r)$ and $G_l(r)$ are subject to the Wronskian conditions (7.36).

For charged-particle scattering, $F_l(r)$ and $G_l(r)$ are Coulomb wave functions and cannot be reduced to elementary functions. However, incomplete numerical tables are available (Nat. Bur. Standards, 1952; Christy and Latter, 1948; Bloch *et al.* 1951; Robson, 1958).

For scattering of neutral particles, $F_l(r)$ and $G_l(r)$ can be expressed in terms of the spherical Bessel function $j_l(kr)$ and the spherical Neumann function $n_l(kr)$ respectively (Schiff, 1949, p. 77):

$$F_l(r) = kr j_l(kr), \quad G_l(r) = -kr n_l(kr). \qquad (12.18)$$

We are interested in the solution of equation (12.15) which is asymptotic for large r to the spherical wave form

$$u_l(r) \sim \exp i(kr - \tfrac{1}{2}l\pi - \alpha \ln 2kr). \qquad (12.19)$$

This may be satisfied by defining

$$u_l(r) = \exp{(-i\sigma_l)} [G_l(r) + i F_l(r)]. \qquad (12.20)$$

The general solution of (12.15) may then be written

$$\psi_l(r) = A u_l(r) + B u_l^*(r), \qquad (12.21)$$

where $u_l^*(r)$ is the complex conjugate of $u_l(r)$. Comparison with the asymptotic condition analogous to equation (12.2),

$$\psi_l(r) \sim \frac{1}{2i} i^l(2l+1)\{\eta_l\, e^{i(kr-\frac{1}{2}l\pi-\alpha\ln 2kr)} - e^{-i(kr-\frac{1}{2}l\pi-\alpha\ln 2kr)}\}, \quad (12.22)$$

leads to the values

$$A = \tfrac{1}{2}i^{l-1}(2l+1)\eta_l, \quad B = -A/\eta_l. \quad (12.23)$$

It is useful to relate these quantities to the value of the logarithmic derivatives f_l at the nuclear boundary,

$$f_l = R[\psi_l'(r)/\psi_l(r)]_{r=R}. \quad (12.24)$$

If we write

$$R[u_l'(r)/u_l(r)]_{r=R} = \Delta_l + iS_l, \quad (12.25)$$

then equations (12.24) and (7.36) lead to values for Δ_l and S_l:

$$\Delta_l = R\left[\frac{G_l G_l' + F_l F_l'}{G_l^2 + F_l^2}\right]_{r=R} = -\tfrac{1}{2}RS_l'/S_l, \quad (12.26)$$

$$S_l = \left[\frac{kR}{G_l^2 + F_l^2}\right]_{r=R} = kR/|u_l(R)|^2. \quad (12.27)$$

We define also the phase δ_l of $u_l(r)$ at $r = R$ by

$$e^{2i\delta_l} = u_l(r)/u_l^*(r) = \frac{G_l(r) + iF_l(r)}{G_l(r) - iF_l(r)} e^{-2i\sigma_l}. \quad (12.28)$$

For neutrons, equation (12.28) reduces to

$$\tan \delta_l = F_l(R)/G_l(R) = -j_l(kR)/n_l(kR), \quad (12.29)$$

with $\delta_l \sim (kR - \tfrac{1}{2}l\pi)$ for large R. δ_l has been tabulated for $l = 0(1)20$ and $x = kR = 0(0\cdot1)10$ (Morse, Lowan, Feshbach and Lax, 1945). Expressions for δ_l for the first few values of l are

$$\delta_0 = x, \quad \delta_1 = x - \tfrac{1}{2}\pi + \text{arcot}\, x, \quad \delta_2 = x - \pi + \text{arcot}\,\tfrac{1}{3}(x^2 - 3),$$

$$\delta_3 = x - \tfrac{3}{2}\pi + \text{arcot}\,[x(x^2 - 15)/(6x^2 - 15)].$$

However, values for charged particles must be computed from tables of Coulomb wave functions using equation (12.28).

The phase δ_l may be related to the constant η_l of (12.3):

$$\eta_l = \frac{f_l - \Delta_l + iS_l}{f_l - \Delta_l - iS_l} e^{-2i\delta_l}, \quad (12.30)$$

so that a real f gives $|\eta_l|^2 = 1$ and a zero reaction cross-section. Substitution of equation (12.30) into equation (12.5) leads to the partial elastic scattering cross-section,

$$\sigma_{el}^l = \frac{\pi}{k^2}(2l+1)\,|\,A_{pot}^l + A_{res}^l\,|^2, \qquad (12.31)$$

where A_{pot}^l and A_{res}^l are the amplitudes for external (potential) scattering and for internal (resonance) scattering respectively.

$$A_{pot}^l = e^{2i\delta_l} - 1, \qquad (12.32)$$

$$A_{res}^l = \frac{-2iS_l}{f_l - (\Delta_l + iS_l)}. \qquad (12.33)$$

The corresponding angular distribution is given by $(1/v)\,N_{el}(\theta)\,d\omega$, related to N_{el} in equation (12.4) by $N_{el}(\theta)\,d\omega = N_{el}$. Hence

$$d\sigma_{el}/d\omega = \frac{1}{4k^2}\left|\sum_{l=0}^{\infty}(2l+1)(\eta_l - 1)P_l(\cos\theta)\right|^2 d\omega$$

$$= \frac{1}{4k^2}\left|\sum_{l=0}^{\infty}(2l+1)\,e^{-2i\delta_l}(A_{pot}^l + A_{res}^l)P_l(\cos\theta)\right|^2 d\omega. \qquad (12.34)$$

The amplitude A_{pot}^l describes the effects of outside potentials, the picture being of a hard sphere of radius R surrounded by a nuclear potential at small distances and a Coulomb potential at large distances. A_{res}^l allows for specifically nuclear effects inside the sphere.

The reaction cross-section follows from equations (12.7) and (12.30):

$$\sigma_r^l = \frac{\pi}{k^2}(2l+1)\frac{-4S_l\mathrm{Im}f_l}{|\,f_l - (\Delta_l + iS_l)\,|^2}, \qquad (12.35)$$

where Im denotes 'imaginary part'. It is proportional to S_l as given by equation (12.27), so we define a penetration factor by

$$v_l = 1/(G_l^2 + F_l^2). \qquad (12.36)$$

A small penetration factor means a small reaction cross-section as $S_l = kRv_l$. It is often used in the discussion of nuclear reactions. Methods of calculation of Coulomb penetration factors and wave functions are discussed by Thaler and Biedenharn (1957).

More explicit expressions for the elastic and reaction cross-sections depend on additional assumptions about the behaviour of particles inside the nucleus. The models discussed below are classified mainly according to whether the nucleus is assumed opaque or partially transparent. This may be expressed in terms of the value assumed for the logarithmic derivative f_l at the nuclear boundary. However, none of the models gives the inelastic cross-section for nucleon scattering as distinct from the reaction cross-section. It has been found that at low energies ($\lesssim 10$ MeV.) and high energies ($\gtrsim 100$ MeV.), a medium or heavy nucleus is comparatively transparent to nucleons (cf. § 12.4). Inelastic scattering takes place via compound nucleus and direct interaction processes ((nn'), (pp'), (np') and (pp')) at both low energies (Lamarsh and Feshbach, 1956; Yoshida, 1956) and high energies (cf. § 12.5). However, at intermediate energies, inelastic scattering appears largely due to direct interaction with (and excitation of) target nucleons on the extreme diffuse rim of the nucleus and not to knock-on scattering throughout the nuclear volume (Austern, Butler and McManus, 1953; Elton and Gomes, 1957; Butler, 1957). The nucleus is opaque at these energies and the compound nucleus process unimportant.

12.3. The schematic or continuum model

Cross-sections for nuclear reactions are complicated functions of the energy of the incident particles, involving resonances near energy levels of the target nucleus. However, it is possible to obtain the general trend of the energy dependence of cross-sections averaged over individual fluctuations. At sufficiently high energies these resonances in the cross-sections become weak and unimportant, finally merging their effects. In this region nuclear models should reproduce the actual cross-sections without any averaging.

The schematic model of Feshbach and Weisskopf (1949) is based on four assumptions as follows:

(a) The nucleus has a well defined radius R outside which interaction with incident particles does not occur.

(b) The average kinetic energy of the incident particle on penetrating inside the nucleus is initially $\epsilon + E_0$, where ϵ is the incident energy and E_0 the kinetic energy of intra-nuclear motion. For neutrons and protons, E_0 may be estimated using the Fermi gas model of the nucleus as about 22 MeV. For deuterons and α-particles, an estimate is much more difficult but will be assumed to be of the same order, $E_0 \sim 22$ MeV.

(c) The interaction between the scattered particle and the nucleons of the target nucleus must be strong enough for the probability of decay into other channels to be high. This necessitates the incident particle sharing its energy very rapidly with the component nucleons of the target nucleus. From this follows the Bohr postulate that the mode of decay of the compound nucleus is independent of its mode of formation.

(d) Ideally the energy of the incident particles should be high enough to make it energetically possible for the compound nucleus to decay into many channels; i.e. the residual nucleus may be left in many different excited states. However at lower energies the theory still holds, giving cross-section vs. energy curves averaged over resonance peaks.

Statements (c) and (d) define the upper and lower energy limits of the theory respectively. If the energy is too high, (c) is not satisfied because of the decrease of the interaction cross-section between nucleons of high energy. Nuclei thus become transparent to incident nucleons of sufficiently high energy, say $\epsilon > 70$ MeV. Statement (d) is the condition for a smooth cross-section vs. energy curve, satisfied for $A > 50$ if $\epsilon > 3$ MeV. At lower energies the experimental curve may deviate considerably from the smoothed-out theoretical curve in the neighbourhood of energy levels of the compound nucleus. Statement (a) involves a nuclear radius R, which when the incident particle is a neutron or proton may be computed from

$$R = r_0 A^{\frac{1}{3}} \times 10^{-13} \text{ cm.} \tag{12.37}$$

If the incident particle is a deuteron or α-particle, allowance must be made for its size, an empirical formula being (Farwell and Wegner, 1955)

$$R = (r_0 A^{\frac{1}{3}} + 1 \cdot 4) \times 10^{-13} \text{ cm.} \tag{12.38}$$

The elastic and total cross-sections for neutral particles are given by equations (12.12) and (12.13) respectively. However, an incident beam of charged particles is deviated by the Coulomb potential (12.14) before reaching the nuclear surface. The particles can interact with the nucleus only if their distance of closest approach is less than the effective nuclear radius $R + \lambda$. If p is the largest impact parameter satisfying this condition, the cross-section for formation of the compound nucleus is p^2. A simple calculation in electrodynamics yields

$$\sigma_r = \pi p^2 = \pi(R+\lambda)^2 \left[1 - \frac{V(R+\lambda)}{\epsilon} \right], \quad \epsilon > V(R+\lambda), \atop = 0, \qquad\qquad\qquad\qquad \epsilon < V(R+\lambda). \right\} \quad (12.39)$$

This classical cross-section vanishes below the Coulomb barrier energy $V(R+\lambda)$, yet reproduces calculated values to within 15% for $\epsilon/V(R) = 1\cdot2$, the accuracy improving for higher energies.

Wave mechanical corrections to equations (12.38) and (12.39) are

(1) Both neutron and charged-particle cross-sections must be corrected for reflexion of the incoming wave at the nuclear boundary due to the sudden change of potential. This reduces the cross-section, particularly at low energies.

(2) Charged particles with energies less than the Coulomb barrier energy can nevertheless leak through to some extent, so that the reaction cross-section is no longer zero as given by equation (12.39).

The corrections (1) and (2) are both described by the transmission factor T, defined as the ratio of the number of particles penetrating to the nuclear surface R to the number of incident particles. The neutron reaction cross-section becomes

$$\sigma_r \approx (R+\lambda)^2\, T, \qquad\qquad (12.40)$$

and the charged-particle reaction cross-section

$$\sigma_r \approx (R+\lambda)^2 \left[1 - \frac{V(R+\lambda)}{\epsilon} \right] T. \qquad (12.41)$$

A rough estimate of T is obtained by assuming the potential changes from $U=0$ to $U=-U_0$ on crossing the nuclear boundary at $r=0$,

$$U(r)=0 \ (r<0), \quad U(r)=-U_0 \ (r>0).$$

Writing $k=(2\mu\epsilon)^{\frac12}/\hbar$, $K=[2\mu(\epsilon+U_0)]^{\frac12}/\hbar$, where μ is the reduced mass of the system, the wave function $\psi(r)$ becomes

$$\psi(r)=A\,e^{ikr}+B\,e^{-ikr}, \quad (r<0),$$
$$\psi(r)=C\,e^{iKr} \qquad\qquad (r>0).$$

As $|A|^2$ and $|B|^2$ represent the number of incident and reflected particles respectively,
$$T = 1 - |B|^2/|A|^2.$$

The continuity of $\psi'(r)/\psi(r)$ at $r = 0$ leads to the value

$$T = 4kK/(K+k)^2. \tag{12.42}$$

For neutrons, the reaction cross-section is

$$\sigma_r \approx \pi(R+\lambda)^2 . 4kK/(K+k)^2, \tag{12.43}$$

which is still only a rough approximation to the more accurate theory. A simple formula cannot be found for charged particles owing to difficulties involved in treating Coulomb wave functions.

The main assumption (c) of the schematic theory may be expressed mathematically by taking the target nucleus as a 'sink' for the incident particles. Strictly speaking, the wave function inside the nucleus is not a function of r alone. However, it may be approximated by a converging wave $\psi(r) \sim e^{-iKr}$, which expresses the fact that the entering particle does not return. This may be written in terms of the boundary condition in the logarithmic derivative (12.24),

$$f_l = -iKR. \tag{12.44}$$

Substitution of equation (12.44) in equation (12.35) gives the reaction cross-section, identical for the schematic theory with the cross-section σ_c for formation of a compound nucleus:

$$\sigma_r^l = \frac{4\pi}{k^2} \sum_{l=0}^{\infty} (2l+1) \frac{S_l KR}{\Delta_l^2 + (KR+S_l)^2}, \tag{12.45}$$

where S_l and Δ_l are defined by equations (12.27) and (12.26) respectively, valid for both neutral and charged particles. For $l = 0$, $\Delta_0 = 0$ and $S_0 = kR$, so that equation (12.45) reduces to

$$\sigma_r^0 = \pi\lambda^2 T, \tag{12.46}$$

where T is given by equation (12.42). This is simply the product of the maximum s-wave cross-section (12.9) by the transmission factor T. However, partial cross-sections for $l \geqslant 1$ do not behave so simply.

The elastic scattering cross-section may be found from equations (12.31), (12.32), (12.33) and (12.44):

$$\sigma_{el} = \frac{\pi}{k^2} \sum_{l=0}^{\infty} (2l+1) \left| \frac{2iS_l}{\Delta_l + i(KR+S_l)} + (e^{2i\delta_l} - 1) \right|^2. \quad (12.47)$$

The wave number K of the incident particle inside the nucleus is related to its value k outside and the wave number K_0 of average intra-nuclear motion by

$$K^2 = K_0^2 + k^2. \quad (12.48)$$

K_0 can be estimated by computing the maximum allowed momentum of a Fermi–Dirac particle restricted to the nuclear volume V (Feshbach, Peaslee and Weisskopf, 1947),

$$K_0 = 2\pi(3A/16\pi V)^{\frac{1}{3}} \approx \frac{1 \cdot 52 \times 10^{13}}{r_0}. \quad (12.49)$$

Assuming $r_0 = 1 \cdot 4$, equation (14.49) gives

$$K_0 \approx 1 \cdot 1 \times 10^{13} \, \text{cm.}^{-1},$$

corresponding to a Fermi energy of 22 MeV.

Extensive calculations have been carried out for the total cross-sections for neutron collisions with a large number of nuclei (Feshbach and Weisskopf, 1949). Good agreement is found in most cases with low energy experiments at 0–1·6 MeV. giving total cross-sections as functions of energy (Bockelman, Peterson, Adair and Barschall, 1949). Results at higher energies (14, 25 MeV.) enable determinations of nuclear radii R, these being read off calculated cross-section curves as in fig. 97 (Amaldi, Bocciarelli, Cacciapuoti and Trabacchi, 1946; Sherr, 1945). The average value of the nucleon radius found (Borst, Ulrich, Osborne and Hasbrouck, 1949) is $1 \cdot 45 \times 10^{-13}$ cm., in good agreement with the usual value $1 \cdot 4$–$1 \cdot 5 \times 10^{-13}$ cm. However, just as in stripping calculations, the nuclear radius R giving best agreement with experiment is sometimes larger than predicted by equations (12.37) or (12.38). This is partly due to nuclei not having clear-cut radii R and partly to the other approximations involved in the schematic theory. Thus to fit total cross-sections for Zr up to 2 MeV., the radii $R = 7 \cdot 8 \times 10^{-13}$ cm. and $r_0 = 1 \cdot 73$ are required, R being $1 \cdot 3 \times 10^{-13}$ cm. larger than predicted by equation (12.37) using

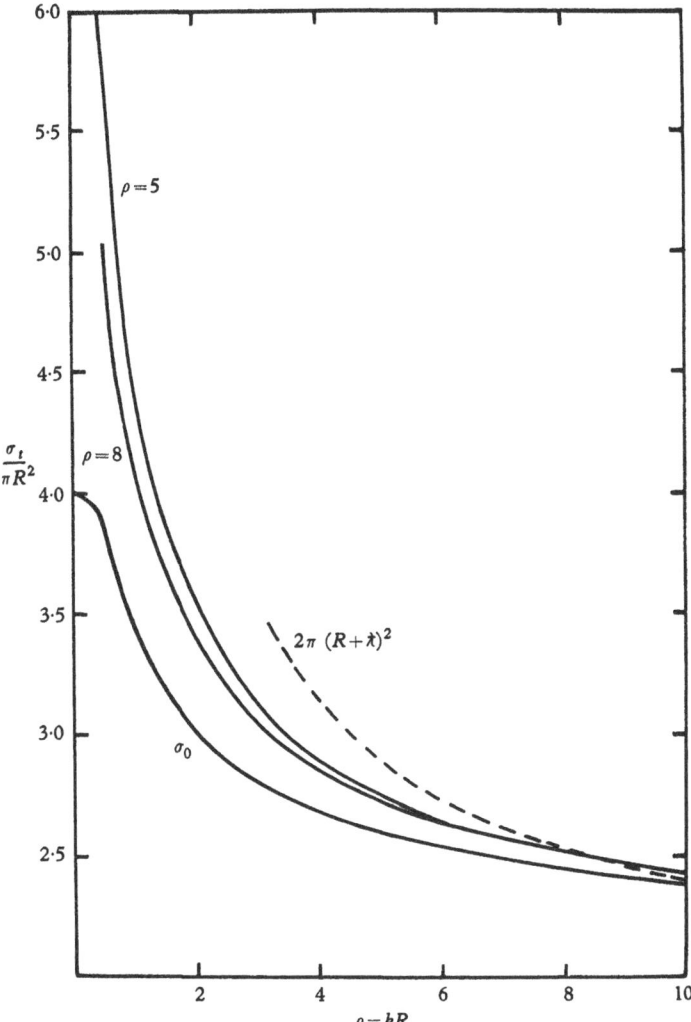

Fig. 97. Total cross-sections for the schematic model as a function of $\rho = kR$, for values of $\rho = 5$ and 8 (Feshbach and Weisskopf, 1949). σ_0 is the total cross-section for an infinitely repulsive sphere of radius R. The broken line gives the approximate behaviour of all these curves for large ρ.

$r_0 = 1\cdot45$ (Feshbach and Weisskopf, 1949). This may be because the nucleus is opaque to low energy nucleons, but partially transparent to higher energy particles. Other nuclei like ^{8}Be, consisting of tightly bound cores, require radii about 20% lower than predicted by equation (12.37).

To obtain consistency with thermal and low energy neutron scattering experiments, it is necessary to drop the assumption of a sharp nuclear boundary (Scott, 1954). Instead a uniform distribution with an outside exponential tail is needed, similar to that found by high energy electron scattering experiments on nuclei. The mean range of this profile turns out to be $r_0 = 1\cdot24$, leading to a Fermi well depth of 42 MeV.

For the limiting case $R \gg \lambda$ or $kR \gg 1$, a simple approximation to the elastic scattering distribution (12.47) for neutrons may be obtained (Bethe and Placzek, 1937). The scattering is mainly shadow scattering and hence largely confined to a forward scattering angle θ of order $\lambda/R = 1/kR$. Provided $\theta \ll 1$ and $kR \gg 1$, the differential cross-section is

$$d\sigma_{el}/d\omega \approx R^2 \left| \frac{J_1(kR)}{\theta} \right|^2. \qquad (12.50)$$

In the typical case of 90 MeV. neutrons elastically scattered by a heavy nucleus of radius $R = 9 \times 10^{-13}$ cm., the first minimum occurs at about $11°$.

12.4. The optical model

The success of the nuclear shell model has cast some doubt on strong coupling methods such as that of the schematic model, which assume an intermediate compound state whenever a nucleon penetrates the nucleus. Shell structure indicates that a nucleon can move freely inside a nucleus without apparently changing the quantum state of the latter. This follows from observations made at the ground state and at low excitation energies. However, it is doubtful whether this is valid also at those excitation energies (~ 8 MeV.) which are created in nuclear reactions with neutrons of a few MeV. In fact at energies of 15 MeV. or more, the interaction between the incident nucleon and the target must be appreciable, for the reaction cross-sections have been found (Phillips,

Davis and Graves, 1952) to be equal to the geometrical cross-sections $\pi(R+\lambda)^2$. Hence at intermediate energies an incident nucleon usually shares its energy with the target nucleus, so that the schematic model is valid.

On the other hand, Serber (1947) has pointed out that at very high energies ($\sim$ 100 MeV.), the mean free path of a nucleon traversing nuclear matter is comparable to the nuclear radius. This transparency effect is strikingly evident in experiments on the scattering of 90 MeV. neutrons by nuclei (Cook *et al.* 1949). These show that the collision radii for light elements lie considerably below values found at 14 MeV. (Amaldi and Cacciapuoti, 1947) and 25 MeV. (Sherr, 1945), whereas radii for heavy elements tend to approach them. This partial transparency at the higher energies is consistent with a mean free path for a 90 MeV. nucleon in nuclear matter of about $4 \cdot 5 \times 10^{-13}$ cm., so that the schematic model is no longer valid (Fernbach, Serber and Taylor, 1949).

A suitable model must involve an interaction such that the nucleon can enter and freely traverse the nucleus, but with a finite probability of absorption less than one to form a compound nucleus. To describe this we decompose the reaction and elastic cross-sections:

$$\sigma_{el} = \sigma_{se} + \sigma_{ce}, \qquad (12.51)$$

$$\sigma_r = \sigma_c - \sigma_{ce}. \qquad (12.52)$$

The shape elastic cross-section σ_{se} does not involve an intermediate compound nucleus, in contrast with the compound elastic cross-section σ_{ce}, for which the incident particle is emitted from the compound nucleus back into the entrance channel. The cross-section σ_c for compound nucleus formation is thus the sum of the reaction and compound elastic cross-sections. Adding equations (12.51) and (12.52), the total cross-section becomes

$$\sigma_{tot} = \sigma_{se} + \sigma_c. \qquad (12.53)$$

The optical model describes the scattering of the nucleon wave by a sphere of material characterized by a refractive index and an absorption coefficient, the latter including the effect of compound elastic, inelastic and exchange scattering. Writing the diffraction cross-section as $\sigma_d = \sigma_{se}$ and the absorption cross-section as $\sigma_a = \sigma_c$, we have

$$\sigma_{tot} = \sigma_d + \sigma_a. \qquad (12.54)$$

The optical model leads directly to expressions for the diffraction and absorption cross-sections σ_d, σ_a and not to the elastic and reaction cross-sections σ_{el}, σ_r as in the schematic theory. For the latter, the distinction was unnecessary, because the interaction was defined as strong enough to make the probability of the incident particle being emitted from the compound nucleus into the entrance channel zero, so that $\sigma_{ce} = 0$.

12.4.1. *Scattering on the optical model.* If the energy of the incident particle is high enough to be in the continuum region, the cross-section becomes a smooth, monotonically decreasing function of energy. We assume that outside the resonance region, $\sigma_{ce} = 0$ (cf. equation (12.82)). Therefore $\sigma_d = \sigma_{el}$ and $\sigma_a = \sigma_r$ as defined by equations (12.5) and (12.7).

The interaction of high energy nucleons with nuclei is analogous to the scattering of light by a conducting glass sphere, e.g. a colloidal suspension of gold particles in a glass bead. Such a medium may be macroscopically described by a complex refractive index,

$$n^2 = \epsilon + i(4\pi\sigma/\omega), \qquad (12.55)$$

where ϵ is the dielectric constant and σ the electric conductivity. This describes the average of a multiple of microscopic events, including scattering processes at each metallic particle.

The analogous description of the scattering and absorption of nucleons by nuclei requires the incident energy to be sufficiently great for many levels to contribute. The optical model as employed by many authors (Feshbach and Weisskopf, 1949; Pasternack and Snyder, 1950; Le Levier and Saxon, 1952; Feshbach, Porter and Weisskopf, 1954) uses a complex square well potential, this being equivalent to a complex refractive index (Bethe, 1940 a). Alternative complex well potentials include the exponential (Bethe, 1940 a), parabolic (Heckrotte, 1954 a) and other well shapes (Mohr and Robson, 1956; Margolis and Troubetzkoy, 1957). The best results have been obtained using a complex square well with an exponential tail, or Woods–Saxon potential (Woods and Saxon, 1954; Melkanoff, Nodvick, Saxon and Woods, 1957).

The introduction of a non-Hermitian potential into the Hamiltonian means that the probability density is no longer conserved. If the potential is

$$V(r) = -V_0 - iW_0 \quad (r < R)$$
$$= 0 \quad\quad\quad (r > R), \quad\quad (12.56)$$

then the continuity equation becomes

$$\partial\rho/\partial t + \nabla \cdot \mathbf{j} = -2\rho W_0/\hbar, \quad\quad (12.57)$$

where $\rho = \psi^*\psi$, $\mathbf{j} = (\hbar/\mu)\,\mathrm{Im}(\psi^*\nabla\psi)$, and μ is the reduced mass of the system. The system appears to absorb incident nucleons at the rate $2\rho W_0/\hbar$ per unit volume, including the effects of inelastic scattering. The complex potential (12·56) gives the diffraction or elastic scattering effect through its real component V_0, representing the average potential energy throughout the nucleus. The absorption effect is similarly represented by the unreal component W_0 and the absorption coefficient by

$$\kappa = 2\rho W_0/\hbar. \quad\quad (12.58)$$

This is equal to the particle density multiplied by the average cross-section for neutron scattering by a particle in the nucleus,

$$\kappa = 2A\sigma/4\pi R^3, \quad\quad (12.59)$$

where in terms of the n-p and n-n cross-sections,

$$\sigma = [Z\sigma_{np} + (A-Z)\sigma_{nn}]/A. \quad\quad (12.60)$$

The wave number inside the nucleus is given by equation (12.48), written as

$$K = k(1 + V_0/\epsilon)^{\frac{1}{2}}, \quad\quad (12.61)$$

where k is the wave number of the incident particle outside the nucleus and ϵ is the energy. The potential V_0 is usually taken about 8 MeV. larger than the energy of the Fermi sphere (22 MeV.), so that $V_0 = 30$ MeV.

At 90 MeV., σ_{np}(free) $= 83$ mb., which must be reduced to allow for exclusion principle effects in the scattering of neutrons by protons bound in a nucleus (Cook et al. 1949). Thus $\sigma_{np} = \frac{2}{3}\sigma_{np}$ (free), which for

$$E = (\hbar^2/\mu)\,K^2 = 90 + 30 = 120 \text{ MeV.},$$

gives $\sigma_{np} = 41 \cdot 5$ mb. (Goldberger, 1948). Taking $\sigma_{nn} = \frac{1}{4}\sigma_{np}$, this leads to $\kappa = 2 \cdot 4 \times 10^{12}$ cm.$^{-1}$ for $Z/A = 0 \cdot 39(U)$. From equation (12.59), we may assume κ varies roughly inversely with energy in the continuum region, say from 10 MeV. up. The mean free path $1/\kappa$ is therefore roughly proportional to energy in this region.

The cross-sections for diffraction and absorption scattering follow directly from the general expressions (12.31) and (12.35) for the elastic and reaction cross-sections respectively. Writing

$$X = kR[1 + (V_0 + iW_0)/\epsilon]^{\frac{1}{2}} \quad (x = kR),$$

a solution at $r = R$ of the wave equation for the complex potential (12.56) is needed satisfying the asymptotic conditions (12.22) and (12.20). For neutral particles,

$$\psi_l(R) = Xj_l(X) \quad (r \leqslant R),$$

$$u_l(R) = xh_l(x) \quad (r \geqslant R),$$

where j_l and h_l are the spherical Bessel and Hankel functions respectively. The logarithmic derivative (12.24) becomes

$$f_l = 1 + j_l'(X)/j_l(X), \tag{12.62}$$

and
$$\Delta_l + iS_l = 1 + h_1'(x)/h_l(x). \tag{12.63}$$

The phase shift δ_l is given by equation (12.29). For full details, see Chamberlain and Garrison (1954).

An equivalent theory valid for high energies may be obtained by pursuing the direct optical analogy. We divide the sphere of radius R representing the nucleus into a number of concentric annular cylinders of length $T(\rho)$ and radius ρ. Two conditions must be satisfied:

(a) An incident nucleon of 'size' λ is not absorbed or scattered until right inside an annular cylinder, i.e. $T > \lambda = 1/k$. At 90 MeV., $\lambda = 0 \cdot 5 \times 10^{-13}$ cm. This means that a boundary layer exists at the surface of a disc in which the absorption coefficient κ and the increment in wave number $k_1 = K - k$ rise to their interior values in a distance greater than $1/k$. There is thus no scattering at the surface of a disc.

(b) To replace the series of annular cylinders by an integration formula over the sphere, we require $T \ll R$, i.e. $\lambda \ll R$ or $kR \gg 1$.

This means the incident energy E must be sufficiently large. Assuming $R = 1 \cdot 45 \times 10^{-13} A^{\frac{1}{3}}$ cm., $A = 125$ gives about 14 discs at 90 MeV., so that the approximation should be valid at this energy.

To illustrate the calculation, consider the scattering from a single disc of radius ρ and thickness T. A beam of unit intensity and amplitude transmitted through the disc has its intensity diminished by absorption by a factor $e^{-\kappa T}$. The transmitted wave has an amplitude and relative phase

$$a = \exp\left(-\tfrac{1}{2}\kappa + ik_1\right) T.$$

The absorption cross-section is $\pi\rho^2$ times the probability that the particle collides with the disc,

$$\sigma_a = \pi\rho^2(1 - |a|^2) = \pi\rho^2(1 - e^{-\kappa T}). \qquad (12.64)$$

The diffraction cross-section follows by noting that in the shadow of the disc, the wave differs from a plane wave by the amplitude $(1 - a)$. This latter represents a scattered wave, its cross-section given by

$$\sigma_d = \pi\rho^2 |1 - a|^2$$
$$= \pi\rho^2(1 - 2e^{-\frac{1}{2}\kappa T}\cos k_1 T + e^{-\kappa T}). \qquad (12.65)$$

The angular dependence of the scattered amplitude is (Mott and Massey, 1949, p. 6)

$$f(\theta) = k \int_0^{\rho} (1 - a) J_0(k\alpha \sin\theta)\, \alpha\, d\alpha$$
$$= (1 - a)\rho J_1(k\rho \sin\theta)/\sin\theta, \qquad (12.66)$$

leading to the differential scattering cross-section

$$d\sigma_d/d\omega = |f(\theta)|^2\, d\omega$$
$$= (\sigma_d/\pi)\,[J_1(k\rho \sin\theta)/\sin\theta]^2. \qquad (12.67)$$

Scattering by a sphere may now be dealt with. The portion of the wave which strikes the sphere at a distance ρ from a line through the centre of the sphere, emerges after travelling a distance $2s$, with $s^2 = R^2 - \rho^2$. Its amplitude after emerging is $a = \exp\left(-\kappa + 2ik_1\right) s$, so in place of equation (12.64), we integrate over a number of concentric annular cylinders:

$$\sigma_a = 2\pi \int_0^{R} (1 - e^{-2\kappa s})\rho\, d\rho$$
$$= \pi R^2 \left[1 - \frac{1}{2\kappa^2 R^2}\{1 - (1 + 2\kappa R)\, e^{-2\kappa R}\}\right]. \qquad (12.68)$$

The diffraction cross-section becomes

$$\sigma_d = 2\pi \int_0^R |\, \mathbf{1} - \exp\left(-\kappa + 2ik_1\right)s\,|^2 \rho\, d\rho$$

$$= \pi R^2 \left[\mathbf{1} + \frac{\mathbf{1}}{2\kappa^2 R^2}\{\mathbf{1} - (\mathbf{1} + 2\kappa R)\, \mathrm{e}^{-2\kappa R}\} \right.$$

$$- \frac{\mathbf{1}}{R^2(\tfrac{1}{4}\kappa^2 + k_1^2)^2}\{(\tfrac{1}{4}\kappa^2 - k_1^2) + \mathrm{e}^{-\kappa R}$$

$$\times [2k_1 R(\tfrac{1}{4}\kappa^2 + k_1^2) + k_1 \kappa]\sin 2k_1 R$$

$$\left. - \mathrm{e}^{-\kappa R}[(\tfrac{1}{4}\kappa^2 - k_1^2) + \kappa R(\tfrac{1}{4}\kappa^2 + k_1^2)]\cos 2k_1 R\} \right]. \quad (12.69)$$

The angular dependence of the scattered amplitude is

$$f(\theta) = k \int_0^R [\mathbf{1} - \exp\left(-\kappa + 2ik_1\right)s]\, J_0(k\rho \sin\theta)\rho\, d\rho. \quad (12.70)$$

This expression cannot be conveniently integrated, but may be expressed as a sum. Putting $l + \tfrac{1}{2} = k$ and using

$$J_0\{(l + \tfrac{1}{2})\sin\theta\} = P_l(\cos\theta),$$

valid for large l and small θ, we find

$$f(\theta) = \tfrac{1}{2}k \sum_{l=0}^{l+\frac{1}{2}<kR} (2l + \mathbf{1})\{\mathbf{1} - \exp\left(-\kappa + 2ik_1\right)s_l\}P_l(\cos\theta), \quad (12.71)$$

where $s_l = \{k^2 R^2 - (l + \tfrac{1}{2})^2\}^{\frac{1}{2}}/k$. This result may also be obtained using the W.K.B. method (Mott and Massey, 1949, p. 122) to find the phase shifts, giving $\delta_l = (k_1 + \tfrac{1}{2}iK)s_l$, which leads direct to equation (12.71).

Comparison with scattering data at 90 MeV. (Coon and Taschek, 1949) enables nuclear radii R to be determined once κ is chosen. The value $\kappa = 2 \cdot 2 \times 10^{12}$ cm.$^{-1}$, corresponding to a mean free path in nuclear matter of $4 \cdot 5 \times 10^{-13}$ cm., gives good agreement with $R = 1 \cdot 4 A^{\frac{1}{3}} \times 10^{-13}$ cm. over a wide range of nuclei (Fernbach, Serber and Taylor, 1949; Culler and Waniek, 1955). The mean free path agrees with that estimated from the high energy n-p cross-section, but the results are not sensitive as κ can be increased considerably without altering R appreciably. The nuclear opacity $\sigma_a/\pi R^2$ varies from $0 \cdot 52$ for Li to $0 \cdot 88$ for U, but σ_d is nearly double these values.

To compare the optical and schematic models, we note the latter uses the logarithmic derivative expression $f_l = -ikR$, which is independent of l and hence leads to an f value for the whole wave function independent of angle. On the other hand, the optical model expression (12.62) is dependent on l, and hence the f value for the whole wave function is apparently dependent on angle. However if the energy is high, then $X = u + iv \gg 1$, so on using the asymptotic formula for the spherical Bessel function in the limit $|X| \gg l$, $f_l \approx v - iu$ $(u, v \gg l)$, which is independent of l. Thus the logarithmic derivative of the whole wave function is at least approximately independent of angle in the optical model. Hence the cross-sections can turn out to be quite similar for the schematic and optical models at high energies (Le Levier and Saxon, 1952; Chase and Rohrlich, 1954).

Fig. 98 shows calculated differential cross-sections for the elastic scattering of 18·6 MeV. neutrons by Al as calculated by the schematic and optical models, together with experimental points. The two curves are very similar, but the optical model is in slightly better agreement with experiment, particularly in reproducing scattering peaks. However the values $r_0 = 1\cdot2$ and $V_0 = 45$ MeV. similar to those found by Scott (1954) give better results. For discussions of recent data, see (Beyster, Walt and Salmi, 1956; Morinaga and Peaslee, 1957). The elastic scattering of α-particles has also been dealt with by the optical model (Igo and Thaler, 1957).

Polarization effects have been treated for the optical model of the nucleus, assuming a complex potential plus an $\mathbf{L} \cdot \mathbf{S}$ interaction, using either the Born approximation (Heckrotte, 1954b; Fernbach, Heckrotte and Lepore, 1955; Sawicki, 1956; Brown, 1957) or the W.K.B. approximation (Fernbach *et al.* 1955; Sternheimer, 1955 a, b). Results with high energy particles indicate that only the latter method gives reasonable results.

12.4.2. *Scattering in the resonance region in the gross structure approximation.* So far the optical model has been applied only to energies in the continuum region where the compound elastic cross-section σ_{ce} is assumed zero and the neutron wavelength is considerably smaller than nuclear dimensions.

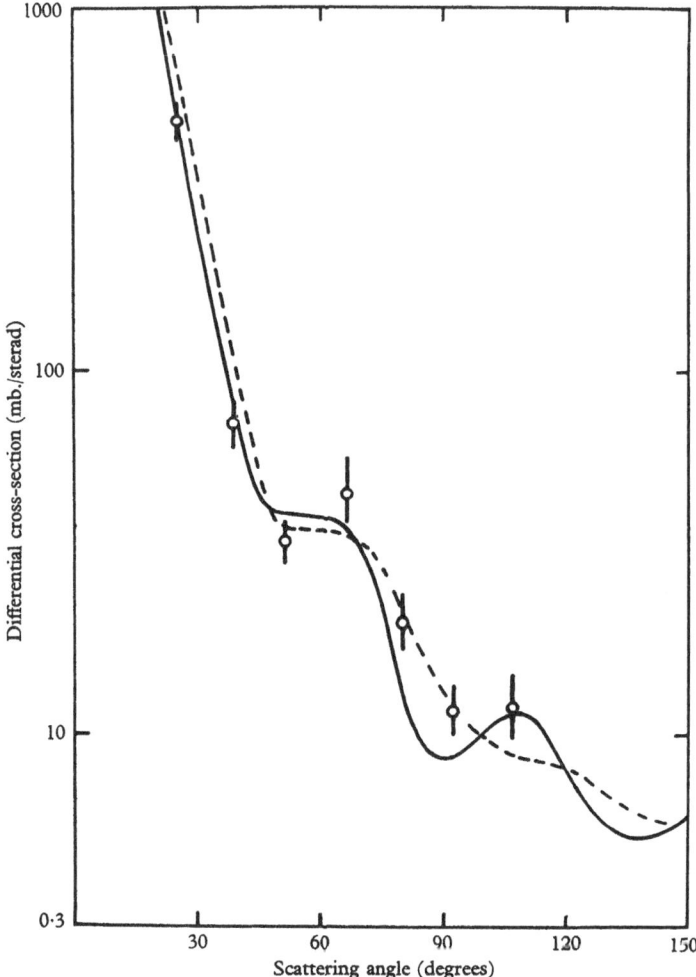

Fig. 98. Differential cross-section vs. angle of scattering on a logarithmic scale (Le Levier and Saxon, 1952) for the elastic scattering of 18·6 MeV. neutrons by Al. The solid curve is due to the optical model and the broken curve to the schematic model. Parameters used are

$$V_0 = 30 \text{ MeV.}, \quad W_0 = 20 \text{ MeV.}, \quad \kappa = 6\cdot2 \times 10^{12} \text{ cm.}^{-1}, \quad R = 1\cdot42A^{\frac{1}{3}} \times 10^{-13} \text{ cm.}$$

Experimental points (Burkig and Wright, 1951) have been normalized to give agreement with the optical model at 26°.

At high energies, the absorption potential W_0 is large, so that writing $W_0 = i\zeta V_0$, equation (12.56) becomes

$$V(r) = -(1 + i\zeta) V_0. \qquad (12.72)$$

At 18 MeV., $V_0 = 30$ MeV. and $W_0 = 20$ MeV., giving $\zeta = 0.67$ and $\kappa = 6.2 \times 10^{12}$ cm.$^{-1}$ (Le Levier and Saxon, 1952). However, the success of the shell model indicates $\zeta \approx 0$ for the ground state of nuclei. In fact, $\zeta = 0.03$ for neutrons in the energy range 0–3 MeV., assuming the empirical value $V_0 = 42$ MeV. and the radius $R = 1.45 A^{\frac{1}{3}} \times 10^{-13}$ cm. (Feshbach et al. 1954).

Some idea of the energy dependence of ζ can be found from equation (12.58), which for $\rho = 1/V$ gives $\zeta = \mathrm{const.}\, e^{\frac{1}{2}} \kappa$. The absorption coefficient κ is a function of energy, rising from a value of about 0.42×10^{12} cm.$^{-1}$ for zero energy neutrons to a maximum in the resonance scattering energy region. In the continuum energy range, the definition (12.59) shows that κ is roughly inversely proportional to energy. These considerations indicate that ζ (and hence the absorption potential W_0) increases at least as fast as the incident energy up to a maximum value in the resonance scattering region, but at higher energies (continuum region), it decreases slowly according to an $\epsilon^{-\frac{1}{2}}$ law. The energy dependence of ζ is discussed by Melkanoff, Moszkowski, Nodvik and Saxon (1955).

The cross-sections in the resonance region fluctuate too rapidly with energy to be described by a simple one-particle potential of type (12.72). Moreover, the neutron wavelength at energies below several MeV. is not small compared to R, again militating against the optical model. However, one solution is to employ cross-sections averaged over the resonances (Feshbach et al. 1954).

Using the notation of § 12.4.1, the partial elastic and reaction cross-sections are

$$\sigma_{el}^l = \pi \lambda^2 (2l + 1) \,|\, 1 - \eta_l \,|^2, \qquad (12.73)$$

$$\sigma_r^l = \pi \lambda^2 (2l + 1)(1 - |\, \eta_l \,|^2), \qquad (12.74)$$

where η_l is the complex reflexion factor of § 12.1. It is a complicated function of the incident energy ϵ, exhibiting rapid fluctuations due to the numerous close-spaced resonances of the compound nucleus. We therefore define the average reflexion factor

$$\bar{\eta}_l = \frac{1}{I} \int_{\epsilon - I/2}^{\epsilon + I/2} \eta_l(\epsilon') \, d\epsilon', \qquad (12.75)$$

which is a smooth function of ϵ if the interval $I \ll \epsilon$. Hence slowly varying functions of ϵ like λ^2 need not be averaged. We also define the average cross-sections:

$$\bar{\sigma}_{\text{el}}^l = \pi\lambda^2(2l+1)\overline{|1-\eta_l|^2}, \tag{12.76}$$

$$\bar{\sigma}_r^l = \pi\lambda^2(2l+1)(1-\overline{|\eta_l|^2}), \tag{12.77}$$

where the bar signifies the average over the interval I.

One can easily verify the following relations:

$$\bar{\sigma}_{\text{el}}^l = \pi\lambda^2(2l+1)\{|1-\bar{\eta}_l|^2 - |\bar{\eta}_l|^2 + \overline{|\eta_l|^2}\}, \tag{12.78}$$

$$\bar{\sigma}_{\text{tot}}^l = \pi\lambda^2(2l+1)\{|1-\bar{\eta}_l|^2 + 1 - |\bar{\eta}_l|^2\}. \tag{12.79}$$

Thus the average total cross-section depends only on $\bar{\eta}_l$.

The average elastic cross-section may be divided into the shape elastic and compound elastic cross-sections σ_{se}^l and σ_{ce}^l according to equation (12.50):

$$\sigma_{\text{el}}^l = \sigma_{\text{se}}^l + \sigma_{\text{ce}}^l, \tag{12.80}$$

where

$$\sigma_{\text{se}}^l = \pi\lambda^2(2l+1)|1-\bar{\eta}_l|^2, \tag{12.81}$$

$$\sigma_{\text{ce}}^l = \pi\lambda^2(2l+1)\{\overline{|\eta_l|^2} - |\bar{\eta}_l|^2\}. \tag{12.82}$$

It is clear that in the continuum energy region, η_l and $|\eta_l|^2$ are equal to their average values so that the compound elastic cross-section vanishes. Also we can combine σ_{ce}^l and $\bar{\sigma}_r^l$ into a new cross-section σ_c^l for compound nucleus formation:

$$\sigma_c^l = \sigma_{\text{ce}}^l + \bar{\sigma}_r^l = \pi\lambda^2(2l+1)(1-|\bar{\eta}_l|^2). \tag{12.83}$$

Thus from equations (12.81) and (12.83), σ_{se}^l and σ_c^l have the form of scattering and reaction cross-sections for a new 'gross structure' problem with phase defined by the slowly varying function $\bar{\eta}_l$. If the energy is high enough to be in the continuum region, then $\bar{\eta}_l = \eta_l$ and the gross structure problem becomes the same as the normal one. The method is quite general, and may be used on cross-sections obtained with any model. In particular, the scattering produced by the complex square well potential (12.56) for the optical model may be found (Feshbach et al. 1954) by averaging the expression (12.30) for η_l, using equations (12.61), (12.62) and (12.63). The Breit–Wigner resonance formulae of § 12.7 may also be averaged in the same way (Feshbach et al. 1954).

12.5. The statistical model for high energy collisions— recoil events

If a nucleus is struck by a very high energy nucleon, particles may be ejected from the nucleus by at least two processes, recoil and evaporation. If the target nucleon receives so much energy that it leaves the nucleus in a time short compared to a nuclear period ($\sim 10^{-21}$ sec.), it is called a recoil particle. However, if the energy of the incident particle is evenly distributed throughout the nucleus by many collisions of the nucleons among one another (compound nucleus), and if particles are ejected only after many nuclear periods, the process is called nuclear evaporation. The latter is treated by the statistical model in § 12.6. Here we consider only events taking place in a time of order 10^{-22} sec., i.e. recoil events.

To treat high energy nuclear reactions, one must consider individual nucleon-nucleon collisions in which a relatively small amount of energy is transferred in a single collision (Serber, 1947). Also account must be taken of the degeneracy of nuclear matter, e.g. exclusion principle effects due to the presence of nucleons other than the struck nucleon. Recoil events have been dealt with on this basis (Goldberger, 1948) using the statistical model of the nucleus (Bethe and Bacher, 1936).

The nucleus is pictured as a mixture of two non-interacting Fermi gases of neutrons and protons, bound in a uniform potential of depth about 30 MeV., the highest filled states being at an energy of about 8 MeV. The maximum Fermi energies of the neutrons and protons are

$$E = (\hbar^2/2m)(3\pi^2 N/V)^{\frac{2}{3}}, \qquad (12.84)$$

where N is the number of neutrons or protons and V is the nuclear volume. The exclusion principle for the incident particle requires that both projectile and target particle be outside the occupied sphere in momentum space after the collision. This sphere has a radius

$$p = \hbar(3\pi^2 N/V)^{\frac{1}{3}}. \qquad (12.85)$$

The most probable momentum transfer in a high energy nucleon-nucleon collision is of order $\hbar/r_0$, where r_0 is the range of nuclear forces. Since $p\hbar/r_0 \sim \frac{1}{2}$, a sizeable fraction of collisions is forbidden by the Pauli principle so that the effective cross-section is decreased. This makes the nucleus more transparent to high energy neutrons, resulting in a longer mean free path.

12.5.1. *The mean free path.* The mean free path of a high energy nucleon inside a nucleus is

$$\Lambda = V/A = 1/(\rho\sigma), \qquad (12.86)$$

where ρ is the nucleon density and σ is the average nucleon-nucleon cross-section (12.60). It satisfies the law

$$p = e^{-x/\Lambda}, \qquad (12.87)$$

where p is the probability of the nucleon penetrating a distance x before undergoing a collision. The mean free path is related to the absorption coefficient κ of equation (12.58) by $\Lambda = 1/\kappa$, so that from § 12.4.2, $\epsilon = 90$ MeV. gives $\Lambda = 4\cdot2 \times 10^{-13}$ cm. for $Z/A = \frac{1}{2}$ (light nuclei) and $\Lambda = 4\cdot8 \times 10^{-13}$ cm. for $Z/A = 0\cdot39$ (U). This includes an exclusion principle correction to the n-p cross-section (cf. § 12.4.1). If this were not allowed for, the mean free paths would be $\Lambda = 3\cdot2 \times 10^{-13}$ cm ($Z/A = \frac{1}{2}$) and $\Lambda = 2\cdot8 \times 10^{-13}$ cm. ($Z/A = 0\cdot39$).

To find the reduced cross-section $\bar{\sigma}$, allowing for exclusion principle effects, in terms of the n-p cross-section σ, consider a neutron with energy ϵ incident on a nucleus. Inside the nucleus it has an energy $E_0 = \epsilon + V$, where V is the Fermi well depth. Let $\mathbf{P}_0$ and $\mathbf{P}_f$ be the initial and final momenta of the incident neutron inside the nucleus in the Lab. system and $\mathbf{p}_0$ and $\mathbf{p}_f$ the corresponding momenta in the CM system. Take also the initial momentum of the target nucleon inside the nucleus as $\mathbf{P}_1$. Then $p_f = p_0 = \frac{1}{2} | \mathbf{P}_0 - \mathbf{P}_1 |$, and the momentum transfer in a collision is $\mathbf{g} = \mathbf{p}_f - \mathbf{p}_0 = \mathbf{P}_f - \mathbf{P}_0$. The cross-section element $I d\omega_{p_f}$ describing a collision between $\mathbf{P}_0$ and $\mathbf{P}_1$ is assumed known:

$$I\, d\omega_{p_f} = I\delta(p_f - p_0)\, d\mathbf{p}_f/p_f^2 = I\delta(p_f - p_0)\, d\mathbf{g}/p_0^2, \qquad (12.88)$$

where $d\omega_{p_f}$ is the element of solid angle about the final projectile momentum and P_1 is fixed at the Fermi value.

The effective total cross-section $\bar{\sigma}$ is found by integrating the product of the relative velocity $| \mathbf{P}_0 - \mathbf{P}_1 |/m$ and the differential cross-section I for free $n\text{-}p$ collisions over the allowed regions of P and dividing by the incident velocity P_0/m:

$$\bar{\sigma} = \frac{1}{p_0} \int d\mathbf{g} \int d\mathbf{P}_1 N(\mathbf{P}_1) \, | \, \mathbf{P}_0 - \mathbf{P}_1 \, | \, I(p_0, \mathbf{p}_0 \cdot \mathbf{p}_f) \frac{1}{p_0^2} \delta(p_0 - p_f), \quad (12.89)$$

where $N(\mathbf{P}_1)$ is the density of target nucleons in momentum space, defined by $m = \int d\mathbf{P}_1 N(\mathbf{P}_1)$. The allowed regions of integration are confined to values of P, g such that both the final momentum vectors of the collision partners lie outside the occupied sphere in momentum space.

Goldberger (1948) evaluated equation (12.89) by following in detail, collision by collision, the passage of a large number of particles through the nucleus until the particles either escaped or lost sufficient energy to be captured. At an incident energy of 90 MeV., employing an experimental $n\text{-}p$ differential cross-section symmetric about $\frac{1}{2}\pi$, $\bar{\sigma} \approx 0.69$, leading to the mean free path values quoted above.

12.5.2. *Differential cross-sections for neutron scattering and energy transfer.* Writing the differential cross-section for scattering of a neutron by a free proton as $I(\mathbf{p}_0 - \mathbf{p}_f)$, that by a proton bound in a nucleus follows similarly to the total cross-section (12.89). However, the integration over $\mathbf{g}$ is not carried out, so taking $d\mathbf{g} = d\mathbf{P}_f$, we obtain the differential cross-section as a function of the momentum of the scattered neutron.

$$I \, d\mathbf{P}_f = \frac{1}{P_0} \int d\mathbf{P}_1 N(\mathbf{P}_1) \, | \, \mathbf{P}_0 - \mathbf{P}_1 \, | \, I(\mathbf{p}_0 - \mathbf{p}_f) \frac{1}{p_0^2} \delta(p_0 - p_f) \, d\mathbf{P}_f. \quad (12.90)$$

The simplest approach takes $I(\mathbf{p}_0 - \mathbf{p}_f)$ as isotropic, but a better approximation uses an empirical formula which fits the experimental angular distribution at 90 MeV.,

$$\sigma(\mathbf{p}_0 - \mathbf{p}_f) = \text{const.}/\{1 \cdot 0714 + (\mathbf{p}_0 - \mathbf{p}_f)^2\}. \quad (12.91)$$

Evaluation of $I d\mathbf{P}_f$ in equation (12.90) for fixed values of scattering angle θ leads to the energy spectra of scattered neutrons, illustrated in fig. 99 for a heavy nucleus. Integration of this with respect to $\mathbf{P}_f$ gives the angular distribution shown in fig. 100, again for a heavy nucleus (Goldberger, 1948). This differs markedly from the

406 NUCLEAR SCATTERING

strong forward scattering expected qualitatively. Forward scattering implies small momentum transfers, which are greatly reduced by the exclusion principle. Thus the distribution of scattering shows a pronounced dip in the forward direction. Similarly the exclusion

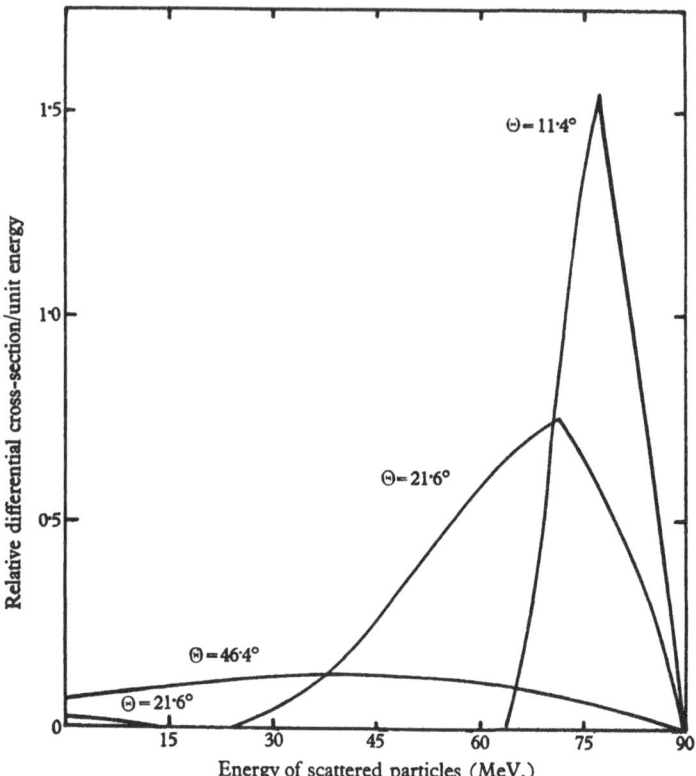

Fig. 99. Energy distribution of particles scattered from a heavy nucleus at fixed Lab. scattering angles (Goldberger, 1948). The energy of the incident neutron is 90 MeV. The discontinuous derivatives at the maxima result from the exclusion principle.

principle discourages large momentum transfers, leading to a back scattering minimum. Inclusion of multiple scattering (Goldberger, 1948) gives a similar result, shown in fig. 100.

The energy transfer E' for a single scatter is defined by the differential cross-section $\tau(\epsilon, E')\,dE'$, found by averaging the differential cross-section over a portion of the Fermi momentum sphere of radius P_1 for the initial state of the struck nucleon:

$$\tau(\epsilon, E')\, dE' = dE' \int I(\mathbf{P}_1, \mathbf{P}_0, E')\, d\mathbf{P}_1 / \tfrac{4}{3}\pi P_1^3. \qquad (12.92)$$

Results are given in fig. 101 for a 100 MeV. neutron. The energy transfer curve shows pronounced dips for very small and very

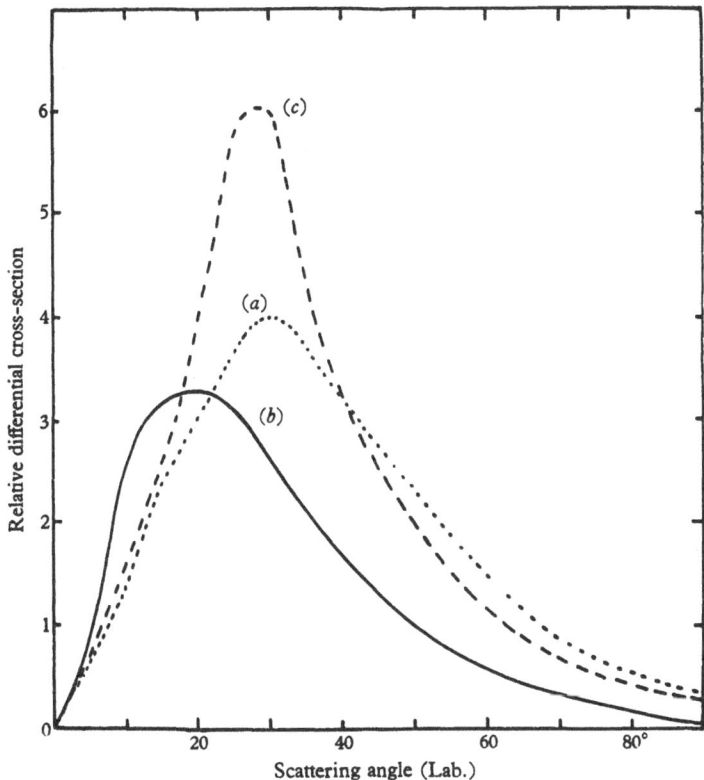

Fig. 100. Angular distribution of 90 MeV. neutrons scattered by a heavy nucleus. Each curve is plotted on a different arbitrary scale (Goldberger, 1948).

(a) Dotted curve: scattering assuming only a single collision inside the nucleus, and employing an isotropic n-p scattering distribution normalized to the total measured cross-section of 83 mb.

(b) Full curve: scattering assuming a single collision inside the heavy nucleus, and employing the forward scattering fit (12·91) for the n-p scattering distribution.

(c) Broken curve: scattering allowing for multiple collisions inside the heavy nucleus and employing the forward scattering n-p distribution (12.91).

large momentum transfers, again illustrating the exclusion principle effect, and is peaked about 16 MeV. (Horning and Baumhoff, 1949). The inclusion of multiple scattering effects leaves the position of the peak unchanged, but makes the larger total energy

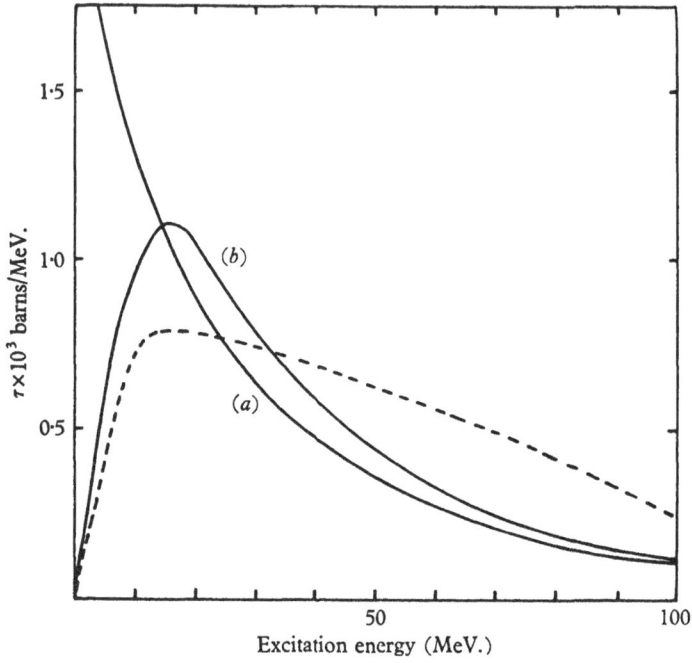

Fig. 101. Full lines: differential cross-section for energy transfer in a collision between an incident neutron of 100 MeV. and a nucleon.
Curve (a): struck nucleus free and at rest.
Curve (b): struck particle within a nucleus.
Both curves employ the Born approximation with a Yukawa interaction and describe single scattering (Horning and Baumhoff, 1949).
Broken line: distribution of excitation energies of the residual nucleus immediately after bombardment of a heavy nucleus by 86·6 MeV. neutrons, given in arbitrary units and taking account of multiple scattering (Goldberger, 1948). The calculations employ the experimental differential scattering cross-section at 90 MeV. and not the Born approximation.

transfers much more probable. The average number of collisions for a neutron entering the nucleus with energy 86·6 MeV. and leaving with $\epsilon > 15$ MeV. is about two. The average excitation energy is 42·5 MeV., or about half the incident energy. This is distributed among the nucleons, the subsequent behaviour of the compound nucleus being described by the evaporation model (§12.6).

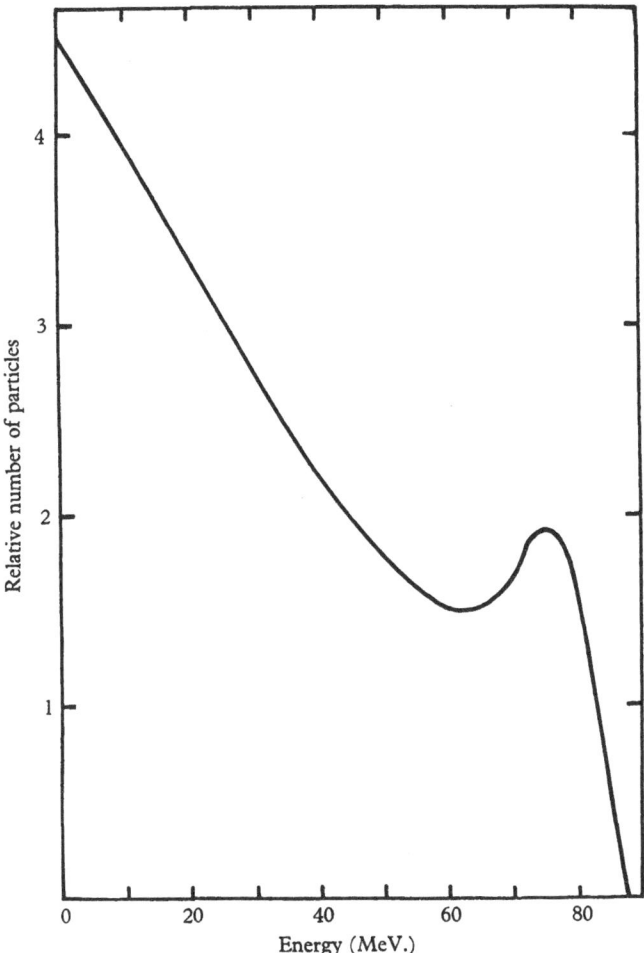

Fig. 102. Energy distribution of particles emerging at all angles immediately after the bombardment of a heavy nucleus with 86·6 MeV. neutrons (Goldberger, 1948).

The energy distribution of particles emerging at all angles immediately after the collision is shown in fig. 102. The peak at the high energy end represents the large number of particles undergoing only one collision before escaping from the nucleus.

12.6. Decay of the compound nucleus according to the evaporation model

The case of energy exchange within a compound nucleus is analogous to the equalization of heat energy of a solid body or liquid, the subsequent expulsion of particles being similar to an evaporation process. These 'delayed' particles are emitted after a time large compared with a nuclear period ($\sim 10^{-21}$ sec.).

Weisskopf (1937) has applied statistical methods to describe the decay of heavy nuclei highly excited by, e.g., collision with very fast neutrons or absorption of γ-rays. This treatment is usually described as employing the *evaporation model* (Serber, 1947; Wolff and Hechrotte, 1948). The individual properties of the separate quantum states are then of no account because of the very small distances between energy levels. Thus statistical information is obtainable by averaging over many quantum states of the same energy, a general qualitative formula resulting for the emission of particles by heavy nuclei.

The simplest reaction involves the collision of a very fast neutron with a heavy nucleus. The probability of formation of a highly excited nucleus A is described by the reaction cross-section σ_r as found by another method such as the schematic model. We suppose that the nucleus A can decay into a nucleus B with emission of a neutron of energy ϵ:

$$A \to B + n. \tag{12.93}$$

The cross-section for this process is related to the reverse reaction of neutron capture by B by the reciprocity theorem (6.65):

$$g_A p_A^2 \, \sigma_{A \to B} = g_B p_B^2 \, \sigma_{B \to A}. \tag{12.94}$$

In this case the momenta p_A and p_B are equal,

$$p_A = p_B = p = \epsilon v / c^2,$$

so that equation (12.94) gives the cross-section for neutron emission in terms of that for neutron capture,

$$\sigma_{A \to B} = \sigma_{B \to A} g_B / g_A, \tag{12.95}$$

where g_B and g_A are statistical weighting factors describing the number of states into which A can decay and the number of states of B into which the neutron can be captured respectively.

Let $\rho_A(E)\,dE$ and $\rho_B(E)\,dE$ be the number of levels of nuclei A and B between E and $E + dE$, the energies being measured from the respective ground states. Thus there are $\rho_B(E_B)\,d\epsilon$ states of the nucleus B into which $A(E_A)$ can decay if the emitted neutron energy lies between ϵ and $\epsilon + d\epsilon$. The number of quantum states in the volume Ω and energy range $d\epsilon$ available to the neutron is

$$g\,\frac{1}{\hbar^3}\,dp = \frac{gm}{2\pi^2\hbar^3}\,p\,d\epsilon,$$

where g denotes the number of spin-states of the particles concerned ($g = 2$ for nucleons, 3 for ^{2}H and 1 for ^{4}He). Hence

$$g_B = \rho_B(E_B)\frac{\Omega gm}{2\pi^2\hbar^3}\,p\,d\epsilon.$$

Similarly $g_A = \rho_A(E_A)\,d\epsilon$ so that equation (12.95) becomes

$$W_n(\epsilon)\,d\epsilon = \sigma(E_A, \epsilon)\frac{\rho_B(E_B)}{\rho_A(E_A)}\frac{gm}{2\pi^2\hbar^3}\,d\epsilon, \tag{12.96}$$

where $W_n(\epsilon)\,d\epsilon$ is the average over all excited states of A with energy near E_A, of the probability per unit time of the decay reaction (12.93). Similarly $W_c(\epsilon)\,d\epsilon$ is the mean probability per unit time of neutrons with energy between ϵ and $\epsilon + d\epsilon$, being captured by B to form the nucleus A with energy between E_A and $E_A + d\epsilon$. Writing

$$\sigma_{A\to B} = \sigma(E_B, \epsilon), \qquad \sigma_{B\to A} = \sigma(E_A, \epsilon),$$

we have $\qquad W_n = \sigma(E_B, \epsilon)\,v/\Omega, \quad W_c = \sigma(E_A, \epsilon)\,v/\Omega.$

We now define the entropy $S(E)$ of a nucleus with energy between E and $E + dE$ by the logarithm of the level density,

$$S_A(E) = \ln\rho_A(E), \quad S_B(E) = \ln\rho_B(E). \tag{12.97}$$

Equation (12.96) becomes

$$W_n(\epsilon)\,d\epsilon = \sigma(E_A, \epsilon)\frac{gm\epsilon}{\pi^2\hbar^3}\cdot\exp\{S_B(E_A - E_0 - \epsilon) - S_B(E_A)\}, \tag{12.98}$$

where E_0 is the binding energy of the emitted particle (kinetic energy ϵ) to the nucleus A.

If we assume $E_A - E_0 \gg \epsilon$ and $S_A(E) = S_B(E)$, then a Taylor theorem expansion is valid,

$$S_B(E_A - E_0 - \epsilon) = S_A(E_A - E_0) - \epsilon/T_B(E_A - E_0) - f(\epsilon), \tag{12.99}$$

where $f(\epsilon)$ contains terms of second order and $T(E)$ is defined by

$$dS/dE = 1/T(E). \qquad (12.100)$$

$T(E)$ has the dimensions of energy and is equal to the product of ordinary temperature by the Boltzmann constant k. The nuclear temperature $T_B(E)$ is the temperature at which E is the most probable energy of the nucleus in thermodynamic equilibrium. The decay cross-section of A becomes

$$\sigma(E_B, \epsilon)\, d\epsilon = \frac{1}{v} W_n(\epsilon)\, d\epsilon$$

$$= \sigma(E_A, \epsilon)\frac{gm}{\pi^2\hbar^3 v}\exp\{S_B(E_A) - S_A(E_A)\}$$

$$\times \epsilon \exp\{-\epsilon/T_B(E_A - E_0)\}\, e^{-f(\epsilon)}\, d\epsilon, \qquad (12.101)$$

where $T(E)$ is given in terms of the level density by

$$T(E) = \rho(E)\Big/\frac{d\rho(E)}{dE}. \qquad (12.102)$$

If in addition we assume

$$S_B(E_A - E_0) - S_A(E_A) = -E_0^*/T_A(E_A), \qquad (12.103)$$

where E_0^* is an energy equal to E_0 only if $E_0 \ll E_A$ and $S_A(E) = S_B(E)$, then equation (12.101) reduces to a form similar to the evaporation formula. The differences are the factor $e^{-f(\epsilon)}$ and the fact that the temperature $T_B(E_A - E_0)$ refers not to the evaporating nucleus $A(E_A)$, but to the final nucleus $B(E_A - E_0)$. This is due to the neutron emission causing a considerable cooling, so that its energy distribution corresponds to the temperature of the final nucleus.

The total probability of neutron emission is found by integration of $W_n(\epsilon)$. The neutron width Γ_n of the emitting level of the nucleus A (that part of the level breadth arising from neutron emission) follows by multiplying by $\hbar$:

$$\Gamma_n = \hbar\int W_n(\epsilon)\, d\epsilon = \bar{\sigma}\frac{gm}{\pi^2\hbar^3}T_B^2(E_A - E_0)\cdot\exp\{S_B(E_A - E_c) - S_A(E_A)\},$$

$$(12.104)$$

where $\bar{\sigma}$ is the mean free path of $\sigma(E_A, \epsilon)\, e^{-f(\epsilon)}$ averaged over the Maxwell distribution:

$$\bar{\sigma} = \int \epsilon\sigma(E_A, \epsilon)\, e^{-\epsilon/T_B - f(\epsilon)}\, d\epsilon\Big/\epsilon\, e^{-\epsilon/T}\, d\epsilon. \qquad (12.105)$$

To estimate $f(\epsilon)$, we observe from equation (12.102) that dS/dE plotted against E is a curve concave downwards. From equation (12.99), $f(o) = o$, $f(\epsilon) > o$ and $f(\epsilon)$ increases monotonically with ϵ. The relative energy distribution of the emitted neutrons is

$$W_n(\epsilon) = \text{const. } \sigma(E_A, \epsilon)\, e^{-\epsilon/T_B} . e^{-f(\epsilon)}. \qquad (12.106)$$

This possesses a maximum displaced towards higher energies and has a smaller breadth than the evaporation formula obtained by putting $f(\epsilon) = o$. A similar argument applied to equation (12.100), assuming $S_A(E) \sim S_B(E)$, leads to the result $E_0^* > E_0$.

In order to calculate the nuclear temperature T from equation (12.102), we require the level density $\rho(E)$, although this can only be found roughly. The nucleus is considered as a degenerate gas with strong interaction forces between the particles. Also E must increase monotonically with T, say $E = T^n/a$, where a is a constant If the temperature is high enough to excite all degrees of freedom (classical limiting case), then $n = 1$. For relatively low temperatures, $n = 4$ for solid bodies and 2 for degenerate gases. Experience with metallic electrons shows that the law $E = T^2/a$ holds in spite of very strong interactions between electrons, so we assume $n = 2$ for nuclei also. Integration of equation (12.102) and use of equation (12.97) leads to the expressions

$$S(E) = 2(E/a)^{\frac{1}{2}} + \text{const.}, \qquad (12.107)$$

$$\rho(E) = \text{const. } \exp\{2(E/a)^{\frac{1}{2}}\}. \qquad (12.108)$$

The value of the constant a may be found if the level density $\rho(E)$ is known for two values E_1, E_2:

$$a = 4\left\{\frac{E_1^{\frac{1}{2}} - E_2^{\frac{1}{2}}}{\ln \rho(E_1) - \ln \rho(E_2)}\right\}^2. \qquad (12.109)$$

Investigation of γ- and α-ray spectra and of resonances in nucleon scattering enables rough estimates of a to be made. For heavy nuclei, $a \sim o \cdot 1$ MeV., increasing slowly with mass number A. Hence

$$T(E) = (aE)^{\frac{1}{2}}, \quad o \cdot o8 < a < o \cdot 5, \qquad (12.110)$$

the lower limit applying for $A = 230$ and the upper limit for $A = 60$. These expressions are only qualitative and are inapplicable to a region containing only a few levels.

12.7. Resonance scattering

Experiments on the interaction of α-particles and protons with nuclei reveal scattering and disintegration maxima at certain energies. Many of these resonances are very sharp, but even more striking effects occur with both slow and fast neutrons.

The α-particle and proton resonances were formerly interpreted from a one-particle viewpoint, the charged incident particle moving behind the Coulomb barrier in the potential field of the nucleus and having certain virtual energy levels in this field. However this picture proved untenable because of neutron resonance phenomena. Bohr pointed out that the solution lay in considering nuclear dynamics as a many-body problem.

In the one-particle picture, the distance between adjacent energy levels is smaller than, but still comparable with, nuclear binding energies ($\sim$ a few hundred keV.), and their width is of the same order as their distance apart.

In the many-body picture, the level separation for neutron resonance phenomena decreases very rapidly with both increasing excitation energy and increasing mass number (Bohr, 1936). The levels are also very much sharper than in the one-particle picture, and both level widths and spacing are negligibly small compared to nuclear binding energies except for very light nuclei. The latter have level spacings D as large as a few hundred keV., but nuclei with $A \geqslant 100$, possessing excitation energies E just sufficient for a dissociation into a neutron and a residual nucleus ($E \sim 8$ MeV.), have $D \sim 10$ eV. For $A < 200$, the corresponding values are $D \sim 10$ eV., $E \sim 5$ MeV. For $A < 100$, the level spacing increases very rapidly.

Proton capture resonances have been detected only for very light nuclei, $D \sim 10 - 100$ keV., $E \sim 10$ MeV. Capture resonances are not observed with heavier nuclei because the energy needed to get the proton over the Coulomb barrier is large, resulting in an excitation energy so large that the resonance levels overlap. The few resonances observed using α-particles as projectiles lead to reactions usually resulting in emission of protons or neutrons.

The cross-section for a nuclear reaction as a function of incident particle energy may be expressed by a dispersion formula similar to that for the scattering of light. Consider a particle P of energy ϵ_p incident on a nucleus A in a state p, forming a compound nucleus C with a number of energy levels E_r. A particle Q of energy ϵ_Q is emitted, leaving a residual nucleus B in a state q:

$$A + P \to C \to B + Q. \tag{12.111}$$

Equation (12.104) is subject to the energy conservation rule

$$E_{A_p} + E_P + \epsilon_P = E_{B_q} + E_Q + \epsilon_Q, \tag{12.112}$$

where E_{A_p} is the internal energy of nucleus A in state p and E_p the internal energy of the incident particle, etc.

Equation (12.111) describes a second order process which may be computed using Dirac's second order perturbation theory. This was first carried out by Breit and Wigner (1936) for a single level and later generalized to the case of many levels (Bethe and Placzek, 1937; Bethe, 1937; Blatt and Weisskopf, 1952). However, the strong interaction between the incident particle and the nucleus renders the perturbation proof of the dispersion formulae invalid except as a very rough approximation. Several authors (Kapur and Peierls, 1938; Siegert, 1939; Breit, 1946) have modified this approach, showing that the dispersion formulae hold rigorously if the widths of the resonance levels contributing to the scattering are much smaller than the level separation. If the level widths are less than the level separation, the formulae are still approximately valid.

However the physical meaning of compound states with level widths greater than the level separation is questionable, although these must occur at higher energies. The proof of the dispersion formulae in this case and the definition and properties of levels far off resonance has lead to great difficulties which perturbation theories seem unable to resolve. The levels far off resonance contribute largely to the background between resonances and are responsible, partly or fully, for the potential scattering.

An alternative approach (Feshbach *et al.* 1947) avoids these difficulties by expressing the reaction and scattering cross-sections in terms of the theory of § 12.2. We have from equations (12.26) and (12.27),

$$\Delta_l = -\tfrac{1}{2} x S_l'/S_l = x \,|\, u_l \,|'/|\, u_l \,|, \quad S_l = x/|\, u_l(R) \,|^2, \tag{12.113}$$

where the derivative is with respect to $x = kR$ and u_l is defined by equation (12.19). Using equation (12.113) for the cross-section expressions (12.30) and (12.34) and taking

$$f_l = g_l - i h_l, \qquad (12.114)$$

$$\sigma_r^l = \frac{4\pi}{k^2}(2l+1)\frac{h_l S_l}{(h_l + S_l)^2 + (g_l - \Delta_l)^2}, \qquad (12.115)$$

$$\sigma_{\text{el}}^l = \frac{4\pi}{k^2}(2l+1)\left| \frac{S_l}{i(h_l + S_l) - (g_l - \Delta_l)} + e^{i\delta}\sin\delta \right|^2, \qquad (12.116)$$

where g_l, h_l and u_l are evaluated at $x = kR$.

12.7.1. *The Breit–Wigner dispersion formulae.* In order to find the value of f_l appropriate to resonance scattering, the physical meaning of the latter must be considered. The wave function ψ_l inside the nuclear boundary may be written as the sum of an ingoing and an outgoing wave,

$$\psi_l \sim \exp(-iKr) + b_l \exp(iKr) \quad (r < R), \qquad (12.117)$$

where b_l is the (complex) amplitude of the outgoing wave, subject to $|b_l| \leqslant 1$. K is defined by equations (12.48) and (12.61) and corresponds to the energy of the incident particle inside the nucleus, given as the sum of the Fermi energy and the particle binding energy (~ 30 MeV.). The limiting case $|b_l|^2 = 1$ leads to pure elastic scattering and $b_l = 0$ to the schematic theory of §14.3, the latter expressing the condition that the incident particle does not return. The intermediate case follows by writing

$$b_l = \exp(2i\zeta_l)\exp(-2q_l), \qquad (12.118)$$

where the phase ζ_l and q_l are both real functions of the incident energy ϵ. b_l is similar to the scattered amplitude a for the high energy optical model (§12.4.1). The wave function becomes

$$\psi_l \sim C\cos(Kr + \zeta_l + iq_l) \quad (r < R), \qquad (12.119)$$

so that $\qquad f_l = R[\psi'_l/\psi_l]_{r=R} = -KR\tan[Z_l(\epsilon) + iq_l], \qquad (12.120)$

where $\qquad Z_l(\epsilon) = Kr + \zeta_l(\epsilon). \qquad (12.121)$

Equation (12.120) is similar to equation (12.62) for f_l on the optical model, using a complex potential. However, a different pair of phase and absorption coefficients ζ_l and q_l for each l value enables fitting of resonance levels for each l also, whereas the optical model parameters are fixed and cannot 'follow' the resonances for different l.

Both K and $\zeta_l(\epsilon)$ increase smoothly with energy so that for a fixed value of r, $Z_l(\epsilon)$ should be a smooth, monotonically increasing function of energy ϵ. The wave function (12.112) must be joined smoothly at the nuclear boundary to the external wave function, so f_l is continuous at $r = R$. It is simplest to consider s-wave scattering of neutrons, for which the external wave function is

$$\psi_0 = A \sin(kr + \delta_0). \qquad (12.122)$$

We have to join smoothly at $r = R$ a periodic function of amplitude C and high wave number K to one of amplitude A and low wave number k (assuming the incident energy is in the energy region $\epsilon < 10$ MeV. where resonance scattering is possible). The ratio of the amplitudes at $r = R$ is

$$C/A = (k/K)(K^2 + f_0^2)^{\frac{1}{2}}/(k^2 + f_0^2)^{\frac{1}{2}}, \qquad (12.123)$$

where $f_0 = k \cot(kR + \delta_0) = -K \tan(Z(\epsilon) + iq)$.

In general, $C/A \sim k/K$ so that the amplitude of the wave inside the nucleus is small compared to that outside the nucleus. The exception occurs if ψ has a maximum or minimum at or near $r = R$, $\psi_0'(R) = 0$ or $f_0 = 0$, in which case $C = A$. If this holds, the incident particle has a high probability of penetrating the nucleus. In general the condition of zero gradient ($f_l = 0$) defines a resonance energy. From equation (12.120), this requires $Z_l(\epsilon_s) = n\pi$ (n any integer) and $q = 0$. Thus the series of formal resonance energies ϵ_s is defined by

$$f_l(\epsilon_s, q = 0) = f_l^0(\epsilon_s) = -KR \tan Z_l(\epsilon_s) = 0. \qquad (12.124)$$

Expansion of f_l as a Taylor's series in q and in ϵ in the neighbourhood of the resonance ϵ_s gives to first order

$$
\begin{aligned}
f_l(\epsilon) &\approx f_l(\epsilon_s, q_l = 0) + (\epsilon - \epsilon_s)\left(\frac{\partial f_l^0}{\partial \epsilon}\right)_s + q\left(\frac{\partial f_l^0}{\partial q_l}\right)_s \\
&= (\epsilon - \epsilon_s)\left(\frac{\partial f_l^0}{\partial \epsilon}\right)_s - iq_l KR, \qquad (12.125)
\end{aligned}
$$

where $(\partial f_l^0/\partial \epsilon)_r$ denotes the derivative of f_l with respect to ϵ at $\epsilon = \epsilon_r$ and $q = 0$. From equation (12.114),

$$g_l = (\epsilon - \epsilon_s)\left(\frac{\partial f_l^0}{\partial \epsilon}\right)_s, \quad h_l = qKR. \tag{12.126}$$

These expressions may be substituted in equations (12.115) and (12.116) to obtain the reaction and scattering cross-sections. We define the particle width Γ_α^s, the reaction width Γ_r^s and the total width Γ^s by expressions valid in the neighbourhood of the resonances:

$$\Gamma_\alpha^s = -2S_l/(\partial f_l^0/\partial \epsilon)_s, \quad \Gamma_r^s = -2h_l/(\partial f_l^0/\partial \epsilon)_s, \tag{12.127}$$

$$\Gamma^s = \Gamma_\alpha^s + \Gamma_r^s. \tag{12.128}$$

The term 'particle width' follows from its association with the incident particle energy via $S_l = kR/|u_l(r)|^2$, and the term 'reaction width' with the relation between h_l and the internal energy of the nucleus via $h_l = qKR$. If the Hamiltonian H_0 describing pure scattering corresponds to the value $f_l(\epsilon_s, q = 0)$, then radiative absorption is allowed for by an imaginary addition to H_0:

$$H\Psi = (H_0 + H')\Psi, \quad H' = -\tfrac{1}{2}i\Gamma_\alpha.$$

This describes the characteristic time dependence

$$|\Psi|^2 \sim |\exp\{-i(\epsilon_s - \tfrac{1}{2}i\Gamma_\alpha)\,t\}|^2 = \exp(-\Gamma_\alpha t).$$

The solution of H needed to determine $f_l(E)$ is given by

$$H\Psi = \epsilon_s\Psi, \quad \text{i.e.} \quad H_0\Psi = (\epsilon_s + \tfrac{1}{2}i\Gamma_\alpha)\Psi.$$

Then $\quad f_l(\epsilon_s) = f_l^0(\epsilon_s + \tfrac{1}{2}i\Gamma_\alpha) \approx f_l^0(\epsilon_s) + i(\tfrac{1}{2}\Gamma_\alpha^s)(\partial f_l^0/\partial \epsilon)_s. \tag{12.129}$

The imaginary part $-h_l$ of $f_l(\epsilon_s)$ becomes

$$h_l = -(\tfrac{1}{2}\Gamma_\alpha^s)(\partial f_l^0/\partial \epsilon)_s,$$

which agrees with the definition (12.127).

Introducing the actual resonance energy

$$\epsilon_{sl}' = \epsilon_s + \Delta_l(\epsilon)/(\partial f_l^0/\partial \epsilon)_s, \tag{12.130}$$

the partial reaction and elastic cross-sections follow from equations (12.115), (12.116), (12.125) and (12.128):

$$\sigma_r^l = \frac{4\pi}{k^2}(2l+1)\frac{\Gamma_\alpha^s \Gamma_r^s}{(\epsilon - \epsilon'_{sl})^2 + \frac{1}{4}(\Gamma_l^s)^2}, \qquad (12.131)$$

$$\sigma_{el}^l = \frac{4\pi}{k^2}(2l+1)\left|\frac{\frac{1}{2}\Gamma_\alpha^s}{\epsilon - \epsilon'_{sl} + \frac{1}{2}i\Gamma_l^s} + e^{i\delta}\sin\delta\right|^2. \qquad (12.132)$$

Equation (12.131) for the reaction cross-section has the typical dispersion form, with a strong maximum at $\epsilon = \epsilon'_s$ of half-width Γ^s. The elastic cross-section (12.132) has a similar resonance scattering peak, partially masked by potential scattering.

Equations (12.131) and (12.132) are not the same as the second order perturbation expressions of Breit and Wigner, for they do not include contributions from resonance levels other than the nearest to the incident energy. This is certainly valid for energies near resonance where contributions of other levels are small in comparison and avoids difficulties mentioned earlier. However, the contributions of other levels may be taken into account if so desired. From equations (12.31) to (12.35), the total cross-sections become

$$\sigma_{el}^l = \frac{\pi}{k^2}(2l+1)\left|-2iS_l g_l(\epsilon) + e^{2i\delta_l} - 1\right|^2, \qquad (12.133)$$

$$\sigma_r^l = \frac{\pi}{k^2}(2l+1)\left[-4S_l h_l |g_l(\epsilon)|^2\right], \qquad (12.134)$$

where

$$g_l(\epsilon) = \frac{1}{f_l - (\Delta_l + iS)_l} = \sum_s \frac{C_s}{P_s - \epsilon}. \qquad (12.135)$$

Here the resonance levels are defined by the poles P_{sl} of the function $g_l(E)$ and the C_s are residues:

$$f_l(P_s) = \Delta_l(P_s) + iS_l(P_s), \qquad (12.136)$$

$$C_s = 1/(\partial f_l^0 / \partial \epsilon)_{P_s}. \qquad (12.137)$$

From equations (12.136), (12.127) and (12.129),

$$f_l(P_s) = \Delta_l(P_s) - i\frac{1}{2}(\partial f_l / \partial \epsilon)_{P_s} = f_l^0(P_s) + \frac{1}{2}i(\partial f_l^0 / \partial \epsilon)_{P_s}.$$

Neglecting second order terms, this becomes

$$f_l^0(P_s) - \Delta_l(P_s) = -\frac{1}{2}i(\Gamma_\alpha^s + \Gamma_r^s)(\partial f_l^0 / \partial \epsilon)_{P_s},$$

or employing the definition (12.130),

$$f_l^0(P_s)/(\partial f_l^0/\partial \epsilon)_{P_s} - (\epsilon_s' - \epsilon_s) = -\tfrac{1}{2}i(\Gamma_\alpha^s + \Gamma_r^s). \qquad (12.138)$$

Also to first order, using equation (12.124),

$$f_l^0(P_s) = f_l^0(\epsilon_s) + (P_s - \epsilon_s)(\partial f_l^0/\partial \epsilon)_{\epsilon_s} = (P_s - \epsilon_s)(\partial f_l/\partial \epsilon)_{\epsilon_s}. \qquad (12\cdot139)$$

Combining equations (12.138) and (12.139) leads to

$$P_s = \epsilon_s - \tfrac{1}{2}i(\Gamma_{\alpha l}^s + \Gamma_{rl}^s). \qquad (12.140)$$

The residues of equation (12.137) are approximated to first order by

$$C_s \approx -\tfrac{1}{2}\Gamma_r^s/h_l. \qquad (12.141)$$

Hence $$g_l(\epsilon) = \sum_s \frac{\tfrac{1}{2}\Gamma_r^s/h_l}{\epsilon - \epsilon_{sl} + \tfrac{1}{2}i(\Gamma_{\alpha l}^s + \Gamma_{rl}^s)}. \qquad (12.142)$$

Substitution of equations (12.142) and (12.127) into equations (12.133) and (12.134) gives the scattering formulae

$$\sigma_{el}^l = \frac{4\pi}{k^2}(2l+1)\left|\tfrac{1}{2}\sum_s \frac{\Gamma_\alpha^s}{\epsilon - \epsilon_{sl} + \tfrac{1}{2}i(\Gamma_{\alpha l}^s + \Gamma_{rl}^s)} + e^{i\delta}\sin\delta\right|^2, \qquad (12.143)$$

$$\sigma_r^l = \frac{\pi}{k^2}(2l+1)\left|\sum_s \frac{\Gamma_\alpha^s \Gamma_r^s}{\epsilon - \epsilon_{sl} + \tfrac{1}{2}i(\Gamma_{\alpha l}^s + \Gamma_{rl}^s)}\right|^2, \qquad (12.144)$$

the summation being over all resonance levels ϵ_s, however far from the incident energy. The cross-sections are identical with the Breit–Wigner formulae in all respects except for a phase factor $e^{i\phi_s}$ missing in the summation for σ_r^l (Feshbach et al. 1947; Kapur and Peierls, 1938).

12.7.2. *The absorption and scattering cross-section of neutrons.* For neutron scattering, $\Gamma_\alpha^s = \Gamma_n^s$, where the level width is given approximately by equations (12.127) and (12.124):

$$\Gamma_{nl}^s = \frac{2S_l}{KR}\bigg/(\partial Z_l/\partial \epsilon)_s, \quad S_l = kRv_l$$

or $$\Gamma_{nl}^s = \frac{2k}{K}v_l/(\partial Z_l/\partial \epsilon)_s. \qquad (12.145)$$

The resonance energies ϵ_s are defined by equation (12.124),

$$Z_l(\epsilon_s) = n\pi, \quad (n \text{ an integer}). \qquad (12.146)$$

The function $f(\epsilon)$ is qualitatively determined by assuming $Z(\epsilon)$ increases smoothly from $n\pi$ to $(n+1)\pi$ when the energy ϵ goes from one resonance to another (Feshbach *et al.* 1947),

$$(\partial Z_l/\partial \epsilon)_s = \pi/D_l^*, \tag{12.147}$$

where $D_l^* \sim D_l$, the average level separation. The neutron level width becomes

$$\Gamma_{nl}^s = \frac{2k}{K} \frac{D_l^*}{\pi} v_l, \tag{12.148}$$

where for a random level distribution, D_l^* is related to D_l by (Fermi *et al.* 1947)

$$D_l^*/D_l = \tfrac{1}{2}. \tag{12.149}$$

The penetration factor v_l defined by equation (12.36) is given by

$$1/v_l = F_l^2 + G_l^2 = x^2 \,|\, h_l^{(1)}(x)\,|^2, \tag{12.150}$$

where $h_l^{(1)}(x)$ is the spherical Hankel function. Values of v_l for $l = 0\text{–}3$ are

$$\left.\begin{aligned}
v_0 &= 1, \quad v_1 = x^2/(1+x^2), \quad v_2 = x^4/(9+3x^2+x^4),\\
v_3 &= x^6/(225+45x^2+6x^4+x^6),
\end{aligned}\right\} \tag{12.151}$$

where $x = kR$. The phase shifts $\delta_l(x)$ are given by equation (12.29). In particular, the neutron level width for s-waves is

$$\Gamma_{n0}^s = \frac{2k}{K} \frac{D_l^*}{\pi}. \tag{12.152}$$

Analysis of neutron absorption s-wave cross-sections indicates $\Gamma_n \sim 10^{-3}\sqrt{\epsilon_s}$ eV. This value gives rise to $D_l^* \sim 10$ eV., which seems reasonable for the resonance level distance.

Consider the average of σ_r^l over a small energy interval containing many levels. This is the cross-section actually observed if the beam is not sufficiently monoenergetic to distinguish resonances.

$$\begin{aligned}
\langle \sigma_r^l \rangle_{\mathrm{av}} &= \frac{\pi}{k^2}(2l+1)\frac{1}{S}\sum_s \frac{1}{D_l}\int_{\epsilon=\epsilon_s-\frac{1}{2}D_l}^{\epsilon=\epsilon_s+\frac{1}{2}D_l} \frac{\Gamma_{nl}^s \Gamma_{rl}^s}{(\epsilon-\epsilon_s)^2 + \frac{1}{4}(\Gamma_l^s)^2}\,d\epsilon \\
&= \frac{\pi}{k^2}(2l+1)\frac{1}{D_l}\int_{\eta=-D_l}^{\eta=D_l} \frac{\Gamma_{nl}\Gamma_{rl}}{\eta^2+\frac{1}{4}(\Gamma_l)^2}\,d\eta = \frac{\pi}{k^2}(2l+1)\frac{1}{D_l}\cdot\frac{2\pi\Gamma_{nl}\Gamma_{rl}}{\Gamma_l} \\
&= \frac{4\pi}{kK}\frac{D_l^*}{D_l}(2l+1)\,v_l\frac{\Gamma_{rl}}{\Gamma_l}, \tag{12.153}
\end{aligned}$$

where Γ_{nl}, Γ_{rl} and Γ_l are the average values of Γ_{nl}^s, Γ_{rl}^s and Γ_l^s respectively over the energy interval ΔE, and D_l is the average level distance. In evaluating the integral in (12.153), D_l/Γ_l is assumed large, $\arctan D_l/\Gamma_l \approx \frac{1}{2}\pi$, which is the condition for validity of the resonance scattering expressions.

For very low energies, the inequality $\Gamma_r \gg \Gamma_n$ holds. Since for heavy nuclei $(A > 100)$, $\Gamma_r \sim 0.1$ eV. and $D_0^* \sim 10$ eV., we find from equation (12.148) that $\Gamma_r = \Gamma_n$ for $E \sim 7$ keV. Hence for s-wave scattering

$$\langle \sigma_r^0 \rangle_{av} = \frac{4}{kK}\frac{D_0^*}{D_0} \approx \frac{250}{(E_{ev.})^{\frac{1}{2}}} \text{ barns}, \quad E \ll 7 \text{ keV}., \quad A > 100. \quad (12.154)$$

Equation (12.154) does not apply to lighter nuclei because the larger D_l involved restricts the limits of validity to even lower energies, making it impossible to take an average over many levels.

The reaction cross-section may be simplified further if the ratio Γ_{rl}/D_l^* and the wave number K are assumed independent of l and ϵ,

$$\frac{\Gamma_{nl}}{\Gamma_{rl}} = \frac{\Gamma_{n0}}{D_0^*}v_l\frac{D_l^*}{\Gamma_{rl}} = \frac{\Gamma_{n0}}{D_0^*}v_l\frac{D_0^*}{\Gamma_{r0}} = \frac{\Gamma_{n0}}{\Gamma_{r0}}v_l = \frac{x}{x_0}v_l,$$

where x_0 is the value of $x = kR$ for which

$$\Gamma_{n0} = \Gamma_{r0}, \quad x_0 = \frac{\pi}{2}\frac{KR}{D_0^*}\Gamma_{r0}.$$

The total reaction cross-section is obtained by summing equation (12.153) over l:

$$\langle \sigma_r \rangle_{av} = \frac{4\pi}{kK}\sum_{l=0}^{\infty}\frac{D_l^*}{D_l}(2l+1) \cdot 1 \bigg/ \left(\frac{1}{v_l} + \frac{x}{x_0}\right). \quad (12.155)$$

The assumption that Γ_{rl}/D_l^* is independent of energy breaks down when inelastic scattering or fission sets in. This increases Γ_{rl} sharply at higher energies, but is unimportant for $\epsilon < 0.4$ MeV.

The elastic cross-section has a characteristic shape shown in fig. 103, calculated for the In resonance energy using the value $D_0^* = 25$ eV., which gives rise to the observed neutron width of 3×10^{-3} eV. in equation (12.152). The depression at low energy and the high shoulder at the higher energy side are due to destructive and constructive interference of the potential and resonance scattering amplitudes. The extensions of these features in the energy scale are much larger than the width Γ_l of the line, being of order $2D/KR$ or about 20 % of D for heavy nuclei.

The average elastic cross-section is found similarly to the reaction cross-section. The average over many resonances contained within a narrow energy interval E is again the magnitude

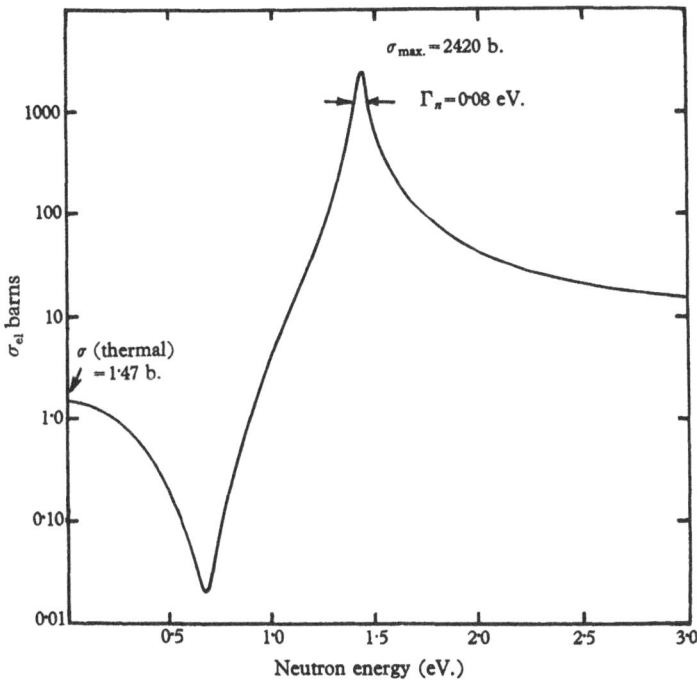

Fig. 103. Elastic scattering cross-section σ_{el} plotted on a logarithmic scale as a function of neutron energy in the neighbourhood of a resonance (Feshbach *et al.* 1947). Numerical values used are those for the indium resonance, $D^* = 25$ eV., $\Gamma_n = 3 \times 10^{-3}$ eV.

which is measured most readily. Expressed in terms of average level widths,

$$\langle\sigma_{el}^l\rangle_{av} = \frac{4\pi}{k^2}(2l+1)\frac{1}{D_l}\int_{\epsilon_s-\frac{1}{2}D_l}^{\epsilon_s+\frac{1}{2}D_l}\left|\frac{\frac{1}{2}\Gamma_{nl}}{(\epsilon-\epsilon_s)+\frac{1}{2}i\Gamma_l}+e^{i\delta_l}\sin\delta_l\right|^2 d\epsilon$$

$$= \frac{4\pi}{k^2}(2l+1)\left[\left(1-\frac{\pi}{D_l}\Gamma_{nl}\right)\sin^2\delta_l+\frac{\pi}{2D_l}\frac{\Gamma_{nl}^2}{\Gamma_l}\right]. \quad (12.156)$$

Hence

$$\langle\sigma_{\text{el}}\rangle_{\text{av}} = \sigma_0 + \frac{4\pi}{k^2}\frac{x}{X}\sum_l \frac{D_l^*}{D_l}(2l+1)v_l\left[\left(1 + \frac{1}{v_l}\frac{x_0}{x}\right)^{-1} - 2\sin^2\delta_l\right],$$

(12.157)

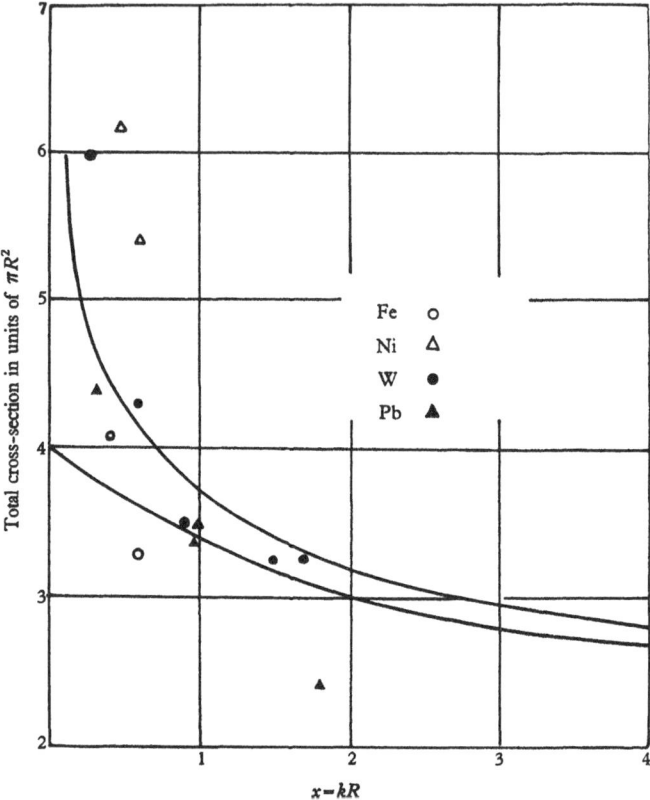

Fig. 104. Total cross-section for neutrons as a function of kR, plotted in units of πR^2. σ_{tot} is the total cross-section for In $(A = 115)$ and σ_0 is the cross-section for scattering by an impenetrable sphere of radius $R = 1\cdot5 \times 10^{-13}A^{\frac{1}{3}}$ cm., assuming $D^*/D = \frac{1}{2}$ (Feshbach et al. 1947). Experimental points are due to Fields et al. (1947).

where $X = KR$ and σ_0 is the scattering cross-section of an impenetrable sphere of radius R,

$$\sigma_0 = \frac{4\pi}{k^2}\sum_l (2l+1)\sin^2\delta_l.$$

(12.158)

σ_0 is $4\pi R^2$ for small energies $(kR \ll 1)$ and approaches $2\pi R^2$ asymptotically for large energies $(kR \gg 1)$.

The quantity usually measured is not $\langle \sigma_{el} \rangle_{av}$ but the total cross-section,

$$\langle \sigma_{tot} \rangle_{av} = \langle \sigma_r \rangle_{av} + \langle \sigma_{el} \rangle_{av}$$

$$= \sigma_0 + \frac{2\pi^2}{k^2 D} \sum_{l=0}^{\infty} (2l+1) \Gamma_{nl} \cos 2\delta_l$$

or $$\langle \sigma_{tot} \rangle_{av} = \sigma_0 + \frac{4\pi}{k^2} \frac{D^*}{D} \cdot \frac{x}{X} \sum_{l=0}^{\infty} (2l+1) v_l \cos 2\delta_l, \qquad (12.159)$$

where we assume D_l^* is independent of l,

$$D_l^*/D_l = D^*/D \sim \tfrac{1}{2}.$$

Note that $\langle \sigma_{tot} \rangle_{av}$ is independent of x_0.

The total nuclear cross-section is always greater than σ_0. For low energies, only the $l=0$ term of (12.159) is used, so assuming $x \ll 1$, the total cross-section becomes

$$\langle \sigma_{tot} \rangle_{av} = 4\pi R^2 \left[1 + \frac{D^*}{D} \cdot \frac{1}{Xx} \right]. \qquad (12.160)$$

The excess of $\langle \sigma_{tot} \rangle_{av}$ over $4\pi R^2$ for small x is large since the $1/x$ term becomes very large $(1/v$ law of absorption$)$. For larger energies, several values of l contribute to the cross-section. In the limit the positive and negative terms cancel so that $\langle \sigma_{tot} \rangle_{av} \rightarrow \sigma_0$ $(k^2 \rightarrow \infty)$.

Fig. 104 shows measurements of the total cross-section of several elements for neutrons of 20–800 keV. energy (Fields *et al.* 1947; Warren *et al.* 1947). Theoretical curves show good general agreement for elements with $A \sim 100$.

Charged-particle scattering may be treated in a similar fashion, using equations (12.131) and (12.132). Detailed accounts of applications are available elsewhere (Bethe, 1937; Blatt and Weisskopf, 1952).

426

KINEMATICS OF ELASTIC COLLISIONS AND CONVERSION FORMULAE TO CENTRE OF MASS SYSTEM

We shall introduce here the momentum and energy relationships for the interaction illustrated in fig. A. 1 and derive expressions connecting the physical quantities.

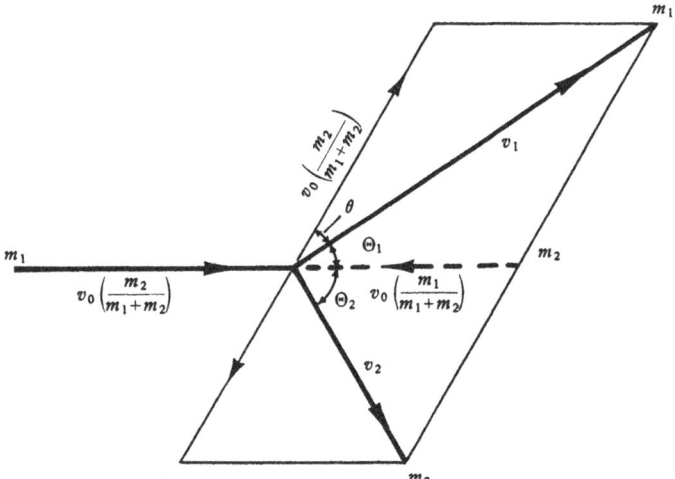

Fig. A. 1. Elastic scattering of mass m_1 by mass m_2, shown in both CM and Lab. co-ordinates. The dark lines represent the paths of the particles in the Lab. system.

Define $m_1 = $ mass of incident (scattered) particle.

$m_2 = $ mass of target (recoil) particle.

$\Theta_1 = $ angle of scatter of m_1.

$\Theta_2 = $ angle of recoil of m_2.

$v_0 = $ incident velocity of m_1 (corresponding kinetic energy E_0 and momentum $\mathbf{p}_0$).

$v_1 = $ velocity of m_1 after scattering (kinetic energy E_1, momentum $\mathbf{p}_1$).

$v_2 = $ velocity of recoil of m_2 (kinetic energy E_2, momentum $\mathbf{p}_2$).

These symbols specify quantities referred to the Lab. system, the motions of the particles in this system being indicated in fig. A. 1 by heavy lines. Conservation of momentum requires that

$$\mathbf{p_0} = \mathbf{p_1} + \mathbf{p_2}. \qquad (A. 1)$$

Since the collision is elastic, kinetic energy is conserved:

$$E_0 = E_1 + E_2. \qquad (A. 2)$$

All the important relationships can be derived from these equations but the calculation is much simplified by referring the collision to the CM system in which there is no net momentum, $\Sigma \mathbf{p} = 0$. The motions of the particles in this system are represented by light lines in fig. A. 1.

We shall first of all carry through the calculation for the case of non-relativistic collisions ($v_0 \ll c$), which is more readily visualized, then undertake the general case of relativistic collisions.

Non-relativistic treatment

The mutual velocity of the masses must remain v_0 in the CM system, hence their respective velocities relative to this system are $v_0 m_2/(m_1 + m_2)$ and $v_0 m_1/(m_1 + m_2)$. The collision produces a change of direction (the distinction between 'scattering' and 'recoiling' is meaningless in the CM system) of angle θ.

After the collision the trajectories of the particles must remain collinear to conserve momentum, and their velocities must remain unchanged in order to preserve the same total kinetic energy. Transformation to the Lab. system is effected by vector addition of the velocity of the centre of mass relative to m_2, i.e. $v_0 m_1/(m_1 + m_2)$. The following relationships can then be derived from inspection of fig. A. 1:

$$\theta = (\pi - 2\Theta_2), \qquad (A. 3)$$

$$v_0 m_2/(m_1 + m_2) . \sin \Theta_1 = v_0 m_1/(m_1 + m_2) . \sin (\theta - \Theta_1),$$

whence

$$\sin (\theta - \Theta_1) = \frac{m_1}{m_2} \sin \Theta_1 = k \sin \Theta_1. \qquad (A. 4)$$

Also $\quad \tan \Theta_1 = \left(v_0 \frac{m_2}{m_1 + m_2} \sin \theta \right) \Big/ \left(v_0 \frac{m_1}{m_1 + m_2} + v_0 \frac{m_2 \cos \theta}{m_1 + m_2} \right),$

so that $\quad \tan \Theta_1 = \dfrac{m_2 \sin \theta}{(m_1 + m_2 \cos \theta)} = \dfrac{\sin \theta}{(k + \cos \theta)}, \qquad (A. 5)$

where k is written for the mass ratio m_1/m_2.

Two useful formulae connecting the masses and scattering angles in the Lab. system follow from the above equations. From equation (A. 4),

$$k = \frac{\sin(\theta - \Theta_1)}{\sin \Theta_1} = \frac{\sin\{\pi - (2\Theta_2 + \Theta_1)\}}{\sin \Theta_1},$$
$$k = \frac{\sin(2\Theta_2 + \Theta_1)}{\sin \Theta_1}. \qquad\qquad\qquad \left.\right\} \quad (A.6)$$

In the case $k > 1$, Θ_2 is double-valued for $\Theta_1 < \Theta_{1,\max}$ (equation (A. 8)) corresponding to the two solutions $|\arcsin(k \sin \Theta_1)|$ and $\pi - |\arcsin(k \sin \Theta_1)|$. Hence for a particular angle of scattering Θ_1, the conjugate angle Θ_2 depends only on the mass ratio. Another form of the equation is

$$\tan \Theta_1 = \{m_2 \sin(\pi - 2\Theta_2)\}/\{m_1 + m_2 \cos(\pi - 2\Theta_2)\}$$
$$= \frac{\sin 2\Theta_2}{(k - \cos 2\Theta_2)}, \qquad\qquad (A.7)$$

which is more convenient when Θ_2 is known and Θ_1 required. The behaviour of Θ_1 vs. Θ_2 is illustrated in fig. A. 2 for several k values.

When $m_1 < m_2$, m_1 can be scattered through all angles Θ_1 from 0 to 180°, while m_2 recoils at angles 90 to 0°. For the special case $m_1 = m_2$, it is obvious that $\theta = 2\Theta_1$ and that $(\Theta_1 + \Theta_2) = 90°$. For unequal masses, $(\Theta_1 + \Theta_2)$ is not constant. For $m_1 > m_2$, Θ_2 can vary from 90 to 0° as before but Θ_1 approaches a definite maximum determined by the mass ratio, then decreases again. The maximum angle of scatter is

$$\Theta_{1,\max} = \arcsin k^{-1}. \qquad\qquad (A.8)$$

Velocity and energy relations follow directly from fig. A. 1:

$$\frac{v_1}{v_0} = \frac{\sin \Theta_2}{\sin(\Theta_1 + \Theta_2)}, \qquad\qquad (A.9)$$

$$\frac{v_2}{v_0} = \frac{k \sin \Theta_1}{\sin(\Theta_1 + \Theta_2)} = \frac{2m_1 \cos \Theta_2}{m_1 + m_2}, \qquad\qquad (A.10)$$

$$\frac{E_1}{E_0} = \left\{ \frac{\sin \Theta_2}{\sin(\Theta_1 + \Theta_2)} \right\}^2, \qquad\qquad (A.11)$$

$$\frac{E_2}{E_0} = k \left\{ \frac{\sin \Theta_1}{\sin(\Theta_1 + \Theta_2)} \right\}^2 = \frac{4m_1 m_2 \cos^2 \Theta_2}{(m_1 + m_2)^2},$$

whence
$$E_2 = E_{\max} \cos^2 \Theta_2. \qquad\qquad (A.12)$$

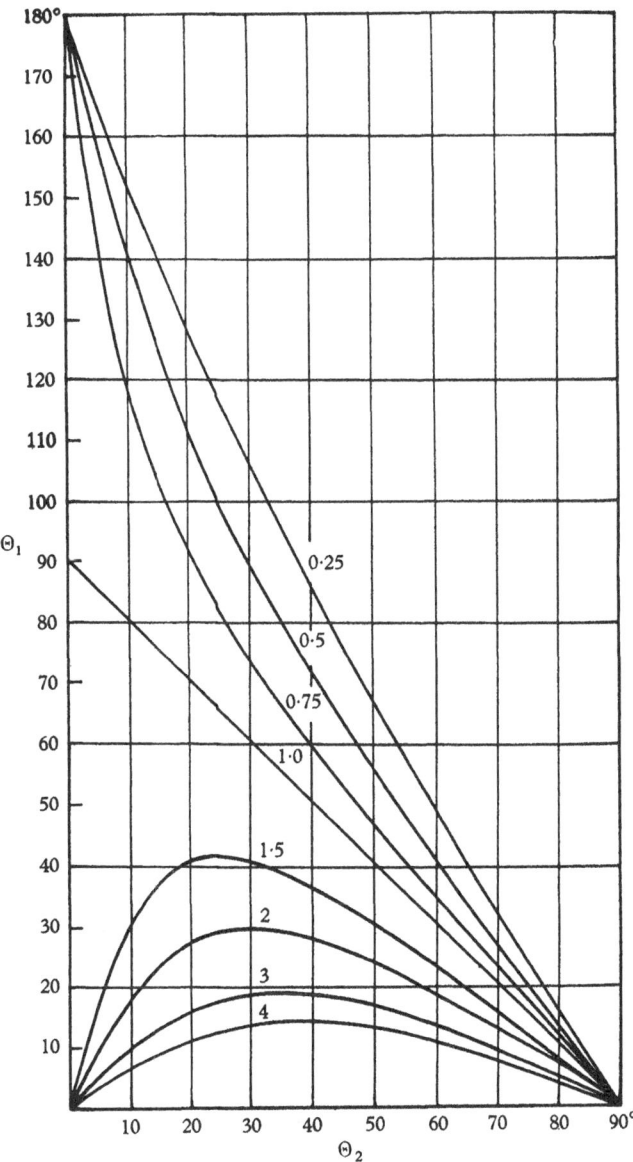

Fig. A. 2. Relationship between angles of scattering Θ_1 and recoil Θ_2 in the Lab. system for various k values ($k = m_1/m_2$).

$E_{\max}$ is the maximum energy which can be transferred to the struck particle. Also

$$\frac{E_2}{E_1} = k \left\{\frac{\sin \Theta_1}{\sin \Theta_2}\right\}^2. \tag{A. 13}$$

When $k > 1$, the two values of E_1 follow from equation (A. 11) by inserting the two Θ_2 values derived from equation (A. 6).

Part of the kinetic energy is associated with the motion of the CM of the whole system:

$$E_{\mathrm{CM}} = \tfrac{1}{2}(m_1 + m_2) \left(\frac{v_0 m_1}{m_1 + m_2}\right)^2 = E_0 \left(\frac{m_1}{m_1 + m_2}\right). \tag{A. 14}$$

Thus, the energy associated with the relative motion of two particles is $(E_0 - E_{\mathrm{CM}})$ or

$$E_R = E_0 \left(\frac{m_2}{m_1 + m_2}\right) = \tfrac{1}{2}\mu v_0^2, \tag{A. 15}$$

where μ is the reduced mass of the system, $m_1 m_2/(m_1 + m_2)$. In terms of energy, two collisions are identical, regardless of which is the projectile and which the target, if the quantity $E_0 m_2/(m_1 + m_2)$ is the same in each case.

Consider now the conversion of cross-sections from Lab. system to CM co-ordinates. The total cross-section is invariant with respect to transformation of co-ordinate system, i.e. the physical process does not depend on how it is observed. It follows that

$$\sigma(\theta)\, d\omega = \sigma(\Theta)\, d\Omega, \tag{A. 16}$$

from which $\sigma(\theta)$ can be expressed in terms of $\sigma(\Theta)$ and the collision angles. Two cases must be considered, depending on whether the cross-section for scattering or that for recoil is measured in the Lab. system.

(i) When the cross-section for scattering $\sigma(\Theta_1)$ has been measured in the Lab. system, we express elements of solid angle in terms of scattering angles. Since the scattering is axially symmetric,

$$\sigma(\theta)\, 2\pi\, d(\cos \theta) = \sigma(\Theta_1)\, 2\pi\, d(\cos \Theta_1),$$

whence
$$\sigma(\theta) = \sigma(\Theta_1) \left(\frac{\sin \Theta_1}{\sin \theta}\right) \frac{d\Theta_1}{d\theta}.$$

Differentiating equation (A. 4) and substituting for k from the same equation leads to

$$\frac{d\Theta_1}{d\theta} = \frac{\sin \Theta_1 \cos(\theta - \Theta_1)}{\sin \theta},$$

so that
$$\sigma(\theta) = \sigma(\Theta_1) \left(\frac{\sin \Theta_1}{\sin \theta}\right)^2 \cos(\theta - \Theta_1). \qquad \text{(A. 17)}$$

For the special case $m_1 = m_2$ this simplifies considerably. Then $\theta = 2\Theta_1$ and it follows that

$$\sigma(\theta) = \sigma(\Theta_1)/4 \cos \Theta_1. \qquad \text{(A. 18)}$$

(ii) When the cross-section $\sigma(\Theta_2)$ for recoil is measured, a corresponding treatment gives

$$\sigma(\theta) = \sigma(\Theta_2) \left(\frac{\sin \Theta_2}{\sin \theta}\right) \frac{d\Theta_2}{d\theta}.$$

By equation (A. 3), $d\Theta_2/d\theta = 0.5$ and the ratio of sines is $(2 \cos \Theta_2)^{-1}$. Hence
$$\sigma(\theta) = \sigma(\Theta_2)/4 \cos \Theta_2, \qquad \text{(A. 19)}$$

which must lead to the same $\sigma(\theta)$ value as equation (A. 17) if Θ_1 and Θ_2 are conjugate angles.

Relativistic treatment

Consider the same collision using m_1 and m_2 for rest masses of the colliding bodies. Subscripts 0, 1 and 2 will be used to denote quantities associated with the two masses in the same sense as listed earlier. Subscript c indicates that the quantity is referred to the CM system, in which total momentum is zero.

Considerable simplification results in the form of relativistic equations by introducing
$$\gamma = (1 - \beta^2)^{-\frac{1}{2}}, \qquad \text{(A. 20)}$$

where $\beta = v/c$, v being the particle velocity and c the velocity of light. The symbol T is used in this section for kinetic energy to distinguish it from the total energy $E \; (= T + mc^2)$.

The three co-ordinates of momentum p_x, p_y and p_z multiplied by c, together with E form a four-vector for which the familiar Lorentz transformation equations hold. If the CM system has velocity $\beta_c c$ in the X direction relative to the Lab. system (i.e. relative to m_2, $-\beta_c c$ specifying the velocity of m_2 relative to the CM) and γ_c is related to it by equation (A. 20), then p and E are connected as follows:

X components
$$\begin{cases} cp_c = \gamma_c(cp - \beta_c E), & \text{(A. 21\,}a) \\ cp = \gamma_c(cp_c + \beta_c E_c). & \text{(A. 21\,}b) \\ E_c = \gamma_c(E - \beta_c cp), & \text{(A. 22\,}a) \\ E = \gamma_c(E_c + \beta_c cp_c). & \text{(A. 22\,}b) \end{cases}$$

Y and Z components $cp_c = cp,$ (A. 23)

whence $c^2\mathbf{p}^2 - E^2 = \text{constant},$ (A. 24)

independent of the frame of reference.

Relativistic momentum and energy may be expressed in terms of γ thus:

$$p = m\beta c\gamma = mc(\gamma^2 - 1)^{\frac{1}{2}},$$ (A. 25)

$$E = mc^2\gamma = c^2p^2 + m^2c^4,$$ (A. 26)

$$T = mc^2(\gamma - 1).$$ (A. 27)

Conservation of momentum requires that in the CM system the bodies have equal and opposite momenta, i.e.

$$m_1 c(\gamma_{c1}^2 - 1)^{\frac{1}{2}} = m_2 c(\gamma_c^2 - 1)^{\frac{1}{2}},$$

where γ_{c1} refers to the value of γ in the CM system for mass m_1, whence

$$m_1^2(\gamma_{c1}^2 - 1) = m_2^2(\gamma_c^2 - 1).$$ (A. 28)

Since the collision is elastic, by definition γ_{c1} and γ_c will also apply to m_1 and m_2 respectively after collision. Now apply transformation (A. 21 a) to get the momentum of m_1 in the CM system:

$$m_1 c^2(\gamma_{c1}^2 - 1)^{\frac{1}{2}} = \gamma_c[m_1 c^2(\gamma_0^2 - 1)^{\frac{1}{2}} - \beta_c m_1 c^2\gamma_0],$$

where γ_0 applies to m_1 in the Lab. system before collision. Hence

$$(\gamma_{c1}^2 - 1)^{\frac{1}{2}} = \gamma_c(\gamma_0^2 - 1)^{\frac{1}{2}} - \gamma_0(\gamma_c^2 - 1)^{\frac{1}{2}}.$$ (A. 29)

We can now extract γ_{c1} and γ_c, which express CM motions of the particles, in terms of quantities measured in the Lab. system. From equations (A. 28) and (A. 29) we obtain two expressions for $(\gamma_{c1}^2 - 1)^{\frac{1}{2}}$. Equating them and squaring both sides gives

$$(\gamma_c - 1) = \frac{\gamma_c^2(\gamma_0^2 - 1)}{(k^{-1} + \gamma_0)^2},$$

where k is again the mass ratio m_1/m_2. Solving this for γ_c gives

$$\gamma_c = \frac{(1 + k\gamma_0)}{(1 + 2k\gamma_0 + k^2)^{\frac{1}{2}}}.$$ (A. 30)

Putting this expression back into equation (A. 28) and extracting γ_{c1}:

$$\gamma_{c1} = \frac{(\gamma_0 + k)}{(1 + 2k\gamma_0 + k^2)^{\frac{1}{2}}},$$ (A. 31)

from which the velocities of m_1 and m_2 in the CM system may be obtained by using equation (A. 20).

In the special case of equal masses, $k = 1$, and

$$\gamma_c = \gamma_{c1} = \left(\frac{1 + \gamma_0}{2}\right)^{\frac{1}{2}} \qquad (A.\,32)$$

We now obtain the relationships between CM and Lab. system angles. By equations (A. 21 b) and (A. 23), components of momentum are

$$p_{2x} = \gamma_c\left(-p_c\cos\theta + \frac{\beta_c}{c}E_{c2}\right), \qquad (A.\,33)$$

$$p_{2y} = -p_c\sin\theta. \qquad (A.\,34)$$

(Since m_1 was defined to be incident in the $+X$ direction the momentum of m_2 in the CM system is negative, which accounts for the change of sign.) The angle of recoil in the Lab. system Θ_2 is related to the momentum components by $\tan\Theta_2 = (p_{2y}/p_{2x})$. Inserting expressions (A. 33) and (A. 34) and substituting for p_c and E_{c2} according to (A. 25) and (A. 26) leads to

$$\tan\Theta_2 = \frac{-m_2\gamma_c c\beta_c\sin\theta}{\gamma_c\left(-m_2\gamma_c c\beta_c\cos\theta + \dfrac{\beta_c}{c}m_2c^2\gamma_c\right)},$$

whence
$$\tan\Theta_2 = \frac{-\cot\tfrac{1}{2}\theta}{\gamma_c}. \qquad (A.\,35)$$

The significance of the negative sign here is merely that m_2 must recoil on the opposite side of the collision axis to m_1.

To get the angle of scattering Θ_1 in the Lab. system in terms of θ we proceed similarly. The X and Y components of momentum are

$$p_{1x} = \gamma_c\left(p_c\cos\theta + \frac{\beta_c}{c}E_{c1}\right)$$

and
$$p_{1y} = p_c\sin\theta.$$

Taking (p_{1y}/p_{1x}) as before, one finds

$$\tan\Theta_1 = \frac{\sin\theta}{\gamma_c\left\{\cos\theta + k\left(\dfrac{\gamma_{c1}}{\gamma_c}\right)\right\}} \qquad (A.\,36)$$

A more convenient form, if θ is required in terms of Θ_1, follows from rearranging equation (A. 36):

$$\gamma_c \cos\theta \tan\Theta_1 + k\gamma_{c1} \tan\Theta_1 = (1 - \cos^2\theta)^{\frac{1}{2}}.$$

Squaring and introducing γ_c and γ_{c1} from equations (A. 30) and (A. 31) leads to a quadratic equation in $\cos\theta$, whence

$$\cos\theta = \frac{-\gamma_c \gamma_{c1} k \tan^2\Theta_1 \pm [(1 - k^2)\tan^2\Theta_1 + 1]^{\frac{1}{2}}}{1 + \gamma_c^2 \tan^2\Theta_1}. \qquad \text{(A. 37)}$$

As in the non-relativistic case (equation (A. 8)), there is an upper limit to the scattering angle, which occurs when the radical is zero, i.e. $\tan^2\Theta_{1,\max} = (k^2 - 1)^{-1}$, or $\sin\Theta_{1,\max} = k^{-1}$. For the case $k < 1$, one root of equation (A. 37) must be rejected. If $\Theta_1 < 90°$, the more positive root must be taken and if $\Theta_1 > 90°$, the more negative.

A useful relationship between Lab. system angles follows from equations (A. 35) and (A. 36). Writing $\sin\theta = 2\sin\frac{1}{2}\theta \cos\frac{1}{2}\theta$ and $\cos\theta = 1 - 2\sin^2\frac{1}{2}\theta$, one gets from the first equation:

$$\sin\theta = \frac{-2\gamma_c \tan\Theta_2}{1 + \gamma_c^2 \tan^2\Theta_2} \quad \text{and} \quad \cos\theta = \frac{\gamma_c^2 \tan^2\Theta_2 - 1}{1 + \gamma_c^2 \tan^2\Theta_2}.$$

Now put these into the second equation. On simplification, the following relation is found:

$$\tan\Theta_1 = \frac{-2\tan\Theta_2}{[(\gamma_c^2 + k\gamma_{c1}\gamma_c)\tan^2\Theta_2 + k(\gamma_{c1}/\gamma_c) - 1]}. \qquad \text{(A. 38)}$$

Alternatively, (A. 38) may be expressed in terms of Lab. system quantities only, by using expressions (A. 30) and (A. 31) for γ_c and γ_{c1}:

$$\tan\Theta_1 = \frac{-2\tan\Theta_2(1 + k\gamma_0)}{[(1 + k\gamma_0)^2 \tan^2\Theta_2 + k^2 - 1]}. \qquad \text{(A. 39)}$$

For the important case of equal masses, $\gamma_c = \gamma_{c1}$ and

$$\tan\Theta_1 \tan\Theta_2 = -\gamma_c^{-1} \quad \text{or} \quad -2/(1 + \gamma_0). \qquad \text{(A. 40)}$$

The angle between scattered and recoil particles is $90°$ at low energies and falls only slightly below this until very high energies are reached. Writing

$$\Phi = (\Theta_1 + \Theta_2)$$

and

$$\tan\Phi = (\tan\Theta_1 + \tan\Theta_2)/(1 - \tan\Theta_1 \tan\Theta_2),$$

values for $\tan \Theta_1$ and $\tan \Theta_2$ can be substituted from equations
(A. 36) and (A. 35), which gives

$$\tan \Phi = 2/(\beta_c^2 \gamma_c \sin \theta). \tag{A. 41}$$

For $\theta = 90°$, the value of Φ is $88° 30' 50''$ at 100 MeV. and falls to
$77° 51' 30''$ at 1 BeV.

Consider now the energies of m_1 and m_2 after collision. By
applying equation (A. 22 b), the energy of m_2 in the Lab. system is

$$E_2 = \gamma_c(E_{c2} - \beta_c c p_c \cos \theta). \tag{A. 42}$$

Hence the kinetic energy expressed in terms of CM scattering
angle is

$$T_2 = \gamma_c(m_2 c^2 \gamma_c - c^2 \beta_c^2 m_2 \gamma_c \cos \theta) - m_2 c^2$$
$$= m_2 c^2 [(\gamma_c^2 - 1)(1 - \cos \theta)]. \tag{A. 43}$$

In order to get T_2 in terms of Lab. system angles, we use equation
(A. 35) again, leading to

$$\cos \theta = \frac{\gamma_c^2 - \gamma_c^2 \cos^2 \Theta_2 + \cos^2 \Theta_2}{\gamma_c^2 - \gamma_c^2 \cos^2 \Theta_2 - \cos^2 \Theta_2}.$$

Substitute this in (A. 43) and replace γ_c by equation (A. 30). The
result is

$$T_2 = 2 m_2 c^2 \left[\frac{p_0^2 c^2 \cos^2 \Theta_2}{(E_0 + m_2 c^2)^2 - p_0^2 c^2 \cos^2 \Theta_2} \right]. \tag{A. 44}$$

The maximum recoil energy occurs when $\Theta_2 = 0$ which applies, of
course, to the low energy case also. The maximum fraction of
energy which can be transferred from the incident particle is
(writing p_0 and E_0 in terms of γ_0)

$$\frac{T_{2, \max}}{T_0} = \frac{2(\gamma_0 + 1)}{k + k^{-1} + 2\gamma_0}. \tag{A. 45}$$

In the non-relativistic limit ($\gamma_0 \simeq 1$), this takes the familiar form

$$\frac{T_{2, \max}}{T_0} = \frac{4}{k + k^{-1} + 2} = \frac{4 m_1 m_2}{(m_1 + m_2)^2}, \tag{A. 46}$$

but for the relativistic case ($\gamma_0 \gg 1$), the fraction tends to unity.

The relationship between scattering angle and energy of m_1
follows from equation (A. 22 b):

$$E_1 = \gamma_c(E_{c1} + \beta_c c p_{c1} \cos \theta). \tag{A. 47}$$

Writing E_{c1}, p_{c1} and β_c in terms of m_1, γ_c and γ_{c1} gives

$$E_1 = m_1 c^2 [\gamma_c \gamma_{c1} + (\gamma_c^2 - 1)^{\frac{1}{2}} (\gamma_{c1}^2 - 1)^{\frac{1}{2}} \cos \theta] \qquad (A.48)$$

and $T_1 = E_1 - m_1 c^2$. When $k > 1$, the two values of $\cos \theta$ (equation (A.37)) must be used. Alternatively, knowing T_2, T_1 may be calculated from conservation of kinetic energy, since $T_1 + T_2 = T_0$.

Consider now the transformation of Lab. system cross-sections to CM co-ordinates. We again use the general equation (A.16), determining first the conversion factor for the case in which the cross-section for recoil has been measured. This is the more useful quantity in many high energy experiments, including, in particular, all cases of neutron scattering:

$$\sigma(\theta) = \sigma(\Theta_2) \left| \frac{d \cos \Theta_2}{d \cos \theta} \right|. \qquad (A.49)$$

Using equation (A.35), one obtains $\cos \theta = 1 - 2(1 + \gamma_c^2 \tan^2 \Theta_2)^{-1}$, which on differentiation and substitution back into equation (A.49), leads to

$$\frac{\sin \Theta_2 \, d\Theta_2}{\sin \theta \, d\theta} = \frac{\cos^4 \Theta_2 + 2\gamma_c^2 \sin \Theta_2 \cos \Theta_2 + \gamma_c^4 \sin^4 \Theta_2}{4\gamma_c^2 \cos \Theta_2}.$$

Introducing $\gamma_c^2 = (1 - \beta_c^2)^{-1}$ and simplifying gives

$$\sigma(\theta) = \sigma(\Theta_2) \left[\frac{\gamma_c^2 (1 - \beta_c^2 \cos^2 \Theta_2)^2}{4 \cos \Theta_2} \right]. \qquad (A.50)$$

The corresponding conversion factor to apply to the cross-section for scattering can be obtained in several different forms. That given below is probably the simplest algebraically. Corresponding to equation (A.49), we have

$$\sigma(\theta) = \sigma(\Theta_1) \left| \frac{d (\cos \Theta_1)}{d (\cos \theta)} \right|. \qquad (A.51)$$

By differentiating equation (A.36) with respect to θ and substituting for Θ_1 in terms of θ one readily obtains

$$\sigma(\theta) = \sigma(\Theta_1) \left[\frac{(\gamma_c + k\gamma_{c1} \cos \theta)}{\{\sin^2 \theta + (\gamma_c \cos \theta + k\gamma_1)^2\}^{\frac{3}{2}}} \right]. \qquad (A.52)$$

THE SCATTERING PHASE SHIFT FOR COLLISIONS AT ZERO ENERGY

The wave equation describing the elastic collision of two bodies is

$$\left[\frac{d^2}{dr^2} + k^2 - \frac{l(l+1)}{r^2} - U(r) \right] \phi_l(r) = 0, \qquad (B.1)$$

where $\phi_l(r) \sim \sin(kr - \frac{1}{2}l\pi + \eta(k))$ for $r \to \infty$, and $\eta(k)$ is the scattering phase shift, defined as continuous in k and such that $\eta(k) \to 0$ as $k \to \infty$. A general theorem may be stated, viz. $\eta(0) = n\pi$, where n is the number of composite bound states possible for the two-body system (Levinson, 1949; Swan, 1955). Thus for n-p scattering, triplet and singlet s-phases $\eta(0) = \pi$ and 0 respectively are obtained, corresponding to the triplet ground state and virtual singlet state of the deuteron. The physical picture is that a potential strong enough to produce a bound state 'pulls in' the incident wave by a distance increasing to one-half wavelength ($\eta(0) = \pi$) in the limit of very small energies. If the potential is too weak for a bound state to be possible, the wave is not pulled in, giving $\eta(0) = 0$.

To prove $\eta(0) = n\pi$ for s-states, consider

$$\left[\frac{d^2}{dr^2} + k^2 - U(r) \right] \phi(r) = 0. \qquad (B.2)$$

We choose solutions of (B.2) with the asymptotic properties

$$f(k, r) \sim e^{-ikr}, \quad f(-k, r) \sim e^{ikr} \quad \text{for} \quad r \to \infty.$$

Denote $\qquad f(k, 0) = f(k), \quad f(-k, 0) = f(-k). \qquad (B.3)$

Then $\qquad \phi(r) = f(k, r) - \dfrac{f(k)}{f(-k)} f(-k, r)$

is a solution of equation (B.2) possessing the properties

$$\phi(0) = 0, \quad \phi(r) \sim e^{-ikr} - \frac{f(k)}{f(-k)} e^{ikr}. \qquad (B.4)$$

Put $f(k) = |f(k)| e^{i\eta(k)}$; then from equation (B. 3) we find

$$f(-k^*, r) = f^*(k, r), \quad f(k, r) = f^*(-k^*, r).$$

$$f(-k) = f^*(k^*) = f(k^*) e^{-i\eta(k^*)}$$

$$= |f(k)| e^{-i\eta(k)} \quad \text{for real } k. \tag{B. 5}$$

Therefore $\phi(r) \sim e^{-ikr} - e^{2i\eta(k)} e^{ikr}$ for $r \to \infty.$ (B. 6)

Thus $\eta(k)$ is the scattering phase for real k provided $\eta(k) \to 0$ as $k \to \infty$.

A bound state is defined by the condition

$$k = -i\gamma, \quad \phi(r) \to e^{-\gamma r} \quad \text{for} \quad r \to \infty,$$

i.e. $e^{2i\eta(k)} = f(k)/f(-k) = 0$

or $f(k) = 0.$ (B. 7)

(Alternatively a pole of $f(-k)$ would give a zero S-matrix (Jost, 1947). However, these 'redundant zeros' are non-physical and cannot represent bound states as they are independent of the potential well parameters. This also follows on the assumption that H is Hermitian.) Actually zeros of $f(k)$ are restricted along the unreal axis of k in the negative complex k-plane, $\text{Im}(k) \leqslant 0$ (fig. B. 1). From equation (B. 5), a zero of $f(k)$ for $\text{Im}(k) \leqslant 0$ would have to be accompanied by a zero of $f(-k)^*$, giving zeros at $k = \pm k_0 - i\gamma$ say. This leads to $\phi(r) \sim e^{-\gamma r} e^{\pm ik_0 r}$, corresponding to solutions for bound states with complex binding energy k^2. However the probability density $\phi^* \phi \sim e^{-2\gamma r}$ indicates binding energies dependent on γ alone. Put more generally, zeros of $f(k)$ for complex k with $\text{Im}(k) \leqslant 0$, contradict the Hermitian property of the energy matrix, unless $k = -i\gamma$.

Equation (B. 2) may be put in the integral equation form

$$f(k, r) = e^{-ikr} + \int_r^\infty \frac{\sin k(r'-r)}{k} U(r') f(k, r') \, dr'. \tag{B. 8}$$

Put $f(k, r) = \sum_{n=0}^{\infty} f_n(k, r),$ (B. 9)

where $f_{n+1}(k, r) = \int_r^\infty \frac{\sin k(r'-r)}{k} U(r') f_n(k, r') \, dr'.$ (B. 10)

Then

$$| f_{n+1}(k,r) | < \left| \frac{k^n U_0^n\, e^{-(n\lambda+ik)r}}{n!\, \lambda^n(\lambda+2ik)(2\lambda+2ik)\ldots(n\lambda+2ik)} \right|, \quad \text{(B.11)}$$

where $| U(r) | < U_0\, e^{-\lambda r}$ for $\lambda < \epsilon$, so that (B.9) is absolutely convergent for all complex k, as $n^4\lambda^2 + 4n^2k^2 > \dfrac{1}{\lambda^2}\, U_0^2\, e^{-2\lambda r}$ for n sufficiently large. Hence $f(k,r)$ and therefore $f(k)$ are analytic everywhere in the complex k-plane. In particular, $f(k)$ has no poles.

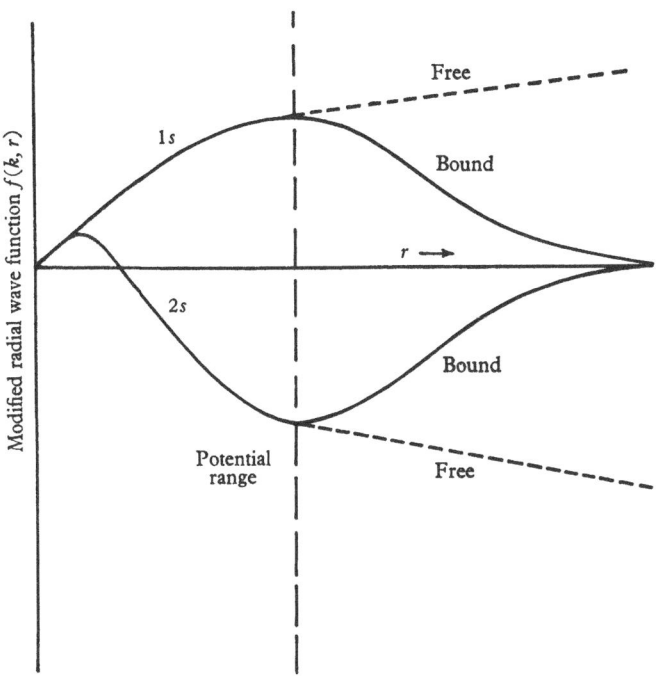

Fig. B.1. Bound and free wave functions for 1s and 2s states.

Consider the following integral over the semicircular contour C of radius in fig. B.2:

$$\int_C \frac{f'(k)}{f(k)}\, dk = [\ln f(k)]_c$$

$$= \ln[f(0_+)/f(0_-)] - \ln[f(\kappa)/f(-\kappa)] + \int_0^{-\pi} f_\theta'(\kappa\, e^{i\theta})/f(\kappa\, e^{i\theta})\, d\theta$$

$$= 2i\eta(0) - 2i\eta(\kappa) + \epsilon(\kappa) \quad \text{from equation (B.7).}$$

Then $\qquad \lim_{\kappa \to \infty} \int_C \dfrac{f'(k)}{f(k)}\, dk = 2i\eta(0) - 2i\eta(\infty) = 2i\eta(0),$ $\qquad$ (B. 12)

provided $\eta(k)$ is single-valued and continuous in C, and $f(k)$ is sufficiently small on the semicircular part of C.

However, if there are n composite bound states, then from equation (B. 7), $f(k)$ has n zeros for $\mathrm{Im}(k) < 0$, making

$$\int_C \frac{f'(k)}{f(k)}\, dk = 2\pi i n.$$ $\qquad$ (B. 13)

Equating equations (B. 12) and (B. 13),

$$\eta(0) = n\pi.$$ $\qquad$ (B. 14)

In the case of a zero of $f(k)$ at $k = 0$ (system just bound),

$$\left.\begin{aligned}\int_C \frac{f'(k)}{f(k)}\, dk &= 2\pi i(n + \tfrac{1}{2}), \\[4pt] \eta(0) &= (n + \tfrac{1}{2})\,\pi.\end{aligned}\right\}$$ $\qquad$ (B. 15)

Collision phenomena involving three or more particles, such as n-d or e-H scattering, lead to integro-differential equations instead of equation (B. 1), provided the Pauli exclusion principle and/or exchange effects are allowed for:

$$\left[\frac{d^2}{dr^2} + k^2 - \frac{l(l+1)}{r^2} - U(r)\right] f_1(r) = \lambda \int_0^\infty K(r, r')\, f_l(r')\, dr'.$$ $\quad$ (B. 16)

The procedure of § B. 1 is not now useful, as $f(k)$ is no longer analytic and may possess poles, each of which contributes $-\pi$ to the contour integral (B. 13). However, the contribution of these poles to $\eta(0)$ is balanced out by an equal number of zeros of $f(k)$ which do not represent bound states, being independent of $U(r)$ and λ. It may be proved that

$$\eta(0) = (n + m)\,\pi,$$ $\qquad$ (B. 17)

where n is the number of possible composite bound states and m the number of states from which the incident particle is excluded by the Pauli principle (Swan, 1955). Thus electron scattering by the rare gases He, Ne, A, Kr and Xe gives $\eta(0) = \pi,\ 2\pi,\ 3\pi,\ 4\pi$ and 5π corresponding to the number of full electron shells (excluded states) in each case.

Each excluded state on the Pauli principle is due to parallel spins of two identical particles in the same quantum state. The exclusion reveals itself by the vanishing of the total wave function for the system if the scattered wave function is put equal to the excluded bound state wave function. Thus for triplet e-H scattering,

$$\Psi(12) = \phi(1)\,F(2) - \phi(2)\,F(1)$$

vanishes for $F(r) = \phi(r)$, the H bound state wave function.

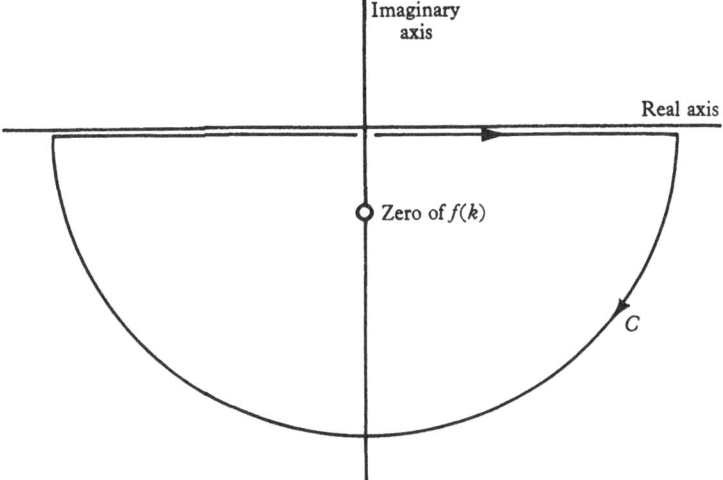

Fig. B.2. The contour C in the k-plane.

In general, each excluded state contributes a bound state type solution asymptotic to $e^{-\gamma r}$, which is independent of $U(r)$ and λ. Each corresponds to a zero of $f(k)$, contributing π to equation (B.13) and hence to $\eta(0)$. This may be seen pictorially using fig. B.2. The wave function inside the potential well is roughly independent of whether a bound or free state exists if the energy of the latter is very small. If the incident particle is excluded from the $1s$ state, it must be put into the $2s$ state, which has an extra node corresponding to a phase shift π. Thus triplet e-H scattering corresponds to a virtual $1s\,2s$ state of H^- as the $(1s)^2$ state is forbidden, giving $\eta(0) = \pi$. These considerations are easily generalized to more complicated cases.

REFERENCES

AAMODT, R. L., PETERSON, V. and PHILLIPS, R. (1949). *University of California Radiation Laboratory Report*, UCRL-526; also (1952), *Phys. Rev.* **88**, 739.

ABRAHAM, G. (1956). *Nuc. Phys.* **1**, 415.

ABRAHAM, G., COHEN, L. and ROBERTS, A. S. (1955). *Proc. Phys. Soc.* A, **68**, 265.

ADAIR, R. K. (1952). *Phys. Rev.* **86**, 155.

ADAIR, R. K., DARDEN, S. E. and FIELDS, R. E. (1954). *Phys. Rev.* **96**, 503.

AGENO, M., AMALDI, E., BOCCIARELLI, D. and TRABACCHI, G. C. (1943), *Nuovo. Cim.* **1**, 253.

AGENO, M., AMALDI, E., BOCCIARELLI, D. and TRABACCHI, G. C. (1947). *Phys. Rev.* **71**, 20.

AJZENBERG, F. (1951). *Phys. Rev.* **82**, 43.

AJZENBERG, F. and LAURITSEN T. (1955). *Rev. Mod. Phys.* **27**, 77.

ALLEN, A. J., NECHAJ, J. F., SUN, K. H. and JENNINGS, B. (1951). *Phys. Rev.* **81**, 536.

ALLEN, K. W. and ALMQVIST, E. (1953). *Rev. Sci. Instr.* **24**, 70.

ALLRED, J. C., ARMSTRONG, A. H., HUDSON, A. M., POTTER, R. M., ROBINSON, E. S., ROSEN, L. and STOVALL, E. J. (1952a). *Phys. Rev.* **88**, 425.

ALLRED, J. C., ARMSTRONG, A. H., BONDELID, R. O. and ROSEN, L. (1952b). *Phys. Rev.* **88**, 433.

ALLRED, J. C., ARMSTRONG, A. H. and ROSEN, L. (1953). *Phys. Rev.* **91**, 90.

ALLRED, J. C., ROSEN, L., TALLMADGE, F. K. and WILLIAMS, J. H. (1951). *Rev. Sci. Instr.* **22**, 191.

ALSTON, M. H., GREIVE, A. V., EVANS, W. H., GREEN, L. L. and WILLMOTT, J. C. (1954). *Proc. Phys. Soc.* A, **67**, 657.

ALTAR, W. and GARBUNY, M. (1949). *Phys. Rev.* **76**, 496; *ibid.* (1947), **72**, 528.

ALVAREZ, L. W., BRADNER, H., FRANCK, J. V., GORDON, H., GOW, J. D., MARSHALL, L. G., OPPENHEIMER, F., PANOFSKY, W. K. H., RICHMAN, C. and WOODYARD, J. R. (1955). *Rev. Sci. Instr.* **26**, 111.

AMALDI, E., BOCCIARELLI, D., CACCIAPUOTI, B. N. and TRABACCHI, G. C. (1946). *Nuovo. Cim.* **3**, 15, 203.

AMALDI, E. and CACCIAPUOTI, B. N. (1947). *Phys. Rev.* **71**, 739.

ANDERSON, H. L. (1948). *Preliminary Report No. 3, Nuclear Science Series*, National Research Council, Washington, D.C.

ANDERSON, H. L., FERMI, E. and MARSHALL, L. (1946). *Phys. Rev.* **70**, 815.

ANDERSON, H. L., FERMI, E., WATTENBERG, A., WEIL, G. L. and ZINN, W. H. (1947). *Phys. Rev.* **72**, 16.

AOKI, H. (1939). *Proc. Phys.-Math. Soc., Japan*, **21**, 232.

ARGO, H. V., TASCHEK, R. F., AGNEW, H. M., HEMMENDINGER, A. and LELAND, W. T. (1952). *Phys. Rev.* **87**, 612.

ARNOLD, W. R., PHILLIPS, J. A., SAWYER, G. A., STOVALL, E. J. and
  TUCK, J. L. (1954). *Phys. Rev.* 93, 483.
ARON, W. A., HOFFMAN, B. G. and WILLIAMS, F. C. (1949). *University
  of California Radiation Laboratory Report,* UCRL-121.
ASHKIN, J. and WU, T. Y. (1948). *Phys. Rev.* 73, 973.
AUGER, P. and MONOD-HERZEN, G. (1933). *C.R. Acad. Sci., Paris,*
  196, 1102.
AUSTERN, N., BUTLER, S. T. and McMANUS, H. (1953). *Phys. Rev.* 92,
  350.
BAILEY, C. L., BENNETT, W. E., BERGSTRALH, T., NUCKOLLS, R. G.,
  RICHARDS, H. T. and WILLIAMS, J. H. (1946). *Phys. Rev.* 70, 583.
BAKER, C. P. and BACHER, R. F. (1941). *Phys. Rev.* 59, 332.
BARKAS, W. H. and WHITE, M. G. (1939). *Phys. Rev.* 56, 288.
BARKER, F. C. (1956). *Atomic Energy Research Establishment Report,*
  A.E.R.E. T/R 1791.
BARKER, F. C. and PEIERLS, R. E. (1949). *Phys. Rev.* 75, 312.
BARSCHALL, H. H. and KANNER, M. H. (1940). *Phys. Rev.* 58, 590.
BARSCHALL, H. H., ROSEN, L., TASCHEK, R. F. and WILLIAMS, J. H.
  (1952). *Rev. Mod. Phys.* 24, 1.
BARSCHALL, H. H. and TASCHEK, R. H. (1949). *Phys. Rev.* 75, 1819.
BASHKIN, S., MOORING, F. P. and PETREE, B. (1951). *Phys. Rev.* 82, 378.
BASKIR, E., HAFNER, E. M., ROBERTS, A. and TINLOT, J. H. (1957).
  Phys. Rev. 106, 564.
BATSON, A. P., COOPER, P. N. and RIDDIFORD, L. (1956). Private com-
  munication.
BATSON, A. P. and RIDDIFORD. L. (1956). *Proc. Roy. Soc.* A, 236, 175.
BEECK, O. (1935). *Rev. Sci. Instr.* 6, 399.
BEIDUK, F. M., PRUETT, J. R. and KONOPINSKI, E. J. (1950). *Phys. Rev.*
  77, 622.
BEISER, A. (1952). *Rev. Mod. Phys.* 24, 273.
BELL, R. E. (1954). *Ann. Rev. Nuclear Sci.* 4, 93.
BENNETT, W. E., BONNER, T. W., MANDEVILLE, C. E. and WATT, B. E.
  (1946). *Phys. Rev.* 70, 882.
BENVENISTE, J. and ZENGER, J. (1954). *University of California Radiation
  Laboratory Report,* UCRL-4266.
BERETTA, L., CLEMENTEL, E. and VILLI, C. (1955). *Phys. Rev.* 98, 1526.
BERNSTEIN, S., ROBERTS, L. D., STANFORD, C. P., DABBS, J. W. T. and
  STEPHENSON, T. E. (1954). *Phys. Rev.* 94, 1243.
BETHE, H. A. (1937). *Rev. Mod. Phys.* 9, 69.
BETHE, H. A. (1940a). *Phys. Rev.* 57, 1125.
BETHE, H. A. (1940b). *Phys. Rev.* 57, 260, 390.
BETHE, H. A. (1949). *Phys. Rev.* 76, 38.
BETHE, H. A. and MORRISON, P. (1956). *Elementary Nuclear Theory,*
  2nd ed. J. Wiley and Sons, Inc., N.Y.
BETHE, H. A. and BACHER, R. (1936). *Rev. Mod. Phys.* 8, 83.
BETHE, H. A. and LONGMIRE, C. (1954). *Phys. Rev.* 77, 647.
BETHE, H. A. and PLACZEK, G. (1937). *Phys. Rev.* 51, 450.
BETHE, H. A. and PLACZEK, G. (1940). *Phys. Rev.* 57, 1075.
BEYSTER, J. R., WALT, M. and SALMI, E. W. (1956). *Phys. Rev.* 104, 1319.

BHATIA, A. B., HUANG, K., HUBY, R. and NEWNS, H. C. (1952). *Phil. Mag.* **43**, 485.

BIEDENHARN, L. C., BOYER, K. and GOLDSTEIN, M. (1956). *Phys. Rev.* **104**, 383.

BIRAM, M. B. (1954). *Atomic Energy Research Establishment Document*, A.E.R.E., T/R1523.

BIRAM, M. B. and TAIT, J. H. (1950). *Atomic Energy Research Establishment Document*, A.E.R.E., T/R563.

BIRGE, R. W. and KRUSE, U. E. (1951). *Phys. Rev.* **81**, 649.

BIRGE, R. W., KRUSE, U. E. and RAMSEY, N. F. (1951). *Phys. Rev.* **83**, 274.

BIRNBAUM, W., CRANDALL, W. E., MILLBURN, G. P. and PYLE, R. V. (1955). *University of California Radiation Laboratory Report*, UCRL-2756.

BISHOP, G. R., WESTHEAD, J. M., PRESTON, G. and HALBAN, H. (1952). *Nature, Lond.* **170**, 113.

BLAIR, J. M., FREIER, G., LAMPI, E. E., SLEATOR, W. and WILLIAMS, J. H. (1948). *Phys. Rev.* **74**, 553.

BLATT, J. M. (1948). *Phys. Rev.* **77**, 647.

BLATT, J. M. and JACKSON, J. D. (1949). *Phys. Rev.* **76**, 18.

BLATT, J. M. and WEISSKOPF, V. F. (1952). *Theoretical Nuclear Physics*, 1st ed. J. Wiley and Sons, Inc., N.Y.

BLEANEY, B. (1951). *Proc. Phys. Soc.* A, **64**, 315; also (1951), *Phil. Mag.* **42**, 441.

BLEWETT, J. P. (1954). *Ann. Rev. Nuclear Sci.* **4**, 1.

BLIN-STOYLE, R. J. (1951). *Proc. Phys. Soc.* A, **64**, 700.

BLOCH, F. (1936). *Phys. Rev.* **50**, 259; *ibid.* (1937), **51**, 994.

BLOCH, I., HULL, M. H., BROYLES, A. A., BOURICIOUS, W. G., FREEMAN, B. E. and BREIT, G. (1950). *Phys. Rev.* **80**, 553.

BLOCH, I., HULL, M. H., BROYLES, A. A., BOURICIOUS, W. G., FREEMAN, B. E. and BREIT, G. (1951). *Rev. Mod. Phys.* **23**, 147.

BLOCK, M. M., HARTH, E. M., COCCONI, V. J., HART, E., FOWLER, W. B., SHUTT, R. P., THORNDIKE, A. M. and WHITTEMORE, E. M. (1956). *Phys. Rev.* **103**, 1484.

BLOEMBERGEN, N., PURCELL, E. M. and POUND, R. V. (1948). *Phys. Rev.* **73**, 679.

BLOSSER, H. G. and HANDLEY, T. H. (1955). *Phys. Rev.* **100**, 1340.

BOCKELMAN, C. K., PETERSON, R. E., ADAIR, R. K. and BARSCHALL, H. H. (1949). *Phys. Rev.* **76**, 277.

BOGACHEV, N. P. and VZOROV, I. K. (1954). *Dokl. Akad. Nauk. S.S.S.R.* **99** (6), 931 (A.E.R.E. Lib./Trans. 541).

BOHR, N. (1936). *Nature, Lond.* **137**, 344.

BONDELID, R. O., BRADEN, C. H., BATTAT, M. E. and BOHLMAN, P. (1952). *Phys. Rev.* **87**, 699.

BONNER, T. W. and BUTLER, J. W. (1951). *Phys. Rev.* **83**, 1091.

BONNER, T. W., FERRELL, R. A. and RINEHART, M. C. (1952). *Phys. Rev.* **87**, 1032.

BORST, L. B. and HARKINS, W. D. (1940). *Phys. Rev.* **57**, 659.

BORST, L. B. and SAILOR, V. L. (1953). *Rev. Sci. Instrum.* **24**, 141.

BORST, L. B., ULRICH, A. J., OSBORNE, C. L. and HASBROUCK, H. (1946). *Phys. Rev.* **70**, 108.

BRADEN, C. H. (1951). *Phys. Rev.* **84**, 762.
BRANSDEN, B. H. (1950). *Proc. Roy. Soc.* A, **209**, 380.
BRANSDEN, B. H. (1952). *Proc. Phys. Soc.* A, **65**, 972.
BRANSDEN, B. H. and BURHOP, E. H. S. (1950). *Proc. Phys. Soc.* A, **63**, 1337.
BRANSDEN, B. H. and McKEE, J. S. C. (1954). *Phil. Mag.* **45**, 869.
BRANSDEN, B. H., ROBERTSON, H. H. and SWAN, P. (1956). *Proc. Phys. Soc.* A, **69**, 877.
BREIT, G. (1946). *Phys. Rev.* **69** 472.
BREIT, G. (1955). *Phys. Rev.* **99**, 1581.
BREIT, G. (1957). *Phys. Rev.* **106**, 314.
BREIT, G., CONDON, E. V. and PRESENT, R. D. (1936). *Phys. Rev.* **50**, 825.
BREIT, G., EHRMAN, J. B. and HULL, M. H. (1955). *Phys. Rev.* **97**, 1047.
BREIT, G., KITTEL, C. and THAXTON, H. M. (1940). *Phys. Rev.* **57**, 255.
BREIT, G., THAXTON, H. M. and EISENBUD, L. (1939). *Phys. Rev.* **55**, 1018.
BREIT, G. and WIGNER, E. (1936). *Phys. Rev.* **49**, 519, 642.
BRETSCHER, E., FRENCH, A. P. and SEIDEL, F. G. P. (1948). *Phys. Rev.* **73**, 815.
BRIGGS, G. H. (1954). *Rev. Mod. Phys.* **26**, 1.
BROCKMAN, K. W. (1956). *Phys. Rev.* **102**, 391.
BROLLEY, J. E., COON, J. H. and FOWLER, J. L. (1951). *Phys. Rev.* **82**, 190.
BROLLEY, J., FOWLER, J. L. and STOVALL, E. J. (1951). *Phys. Rev.* **82**, 502.
BROWN, A. B., SNYDER, C. W., FOWLER, W. A. and LAURITSEN, C. C. (1951), *Phys. Rev.* **82**, 159.
BROWN, G. E. (1957). *Proc. Phys. Soc.* A, **70**, 361.
BRUECKNER, K. (1953). *Phys. Rev.* **89**, 834.
BRUECKNER, K., HARTSOUGH, W., HAYWARD, E. and POWELL, W. M. (1949) *Phys. Rev.* **75**, 555.
BRUGGER, R. M., BONNER, T. W. and MARION, J. B. (1955). *Phys. Rev.* **100**, 84.
BRUSSEL, M. K. and WILLIAMS, J. H. (1957). *Phys. Rev.* **106**, 286.
BUCKINGHAM, R. A., HUBBARD, S. J. and MASSEY, H. S. W. (1952). *Proc. Roy. Soc.* A, **211**, 183.
BUCKINGHAM, R. A. and MASSEY, H. S. W. (1941). *Proc. Roy. Soc.* A, **179**, 123.
BURHOP, E. H. S. and MASSEY, N. S. W. (1947). *Proc. Roy. Soc.* A, **192**, 156.
BURKE, P. G. and ROBERTSON, H. H. (1957). *Proc. Phys. Soc.* A, **70**, 777.
BURKIG, J. W. and WRIGHT, B. T. (1951). *Phys. Rev.* **82**, 451.
BUTLER, S. T. (1950). *Phys. Rev.* **80**, 1095.
BUTLER, S. T. (1951). *Proc. Roy. Soc.* A, **208**, 559.
BUTLER, S. T. (1952). *Phys. Rev.* **88**, 133.
BUTLER, S. T. (1957). *Phys. Rev.* **106**, 272.
BUTLER, S. T. and HITTMAIR, O. H. (1957). *Nuclear Stripping.* J. Wiley and Sons, Inc., N.Y.
CALDWELL, D. O. and RICHARDSON, J. O. (1955). *Phys. Rev.* **98**, 28.
CAPLEHORN, W. F. and RUNDLE, G. P. (1951). *Proc. Phys. Soc.* A, **64**, 546.
CARO, D. E., MARTIN, L. H. and ROUSE, J. L. (1955). *Aust. J. Phys.* **8**, 306.
CARROLL, H. (1941). *Phys. Rev.* **60**, 702.

CASE, K. M. and PAIS, A. (1950). *Phys. Rev.* **80**, 203.
CASIMIR, H. (1936). *Physica*, **3**, 936.
CASSELS, J. M., STAFFORD, G. H. and PICKAVANCE, T. C. (1951). *Nature, Lond.* **168**, 556.
CATALA, J. and GIBSON, W. M. (1951). *Nature, Lond.* **167**, 551.
CESTER, R., HOANG, T. F. and KERNAN, A. (1956). *Phys. Rev.* **103**, 1443.
CHADWICK, J., MAY, A. N., PICKAVANCE, T. G. and POWELL, C. F. (1944). *Proc. Roy. Soc.* A, **183**, 1.
CHAMBERLAIN, O. and CLARK, D. D. (1956). *Phys. Rev.* **102**, 473.
CHAMBERLAIN, O., DONALDSON, R., SEGRÈ, E., TRIPP, R., WIEGAND, C. and YPSILANTIS, T. (1954). *Phys. Rev.* **95**, 850.
CHAMBERLAIN, O. and GARRISON, J. D. (1954). *Phys. Rev.* **95**, 1349.
CHAMBERLAIN, O. and GARRISON, J. D. (1956). *Phys. Rev.* **103**, 1860.
CHAMBERLAIN, O., PETTENGILL, G., SEGRÈ, E. and WIEGARD, C. (1954). *Phys. Rev.* **93**, 1424.
CHAMBERLAIN, O., SEGRÈ, E., TRIPP, R. D., WIEGAND, C. and YPSILANTIS, T. (1957). *Phys. Rev.* **105**, 288.
CHAMBERLAIN, O., SEGRÈ, E. and WIEGAND, C. (1951). *Phys. Rev.* **83**, 923.
CHAMBERLAIN, O. and STERN, M. O. (1954). *Phys. Rev.* **94**, 666.
CHAMBERLAIN, O. and WIEGAND, C. (1950). *Phys. Rev.* **79**, 81.
CHAMPION, F. C. and POWELL, C. F. (1944). *Proc. Roy. Soc.* A, **183**, 64.
CHASE, D. M. and ROHRLICH, F. (1954). *Phys. Rev.* **94**, 81.
CHEN, F. F., LEAVITT, C. P. and SHAPIRO, A. M. (1956). *Phys. Rev.* **103**, 211.
CHESTON, W. B. (1954). *Phys. Rev.* **96**, 1590.
CHEW, G. F. (1948). *Phys. Rev.* **74**, 809.
CHEW, G. F. (1951 a). *Phys. Rev.* **84**, 710.
CHEW, G. F. (1951 b). *Phys. Rev.* **84**, 1057.
CHEW, G. F. and GOLDBERGER, M. L. (1949). *Phys. Rev.* **75**, 1466.
CHEW, G. F. and STEINBERGER, J. (1950). *Phys. Rev.* **78**, 497.
CHEW, G. F. and WICK, G. C. (1952). *Phys. Rev.* **85**, 636.
CHIH, C. Y. and POWELL, W. M. (1957). *Phys. Rev.* **106**, 539.
CHRISTIAN, R. S. (1949). *Phys. Rev.* **75**, 1675.
CHRISTIAN, R. S. (1952). *Rep. Progr. Phys.* **15**, 68.
CHRISTIAN, R. S. and GAMMEL, J. L. (1953). *Phys. Rev.* **91**, 100.
CHRISTIAN, R. S. and HART, E. W. (1950). *Phys. Rev.* **77**, 441.
CHRISTIAN, R. S. and NOYES, H. P. (1950). *Phys. Rev.* **79**, 85.
CHRISTY, R. F. and LATTER, R. (1948). *Rev. Mod. Phys.* **20**, 185.
CHU, E. L. and SCHIFF, L. J. (1953). *Ann. Rev. Nuclear Sci.* **2**, 79.
CLAASSEN, R. S., BROWN, R. J. S., FREIER, G. D. and STRATTON, W. R. (1951). *Phys. Rev.* **82**, 589; *ibid.* (1952), **88**, 253.
CLADIS, J. B., HADLEY, J. and HESS, W. N. (1952). *Phys. Rev.* **86**, 110.
CLEMENTEL, F. (1951). *Nuovo Cim.* **8**, 185.
CLEMENTEL, E. and VILLI, C. (1955). *Nuovo Cim.* **1**, 1273.
COCKROFT, A. L. and CURRAN, S. C. (1951). *Rev. Sci. Instr.* **22**, 37.
COCKCROFT, J. D., DUCKWORTH, J. C. and MERRISON, A. W. (1949). *Nature, Lond.* **163**, 869.
COHEN, B. L. (1951). *Nucleonics*, **8**, no. 2, 29.
COLLINS, E. R., McKENZIE, C. D. and RAMM, C. A. (1953). *Proc. Roy. Soc.* A, **216**, 219.

REFERENCES 447

Cook, L. (1951). *Rev. Sci. Instr.* **22**, 1006.
Cook, L. J., McMillan, E. M., Peterson, J. M. and Sewell, D. C. (1947). *Phys. Rev.* **72**, 1264.
Cook, L. J., McMillan, E. M., Peterson, J. M. and Sewell, D. C. (1949), *Phys. Rev.* **75**, 7.
Coon, J. H. and Barschall, H. H. (1946). *Phys. Rev.* **70**, 592.
Coon, J., H. Davis, R. and Barschall, H. (1946). *Phys. Rev.* **70**, 104.
Coon, J. H., Bockelman, C. K. and Barschall, H. H. (1951). *Phys. Rev.* **81**, 33.
Coon, J. H. and Taschek, R. F. (1949). *Phys. Rev.* **76**, 710.
Cooper, D. I., Frisch, D. H. and Zimmerman, R. L. (1954). *Phys. Rev.* **94**, 1209.
Coor, T., Hill, D. A., Hornyak, W. F., Smith, L. W. and Snow, G. (1955). *Phys. Rev.* **98**, 1369.
Cork, B. (1952). *University of California Radiation Laboratory Report,* UCRL-1673.
Cork, B. (1955). *Rev. Sci. Instr.* **26**, 210.
Cork, B. and Hartsough, W. (1954a). *Phys. Rev.* **94**, 1300.
Cork, B. and Hartsough, W. (1954b). *Phys. Rev.* **96**, 1267.
Cork, B., Johnston, L. and Richman, C. (1950). *Phys. Rev.* **79**, 71.
Cork, B. and Wenzel, W. A. (1955). *University of California Radiation Laboratory Report,* UCRL-3223.
Cork, B. and Zajec, E. (1953). *University of California Radiation Laboratory Report,* UCRL-2182; also (1953), *Phys. Rev.* **92**, 853.
Craggs, J. D. and Meek, J. M. (1954). *High Voltage Laboratory Technique.* Butterworths Scientific Publications, London.
Cranberg, L., Aiello, W. P., Beauchamp, R. K., Lang, H. J. and Levin, J. S. (1957). *Rev. Sci. Instr.* **28**, 84.
Cranberg, L., Beauchamp, R. K. and Levin, J. S. (1957). *Rev. Sci. Instr.* **28**, 89.
Critchfield, C. L. and Dodder, D. C. (1949a). *Phys. Rev.* **75**, 419.
Critchfield, C. L. and Dodder, D. C. (1949b). *Phys. Rev.* **76**, 602.
Cross, W. G. (1952). *Phys. Rev.* **87**, 223.
Culler, V. and Waniek, R. W. (1955). *Phys. Rev.* **99**, 740.
Cumming, J. B., Swartz, C. A. and Friedlander, G. (1956). *Bull. Amer. Phys. Soc.* II, **1**, no. 4 (W11).
Curran, S. C. (1953). *Luminescence and the Scintillation Counter.* Butterworths Scientific Publications, London.
Curtis, B. R., Fowler, J. L. and Rosen, L. (1949). *Rev. Sci. Instr.* **20**, 388.
Curtis, L. F. and Carson, A. (1949). *Phys. Rev.* **76**, 1412.
Cushman, B. E. (1951). *University of California Radiation Laboratory Report,* UCRL-1238 (Bibliography).
Dabbs, J. W. T., Roberts, L. D. and Bernstein, S. (1955). *Phys. Rev.* **98**, 1512.
Dabbs, J. W. T., Roberts, L. D. and Parker, G. W. (1956). *Bull. Amer. Phys. Soc.* II, **1**, no. 4 (R9).
Daitch, P. B. and French, J. B. (1952). *Phys. Rev.* **87**, 900.
Dalitz, R. H. (1951). *Proc. Roy. Soc. A,* **206**, 509.

DALITZ, R. H. (1952). *Proc. Phys. Soc.* A, **65**, 175.

DALITZ, R. H. (1953). *Proc. Phys. Soc.* A, **66**, 28.

DANCOFF, S. M. (1940). *Phys. Rev.* **58**, 326.

DANCOFF, S. M. (1947). *Phys. Rev.* **72**, 1017.

DANCOFF, S. M. and INGLIS, D. (1936). *Phys. Rev.* **50**, 784.

DANIELS, J. M., GRACE, M. A. and ROBINSON, F. N. H. (1951). *Nature, Lond.* **168**, 780.

DARBY, J F. and SWAN, J. B. (1948). *Aust. J. Sci. Res.* A, **1**, 18.

DAVENPORT, P., JEFFRIES, J., OWEN, M., PRICE, F. V. and ROAF, D. (1953). *Proc. Roy. Soc.* A, **216**, 66.

DE BORDE, H. H. and MASSEY, H. S. W. (1955). *Proc. Roy. Soc.* A, **68**, 769.

DE CARVALHO, M. G., MARSHALL, J. and MARSHALL, L. (1954). *Phys. Rev.* **96**, 1081.

DEE, P. I. and GILBERT, C. W. (1937). *Proc. Roy. Soc.* A, **163**, 265.

DE HOFFMAN, F. (1950). *Phys. Rev.* **78**, 216.

DE PANGHER, J. (1954). *Phys. Rev.* **95**, 578.

DE PANGHER, J. (1955). *Phys. Rev.* **99**, 1447.

DEARNLEY, I. H., OXLEY, C. L. and PERRY, J. E. (1948). *Phys. Rev.* **73**, 1290.

DEUTSCH, M. (1948). *Nucleonics*, **2**, no. 3, 58.

DICKINSON, W. C. and DODDER, D. C. (1950). *Los Alamos Scientific Laboratory Report*, La-1182; also (1953), *Rev. Sci. Instr.* **24**, 428.

DICKSON, J. M. and SALTER, D. C. (1954). *Nature, Lond.* **173**, 946.

DODDER, D. C. and GAMMEL, J. L. (1952). *Phys. Rev.* **88**, 520.

DONALDSON, R. E. and BRADNER, H. (1955). *Phys. Rev.* **99**, 892.

DUKE, P. J., LOCK, W. O., MARCH, P. V., GIBSON, W. M., McEWEN, J. G., HUGHES, I. S. and MUIRHEAD, H. (1957). *Phil. Mag.* **14**, 204.

DUNNING, J. R., PEGRAM, G. B., FINK, G. A., MITCHELL, D. P. and SEGRÈ, E. (1935). *Phys. Rev.* **48**, 704.

DZHELEPOV, V. P. and KAZARINOV, U. M. (1954). *Dokl. Akad. Nauk. S.S.S.R.* **99**, no. 6, 939 (A.E.R.E. Lib./Trans, 542).

DZHELEPOV, V. P., KAZARINOV, U. M., GOLOVIN, B. M., FLJAGIN, V. B. and SATAROV, V. I. (1956). *Nuovo Cim.* **3**, 61.

DZHELEPOV, V. P., MOSKALIEV, V. I. and MIEDVIEV, S. V. (1955). *Dokl. Akad. Nauk. S.S.S.R.* **104**, 380.

EBEL, M. E. and HULL, M. H. (1955). *Phys. Rev.* **99**, 1596.

ELMORE, W. C. and SANDS, M. L. (1949). *Electronics: Experimental Techniques*. McGraw-Hill Book Co. Inc., N.Y.

ELTON, L. R. B. and GOMES, L. C. (1957). *Phys. Rev.* **105**, 1027.

ENGE, H. A. and GRAVE, A. (1956). *Rev. Sci. Instr.* **27**, 1078.

ENNIS, M. E. and HEMMENDINGER, A. (1954). *Phys. Rev.* **95**, 772.

ERIKSEN, E., FOLDY, L. L. and RARITA, W. (1956). *Phys. Rev.* **103**, 781.

ERIKSSON, T. (1956). *Nuc. Phys.* **2**, 91.

ERSKINE, G. A. and MASSEY, H. S. W. (1941). *Proc. Roy. Soc.* A, **212**, 521.

EVANS, R. D. (1955). *The Atomic Nucleus.* McGraw-Hill Book Co., Inc., N.Y.

FAMULARO, K. F., BROWN, R. J. S., HOLMGREN, H. D. and STRATTON, T. F. (1954). *Phys. Rev.* **93**, 928.

FARWELL, G. W. and WEGNER, H. E. (1955). *Phys. Rev.* 95, 1212.
FELD, B. T. (1951). *Nucleonics*, 9, no. 4, 51.
FELD, B. T. (1953). *Experimental Nuclear Physics*, vol. II, E. Segre (ed.).
 J. Wiley and Sons, Inc., N.Y.
FERMI, E. (1936). *Ric. Sci.* (VII), 2, 13.
FERMI, E. and MARSHALL, L. (1947). *Phys. Rev.* 71, 666.
FERMI, E., MARSHALL, J. and MARSHALL, L. (1947). *Phys. Rev.* 72, 193.
FERNBACH, S., HECKROTTE, W. and LEPORE, J. V. (1955). *Phys. Rev.*
 97, 1059.
FERNBACH, S., SERBER, R. and TAYLOR, T. B. (1949). *Phys. Rev.* 75,
 1352.
FESHBACH, H., PEASLEE, D. C. and WEISSKOPF, V. F. (1947). *Phys. Rev.*
 71, 145.
FESHBACH, H., PORTER, C. E. and WEISSKOPF, V. F. (1954). *Phys. Rev.*
 96, 448.
FESHBACH, H. and SCHWINGER, J. (1951). *Phys. Rev.* 84, 194.
FESHBACH, H. and WEISSKOPF, V. F. (1949). *Phys. Rev.* 76, 1550.
FIELDS, R., RUSSELL, B., SACHS, D. and WATTENBERG, A. (1947). *Phys.
 Rev.* 71, 508.
FISCHER, D. and GOLDHABER, G. (1954). *Phys. Rev.* 95, 1350.
FLUGGE, S. (1938). *Z. Phys.* 108, 545.
FOLDY, L. L. and ERIKSEN, E. (1954). *Phys. Rev.* 95, 1048.
FOLDY, L. L. and ERIKSEN, E. (1955). *Phys. Rev.* 98, 775.
FOSTER, F. L., DEWEY, D. R. and GALE, A. J. (1953). *Nucleonics*, II,
 no. 10, 14.
FOWLER, J. L. and BROLLEY, J. E. (1956). *Rev. Mod. Phys.* 28, 103.
FOWLER, J. L. and SLYE, J. M. (1950). *Phys. Rev.* 77, 787.
FOWLER, W. A. and LAURITSEN, C. C. (1949). *Phys. Rev.* 76, 314.
FOWLER, W. B., SHUTT, R. P., THORNDIKE, A. M. and WHITTEMORE,
 W. L. (1954). *Phys. Rev.* 95, 1026.
FOWLER, W. B., SHUTT, R. P., THORNDIKE, A. M. and WHITTEMORE, W. L.
 (1956), *Phys. Rev.* 103, 1479.
FOWLER, W. B., SHUTT, R. P., THORNDIKE, A. M., WHITTEMORE, W. L.,
 COCCONI, V. T., HART, E., BLOCHS, M. M., HARTH, E. M., FOWLER,
 E. C., GARRISON, J. D. and MORRIS, T. W. (1956). *Phys. Rev.* 103,
 1489.
FOX, R., LEITH, C., WOUTERS, L. and MACKENZIE, K. R. (1950). *Phys.
 Rev.* 80, 23.
FOX, R. (1950). *University of California Radiation Laboratory Report*,
 UCRL-867.
FRANCIS, N. C. and WATSON, K. M. (1954). *Phys. Rev.* 93, 313.
FRANK, R. M. and GAMMEL, J. L. (1954). *Phys. Rev.* 93, 463.
FRANK, R. M. and GAMMEL, J. L. (1955). *Phys. Rev.* 98, 1204.
FREEMANTLE, R. G., GROTDAL, T., GIBSON, W. M., MCKEAGUE, R.,
 PROWSE, D. J. and ROBLAT, J. (1954). *Phil. Mag.* 45, 108.
FREIER, G., LAMPI, E., SLEATOR, W. and WILLIAMS, J. H. (1949). *Phys.
 Rev.* 75, 1345.
FRIEDMAN, F. L. and TOBOCMAN, W. (1952). *Phys. Rev.* 87, 208.
FRÖBERG, C. (1955). *Rev. Mod. Phys.* 27, 399.

FROHLICH, H., RAMSEY, W. and SNEDDON, I. (1946). *Cambr. Conf.* p. 166.

FULBRIGHT, H. W., BROMLEY, D. A. and BRUNER, J. A. (1955). *Bull. Amer. Phys. Soc.* **30**, no. 3 (T 11).

Fundamental Mechanisms of Photographic Sensitivity, University of Bristol Symposium. Butterworths Scientific Publications, London, 1951.

GAERTTNER, E. R., PARDUE, L. A. and STREIB, J. F. (1939). *Phys. Rev.* **56**, 656.

GALONSKY, A. and JUDISH, J. P. (1955). *Phys. Rev.* **100**, 121.

GAMMEL, J. L., CHRISTIAN, R. S. and THALER, R. M. (1956). *Phys. Rev.* **105**, 311.

GAMOW, G. and CRITCHFIELD, C. L. (1949). *Theory of Atomic Nucleus and Nuclear Energy Sources.* Clarendon Press, Oxford.

GARDNER, J. H. and PURCELL, E. M. (1949). *Phys. Rev.* **76**, 1262.

GARREN, A. (1953). *Phys. Rev.* **92**, 213. Erratum, 1587.

GEISSLER, G. (1956). *Ann. Phys.* **18**, 125.

GELERNTER, H. (1957). *Phys. Rev.* **105**, 1068.

GIBBONS, J. H., MACKLIN, R. L. and SCHMITT, H. W. (1955). *Phys. Rev.* **100**, 167.

GIBSON, W. M., PROWSE, D. J. and ROTBLAT, J. (1954). *Nature, Lond.* **173**, 1180.

GLASER, D. A. and RAHM, D. C. (1955). *Phys. Rev.* **97**, 474.

GLASHOW, S. and SELOVE, W. (1956). *Phys. Rev.* **102**, 200.

GLASSTONE, S. and EDLUND, M. C. (1952). *The Elements of Nuclear Reactor Theory.* D. Van Nostrand Co., Inc., N.Y.

GLUCKSTERN, R. L. and BETHE, H. A. (1951). *Phys. Rev.* **81**, 761.

GOLDBERGER, M. L. (1948). *Phys. Rev.* **74**, 1269.

GOLDFARB, L. J. B. and FELDMAN, D. (1952). *Phys. Rev.* **88**, 1099.

GOLDING, E. W. (1942). *Electrical Measurements and Measuring Instruments.* Pitman and Sons. Ltd., London.

GOLDSCHMIDT-CLERMONT, Y. (1953). *Ann. Rev. Nuclear Sci.* **3**, 141.

GORDON, W. (1928). *Z. Phys.* **48**, 180.

GORTER, C. J. (1948). *Physica,* **14**, 504.

GRIFFITH, G. L., REMLEY, M. E. and KRUGER, P. G. (1950). *Phys. Rev.* **79**, 443.

GROSSKREUTZ, J. C. and MATHER, K. B. (1949). *Bull. Amer. Phys. Soc.* **24**, no. 7 (H 5); also (1950). *Phys. Rev.* **77**, 580.

GUERNSEY, G., MOTT, G. and NELSON, B. K. (1952). *Phys. Rev.* **88**, 15.

HADLEY, J., KELLY, E., LEITH, C., SEGRÈ, E., WIEGAND, C. and YORK, H. (1949). *Phys. Rev.* **75**, 351.

HAFNER, E. M., HORNYAK, W. F., FALK, C. E., SNOW, G. and COOR, T. (1953). *Phys. Rev.* **89**, 204.

HAFSTAD, L. R., HEYDENBURG, N. P. and TUVE, M. A. (1938). *Phys. Rev.* **53**, 239.

HALL, H. H. (1954). *Phys. Rev.* **95**, 424.

HALL, H. H. and POWELL, J. L. (1953). *Phys. Rev.* **90**, 912.

HALL, T. A. and KOONTZ, P. G. (1947). *Phys. Rev.* **72**, 196.

HALPERN, O. and HOLSTEIN, T. (1941). *Phys. Rev.* **59**, 960.

HAMERMESH, M. and SCHWINGER, J. (1946). *Phys. Rev.* **69**, 145.

HAMOUDA, I., HALTER, J. and SCHERRER, P. (1950). *Phys. Rev.* 79, 539.
HANSON, A. O. (1949). *Phys. Rev.* 75, 1794.
HANSON, A. O., TASCHEK, R. F. and WILLIAMS, J. H. (1949). *Rev. Mod. Phys.* 21, 635.
HARKINS, W. D., KAMEN, M. D., NEWSON, H. W. and GANS, D. M. (1936). *Phys. Rev.* 50, 980.
HARLOW, F. H. and JACOBSON, B. A. (1953). *Phys. Rev.* 92, 766.
HARRIS, S. P. (1950). *Phys. Rev.* 80, 20.
HARTZLER, A. J. and SIEGEL, R. T. (1954). *Phys. Rev.* 95, 185.
HARTZLER, A. J., SIEGEL, R. T. and OPITZ, W. (1954). *Phys. Rev.* 95, 591.
HATCHER, R. D., ARFKEN, G. B. and BREIT, G. (1949). *Phys. Rev.* 75, 1389.
HECKROTTE, W. (1954a). *Phys. Rev.* 95, 1279.
HECKROTTE, W. (1954b). *Phys. Rev.* 94, 1797.
HEIDMANN, J. (1950). *Phil. Mag.* 41, 444.
HEITLER, W. (1954). *The Quantum Theory of Radiation*, 3rd ed. Clarendon Press, Oxford.
HELMHOLTZ, A. C., McMILLAN, E. M. and SEWELL, D. C. (1947). *Phys. Rev.* 72, 1003.
HEMMENDINGER, A. (1956). *Bull. Amer. Phys. Soc.* II, 1, no. 2 (N7).
HEMMENDINGER, A. and ARGO, H. V. (1955). *Phys. Rev.* 98, 70.
HEMMENDINGER, A., JARVIS, G. A. and TASCHEK, R. F. (1949). *Phys. Rev.* 76, 1137.
HERB, R. G., KERST, D. W., PARKINSON, D. B. and PLAIN, G. J. (1939). *Phys. Rev.* 55, 998.
HERB, R. G., SNOWDON, S. C. and SALA, O. (1949). *Phys. Rev.* 75, 246.
HEUSINKVELD, M. and FREIER, G. (1952). *Phys. Rev.* 85, 80.
HEYDENBERG, N. P., HAFSTAD, L. R. and TUVE, M. A. (1939). *Phys. Rev.* 56, 1078.
HEYDENBURG, N. P. and RAMSEY, N. F. (1941). *Phys. Rev.* 60, 42.
HIBDON, C. T. and MUEHLHAUSE, C. O. (1951). *Argonne National Laboratory Report*, ANL-4680.
HILDEBRAND, R. H. (1953). *Phys. Rev.* 89, 1090.
HILL, D. L. (1952). *Phys. Rev.* 87, 1034.
HITTMAIR, O. H. (1955). *Z. Phys.* 142, 219.
HOCHBERG, S., MASSEY, H. S. W., ROBERTSON, H. H. and UNDERHILL, L. H. (1955). *Proc. Phys. Soc.* A, 68, 746.
HOCHBERG, S., MASSEY, H. S. W. and UNDERHILL, L. H. (1954). *Proc. Phys. Soc.* A, 67, 957.
HOCKER, M. (1942). *Physik Zeits.* 43, 236.
HODGSON, E. R., GALLAGHER, J. F. and BOWEY, E. M. (1952). *Proc. Phys. Soc.* A, 65, 992.
HOFMANN, J. A. and STRAUCH, K. (1953). *Phys. Rev.* 90, 449.
HOLM, D. M. and ARGO, H. V. (1956). *Phys. Rev.* 101, 1772.
HONIG, A. (1954). *Phys. Rev.* 96, 234.
HORNING, W. and BAUMHOFF, L. (1949). *Phys. Rev.* 75, 370.
HOROWITZ, J. and MESSIAH, A. (1953a). *Phys. Rev.* 92, 1326.
HOROWITZ, J. and MESSIAH, A. (1953b). *J. Phys. Radium*, 14, 12, 695.

452   REFERENCES

HORWITZ, N., MURRAY, J. J. and CRANDALL, W. E. (1956). *Bull. Amer. Phys. Soc.* II, **1**, no. 4 (W 12).
HU, T. S. and MASSEY, H. S. W. (1949). *Proc. Roy. Soc.* A, **196**, 135.
HUBER, P. and BALDINGER, E. (1952). *Helv. Phys. Acta,* **25**, 435.
HUBY, R. (1952). *Proc. Roy. Soc.* A, **215**, 385.
HUBY, R. (1953). *Prog. Nuc. Phys.* **3**, 177.
HUGHES, D. J. (1953). *Pile Neutron Research.* Addison-Wesley Publishing Co., Inc., Cambridge, Mass.
HUGHES, D. J. and BURGY, M. T. (1949). *Phys. Rev.* **76**, 1413; *ibid.* (1951), **81**, 498.
HUGHES, D. J., BURGY, M. T. and RINGO, G. R. (1950). *Phys. Rev.* **77**, 291.
HUGHES, D. J. and EGGLER, C. (1947). *Phys. Rev.* **72**, 902.
HUGHES, D. J., SPATZ, W. D. B. and GOLDSTEIN, N. (1949). *Phys. Rev.* **75**, 1781.
HULL, M. H. and SAPERSTEIN, A. M. (1954). *Phys. Rev.* **96**, 806.
HUNTOON, R. D., ELLETT, A., BAYLEY, D. S. and VAN ALLEN, J. A. (1940). *Phys. Rev.* **58**, 97.
HURST, D. G. and ALCOCK, N. Z. (1950). *Phys. Rev.* **80**, 117.
IGO, G., CLARK, D. D. and EISBERG, R. M. (1953). *Phys. Rev.* **89**, 879.
IGO, G. and THALER, R. M. (1957). *Phys. Rev.* **106**, 126.
INGLIS, D. R. (1936). *Phys. Rev.* **50**, 783.
INGLIS, D. R. (1953). *Rev. Mod. Phys.* **25**, 390.
ISBEN, H. S. (1952). *Nucleonics,* **10**, no. 3, 10.
JACKSON, J. D. and BLATT, J. M. (1950). *Rev. Mod. Phys.* **22**, 77.
JARVIS, G. A., HEMMENDINGER, A., ARGO, H. V. and TASCHEK, R. F. (1950). *Phys. Rev.* **79**, 929.
JASTROW, R. (1950). *Phys. Rev.* **79**, 389.
JASTROW, R. (1951). *Phys. Rev.* **81**, 165.
JELLEY, J. V. (1953). *Proc. Nuc. Phys.* **3**, 131.
JENSEN, J. H. D., HAXEL, O. and SUESS, H. E. (1948). *Naturwiss.* **35**, 376; *ibid.* (1949), **36**, 153, 155.
JOHNSON, C. H. and BANTA, H. E. (1956). *Rev. Sci. Instr.* **27**, 132.
JOHNSON, C. H. and GALONSKEY, A. (1955). *Phys. Rev.* **100**, 1252.
JOHNSON, V. R., LAUBENSTEIN, M. J. W. and RICHARDS, H. T. (1950). *Phys. Rev.* **77**, 413.
JONES, K. W., DOUGLAS, R. A., McELLISTREM, M. T. and RICHARDS, H. T. (1954). *Phys. Rev.* **94**, 947.
JONES, W. B. (1948). *Phys. Rev.* **74**, 364.
JOST, R. (1947). *Helv. Phys. Act.* **20**, 256.
JUVELAND, A. C. and JENTSCHKE, W. K. (1956a). *Z. Phys.* **144**, 521.
JUVELAND, A. C. and JENTSCHKE, W. K. (1956b). *Bull. Amer. Phys. Soc.* II, **1**, no. 4 (M 10).
KAO, S. K. and CLARK, A. F. (1955). *Phys. Rev.* **99**, 895.
KAPUR, P. L. and PEIERLS, R. (1938). *Proc. Roy. Soc.* A, **166**, 277.
KARR, H. J., BONDELID, R. O. and MATHER, K. B. (1951). *Phys. Rev.* **81**, 37.
KATZ, L. and CAMERON, A. G. W. (1951). *Can. J. Phys.* **29**, 518.
KELLY, E. L., LEITH, C., SEGRÈ, E. and WIEGAND, C. (1950). *Phys. Rev.* **79**, 96.

KERNAN, A. and CESTER, R. (1955). *Bull. Amer. Phys. Soc.* **30**, no. 3, (N6).

KIKUCHI, S. and AOKI, H. (1939). *Phys. Math. Soc. Japan, Proc.* **21**, 75.

KLIGMAN, F. (1940). *J. Exp. and Theor. Phys.* **10**, 15.

KOWARSKI, L. (1955). *Report on Research Reactors.* International Conference on the Peaceful Uses of Atomic Energy, A/Conf. 8/P/946, UNESCO.

KREGER, W. E., JENTSCHKE, W. K. and KRUGER, P. G. (1954). *Phys. Rev.* **93**, 837.

KREGER, W. E., KERMAN, R. O. and JENTSCHKE, W. K. (1952). *Phys. Rev.* **86**, 593.

KRUGER, P. G., SHOUPP, W. E. and STALLMAN, F. W. (1937). *Phys. Rev.* **52**, 678.

KRUGER, P. G., SHOUPP, W. E., WATSON, R. E. and STALLMAN, F. W. (1938). *Phys. Rev.* **53**, 1014.

KRUSE, U. E., TEEM, J. M. and RAMSEY, N. F. (1954). *Phys. Rev.* **94**, 1795.

KURIE, F. N. D. (1933). *Phys. Rev.* **44**, 463.

KURIE, F. N. D. (1948). *Phys. Rev.* **19**, 485.

KURTI, N. and SIMON, F. E. (1935). *Proc. Roy. Soc.* A, **149**, 152.

LAMARSH, J. R. and FESHBACH, H. (1956). *Phys. Rev.* **104**, 1633.

LAMPSON, C. W., MUELLER, D. W. and BARTON, H. A. (1937). *Phys. Rev.* **51**, 1021.

LANDAU, L. (1944). *J. Phys. U.S.S.R.* **8**, 201.

LANDAU, L. and SMORODINSKY, J. (1944). *J. Phys. U.S.S.R.* **8**, 154.

LAUGHLIN, J. S. and KRUGER, P. G. (1948). *Phys. Rev.* **73**, 197.

LAWRENCE, E. O. (1955). *Science,* **122**, 1127.

LAWS, F. A. (1938). *Electrical Measurements.* McGraw-Hill Book Co., Inc., N.Y.

LEVINSON, N. (1949). *K. Danske Vidensk. Selsk., Mat.-fys. Medd.,* **25**, no. 9.

LEITH, C. E. (1950). *Phys. Rev.* **78**, 89.

LE LEVIER, R. E. and SAXON, D. S. (1952). *Phys. Rev.* **87**, 40.

LEPORE, J. V. (1950). *Phys. Rev.* **79**, 137.

LEVINTOV, I. I., MILLER, A. V. and SHAMSHEV, V. N. (1957). *Nuc. Phys.* **3**, 221.

LEVINTOV, I. I., MILLER, A. V., TARUMOV, E. Z. and SHAMSHEV, V. N. (1957). *Nuc. Phys.* **3**, 237.

LEVY, M. M. (1952a). *Phys. Rev.* **88**, 725.

LEVY, M. M. (1952b). *Phys. Rev.* **86**, 806; also *C.R. Acad. Sci., Paris,* **234**, 815, 922, 1255, 1671, 1744.

LEWIS, I. A. D. and WELLS, F. H. (1954). *Millimicrosecond Pulse Techniques.* Pergamon Press, London.

LIVINGSTON, M. S. (1952). *Ann. Rev. Nuclear Sci.* **1**, 157, 163, 169.

LIVINGSTON, M. S. (1954). *High Energy Accelerators.* Interscience Publishers, Inc., N.Y.

LONGLEY, H. J., LITTLE, R. N. and SLYE, J. M. (1952). *Phys. Rev.* **86**, 419.

LORRAIN, P., BÉIQUE, R., GILMORE, P., GIRARD, P. E., BRETON, A. and PICHÉ, P. (1955). *Bull. Amer. Phys. Soc.* **29**, no. 4 (C4).

LOWEN, R. W. (1954). *Phys. Rev.* **96**, 826.
LUBITZ, C. R. and PARKINSON, W. C. (1955). *Rev. Sci. Instr.* **26**, 400.
MANLEY, J. H. and JAKOBSON, M. J. (1954). *Rev. Sci. Instr.* **25**, 368.
MARGOLIS, B. and TROUBETZKOY, E. S. (1957). *Phys. Rev.* **106**, 105.
MARION, J. B., BONNER, T. W. and COOK, C. F. (1955). *Phys. Rev.* **100**, 91.
MARION, J. B., BRUGGER, R. M. and BONNER, T. W. (1955). *Phys. Rev.* **100**, 46.
MARION, J. B. and WEBER, G. (1956). *Phys. Rev.* **102**, 1355; *ibid.* **103**, 167.
MARSHAK, R. E. (1952). *Meson Physics.* McGraw-Hill Book Co. Inc., N.Y.
MARSHALL, J. (1954). *Ann. Rev. Nuclear Sci.* **4**, 141.
MARSHALL, J., MARSHALL, L., NAGLE, D. and SKOLNIK, W. (1954). *Phys. Rev.* **95**, 1020.
MARSHALL, J., MARSHALL, L. and NEDZEL, V. A. (1953). *Phys. Rev.* **92**, 834.
MARSHALL, J., MARSHALL, L. and NEDZEL, V. A. (1955). *Phys. Rev.* **98**, 1513.
MASSEY, H. S. W. (1953). *Proc. Nuc. Phys.* **3**, 235.
MASSEY, H. S. W. and MOISEIWITSCH, B. L. (1951). *Proc. Roy. Soc. A*, **205**, 483.
MATHER, K. B. (1951). *Phys. Rev.* **82**, 133.
MATHER, K. B. (1952). *Phys. Rev.* **88**, 1408.
MATHEWS, P. T. and SALAM, A. (1952). *Phys. Rev.* **86**, 715.
MAY, A. N. and POWELL, G. F. (1947). *Proc. Roy. Soc. A*, **190**, 170.
MAYER, M. G. (1948). *Phys. Rev.* **74**, 235; *ibid.* (1950), **78**, 16, 22.
McCUE, J. J. G. and PRESTON, W. M. (1951). *Phys. Rev.* **84**, 1150.
McINTOSH, J. S., GLUCKSTERN, R. L. and SACK, S. (1952). *Phys. Rev.* **88**, 752.
McKIBBEN, J. L. (1946). *Phys. Rev.* **70**, 101; also Document MDDC-223 (Document Division of U.S.A.E.C., Oak Ridge, Tennessee).
MEAGHER, R. H. (1950). *Phys. Rev.* **78**, 667.
MEIER, R. W., SCHERER, R. and TRUMPY, G. (1954). *Helv. Phys. Act.* **27**, 577.
MEITNER, L. and PHILIPP, K. (1934). *Z. Phys.* **87**, 484.
MELKANOFF, M. A., MOSZKOWSKI, S. A., NODVIK, J. S. and SAXON, D. S. (1955). *Phys. Rev.* **101**, 507.
MELKANOFF, M. A., NODVIK, J. S., SAXON, D. S. and WOODS, R. D. (1957). *Phys. Rev.* **106**, 793.
MELKONIAN, E. (1949). *Phys. Rev.* **76**, 1744.
MERRISON, A. W. and WIBLIN, E. R. (1951). *Nature, Lond.* **167**, 346.
MESCERJAKOV, M. G., BOGACEV, N. P. and NEGANOV, B. S. (1956). *Nuovo Cim.* **3**, 119.
MESHCHEVYOKOV, M. G., NEGANOV, B. S., SOROKO, L. M. and VZOROV, I. K. (1954). *Dokl. Akad. Nauk. S.S.S.R.* **99**, 959.
MIDDLETON, R., EL-BEDEWI, F. A. and TAI, C. T. (1953). *Proc. Phys. Soc. A*, **66**, 95.
MILLBURN, G. P., MOYER, B. J., TAI, Y. and KAPLAN, S. N. (1955). *Bull. Amer. Phys. Soc.* **30**, no. 8 (C5).

MILLER, R. D., SEWELL, D. C. and WRIGHT, R. W. (1951). *Phys. Rev.* **81**, 374.

MONTALBETTI, R., KATZ, L. and GOLDENBERG, J. (1953). *Phys. Rev.* **91**, 659.

MOHR, C. B. O. and ROBSON, B. A. (1956). *Proc. Phys. Soc.* A, **69**, 365.

MOORE, M. J. (1955). *Nature, Lond.*, **175**, 1012.

MORINAGA, H. and PEASLEE, D. C. (1957). *Nuc. Phys.* **3**, 115.

MORRIS, T. W., FOWLER, E. C. and GARRISON, J. D. (1956). *Phys. Rev.* **103**, 1472.

MORRIS, T. W., GARRISON, J. D., FOWLER, E. C., FOWLER, W. B., SHUTT, R. P., THRONDIKE, A. M. and WHITTEMORE, W. L. (1955). *Bull. Amer. Phys. Soc.* **30**, no. 1 (L5).

MORSE, P. M., LOWAN, A. N., FESHBACH, H. and LAX, M. (1945). U.S. Navy, Department of Research and Inventions, Report 62, 1 R.

MOTT, N. F. and MASSEY, H. S. W. (1949). *The Theory of Atomic Collisions*, 2nd ed. Clarendon Press, Oxford.

MOTT, W. E., SUTTON, R. B., FOX, J. G. and KANE, J. A. (1953). *Phys. Rev.* **90**, 712.

NADI, M. E. (1957). *Proc. Phys. Soc.* A, **70**, 62.

NATHANS, R. and HALPERN, J. (1954). *Phys. Rev.* **93**, 437.

NATIONAL BUREAU OF STANDARDS (1952). *Tables of Coulomb Wave Functions*, vol. 1, Applied Mathematics Series, 17.

NISHIJIMA, K. (1951). *Prog. Theor. Phys.* **6**, 815.

NUCKOLLS, R. G., BAILEY, C. L., BENNETT, W. E., BERGSTRALH, T., RICHARDS, H. T. and WILLIAMS, J. H. (1946). *Phys. Rev.* **70**, 805.

OCHIAI, K. (1937). *Phys. Rev.* **52**, 1221.

OHNUMA, S. and FELDMAN, D. (1956). *Phys. Rev.* **102**, 1641.

OKAI, S. and SANO, M. (1956). *Prog. Theor. Phys.* **15**, 203.

O'NEILL, G. K. (1954). *Phys. Rev.* **95**, 1235.

OPPENHEIMER, J. R. and PHILLIPS, M. (1935). *Phys. Rev.* **48**, 500.

OSWALD, L. O. (1957). *Rev. Sci. Instr.* **28**, 80.

OWEN, G. E. and MADANSKY, L. (1957). *Phys. Rev.* **105**, 1766.

OXLEY, C. L. and SCHAMBERGER, R. D. (1952). *Phys. Rev.* **85**, 416.

OXLEY, C. L., CARTWRIGHT, W. F. and ROUVINA, J. (1954). *Phys. Rev.* **93**, 806.

PADFIELD, D. (1949). *Nature, Lond.* **163**, 22.

PALEVSKY, H., SWANK, R. K. and GRENCHIK, R. (1947). *Rev. Sci. Instr.* **18**, 298.

PANOFSKY, W. K. H. and FILLMORE, F. L. (1950). *Phys. Rev.* **79**, 57.

PASTERNACK, S. and SNYDER, H. S. (1950). *Phys. Rev.* **80**, 921.

PEASLEE, D. C. (1948). *Phys. Rev.* **74**, 1001.

PEASLEE, D. C. (1957). *Nuc. Phys.* **3**, 255.

PECK, R. A. and EUBANK, H. P. (1955). *Rev. Sci. Instr.* **26**, 444.

PERRY, J. E. and BAME, S. J. (1955). *Phys. Rev.* **99**, 1368.

PHILLIPS, M., DAVIS, L. and GRAVES, A. C. (1952). *Phys. Rev.* **88**, 600.

PICCIONI, O., CLARK, D., COOL, R., FRIEDLANDER, G. and KASSNER, D. (1955). *Rev. Sci. Instr.* **26**, 232.

Poss, H. L. (1949). *The Properties of Atomic Nuclei.* I. *Spins, Magnetic Moments and Electric Quadrupole Moments.* Brookhaven National Laboratory.

Poss, H. L., Salant, E. O., Snow, G. A. and Yuan, L. C. L. (1952). *Phys. Rev.* 87, 11.

Pound, R. V. (1949), *Phys. Rev.* 76, 1410.

Powell, C. F., Heitler, H. and Champion, F. C. (1940). *Nature, Lond.* 146, 716.

Powell, C. F. and Occhialini, G. P. S. (1947). *Nuclear Physics in Photographs.* Clarendon Press, Oxford.

Powell, W. M. (1953). Unpublished.

Pratt, W. W. (1955). *Phys. Rev.* 97, 131.

Proc. International Conference on the Peaceful Uses of Atomic Energy (1956), vol. II. United Nations, N.Y.

Putnam, T. M. (1952). *Phys. Rev.* 87, 932.

Putnam, T. M., Brolley, J. E. and Rosen, L. (1956). *Phys. Rev.* 104, 1303.

Ragan, G. L., Kanne, W. R. and Taschek, R. F. (1941). *Phys. Rev.* 60, 628.

Rainwater, J. and Havens, W. W. (1946). *Phys. Rev.* 70, 136, 154; also (1947), 71, 65.

Ralph, D. C. and Dunnan, F. F. (1954). *Bull. Amer. Phys. Soc.* 29, no. 7, (X4).

Ralph, D. C., Worthington, H. R. and Herb, R. G. (1950). Unpublished; see Jackson and Blatt (1950).

Randle, T. C., Skyrme, D. M., Snowden, M., Taylor, A. E., Uridge, F. and Wood, E. (1956). *Proc. Phys. Soc.* A, 69, 760.

Randle, T. C., Taylor, A. E. and Wood, E. (1952). *Proc. Roy. Soc.* A, 213, 392.

Rarita, W. and Schwinger, J. S. (1941*a*). *Phys. Rev.* 59, 436.

Rarita, W. and Schwinger, J. S. (1941*b*). *Phys. Rev.* 59, 556.

Rich, M. and Madey, R. (1954). *University of California Radiation Laboratory Report*, UCRL-2301.

Richards, H. T., Smith, R. V. and Browne, C. P. (1950). *Phys. Rev.* 80, 524.

Ridenour, L. N. and Henderson, W. J. (1937). *Phys. Rev.* 52, 889.

Ringo, G. R., Burgy, M. T. and Hughes, D. J. (1951). *Phys. Rev.* 84, 1160.

Robson, B. A. (1958). University of Melbourne, private communication.

Rodgers, F. A., Leiter, H. A. and Kruger, P. G. (1950). *Phys. Rev.* 78, 656.

Rose, M. E. (1949). *Phys. Rev.* 75, 213.

Rosen, L. and Allred, J. C. (1951). *Phys. Rev.* 82, 777.

Rosen, L. and Allred, J. C. (1952). *Phys. Rev.* 88, 431.

Rosenfeld, A. H. (1954). *Phys. Rev.* 96, 139.

Rosenfeld, L. (1948). *Nuclear Forces*, 1st ed. North Holland Publishing Co., Amsterdam.

Rossi, B. B. (1952). *High Energy Particles.* Prentice-Hall, Inc., N.Y.

Rossi, B. B. and Staub, H. B. (1949). *Ionization Chambers and Counters*, McGraw-Hill Book Co. Inc., N.Y.
Rotblat, J. (1951). *Nature, Lond.* **167**, 550; ibid. (1950), **165**, 387.
Royden, H. N. and Caldwell, D. O. (1956). *Rev. Sci. Instr.* **27**, 91.
Russell, B., Sachs, D., Wattenberg, A. and Fields, R. (1948). *Phys. Rev.* **73**, 545.
Sack, S., Biedenharn, L. C. and Breit, G. (1954). *Phys. Rev.* **93**, 321.
Sailor, V. L., Foote, H. L., Landon, H. H. and Wood, R. E. (1956). *Rev. Sci. Instr.* **27**, 26.
Salant, E. and Ramsay, N. (1940). *Phys. Rev.* **57**, 1075.
Salpeter, E. E. (1951). *Phys. Rev.* **82**, 60.
Sanada, J. and Yamabe, S. (1950). *Phys. Rev.* **80**, 750.
Sawicki, J. (1956). *Phys. Rev.* **104**, 1441.
Sawicki, J. (1957). *Phys. Rev.* **106**, 172.
Sawyer, R. B., Wollan, E. O., Bernstein, S. and Peterson, K. C. (1947). *Phys. Rev.* **72**, 109.
Schiff, L. I. (1937). *Phys. Rev.* **52**, 149.
Schiff, L. I. (1949). *Quantum Mechanics*, 1st ed. McGraw-Hill Book Co., Inc., N.Y.
Schamberger, R. D. (1951). *Phys. Rev.* **83**, 1276.
Schamberger, R. D. (1952). *Phys. Rev.* **85**, 424.
Schwinger, J. (1940). *Phys. Rev.* **58**, 1004.
Schwinger, J. (1946). *Phys. Rev.* **69**, 681.
Schwinger, J. (1948). *Phys. Rev.* **73**, 407.
Schwinger, J. (1950). *Phys. Rev.* **78**, 135.
Schwinger, J. and Teller, E. (1937). *Phys. Rev.* **52**, 286.
Scintillation Counter Symposium (1952). Washington, D.C. (See *Nucleonics*, April–August, 1952.)
Scintillation Counter Symposium (1954). Washington, D.C. (See *Nucleonics*, March 1954.)
Scintillation Counter Symposium (1956). Washington, D. C. (See *Nucleonics*, April 1956.)
Scott, J. M. C. (1954). *Phil. Mag.* **45**, 441.
Seagrave, J. D. (1953). *Phys. Rev.* **92**, 1222.
Seagrave, J. D. (1955). *Phys. Rev.* **97**, 757.
Seagrave, J. D. and Cranberg, L. (1957). *Phys. Rev.* **105**, 1816.
Seagrave, J. D. and Henkel, R. L. (1955). *Phys. Rev.* **98**, 666.
Seidl, F. G. P., Hughes, D. J., Palevsky, H., Levin, J. S., Kato, W. Y. and Sjostrand, N. G. (1954). *Phys. Rev.* **95**, 476.
Selove, W. (1951). *Phys. Rev.* **84**, 869; also (1952). *Rev. Sci. Instr.* **23**, 350.
Selove, W., Strauch, K. and Titus, F. (1953). *Phys. Rev.* **92**, 724.
Serber, R. (1947). *Phys. Rev.* **72**, 1008.
Shapiro, A. M., Leavitt, C. P. and Chen, F. F. (1954). *Phys. Rev.* **95**, 663.
Shaw, D. F. (1955). *Proc. Phys. Soc. A*, **68**, 43.
Sherr, R. (1945). *Phys. Rev.* **68**, 240.
Sherr, R., Blair, J. M., Kratz, H. R., Bailey, C. L. and Taschek, R. F. (1947). *Phys. Rev.* **72**, 662.

SHOUPP, W. E., JENNINGS, B. and JONES, W. (1949). *Phys. Rev.* **76**, 502.
SHULL, C. G., WOLLAN, E. O., MORTON, G. H. and DAVIDSON, W. L. (1948). *Phys. Rev.* **73**, 842.
SHULL, F. B., MacFARLAND, C. E. and BRETSCHER, M. M. (1954). *Rev. Sci. Instr.* **25**, 364.
SIEGEL, R. T., HARTZLER, A. J. and LOVE, W. A. (1956). *Phys. Rev.* **101**, 838.
SIEGERT, A. J. F. (1939). *Phys. Rev.* **56**, 750.
SIGNEL, P. S. and MARSHAK, R. E. (1957). *Phys. Rev.* **106**, 832.
SIMMONS, J. E. (1956). *Phys. Rev.* **104**, 416.
SJÖLANDER, A. and KÖHLER, S. (1954). *Arkiv. för Fysik* **8**, 521.
SLATER, J. C. (1952). *Ann. Rev. Nuclear Sci.* **1**, 199.
SLEATOR, W. (1947). *Phys. Rev.* **72**, 207.
SMITH, L. W. and KRUGER, P. G. (1951). *Phys. Rev.* **83**, 1137.
SMITH, L. W., McREYNOLDS, A. W. and SNOW, G. (1955). *Phys. Rev.* **97**, 1186.
SNOWDEN, M. (1953). *Prog. Nuc. Phys.* **3**, 1.
SOODAK, H. and CAMPBELL, E. C. (1950). *Elementary Pile Theory.* J. Wiley and Sons, Inc., N.Y.
STAHL, R. H. and RAMSEY, N. F. (1954). *Phys. Rev.* **96**, 1310.
STAPP, H. P., YPSILANTIS, T. J. and METROPOLIS, N. (1957). *Phys. Rev.* **105**, 302.
STAUB, H. and STEPHENS, W. E. (1939). *Phys. Rev.* **55**, 131.
STAUB, H. and TATEL, H. (1940). *Phys. Rev.* **58**, 820.
STAUB, H. (1953). *Experimental Nuclear Physics*, vol. 1, E. Segrè, (ed.). J. Wiley and Sons, Inc., N.Y.
STEIGERT, F. E. and SAMPSON, M. B. (1953). *Phys. Rev.* **92**, 660.
STELSON, P. H. and PRESTON, W. M. (1951). *Phys. Rev.* **82**, 655; *ibid.* (1951), **83**, 469.
STERN, M. O. and BLOOM, A. L. (1951). *Phys. Rev.* **83**, 178.
STERNHEIMER, R. M. (1955a). *Phys. Rev.* **97**, 1314.
STERNHEIMER, R. M. (1955b). *Phys. Rev.* **100**, 886.
STEWARD, L. (1955). *Phys. Rev.* **98**, 740.
STRAUCH, K. (1955). *Phys. Rev.* **99**, 150.
STRONG, J. S. (1945). *Procedures in Experimental Physics.* Prentice-Hall, Inc., N.Y.
SUGIE, A., HODGSON, P. E. and ROBERTSON, H. H. (1957). *Proc. Phys. Soc.* A, **70**, 1.
SUTTON, R. B., FIELDS, T. H., FOX, J. G., KANE, J. A., MOTT, W. E. and STALLWOOD, R. A. (1955). *Phys. Rev.* **97**, 783.
SUTTON, R. B., HALL, T., ANDERSON, E. E., BRIDGE, H. S., DE WIRE, J. W., LONG, E. A., SNYDER, T. and WILLIAMS, R. W. (1947). *Phys. Rev.* **72**, 1147.
SWAN, P. (1953). *Proc. Phys. Soc.* A, **66**, 238, 740.
SWAN, P. (1955). *Proc. Roy. Soc.* A, **228**, 10.
SWANK, R. K. (1954). *Ann. Rev. Nuclear Sci.* **4**, 111.
SWANSON, D. R. (1953). *Phys. Rev.* **89**, 749.
SWARTZ, C. (1952). *Phys. Rev.* **85**, 73.

SWEETMAN, D. R. (1955). Birmingham University, private communication.

SYMON, K. R. (1948). Thesis, Harvard University; also Rossi (1952).

TAKANO, Y. and HULL, M. H. (1957). *Phys. Rev.* **106**, 321.

TAMOR, S. (1955). *Phys. Rev.* **97**, 1077.

TANNENWALD, P. E. (1953). *Phys. Rev.* **89**, 508.

TASCHEK, R. F., JARVIS, G. A., HEMMENDINGER, A., EVERHART, G. G. and GITTINGS, H. T. (1949). *Phys. Rev.* **75**, 1361.

TATEL, H. (1942). *Phys. Rev.* **61**, 450.

TAYLOR, A. E. and WOOD, E. (1956). *Proc. Phys. Soc.* A, **69**, 645.

TAYLOR, C. J., JENTSCHKE, W. K., REMLEY, M. E., EBY, F. S. and KRUGER, P. G. (1951). *Phys. Rev.* **84**, 1034.

TAYLOR, C. J. BRATENAHL, A., HERNANDEZ, H. P., PETERSON, J. M., SMITH, B. H., STAHL, R. H., GALVAN, J. A. and MULLINS, R. K. (1956). *Bull. Amer. Phys. Soc.* II,I, no. 4 (G9-G11).

TAYLOR, H. M. (1932). *Proc. Roy. Soc.* A, **136**, 605.

TAYLOR, T. I. and HAVENS, W. W. (1949). *Nucleonics*, **5**, no. 6, 4.

THALER, R. M. and BIEDENHARN, L. C. (1957). *Nuc. Phys.* **3**, 207.

THORNDIKE, A. M. and WOTRING, A. W. (1951). *Phys. Rev.* **82**, 295.

TOBOCMAN, W. and KALOS, M. (1955). *Phys. Rev.* **97**, 132.

TOMONAGA, S. (1947). *Prog. Theor. Phys.* **2**, 6.

TOWLER, O. A. (1952). *Phys. Rev.* **85**, 1024.

TRIPP, R. D. (1955). *University of California Radiation Laboratory Report*, UCRL-2975.

TROESCH, A. and VERDE, M. (1951). *Helv. Phys. Acta*, **24**, 39.

TUVE, M. A., HEYDENBURG, N. P. and HAFSTAD, L. R. (1936). *Phys. Rev.* **50**, 806.

VAN DER SPUY, E. (1956). *Nuc. Phys.* **1**, 381.

VAN PATTER, D. M. and WHALING, W. (1954). *Rev. Mod. Phys.* **26**, 402.

VAN RENNES, A. B. (1952). *Nucleonics*, **10**, no. 7, 20; no. 8, 22; no. 9, 32; no. 10, 50.

VARIOUS AUTHORS (1956). *Bull. Amer. Phys. Soc.* II, **1**, no. 1, pp. 55 *et seq.*; pp. 69 *et seq.*

VERDE, M. (1949). *Helv. Phys. Acta*, **24**, 39.

WALLACE, P. R. (1949). *Nucleonics*, **4**, no. 2, 30; *ibid.* (1949), no. 4, 48.

WALLACE, R. (1951). *Phys. Rev.* **81**, 493.

WANTUCH, E. (1951). *Phys. Rev.* **84**, 169.

WARREN, R. E., POWELL, J. L. and HERB, R. G. (1947). *Rev. Sci. Instr.* **18**, 559.

WATT, B. E. (1952). *Phys. Rev.* **87**, 1037.

WATTENBERG, A. (1947). *Phys. Rev.* **71**, 497; also *Preliminary Report*, no. 6, *Nuclear Science Series*. National Research Council, Washington, D.C., 1949.

WEISSKOPF, V. (1937). *Phys. Rev.* **52**, 295.

WELLS, W. H. (1935). *Phys. Rev.* **47**, 591.

WENTZEL, G. (1940). *Helv. Phys. Acta*, **13**, 269.

WENTZEL, G. (1942). *Helv. Phys. Acta*, **15**, 685.

WERNER, A. (1956). *Nuc. Phys.* **1**, 9.

WHALIN, E. A. and REITZ, R. A. (1955). *Rev. Sci. Instr.* **26**, 59.

460   REFERENCES

WHEELER, J. A. (1937). *Phys. Rev.* **52**, 1083.
WHEELER, J. A. and BARSCHALL, H. H. (1940). *Phys. Rev.* **58**, 682.
WHITE, M. G. (1935). *Phys. Rev.* **47**, 573; *ibid.* (1936), **49**, 309.
WHITMORE, B. G. and BAKER, W. B. (1950). *Phys. Rev.* **78**, 799.
WHITTAKER, E. T. and WATSON, G. N. (1953). *Modern Analysis.* Cambridge University Press.
WICK, G. C. (1937). *Physik. Z.* **38**, 403, 689.
WICKERSHAM, A. F. (1954). *University of California Radiation Laboratory Report*, UCRL-2662.
WIEGAND, C. (1948). *Rev. Sci. Instr.* **19**, 790.
WILKINS, J. J. (1951). *Atomic Energy Research Establishment Report*, G/R664.
WILKINS, T. R. (1940). *J. Appl. Phys.* **11**, 35.
WILKINSON, D. H. (1950). *Ionization Chambers and Counters.* Cambridge University Press.
WILLARD, H. B., BAIR, J. K., COHN, H. O. and KINGTON, J. D. (1956). *Bull. Amer. Phys. Soc.* II, **1**, no. 1 (R1).
WILLIAMS, E. J. (1938). *Proc. Roy. Soc.* A, **169**, 521; also (1940), *Phys. Rev.* **58**, 292; and (1945), *Rev. Mod. Phys.* **17**, 217.
WILLIAMS, J. H. and RASMUSSEN, S. W. (1955). *Phys. Rev.* **98**, 56.
WILSON, J. G. (1951). *The Principles of the Cloud Chamber Technique.* Cambridge University Press.
WILSON, R. R. (1947). *Phys. Rev.* **71**, 384.
WILSON, R. R. and CREUTZ, E. C. (1940). *Phys. Rev.* **59**, 916; *ibid.* (1947), **71**, 339.
WILSON, R. R. and CREUTZ, E. C. (1947). *Phys. Rev.* **71**, 339.
WILSON, R. R., LOFGREN, E. J., WRIGHT, B. T. and SHANKLAND, R. S. (1947). *Phys. Rev.* **72**, 1131.
WILSON, R. R. and VOSS, R. G. (1956). *Bull. Amer. Phys. Soc.* II, **1**, no. 1 (UA1).
WOLFE, B., SILVERMAN, A. and DE WIRE, J. W. (1955). *Rev. Sci. Instr.* **26**, 504.
WOLFENSTEIN, L. (1949a). *Phys. Rev.* **75**, 1664.
WOLFENSTEIN, L. (1949b). *Phys. Rev.* **76**, 541.
WOLFENSTEIN, L. (1951). *Phys. Rev.* **82**, 308.
WOLFENSTEIN, L. (1954). *Phys. Rev.* **96**, 1654.
WOLFENSTEIN, L. and ASHKIN, J. (1952). *Phys. Rev.* **85**, 947.
WOLFF, P. and HECHROTTE, W. (1948). *Phys. Rev.* **73**, 264.
WOLLAN, E. O., SHULL, C. G. and KOEHLER, W. C. (1951). *Phys. Rev.* **83**, 700.
WOODS, R. D. and SAXON, D. S. (1954). *Phys. Rev.* **95**, 577.
WORTHINGTON, H. R., McGRUER, J. N. and FINDLEY, D. E. (1953). *Phys. Rev.* **90**, 899.
WOUTERS, L. F. (1951). *Phys. Rev.* **84**, 1069.
WRIGHT, R. W., SAPHIR, G., POWELL, W. M., MAENCHEN, G. and FOWLER, W. B. (1955). *Bull. Amer. Phys. Soc.* **30**, no. 8 (M1).
WRIGHT, S. C. and SCHLUTER, R. A. (1954). *Phys. Rev.* **95**, 639.
WU, T. Y. (1948). *Phys. Rev.* **73**, 934.
WU, T. Y. and ASHKIN, J. (1948). *Phys. Rev.* **73**, 986.

YAGODA, H. (1949). *Radioactive Measurements with Nuclear Emulsions.* J. Wiley and Sons, Inc., N.Y.

YARNELL, J. L., LOVBERG, R. H. and STRATTON, W. R. (1953). *Phys. Rev.* 90, 292.

YEATER, M. L., GAERTTNER, E. R. and BALDWIN, G. C. (1953). *Phys. Rev.* 91, 451.

YNTEMA, J. L. and WHITE, M. G. (1954). *Phys. Rev.* 95, 1226.

YOSHIDA, S. (1956). *Proc. Phys. Soc.* A, 69, 668.

YOST, F. L., WHEELER, J. A. and BREIT, G. (1936). *Phys. Rev.* 49, 174.

ZIMMERMAN, E. J., KERMAN, R. O., SINGER, S., KRUGER, P. G. and JENTSCHKE, W. (1954). *Phys. Rev.* 96, 1322.

ZINN, W. H. (1946). *Phys. Rev.* 70, 102.

INDEX